COMPUTATIONAL BIOCHEMISTRY AND BIOPHYSICS

COMPUTATIONAL BIOCHEMISTRY AND BIOPHYSICS

edited by

Oren M. Becker
Tel Aviv University
Tel Aviv, Israel

Alexander D. MacKerell, Jr.
University of Maryland
Baltimore, Maryland

Benoît Roux
Weill Medical College of Cornell University
New York, New York

Masakatsu Watanabe
Wavefunction, Inc.
Irvine, California

MARCEL DEKKER, INC.   NEW YORK • BASEL

ISBN: 0-8247-0455-X

This book is printed on acid-free paper.

Headquarters
Marcel Dekker, Inc.
270 Madison Avenue, New York, NY 10016
tel: 212-696-9000; fax: 212-685-4540

Eastern Hemisphere Distribution
Marcel Dekker AG
Hutgasse 4, Postfach 812, CH-4001 Basel, Switzerland
tel: 41-61-261-8482; fax: 41-61-261-8896

World Wide Web
http://www.dekker.com

The publisher offers discounts on this book when ordered in bulk quantities. For more information, write to Special Sales/Professional Marketing at the headquarters address above.

Current printing (last digit):
10 9 8 7 6 5 4 3 2 1

PRINTED IN THE UNITED STATES OF AMERICA

Foreword

The long-range goal of molecular approaches to biology is to describe living systems in terms of chemistry and physics. Over the last 70 years great progress has been made in applying the quantum mechanical equations representing the underlying physical laws to chemical problems involving the structures and reactions of small molecules. This work was recognized in the awarding of the Nobel Prize in Chemistry to Walter Kohn and John Pople in 1998. Computational studies of mesoscopic systems of biological interest have been attempted only more recently. Classical mechanics is adequate for describing most of the properties of these systems, and the molecular dynamics simulation method is the most important theoretical approach used in such studies. The first molecular dynamics simulation of a protein, the bovine pancreatic trypsin inhibitor (BPTI), was published more than 20 years ago [1]. Although the simulation was ''crude'' by present standards, it was important because it introduced an important conceptual change in our view of biomolecules. The classic view of biopolymers, like proteins and nucleic acids, had been static in character. The remarkable detail evident in the protein crystal structures available at that time led to an image of ''rigid'' biomolecules with every atom fixed in place [2]. The molecular dynamics simulation of BPTI was instrumental in changing the static view of the structure of biomolecules to a dynamic picture. It is now recognized that the atoms of which biopolymers are composed are in a state of constant motion at ordinary temperatures. The X-ray structure of a protein provides the average atomic positions, but the atoms exhibit fluidlike motions of sizable amplitudes about these averages. The new understanding of protein dynamics subsumed the static picture in that the average positions are still useful for the discussion of many aspects of biomolecule function in the language of structural chemistry. The recognition of the importance of fluctuations opened the way for more sophisticated and accurate interpretations of functional properties.

In the intervening years, molecular dynamics simulations of biomolecules have undergone an explosive development and been applied to a wide range of problems [3,4]. Two attributes of molecular dynamics simulations have played an essential role in their increasing use. The first is that simulations provide individual particle motions as a function of time so they can answer detailed questions about the properties of a system, often more easily than experiments. For many aspects of biomolecule function, it is these details that are of interest (e.g., by what pathways does oxygen get into and exit the heme pocket in myoglobin? How does the conformational change that triggers activity of ras p21 take place?). The second attribute is that, although the potential used in the simulations is approximate, it is completely under the user's control, so that by removing or altering specific contributions to the potential, their role in determining a given property can be examined. This is most graphically demonstrated in the calculation of free energy differences by "computer alchemy" in which the potential is transmuted reversibly from that representing one system to another during a simulation [5].

There are three types of applications of molecular dynamics simulation methods in the study of macromolecules of biological interest, as in other areas that use such simulations. The first uses the simulation simply as a means of sampling configuration space. This is involved in the utilization of molecular dynamics, often with simulated annealing protocols, to determine or refine structures with data obtained from experiments, such as X-ray diffraction. The second uses simulations to determine equilibrium averages, including structural and motional properties (e.g., atomic mean-square fluctuation amplitudes) and the thermodynamics of the system. For such applications, it is necessary that the simulations adequately sample configuration space, as in the first application, with the additional condition that each point be weighted by the appropriate Boltzmann factor. The third area employs simulations to examine the actual dynamics. Here not only is adequate sampling of configuration space with appropriate Boltzmann weighting required, but it must be done so as to properly represent the time development of the system. For the first two areas, Monte Carlo simulations, as well as molecular dynamics, can be utilized. By contrast, in the third area where the motions and their development are of interest, only molecular dynamics can provide the necessary information. The three types of applications, all of which are considered in the present volume, make increasing demands on the simulation methodology in terms of the accuracy that is required.

In the early years of molecular dynamics simulations of biomolecules, almost all scientists working in the field received specialized training (as graduate students and/or postdoctoral fellows) that provided a detailed understanding of the power and limitations of the approach. Now that the methodology is becoming more accessible (in terms of ease of application of generally distributed programs and the availability of the required computational resources) and better validated (in terms of published results), many people are beginning to use simulation technology without training in the area. Molecular dynamics simulations are becoming part of the "tool kit" used by everyone, even experimentalists, who wish to obtain an understanding of the structure and function of biomolecules. To be able to do this effectively, a person must have access to sources from which he or she can obtain the background required for meaningful applications of the simulation methodology. This volume has an important role to play in the transition of the field from one limited to specialists (although they will continue to be needed to improve the methodology and extend its applicability) to the mainstream of molecular biology. The emphasis on an in-depth description of the computational methodology will make the volume useful as an introduction to the field for many people who are doing simulations for the first time. They will find it helpful also to look at two earlier volumes on macromolecular simulations [3,4], as well as the classic general text on molecular dynamics [6]. Equally important in the volume is the connection made with X-ray, neutron scattering, and nuclear magnetic resonance experiments, areas in which molecular dynamics simulations are playing an essential role. A number of well-chosen "special topics" involving applications of simulation methods are described. Also, several chapters broaden the perspective of the book by introducing approaches other than molecular dynamics for modeling proteins and their interactions. They make the connection with what many people regard—mistakenly, in my view—as "computational biology." Certainly with the announced completion of a description of the human genome in a coarse-grained sense, the part of computational biology concerned with the prediction of the structure and function of gene products from a knowledge of the polypeptide sequence is an important endeavor. However, equally important, and probably more so in the long run, is the biophysical aspect of computational biology. The first set of Investigators in Computational Biology chosen this year demonstrates that the Howard Hughes Foundation recognized the importance of such biophysical studies to which this volume serves as an excellent introduction.

I am very pleased to have been given the opportunity to contribute a Foreword to this very useful book. It is a particular pleasure for me to do so because all the editors and fifteen of the authors are alumni of my research group at Harvard, where molecular dynamics simulations of biomolecules originated.

REFERENCES

1. JA McCammon, BR Gelin, and M Karplus. Nature 267:585, 1977.
2. DC Phillips. In: RH Sarma, ed. Biomolecular Stereodynamics, II. Guilderland, New York: Adenine Press, 1981, p 497.
3. JA McCammon and S Harvey. Dynamics of Proteins and Nucleic Acids. Cambridge: Cambridge University Press, 1987.
4. CL Brooks III, M Karplus, and BM Pettitt. Proteins: A Theoretical Perspective of Dynamics, Structure, and Thermodynamics. New York: John Wiley & Sons, 1988.
5. For an early example, see J Gao, K Kuczera, B Tidor, and M Karplus. Science 244:1069–1072, 1989.
6. MP Allen and DJ Tildesley. Computer Simulations of Liquids. Oxford: Clarendon Press, 1987.

Martin Karplus
Laboratoire de chimie Biophysique, ISIS
Université Louis Pasteur
Strasbourg, France
and
Department of Chemistry and Chemical Biology
Harvard University
Cambridge, Massachusetts

Preface

The first dynamical simulation of a protein based on a detailed atomic model was reported in 1977. Since then, the uses of various theoretical and computational approaches have contributed tremendously to our understanding of complex biomolecular systems such as proteins, nucleic acids, and bilayer membranes. By providing detailed information on biomolecular systems that is often experimentally inaccessible, computational approaches based on detailed atomic models can help in the current efforts to understand the relationship of the structure of biomolecules to their function. For that reason, they are now considered to be an integrated and essential component of research in modern biology, biochemistry, and biophysics.

A number of books and journal articles reviewing computational methods relevant to biophysical problems have been published in the last decade. Two of the most popular texts, however, were published more than ten years ago: those of McCammon and Harvey in 1987 and Brooks, Karplus, and Pettitt in 1988. There has been significant progress in theoretical and computational methodologies since the publication of these books. Therefore, we feel that there is a need for an updated, comprehensive text including the most recent developments and applications in the field.

In recent years the significant increase in computer power along with the implementation of a wide range of theoretical methods into sophisticated simulation programs have greatly expanded the applicability of computational approaches to biological systems. The expansion is such that interesting applications to important and complex biomolecular systems are now often carried out by researchers with no special training in computational methodologies. To successfully apply computational approaches to their systems of interest, these ''nonspecialists'' must make several important choices about the proper methods and techniques for the particular question that they are trying to address. We believe that a good understanding of the theory behind the myriad of computational methods and techniques can help in this process. Therefore, one of this book's aims is to provide readers with the required background to properly design and implement computational investigations of biomolecular systems. In addition, the book provides the needed information for calculating and interpreting experimentally observed properties on the basis of the results generated by computer simulations.

This book is organized so that nonspecialists as well as more advanced users can benefit. It can serve as both an introductory text to computational biology, making it useful for students, and a reference source for active researchers in the field. We have tried to compile a comprehensive but reasonably concise review of relevant theoretical and computational methods that is self-contained. Therefore, the chapters, particularly in Part I, are ordered so that the reader can easily follow from one topic to the next and be systematically introduced to the theoretical methods used in computational studies of biomolecular systems. The remainder of the book is designed so that the individual parts as well as their chapters can be read independently. Additional technical details can be found in the references listed in each chapter. Thus the book may also serve as a useful reference for both theoreticians and experimentalists in all areas of biophysics and biochemical research.

This volume thus presents a current and comprehensive account of computational methods and their application to biological macromolecules. We hope that it will serve as a useful tool to guide future investigations of proteins, nucleic acids, and biological membranes, so that the mysteries of biological molecules can continue to be revealed.

We are grateful to the many colleagues we have worked with, collaborated with, and grown with over the course of our research careers. The multidimensionality of those interactions has allowed us to grow in many facets of our lives. Special thanks to Professor Martin Karplus for contributing the Foreword of this book and, most important, for supplying the insights, knowledge, and environment that laid the foundation for our scientific pursuits in computational biochemistry and biophysics and led directly to the creation of this book. Finally, we wish to acknowledge the support of all our friends and family.

Oren M. Becker
Alexander D. MacKerell, Jr.
Benoît Roux
Masakatsu Watanabe

Contents

Contributors

Oren M. Becker Department of Chemical Physics, School of Chemistry, Tel Aviv University, Tel Aviv, Israel

Thomas A. Darden Laboratory of Structural Biology, National Institute of Environmental Health Sciences, National Institutes of Health, Research Triangle Park, North Carolina

Roland L. Dunbrack, Jr. Institute for Cancer Research, Fox Chase Cancer Center, Philadelphia, Pennsylvania

András Fiser Laboratories of Molecular Biophysics, The Rockefeller University, New York, New York

Steven Hayward School of Information Systems, University of East Anglia, Norwich, England

Fumio Hirata Department of Theoretical Study, Institute for Molecular Science, Okazaki National Research Institutes, Okazaki, Japan

Toshiko Ichiye School of Molecular Biosciences, Washington State University, Pullman, Washington

Shigeki Kato Department of Chemistry, Kyoto University, Kyoto, Japan

Paul D. Lyne Computer Aided Drug Design, Biogen, Inc., Cambridge, Massachusetts

Alexander D. MacKerell, Jr. School of Pharmacy, University of Maryland, Baltimore, Maryland

Alexey K. Mazur Institut de Biologie Physico-Chimique, CNRS, Paris, France

Francisco Melo Laboratories of Molecular Biophysics, The Rockefeller University, New York, New York

Michael Nilges Structural and Computational Biology Program, European Molecular Biology Laboratory, Heidelberg, Germany

Lennart Nilsson Department of Biosciences at NOVUM, Karolinska Institutet, Huddinge, Sweden

Benoît Roux Department of Biochemistry and Structural Biology, Weill Medical College of Cornell University, New York, New York

Andrej Šali Laboratories of Molecular Biophysics, The Rockefeller University, New York, New York

Roberto Sánchez Laboratories of Molecular Biophysics, The Rockefeller University, New York, New York

Hirofumi Sato Department of Theoretical Study, Institute for Molecular Science, Okazaki National Research Institutes, Okazaki, Japan

Thomas Simonson Laboratory for Structural Biology and Genomics, Centre National de la Recherche Scientifique, Strasbourg, France

Jeremy C. Smith Lehrstuhl für Biocomputing, Interdisziplinäres Zentrum für Wissenschaftliches Rechnen der Universität Heidelberg, Heidelberg, Germany

John E. Straub Department of Chemistry, Boston University, Boston, Massachusetts

Seiichiro Ten-no Graduate School of Information Science, Nagoya University, Nagoya, Japan

Douglas J. Tobias Department of Chemistry, University of California at Irvine, Irvine, California

Alexander Tropsha Laboratory for Molecular Modeling, University of North Carolina at Chapel Hill, Chapel Hill, North Carolina

Owen A. Walsh Physical and Theoretical Chemistry Laboratory, Oxford University, Oxford, England

Masakatsu Watanabe* Moldyn, Inc., Cambridge, Massachusetts

Weifan Zheng Laboratory for Molecular Modeling, University of North Carolina at Chapel Hill, Chapel Hill, North Carolina

---

* *Current affiliation*: Wavefunction, Inc., Irvine, California.

1

Introduction

Oren M. Becker
Tel Aviv University, Tel Aviv, Israel

Alexander D. MacKerell, Jr.
University of Maryland, Baltimore, Maryland

Benoît Roux
Weill Medical College of Cornell University, New York, New York

Masakatsu Watanabe*
Moldyn, Inc., Cambridge, Massachusetts

I. INTRODUCTION

The first hints of the chemical basis of life were noted approximately 150 years ago. Leading up to this initial awareness were a series of insights that living organisms comprise a hierarchy of structures: organs, which are composed of individual cells, which are themselves formed of organelles of different chemical compositions, and so on. From this realization and the observation that nonviable extracts from organisms such as yeast could by themselves catalyze chemical reactions, it became clear that life itself was the result of a complex combination of individual chemicals and chemical reactions. These advances stimulated investigations into the nature of the molecules responsible for biochemical reactions, culminating in the discovery of the genetic code and the molecular structure of deoxyribonucleic acid (DNA) in the early 1950s by Watson and Crick [1]. One of the most fascinating aspects of their discovery was that an understanding of the mechanism by which the genetic code functioned could not be achieved until knowledge of the three-dimensional (3D) structure of DNA was attained. The discovery of the structure of DNA and its relationship to DNA function had a tremendous impact on all subsequent biochemical investigations, basically defining the paradigm of modern biochemistry and molecular biology. This established the primary importance of molecular structure for an understanding of the function of biological molecules and the need to investigate the relationship between structure and function in order to advance our understanding of the fundamental processes of life.

As the molecular structure of DNA was being elucidated, scientists made significant contributions to revealing the structures of proteins and enzymes. Sanger [2] resolved the

---

* *Current affiliation*: Wavefunction, Inc., Irvine, California.

primary sequence of insulin in 1953, followed by that of an enzyme, ribonuclease A, 10 years later. The late 1950s saw the first high resolution 3D structures of proteins, myoglobin and hemoglobin, as determined by Kendrew et al. [3] and Perutz et al. [4], respectively, followed by the first 3D structure of an enzyme, lysozyme, by Phillips and coworkers [5] in 1965. Since then, the structures of a very large number of proteins and other biological molecules have been determined. There are currently over 10,000 3D structures of proteins available [6] along with several hundred DNA and RNA structures [7] and a number of protein–nucleic acid complexes.

Prior to the elucidation of the 3D structure of proteins via experimental methods, theoretical approaches made significant inroads toward understanding protein structure. One of the most significant contributions was made by Pauling and Corey [8] in 1951, when they predicted the existence of the main elements of secondary structure in proteins, the α-helix and β-sheet. Their prediction was soon confirmed by Perutz [9], who made the first glimpse of the secondary structure at low resolution. This landmark work by Pauling and Corey marked the dawn of theoretical studies of biomolecules. It was followed by prediction of the allowed conformations of amino acids, the basic building block of proteins, in 1963 by Ramachandran et al. [10]. This work, which was based on simple hard-sphere models, indicated the potential of computational approaches as tools for understanding the atomic details of biomolecules. Energy minimization algorithms with an explicit potential energy function followed readily to assist in the refinement of model structures of peptides by Scheraga [11] and of crystal structures of proteins by Levitt and Lifson [12].

The availability of the first protein structures determined by X-ray crystallography led to the initial view that these molecules were very rigid, an idea consistent with the lock-and-key model of enzyme catalysis. Detailed analysis of protein structures, however, indicated that proteins had to be flexible in order to perform their biological functions. For example, in the case of myoglobin and hemoglobin, there is no path for the escape of O_2 from the heme-binding pocket in the crystal structure; the protein must change structure in order for the O_2 to be released. This and other realizations lead to a rethinking of the properties of proteins, which resulted in a more dynamic picture of protein structure. Experimental methods have been developed to investigate the dynamic properties of proteins; however, the information content from these studies is generally isotropic in nature, affording little insight into the atomic details of these fluctuations [13]. Atomic resolution information on the dynamics of proteins as well as other biomolecules and the relationship of dynamics to function is an area where computational studies can extend our knowledge beyond what is accessible to experimentalists.

The first detailed microscopic view of atomic motions in a protein was provided in 1977 via a molecular dynamics (MD) simulation of bovine pancreatic trypsin inhibitor by McCammon et al. [14]. This work, marking the beginning of modern computational biochemistry and biophysics, has been followed by a large number of theoretical investigations of many complex biomolecular systems. It is this large body of work, including the numerous methodological advances in computational studies of biomolecules over the last decade, that largely motivated the production of the present book.

II. OVERVIEW OF COMPUTATIONAL BIOCHEMISTRY AND BIOPHYSICS

Although the dynamic nature of biological molecules has been well accepted for over 20 years, the extent of that flexibility, as manifested in the large structural changes that biomolecules can undergo, has recently become clearer due to the availability of experimentally determined structures of the same biological molecules in different environments. For example, the enzyme triosephosphate isomerase contains an 11 amino acid residue loop that moves by more than 7 Å following the binding of substrate, leading to a catalytically competent structure [15,16]. In the enzyme cytosine-5-methyltransferase, a loop containing one of the catalytically essential residues undergoes a large conformational change upon formation of the DNA–coenzyme–protein complex, leading to some residues changing position by over 20 Å [17]. DNA, typically envisioned in the canonical B form [18], has been shown to undergo significant distortions upon binding to proteins. Bending of 90° has been seen in the CAP–DNA complex [19], and binding of the TATA box binding protein to the TATAAAA consensus sequence leads to the DNA assuming a unique conformation referred to as the TA form [20]. Even though experimental studies can reveal the end points associated with these conformational transitions, these methods typically cannot access structural details of the pathway between the end points. Such information is directly accessible via computational approaches.

Computational approaches can be used to investigate the energetics associated with changes in both conformation and chemical structure. An example is afforded by the conformational transitions discussed in the preceding paragraph. Conformational free energy differences and barriers can be calculated and then directly compared with experimental results. Overviews of these methods are included in Chapters 9 and 10. Recent advances in techniques that combine quantum mechanical (QM) approaches with molecular mechanics (MM) now allow for a detailed understanding of processes involving bond breaking and bond making and how enzymes can accelerate those reactions. Chapter 11 gives a detailed overview of the implementation and current status of QM/MM methods. The ability of computational biochemistry to reveal the microscopic events controlling reaction rates and equilibrium at the atomic level is one of its greatest strengths.

Biological membranes provide the essential barrier between cells and the organelles of which cells are composed. Cellular membranes are complicated extensive biomolecular sheetlike structures, mostly formed by lipid molecules held together by cooperative noncovalent interactions. A membrane is not a static structure, but rather a complex dynamical two-dimensional liquid crystalline fluid mosaic of oriented proteins and lipids. A number of experimental approaches can be used to investigate and characterize biological membranes. However, the complexity of membranes is such that experimental data remain very difficult to interpret at the microscopic level. In recent years, computational studies of membranes based on detailed atomic models, as summarized in Chapter 21, have greatly increased the ability to interpret experimental data, yielding a much-improved picture of the structure and dynamics of lipid bilayers and the relationship of those properties to membrane function [21].

Computational approaches are now being used to facilitate the experimental determination of macromolecular structures by aiding in structural refinement based on either nuclear magnetic resonance (NMR) or X-ray data. The current status of the application of computational methods to the determination of biomolecular structure and dynamics is presented in Chapters 12 and 13. Computational approaches can also be applied in situations where experimentally determined structures are not available. With the rapid advances in gene technology, including the human genome project, the ability of computational approaches to accurately predict 3D structures based on primary sequence represents an area that is expected to have a significant impact. Prediction of the 3D structures of proteins can be performed via homology modeling or threading methods; various approaches to this problem are presented in Chapters 14 and 15. Related to this is the area of protein folding. As has been known since the seminal experimental refolding studies of ribonuclease A in the 1950s, the primary structure of many proteins dictates their 3D structure [22]. Accordingly, it should be possible "in principle" to compute the 3D structure of many proteins based on knowledge of just their primary sequences. Although this has yet to be achieved on a wide scale, considerable efforts are being made to attain this goal, as overviewed in Chapter 17.

Drug design and development is another area of research where computational biochemistry and biophysics are having an ever-increasing impact. Computational approaches can be used to aid in the refinement of drug candidates, systematically changing a drug's structure to improve its pharmacological properties, as well as in the identification of novel lead compounds. The latter can be performed via the identification of compounds with a high potential for activity from available databases of chemical compounds or via de novo drug design approaches, which build totally novel ligands into the binding sites of target molecules. Techniques used for these types of studies are presented in Chapter 16. In addition to aiding in the design of compounds that target specific molecules, computational approaches offer the possibility of being able to improve the ability of drugs to access their targets in the body. These gains will be made through an understanding of the energetics associated with the crossing of lipid membranes and using the information to rationally enhance drug absorption rates. As evidenced by the recent contribution of computational approaches in the development of inhibitors of the HIV protease, many of which are currently on the market, it can be expected that these methods will continue to have an increasing role in drug design and development.

Clearly, computational and theoretical studies of biological molecules have advanced significantly in recent years and will progress rapidly in the future. These advances have been partially fueled by the ever-increasing number of available structures of proteins, nucleic acids, and carbohydrates, but at the same time significant methodological improvements have been made in the area of physics relevant to biological molecules. These advances have allowed for computational studies of biochemical processes to be performed with greater accuracy and under conditions that allow for direct comparison with experimental studies. Examples include improved force fields, treatment of long-range atom–atom interactions, and a variety of algorithmic advances, as covered in Chapters 2 through 8. The combination of these advances with the exponential increases in computational resources has greatly extended and will continue to expand the applicability of computational approaches to biomolecules.

III. SCOPE OF THE BOOK

The overall scope of this book is the implementation and application of available theoretical and computational methods toward understanding the structure, dynamics, and function of biological molecules, namely proteins, nucleic acids, carbohydrates, and membranes. The large number of computational tools already available in computational chemistry preclude covering all topics, as Schleyer et al. are doing in *The Encyclopedia of Computational Chemistry* [23]. Instead, we have attempted to create a book that covers currently available theoretical methods applicable to biomolecular research along with the appropriate computational applications. We have designed it to focus on the area of biomolecular computations with emphasis on the special requirements associated with the treatment of macromolecules.

Part I provides an introduction to the field of computational biochemistry and biophysics for nonspecialists, with the later chapters in Part I presenting more advanced techniques that will be of interest to both the nonspecialist and the more advanced reader. Part II presents approaches to extract information from computational studies for the interpretation of experimental data. Part III focuses on methods for modeling and designing molecules. Chapters 14 and 15 are devoted to the determination and modeling of protein structures based on limited available experimental information such as primary sequence. Chapter 16 discusses the recent developments in computer-aided drug designs. The algorithms presented in Part III will see expanding use as the fields of genomics and bioinformatics continue to evolve. The final section, Part IV, presents a collection of overviews of various state-of-the-art theoretical methods and applications in specific areas relevant to biomolecules: protein folding (Chapter 17), protein simulation (Chapter 18), chemical process in solution (Chapter 19), nucleic acids simulation (Chapter 20), and membrane simulation (Chapter 21).

In combination, the book should serve as a useful reference for both theoreticians and experimentalists in all areas of biophysical and biochemical research. Its content represents progress made over the last decade in the area of computational biochemistry and biophysics. Books by Brooks et al. [24] and McCammon and Harvey [25] are recommended for an overview of earlier developments in the field. Although efforts have been made to include the most recent advances in the field along with the underlying fundamental concepts, it is to be expected that further advances will be made even as this book is being published. To help the reader keep abreast of these advances, we present a list of useful WWW sites in the Appendix.

IV. TOWARD A NEW ERA

The 1998 Nobel Prize in Chemistry was given to John A. Pople and Walter Kohn for their work in the area of quantum chemistry, signifying the widespread acceptance of computation as a valid tool for investigating chemical phenomena. With its extension to bimolecular systems, the range of possible applications of computational chemistry was greatly expanded. Though still a relatively young field, computational biochemistry and biophysics is now pervasive in all aspects of the biological sciences. These methods have aided in the interpretation of experimental data, and will continue to do so, allowing for the more rational design of new experiments, thereby facilitating investigations in the biological sciences. Computational methods will also allow access to information beyond that obtainable via experimental techniques. Indeed, computer-based approaches for the study of virtually any chemical or biological phenomena may represent the most powerful tool now available to scientists, allowing for studies at an unprecedented level of detail. It is our hope that the present book will help expand the accessibility of computational approaches to the vast community of scientists investigating biological systems.

REFERENCES

1. JD Watson, FHC Crick. Nature 171:737, 1953.
2. F Sanger. Annu Rev Biochem 57:1, 1988.

3. JC Kendrew, G Bodo, MH Dintzis, RG Parrish, H Wyckoff, DC Phillips. Nature 181:622, 1958.
4. MF Perutz, MG Rossmann, AF Cullis, H Muirhead, G Will, ACT North. Nature 185:416, 1960.
5. CCF Blake, DF Koenig, GA Mair, ACT North, DC Phillips, VR Sarma. Nature 206:757, 1965.
6. FC Bernstein, TF Koetzle, GJB Williams, DF Meyer Jr, MD Brice, JR Rodgers, O Kennard, T Shimanouchi, M Tasumi. J Mol Biol 112:535, 1977.
7. HM Berman, WK Olson, DL Beveridge, J Westbrook, A Gelbin, T Demeny, S-H Hsieh, AR Srinivasan, B Schneider. Biophys J 63:751, 1992.
8. L Pauling, RB Corey. Proc Roy Soc Lond B141:10, 1953.
9. MF Perutz. Nature 167:1053, 1951.
10. GN Ramachandran, C Ramakrishana, V Sasisekharan. J Mol Biol 7:95, 1963.
11. HA Scheraga. Adv Phys Org Chem 6:103, 1968.
12. M Levitt, S Lifson. J Mol Biol 46:269, 1969.
13. M Karplus, GA Petsko. Nature 347:631, 1990.
14. JA McCammon, BR Gelin, M Karplus. Nature 267:585, 1977.
15. D Joseph, GA Petsko, M Karplus. Science 249:1425, 1990.
16. DL Pompliano, A Peyman, JR Knowles. Biochemistry 29:3186, 1990.
17. S Klimasauskas, S Kumar, RJ Roberts, X Cheng. Cell 76:357, 1994.
18. W Saenger. Principles of Nucleic Acid Structure. New York: Springer-Verlag, 1984.
19. SC Schultz, GC Shields, TA Steitz. Science 253:1001, 1991.
20. G Guzikevich-Guerstein, Z Shakked. Nature Struct Biol 3:32, 1996.
21. KM Merz Jr, B Roux, eds. Biological Membranes: A Molecular Perspective from Computation and Experiment. Boston: Birkhauser, 1996.
22. CB Anfinsen. Science 181:223, 1973.
23. PvR Schleyer, NL Allinger, T Clark, J Gasteiger, PA Kollman, HF Schaefer III, PR Schreiner, eds. The Encyclopedia of Computational Chemistry. Chichester: Wiley, 1998.
24. CL Brooks III, M Karplus, BM Pettitt. Proteins, A Theoretical Perspective: Dynamics, Structure, and Thermodynamics, Vol 71. New York: Wiley, 1988.
25. JA McCammon, SC Harvey. Dynamics of Proteins and Nucleic Acids. New York: Cambridge University Press, 1987.

2
Atomistic Models and Force Fields

Alexander D. MacKerell, Jr.
University of Maryland, Baltimore, Maryland

I. INTRODUCTION

Central to the success of any computational approach to the study of chemical systems is the quality of the mathematical model used to calculate the energy of the system as a function of its structure. For smaller chemical systems studied in the gas phase, quantum mechanical (QM) approaches are appropriate. The success of these methods was emphasized by the selection of John A. Pople and Walter Kohn as winners of the 1998 Nobel prize in chemistry. These methods, however, are typically limited to systems of approximately 100 atoms or less, although approaches to treat large systems are under development [1]. Systems of biochemical or biophysical interest typically involve macromolecules that contain 1000–5000 or more atoms plus their condensed phase environment. This can lead to biochemical systems containing 20,000 atoms or more. In addition, the inherent dynamical nature of biochemicals and the mobility of their environments [2,3] require that large number of conformations, generated via various methods (see Chapters 3, 4, 6, and 10), be subjected to energy calculations. Thus, an energy function is required that allows for 10^6 or more energy calculations on systems containing on the order of 10^5 atoms.

Empirical energy functions can fulfill the demands required by computational studies of biochemical and biophysical systems. The mathematical equations in empirical energy functions include relatively simple terms to describe the physical interactions that dictate the structure and dynamic properties of biological molecules. In addition, empirical force fields use atomistic models, in which atoms are the smallest particles in the system rather than the electrons and nuclei used in quantum mechanics. These two simplifications allow for the computational speed required to perform the required number of energy calculations on biomolecules in their environments to be attained, and, more important, via the use of properly optimized parameters in the mathematical models the required chemical accuracy can be achieved. The use of empirical energy functions was initially applied to small organic molecules, where it was referred to as molecular mechanics [4], and more recently to biological systems [2,3].

II. POTENTIAL ENERGY FUNCTIONS

A. Potential Energy Functions for the Treatment of Biological Molecules

A potential energy function is a mathematical equation that allows for the potential energy, V, of a chemical system to be calculated as a function of its three-dimensional (3D) structure, R. The equation includes terms describing the various physical interactions that dictate the structure and properties of a chemical system. The total potential energy of a chemical system with a defined 3D structure, $V(R)_{\text{total}}$, can be separated into terms for the internal, $V(R)_{\text{internal}}$, and external, $V(R)_{\text{external}}$, potential energy as described in the following equations.

$$V(R)_{\text{total}} = V(R)_{\text{internal}} + V(R)_{\text{external}} \tag{1}$$

$$V(R)_{\text{internal}} = \sum_{\text{bonds}} K_b(b - b_0)^2 \tag{2}$$

$$+ \sum_{\text{angles}} K_\theta(\theta - \theta_0)^2 + \sum_{\text{dihedrals}} K_\chi[1 + \cos(n\chi - \sigma)]$$

and

$$V(R)\text{external} = \sum_{\substack{\text{nonbonded} \\ \text{atom pairs}}} \left(\varepsilon_{ij} \left[\left(\frac{R_{\text{min},ij}}{r_{ij}} \right)^{12} - \left(\frac{R_{\text{min},ij}}{r_{ij}} \right)^{6} \right] + \frac{q_i q_j}{\varepsilon_D r_{ij}} \right) \tag{3}$$

The internal terms are associated with covalently connected atoms, and the external terms represent the noncovalent or nonbonded interactions between atoms. The external terms are also referred to as interaction, nonbonded, or intermolecular terms.

Beyond the form of Eqs. (1)–(3), which is discussed below, it is important to emphasize the difference between the terms associated with the 3D structure, R, being subjected to the energy calculation and the parameters in the equations. The terms obtained from the 3D structure are the bond lengths, b; the valence angles, θ; the dihedral or torsion angles, χ; and the distances between the atoms, r_{ij}. A diagrammatic representation of two hypothetical molecules in Figure 1 allows for visualization of these terms. The values of these terms are typically obtained from experimental structures generated from X-ray crystallography or NMR experiments (see Chapter 13), from modeled structures (e.g., from homology modeling of a protein; see Chapters 14 and 15), or a structure generated during a molecular dynamics (MD) or Monte Carlo (MC) simulation. The remaining terms in Eqs. (2) and (3) are referred to as the parameters. These terms are associated with the particular type of atom and the types of atoms covalently bound to it. For example, the parameter q, the partial atomic charge, of a sodium cation is typically set to $+1$, while that of a chloride anion is set to -1. Another example is a C—C single bond versus a C=C double bond, where the former may have bond parameters of $b_0 = 1.53$ Å, $K_b = 225$ kcal/(mol · Å^2) and the latter $b_0 = 1.33$ Å, $K_b = 500$ kcal/(mol · Å^2) Thus, different parameters allow for different types of atoms and different molecular connectivities to be treated using the same form of Eqs. (2) and (3). Indeed, it is the quality of the parameters, as judged by their ability to reproduce experimentally, and quantum-mechanically determined target data (e.g., information on selected molecules that the parameters are adjusted to reproduce) that ultimately determines the accuracy of the results obtained from compu-

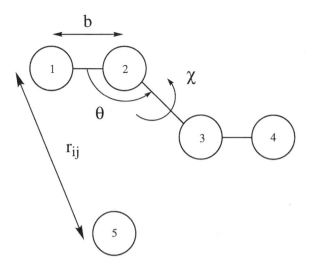

Figure 1 Hypothetical molecules to illustrate the energetic terms included in Eqs. (1)–(3). Molecule A comprises atoms 1–4, and molecule B comprises atom 5. Internal terms that occur in molecule A are the bonds, b, between atoms 1 and 2, 2 and 3, and 3 and 4; angles θ, involving atoms 1–2–3 and atoms 2–3–4, and a dihedral or torsional angle, χ, described by atoms 1–2–3–4. Bonds can also be referred to as 1,2 atom pairs or 1,2 interactions; angles as 1,3 atom pairs or 1,3 interactions; and dihedrals as 1,4 atom pairs or 1,4 interactions. Molecule B is involved in external interactions with all four atoms in molecule A, where the different interatomic distances, r_{ij}, must be known. Note that external interactions (both van der Waals and Coulombic) can occur between the 1,2, 1,3, and 1,4 pairs in molecule A. However, external interactions involving 1,2 and 1,3 interactions are generally not included as part of the external energy (i.e., 1,2 and 1,3 exclusions), but 1,4 interactions are. Often the 1,4 external interaction energies are scaled (i.e., 1,4 scaling) to diminish the influence of these external interactions on geometries, vibrations, and conformational energetics. It should also be noted that additional atoms that could be present in molecule A would represent 1,5 interactions, 1,6 interactions, and so on, and would also interact with each other via the external terms.

tational studies of biological molecules. Details of the parameter optimization process are discussed below.

The mathematical form of Eqs. (2) and (3) represents a compromise between simplicity and chemical accuracy. Both the bond-stretching and angle-bending terms are treated harmonically, which effectively keeps the bonds and angles near their equilibrium values. Bond and angle parameters include b_0 and θ_0, the equilibrium bond length and equilibrium angle, respectively. K_b and K_θ are the force constants associated with the bond and angle terms, respectively. The use of harmonic terms is sufficient for the conditions under which biological computations are performed. Typically MD or MC simulations are performed in the vicinity of room temperature and in the absence of bond-breaking or bond-making events; because the bonds and angles stay close to their equilibrium values at room temperature, the harmonic energy surfaces accurately represent the local bond and angle distortions. It should be noted that the absence of bond breaking is essential for simulated annealing calculations performed at elevated temperatures (see Chapter 13). Dihedral or torsion angles represent the rotations that occur about a bond, leading to changes in the relative positions of atoms 1 and 4 as described in Figure 1. These terms are oscillatory in nature (e.g., rotation about the C — C bond in ethane changes the structure from a low energy staggered conformation to a high energy eclipsed conformation, then back to a low energy staggered conformation, and so on), requiring the use of a sinusoidal function to accurately model them.

In Eq. (2), the dihedral term includes parameters for the force constant, K_χ; the periodicity or multiplicity, n; and the phase, δ. The magnitude of K_χ dictates the height of the barrier to rotation, such that K_χ associated with a double bond would be significantly larger that that for a single bond. The periodicity, n, indicates the number of cycles per 360° rotation about the dihedral. In the case of an sp^3–sp^3 bond, as in ethane, n would equal 3, while the sp^2–sp^2 C=C bond in ethylene would have $n = 2$. The phase, δ, dictates the location of the maxima in the dihedral energy surface allowing for the location of the minima for a dihedral with $n = 2$ to be shifted from 0° to 90° and so on. Typically, δ is equal to 0 or 180, although recent extensions allow any value from 0 to 360 to be assigned to δ [5]. Finally, each torsion angle in a molecule may be treated with a sum of dihedral terms that have different multiplicities, as well as force constants and phases [i.e., the peptide bond can be treated by a summation of 1-fold ($n = 1$) and 2-fold ($n = 2$) dihedral terms with the 2-fold term used to model the double-bonded character of the C—N bond and the 1-fold term used to model the energy difference between the cis and trans conformations]. The use of a summation of dihedral terms for a single torsion angle, a Fourier series, greatly enhances the flexibility of the dihedral term, allowing for more accurate reproduction of experimental and QM energetic target data.

Equation (3) describes the external or nonbond interaction terms. These terms may be considered the most important of the energy terms for computational studies of biological systems. This is because of the strong influence of the environment on the properties of macromolecules as well as the large number of nonbond interactions that occur in biological molecules themselves (e.g., hydrogen bonds between Watson–Crick base pairs in DNA, peptide bond–peptide bond hydrogen bonds involved in the secondary structures of proteins, and dispersion interactions between the aliphatic portions of lipids that occur in membranes). Interestingly, although the proper treatment of nonbond interactions is essential for successful biomolecular computations, it has been shown that the mathematical model required to treat these terms accurately can be relatively simple. Parameters associated with the external terms are the well depth, ε_{ij}, between atoms i and j; the minimum interaction radius, $R_{\min,ij}$; and the partial atomic charge, q_i. Also included is the dielectric constant, ε_D, which is generally treated as equal to 1, the permittivity of vacuum, although exceptions do exist (see below).

The term in square brackets in Eq. (3) is used to treat the van der Waals (VDW) interactions. The particular form in Eq. (3) is referred to as the Lennard-Jones (LJ) 6–12 term. The $1/r^{12}$ term represents the exchange repulsion between atoms associated with overlap of the electron clouds of the individual atoms (i.e., the Pauli exclusion principle). The strong distance dependence of the repulsion is indicated by the 12th power of this term. Representing London's dispersion interactions or instantaneous dipole–induced dipole interactions is the $1/r^6$ term, which is negative, indicating its favorable nature. In the LJ 6-12 equation there are two parameters; The well depth, ε_{ij}, indicates the magnitude of the favorable London's dispersion interactions between two atoms i, j; and $R_{\min,ij}$ is the distance between atoms i and j at which the minimum LJ interaction energy occurs and is related to the VDW radius of an atom. Typically, ε_{ij} and $R_{\min,ij}$ are not determined for every possible interaction pair, i, j; but rather ε_i and $R_{\min,i}$ parameters are determined for the individual atom types (e.g., sp^2 carbon versus sp^3 carbon) and then combining rules are used to create the ij cross terms. These combining rules are generally quite simple, being either the arithmetic mean [i.e., $R_{\mathrm{min},ij} = (R_{\mathrm{min},i} + R_{\mathrm{min},j})/2$] or the geometric mean [i.e., $\varepsilon_{ij} = (\varepsilon_i \varepsilon_j)^{1/2}$]. The use of combining rules greatly simplifies the determination of the ε_i and $R_{\mathrm{min},i}$ parameters.

In special cases the use of combining rules can be supplemented by specific i,j LJ parameters, referred to as off-diagonal terms, to treat interactions between specific atom types that are poorly modeled by the use of combining rules. The final term contributing to the external interactions is the electrostatic or Coulombic term. This term involves the interaction between partial atomic charges, q_i and q_j, on atoms i and j divided by the distance, r_{ij}, between those atoms with the appropriate dielectric constant taken into account. The use of a charge representation for the individual atoms, or monopoles, effectively includes all higher order electronic interactions, such as those between dipoles and quadrupoles. Combined, the Lennard-Jones and Coulombic interactions have been shown to produce a very accurate representation of the interaction between molecules, including both the distance and angle dependencies of hydrogen bonds [6].

Once the 3D structure of a molecule and all the parameters required for the atomic and molecular connectivities are known, the energy of the system can be calculated via Eqs. (1)–(3). First derivatives of the energy with respect to position allow for determination of the forces acting on the atoms, information that is used in the energy minimization (see Chapter 4) or MD simulations (see Chapter 3). Second derivatives of the energy with respect to position can be used to calculate force constants acting on atoms, allowing the determination of vibrational spectra via normal mode analysis (see Chapter 8).

B. All-Atom Versus Extended-Atom Models

Always a limiting factor in computational studies of biological molecules is the ability to treat systems of adequate size for the required amount of simulation time or number of conformations to be sampled. One method to minimize the size of the system is to use extended-atom models versus all-atom models. In extended-atom models the hydrogens are not explicitly represented but rather are treated as part of the nonhydrogen atom to which they are covalently bound. For example, an all-atom model would treat a methyl group as four individual atoms (a carbon and three hydrogens), whereas in an extended-atom model the methyl group would be treated as a single atom, with the LJ parameters and charges adjusted to account for the omission of the hydrogens. Although this approach could be applied for all hydrogens it was typically used only for nonpolar (aliphatic and aromatic) hydrogens; polar hydrogens important for hydrogen bonding interactions were treated explicitly. Extended-atom models were most widely applied for the simulation of proteins in vacuum, where the large number of nonpolar hydrogens yields a significant decrease in the number of atoms compared to all-atom models. However, as more simulations were performed with explicit solvent representation, making the proportion of nonpolar hydrogens in the system much smaller, with ever-increasing computer resources the use of extended-atom models in simulations has decreased. Extended-atom models, however, are still useful for applications where a large sampling of conformational space is required [7].

C. Extensions of the Potential Energy Function

The potential energy function presented in Eqs. (2) and (3) represents the minimal mathematical model that can be used for computational studies of biological systems. Currently, the most widely used energy functions are those included with the CHARMM [8,9], AMBER [10], and GROMOS [11] programs. Two extensions beyond the terms in Eqs. (2) and (3) are often included in biomolecular force fields. A harmonic term for improper dihedrals is often used to treat out-of-plane distortions, such as those that occur with aromatic hydrogens (i.e., Wilson wags). Historically, the improper term was also used to maintain the proper chirality in extended-atom models of proteins (e.g., without the H_α hydrogen, the chirality of amino acids is undefined). Some force fields also contain a Urey–Bradly term that treats 1,3 atoms (the two terminal atoms in an angle; see Fig. 1) with a harmonic bond-stretching term in order to more accurately model vibrational spectra.

Beyond the extensions mentioned in the previous paragraph, a variety of terms are included in force fields used for the modeling of small molecules that can also be applied to biological systems. These types of force fields are often referred to as Class II force fields, to distinguish then from the Class I force fields such as AMBER, CHARMM, and GROMOS discussed above. For example, the bond term in Eq. (2) can be expanded to include cubic and quartic terms, which will more accurately treat the anharmonicity associated with bond stretching. Another extension is the addition of cross terms that express the influence that stretching of a bond has on the stretching of an adjacent bond. Cross terms may also be used between the different types of terms such as bond angle or dihedral angle terms, allowing for the influence of bond length on angle bending or of angle bending on dihedral rotations, respectively, to be more accurately modeled [12]. Extensions may also be made to the interaction portion of the force field [Eq. (3)]. These may include terms for electronic polarizability (see below) or the use of $1/r^4$ terms to treat ion–dipole interactions associated with interactions between, for example, ions and the peptide backbone [13]. In all cases the extension of a potential energy function should, in principle, allow for the system of interest to be modeled with more accuracy. The gains associated with the additional terms, however, are often significant only in specific cases (e.g., the use of a $1/r^4$ term in the study of specific cation–peptide interactions), making their inclusion for the majority of calculations on biochemical systems unwarranted, especially when those terms increase the demand on computational resources.

D.   Alternatives to the Potential Energy Function

The form of the potential energy function in Eqs. (1)–(3) was developed based on a combination of simplicity with required accuracy. However, a number of other forms can be used to treat the different terms in Eqs. (2) and (3). One alternative form used to treat the bond is referred to as the Morse potential. This term allows for bond-breaking events to occur and includes anharmonicity in the bond-stretching surface near the equilibrium value. The ability to break bonds, however, leads to forces close to zero at large bond distances, which may present a problem when crude modeling techniques are used to generate structures [14]. A number of variations in the form of the equation to treat the VDW interactions have been applied. The $1/r^{12}$ term used for modeling exchange repulsion overestimates the distance dependence of the repulsive wall, leading to the use of an $1/r^9$ term [15] or exponential repulsive terms [16]. A more recent variation is the buffered 14-7 form, which was selected because of its ability to reproduce interactions between rare gas atoms [17]. Concerning electrostatic interactions, the majority of potential energy functions employ the standard Coulombic term shown in Eq. (3), with one variation being the use of bond dipoles rather than atom-centered partial atomic charges [16]. As with the extensions to the force fields discussed above, the alternative forms discussed in this paragraph generally do not yield significant gains in accuracy for biomolecular simulations performed in condensed phase environments at room temperature, although for specific situations they may.

III. EMPIRICAL FORCE FIELDS

A. From Potential Energy Functions to Force Fields

Equations (1)–(3) in combination are a potential energy function that is representative of those commonly used in biomolecular simulations. As discussed above, the form of this equation is adequate to treat the physical interactions that occur in biological systems. The accuracy of that treatment, however, is dictated by the parameters used in the potential energy function, and it is the combination of the potential energy function and the parameters that comprises a force field. In the remainder of this chapter we describe various aspects of force fields including their derivation (i.e., optimization of the parameters), those widely available, and their applicability.

B. Overview of Available Force Fields

Currently there a variety of force fields that may, in principle, be used for computational studies of biological systems. Of these force fields, however, only a subset have been designed specifically for biomolecular simulations. As discussed above, the majority of biomolecular simulations are performed with the CHARMM, AMBER, and GROMOS packages. Recent publication of new CHARMM [18–20] and AMBER [21] force fields allows for these to be discussed in detail. Although the forms of the potential energy functions in CHARMM and AMBER are similar, with CHARMM including the additional improper and Urey–Bradley terms (see above), significant philosophical and parameter optimization differences exist (see below). The latest versions of both force fields are all-atom representations, although extended-atom representations are available [22,23].

To date, a number of simulation studies have been performed on nucleic acids and proteins using both AMBER and CHARMM. A direct comparison of crystal simulations of bovine pancreatic trypsin inhibitor show that the two force fields behave similarly, although differences in solvent–protein interactions are evident [24]. Side-by-side tests have also been performed on a DNA duplex, showing both force fields to be in reasonable agreement with experiment although significant, and different, problems were evident in both cases [25]. It should be noted that as of the writing of this chapter revised versions of both the AMBER and CHARMM nucleic acid force fields had become available. Several simulations of membranes have been performed with the CHARMM force field for both saturated [26] and unsaturated [27] lipids. The availability of both protein and nucleic acid parameters in AMBER and CHARMM allows for protein–nucleic acid complexes to be studied with both force fields (see Chapter 20), whereas protein–lipid (see Chapter 21) and DNA–lipid simulations can also be performed with CHARMM.

A number of more general force fields for the study of small molecules are available that can be extended to biological molecules. These force fields have been designed with the goal of being able to treat a wide variety of molecules, based on the ability to transfer parameters between chemical systems and the use of additional terms (e.g., cross terms) in their potential energy functions. Typically, these force fields have been optimized to treat small molecules in the gas phase, although exceptions do exist. Such force fields may also be used for biological simulations; however, the lack of emphasis on properly treating biological systems generally makes them inferior to those discussed in the previous paragraphs. The optimized potential for liquid simulations (OPLS) force field was initially developed for liquid and hydration simulations on a variety of organic compounds [28,29]. This force field has been extended to proteins [30], nucleic acid bases [31], and carbohydrates [32], although its widespread use has not occurred. Some of the most widely used force fields for organic molecules are MM3 and its predecessors [33]. An MM3 force field for proteins has been reported [34]; however, it too has not been widely applied to date.

The consistent force field (CFF) series of force fields have also been developed to treat a wide selection of small molecules and include parameters for peptides. However, those parameters were developed primarily on the basis of optimization of the internal terms [35]. A recent extension of CFF, COMPASS, has been published that concentrates on producing a force field suitable for condensed phase simulations [36], although no condensed phase simulations of biological molecules have been reported. Another force field to which significant effort was devoted to allow for its application to a wide variety of compounds is the Merck Molecular Force Field (MMFF) [37]. During the development of MMFF, a significant effort was placed on optimizing the internal parameters to yield good geometries and energetics of small compounds as well as the accurate treatment of nonbonded interactions. This force field has been shown to be well behaved in condensed phase simulations of proteins; however, the results appear to be inferior to those of the AMBER and CHARMM models. Two other force fields of note are UFF [38] and DREIDING [14]. These force fields were developed to treat a much wider variety of molecules, including inorganic compounds, than the force fields mentioned previously, although their application to biological systems has not been widespread.

It should also be noted that a force field for a wide variety of small molecules, CHARMm (note the small "m," indicating the commercial version of the program and parameters), is available [39] and has been applied to protein simulations with limited success. Efforts are currently under way to extend the CHARMm small molecule force field to make the nonbonded parameters consistent with those of the CHARMM force fields, thereby allowing for a variety of small molecules to be included in computational studies of biological systems.

Although the list of force fields discussed in this subsection is by no means complete, it does emphasize the wide variety of force fields that are available for different types of chemical systems as well as differences in their development and optimization.

C. Free Energy Force Fields

All of the force fields discussed in the preceding sections are based on potential energy functions. To obtain free energy information when using these force fields, statistical mechanical ensembles must be obtained via various simulation techniques. An alternative approach is to use a force field that has been optimized to reproduce free energies directly rather than potential energies. For example, a given set of dihedral parameters in a potential energy function may be adjusted to reproduce a QM-determined torsional potential energy surface for a selected model compound. In the case of a free energy force field, the dihedral parameters would be optimized to reproduce the experimentally observed probability distribution of that dihedral in solution. Because the experimentally determined probability distribution corresponds to a free energy surface, a dihedral energy surface calculated using this force field would correspond to the free energy surface in solution. This allows for calculations to be performed in vacuum while yielding results that, in principle, correspond to the free energy in solution.

The best known of the free energy force fields is the Empirical Conformational Energy Program for Peptides (ECEPP) [40]. ECEPP parameters (both internal and external) were derived primarily on the basis of crystal structures of a wide variety of peptides. Such an approach yields significant savings in computational costs when sampling large numbers of conformations; however, microscopic details of the role of solvent on the biological molecules are lost. This type of approach is useful for the study of protein folding [41,42] as well as protein–protein or protein–ligand interactions [43].

An alternative to obtaining free energy information is the use of potential energy functions combined with methods to calculate the contribution of the free energy of solvation. Examples include methods based on the solvent accessibilities of atoms [44,45], continuum electrostatics–based models [46–49], and the generalized Born equation [50,51]. With some of these approaches the availability of analytical derivatives allows for their use in MD simulations; however, they are generally most useful for determining solvation contributions associated with previously generated conformations. See Chapter 7 for a detailed overview of these approaches.

D. Applicability of Force Fields

Clearly, the wide variety for force fields requires the user to carefully consider those that are available and choose that which is most appropriate for his or her particular application. Most important in this selection process is a knowledge of the information to be obtained from the computational study. If atomic details of specific interactions are required, then all-atom models with the explicit inclusion of solvent will be necessary. For example, experimental results indicate that a single point mutation in a protein increases its stability. Application of an all-atom model with explicit solvent in MD simulations would allow for atomic details of interactions of the two side chains with the environment to be understood, allowing for more detailed interpretation of the experimental data. Furthermore, the use of free energy perturbation techniques would allow for more quantitative data to be obtained from the calculations, although this approach requires proper treatment of the unfolded states of the proteins, which is difficult (see Chapter 9 for more details). In other cases, a more simplified model, such as an extended-atom force field with the solvent treated implicitly via the use of an R-dependent dielectric constant, may be appropriate. Examples include cases in which sampling of a large number of conformations of a protein or peptide is required [7]. In these cases the use of the free energy force fields may be useful. Another example is a situation in which the interaction of a number of small molecules with a macromolecule is to be investigated. In such a case it may be appropriate to treat both the small molecules and the macromolecule with one of the small-molecule-based force fields, although the quality of the treatment of the macromolecule may be sacrificed. In these cases the reader is advised against using one force field for the macromolecule and a second, unrelated, force field for the small molecules. There are often significant differences in the assumptions made when the parameters were being developed that would lead to a severe imbalance between the energetics and forces dictating the individual macromolecule and small molecule structures and the interactions between those molecules. If possible, the user should select a model system related to the particular application for which extensive experimental data are available. Tests of different force fields (and programs) can then be performed to see which best reproduces the experimental data for the model system and would therefore be the most appropriate for the application.

IV. DEVELOPMENT OF EMPIRICAL FORCE FIELDS

As emphasized by the word "empirical" to describe the force fields used for biomolecular computations, the development of these force fields is largely based on the methods and target data used to optimize the parameters in the force field. Decisions concerning these methods and target data are strongly dependent on the force field developer. To a large extent, even the selection of the form of the potential energy function itself is empirical, based on considerations of what terms are and are not required to obtain satisfactory results. Accordingly, the philosophy, or assumptions, used in the development of a force field will dictate both its applicability and its quality. A brief discussion of some of the philosophical considerations behind the most commonly used force fields follows.

A. Philosophical Considerations Behind Commonly Used Force Fields

Step 1 in the development of a force field is a decision concerning its applicability and transferability. The applicability issue was discussed in Section III.D and can be separated, on one level, into force fields for biological molecules and those for small molecules. Applicability also includes the use of explicit solvent representations (i.e., the solvent molecules themselves are included in the simulations), implicit solvent models [i.e., the solvent is included in a simplified, continuum-based fashion, the simplest being the use of a dielectric constant of 78 (for water) versus 1 (for vacuum)], or free energy based force fields. Transferability is concerned with the ability to take parameters optimized for a given set of target data and apply them to compounds not included in the target data. For example, dihedral parameters about a $C-C$ single bond may be optimized with respect to the rotational energy surface of ethane. In a transferable force field those parameters would then be applied for calculations on butane. In a nontransferable force field, the parameters for the $C-C-C-C$ and $C-C-C-H$ dihedrals not in ethane would be optimized specifically by using target data on butane. Obviously, the definition of transferability is somewhat ambiguous, and the extent to which parameters can be transferred is associated with chemical similarity. However, because of the simplicity of empirical force fields, transferability must be treated with care.

Force fields for small molecules are generally considered transferable, the transferability being attained by the use of various cross terms in the potential energy function. Typically, a set of model compounds representing a type of functional group (e.g., azo compounds or bicarbamates) is selected. Parameters corresponding to the functional group are then optimized to reproduce the available target data for the selected model compounds. Those parameters are then transferred to new compounds that contain that functional group but for which unique chemical connectivities are present (see the ethane-to-butane example above). A recent comparison of several of the small-molecule force fields discussed above has shown this approach to yield reasonable results for conformational energies; however, in all cases examples exist of catastrophic failures [52]. Such failures emphasize the importance of user awareness when a force field is being applied to a novel chemical system. This awareness includes an understanding of the range of functional groups used in the optimization of the force field and the relationship of the novel chemical systems to those functional groups. The more dissimilar the novel compound and the compounds included in the target data, the less confidence the user should have in the obtained results. This is also true in the case of bifunctional compounds, where the physical properties of the first functional group could significantly change those of the second group and vice versa. In such cases it is recommended that some tests of the force field be performed via comparison with QM data (see below).

Of the biomolecular force fields, AMBER [21] is considered to be transferable, whereas academic CHARMM [20] is not transferable. Considering the simplistic form of the potential energy functions used in these force fields, the extent of transferability should be considered to be minimal, as has been shown recently [52]. As stated above, the user should perform suitable tests on any novel compounds to ensure that the force field is treating the systems of interest with sufficient accuracy.

Another important applicability decision is whether the force field will be used for gas-phase (i.e., vacuum) or condensed phase (e.g., in solution, in a membrane, or in the crystal environment) computations. Owing to a combination of limitations associated with available condensed phase data and computational resources, the majority of force fields prior to 1990 were designed for gas-phase calculations. With small-molecule force fields this resulted in relatively little emphasis being placed on the accurate treatment of the external interaction terms in the force fields. In the case of the biomolecular force fields designed to be used in vacuum via implicit treatment of the solvent environment, such as the CHARMM Param 19 [6,23] and AMBER force fields [22], care was taken in the optimization of charges to be consistent with the use of an R-dependent dielectric constant. The first concerted effort to rigorously model condensed phase properties was with the OPLS force field [53]. Those efforts were based on the explicit use of pure solvent and aqueous phase computations to calculate experimentally accessible thermodynamic properties. The external parameters were then optimized to maximize the agreement between the calculated and experimental thermodynamic properties. This very successful approach is the basis for the optimization procedures used in the majority of force fields currently being developed and used for condensed phase simulations.

Although while a number of additional philosophical considerations with respect to force fields could be discussed, presentation of parameter optimization methods in the remainder of this section will include philosophical considerations. It is worth reemphasizing the empirical nature of force fields, which leads to the creators of different ones having a significant impact on the quality of the resulting force field even when exactly the same form of potential energy function is being used. This is in large part due to the extensive nature of parameter space. Because of the large number of different individual parameters in a force field, an extensive amount of correlation exists between those parameters. Thus, a number of different combinations of parameters could reproduce a given set of target data. Although additional target data can partially overcome this problem, it cannot eliminate it, making the parameter optimization approach central to the ultimate quality of the force field. It should be emphasized that even though efforts have been made to automate parametrization procedures [54,55], a significant amount of manual intervention is generally required during parameter optimization.

B. Optimization Procedures Used in Empirical Force Fields

Knowledge of the approaches and target data used in the optimization of an empirical force field aids in the selection of the appropriate force field for a given study and acts as the basis for extending a force field to allow for its use with new compounds (see below). In this section some of the general considerations that are involved during the development of a force field are presented, followed by a more detailed description of the parameter optimization procedure.

Presented in Table 1 is a list of the parameters in Eqs. (2) and (3) and the type of target data used for their optimization. The information in Table 1 is separated into categories associated with those parameters. It should be noted that separation into the different categories represents a simplification; in practice there is extensive correlation between the different parameters, as discussed above; for example, changes in bond parameters that affect the geometry may also have an influence on $\Delta G_{solvation}$ for a given model compound. These correlations require that parameter optimization protocols include iterative approaches, as will be discussed below.

Internal parameters are generally optimized with respect to the geometries, vibrational spectra, and conformational energetics of selected model compounds. The equilib-

Table 1 Types and Sources of Target Data Used in the Optimization of Empirical Force Field Parameters

Term	Target data	Source
Internal		
Equilibrium terms, multiplicity, and phase (b_0, θ_0, n, δ)	Geometries	QM, electron diffraction, microwave, crystal survey
Force constants (K_b, K_θ, K_χ)	Vibrational spectra,	QM, IR, Raman
	Conformational properties	QM, IR, NMR, crystal survey
External		
VDW terms (ε_i, $R_{min,i}$)	Pure solvent properties [56] ($\Delta H_{vaporization}$, molecular volume)	Vapor pressure, calorimetry, densities
	Crystal properties ($\Delta H_{sublimation}$ [56] lattice parameters, non-bond distances)	X-ray and neutron diffraction, vapor pressure, calorimetry
	Interaction energies (dimers, rare gas–model compound, water–model compound)	QM, microwave, mass spectrometry
Atomic charges (q_i)	Dipole moments [57]	QM, dielectric permittivity, Stark effect, microwave
	Electrostatic potentials	QM
	Interaction energies (dimers, water–model compound)	QM, microwave, mass spectrometry
	Aqueous solution ($\Delta G_{solvation}$, $\Delta H_{solvation}$, partial molar volume [58])	Calorimetry, volume variations

QM = quantum mechanics; IR = infrared spectroscopy.

rium bond lengths and angles and the dihedral multiplicity and phase are often optimized to reproduce gas-phase geometric data such as those obtained from QM, electron diffraction, or microwave experiments. Such data, however, may have limitations when they are used in the optimization of parameters for condensed phase simulations. For example, it has been shown that the internal geometry of *N*-methylacetamide (NMA), a model for the peptide bond in proteins, is significantly influenced by the environment [59]. Therefore, a force field that is being developed for condensed phase simulations should be optimized to reproduce condensed phase geometries rather than gas-phase values [20]. This is necessary because the form of the potential energy function does not allow for subtle changes in geometries and other phenomena that occur upon going from the gas phase to the condensed phase to be reproduced by the force field. The use of geometric data from a survey of the Cambridge Crystal Database (CSD) [60] can be useful in this regard. Geometries from individual crystal structures can be influenced by non-bond interactions in the crystal, especially when ions are present. Use of geometric data from a survey overcomes this limitation by averaging over a large number of crystal structures, yielding condensed phase geometric data that are not biased by interactions specific to a single crystal. Finally, QM calculations can be performed in the presence of water molecules or with a reaction field model to test whether condensed phase effects may have an influence on the obtained geometries [61].

Optimization of the internal force constants typically uses vibrational spectra and conformational energetics as the primary target data. Vibrational spectra, which comprise the individual frequencies and their assignments, dominate the optimization of the bond and angle force constants. It must be emphasized that both the frequencies and assignments should be accurately reproduced by the force field to ensure that the proper molecular distortions are associated with the correct frequencies. To attain this goal it is important to have proper assignments from the experimental data, often based on isotopic substitution. One way to supplement the assignment data is to use QM-calculated spectra from which detailed assignments in the form of potential energy distributions (PEDs) can be obtained [62]. Once the frequencies and their assignments are known, the force constants can be adjusted to reproduce these values. It should be noted that selected dihedral force constants will be optimized to reproduce conformational energetics, often at the expense of sacrificing the quality of the vibrational spectra. For example, with ethane it is necessary to overestimate the frequency of the $C-C$ torsional rotation in order to accurately reproduce the barrier to rotation [63]. This discrepancy emphasizes the need to take into account barrier heights as well as the relative conformational energies of minima, especially in cases when the force field is to be used in MD simulation studies where there is a significant probability of sampling regions of conformational surfaces with relatively high energies. As discussed with respect to geometries, the environment can have a significant influence on both the vibrational spectra and the conformational energetics. Examples include the vibrational spectra of NMA [20] and the conformational energetics of dimethylphosphate [64], a model compound used for the parametrization of oligonucleotides. Increasing the size of the model compound used to generate the target data may also influence the final parameters. An example of this is the use of the alanine dipeptide to model the protein backbone versus a larger compound such as the alanine tetrapeptide [65].

Optimization of external parameters tends to be more difficult as the quantity of the target data is decreased relative to the number of parameters to be optimized compared to the internal parameters, leaving the solution more undetermined. This increases the problems associated with parameter correlation, thereby limiting the ability to apply automated parameter optimization algorithms. An example of the parameter correlation problem with van der Waals parameters is presented in Table 2, where pure solvent properties for ethane using three different sets of parameters are presented (AD MacKerell Jr, M Karplus, unpublished work). As may be seen, all three sets of LJ parameters presented in Table 2 yield heats of vaporization and molecular volumes in satisfactory agreement with the experimental data, in spite of the carbon R_{min} varying by over 0.5 Å among the three sets. The presence of parameter correlation is evident. As the carbon R_{min} increases and ε values decrease, the hydrogen R_{min} decreases and ε values increase. Thus, it is clear that special care needs to be taken during the optimization of the non-bond parameters to maximize agreement with experimental data while minimizing parameter correlation. Such efforts will yield a force field that is of the highest accuracy based on the most physically reasonable parameters.

Van der Waals or Lennard-Jones contributions to empirical force fields are generally considered to be of less importance than the electrostatic term in contributing to the nonbond interactions in biological molecules. This view, however, is not totally warranted. Studies have shown significant contributions from the VDW term to heats of vaporization of polar-neutral compounds, including over 50% of the mean interaction energies in liquid NMA [67], as well as in crystals of nucleic acid bases, where the VDW energy contributed between 52% and 65% of the mean interaction energies [18]. Furthermore, recent studies on alkanes have shown that VDW parameters have a significant impact on their calculated free energies of solvation [29,63]. Thus, proper optimization of VDW parameters is essential to the quality of a force field for condensed phase simulations of biomolecules.

Significant progress in the optimization of VDW parameters was associated with the development of the OPLS force field [53]. In those efforts the approach of using Monte Carlo calculations on pure solvents to compute heats of vaporization and molecular volumes and then using that information to refine the VDW parameters was first developed and applied. Subsequently, developers of other force fields have used this same approach for optimization of biomolecular force fields [20,21]. Van der Waals parameters may also be optimized based on calculated heats of sublimation of crystals [68], as has been done for the optimization of some of the VDW parameters in the nucleic acid bases [18]. Alternative approaches to optimizing VDW parameters have been based primarily on the use of QM data. Quantum mechanical data contains detailed information on the electron distribution around a molecule, which, in principle, should be useful for the optimization of VDW

Table 2   Ethane Experimental and Calculated Pure Solvent Properties[a]

Lennard Jones parameters[b]		Heat of vaporization[c]	Molecular volume
Carbon	Hydrogen		
3.60/0.190	3.02/0.0085	3.50	90.7
4.00/0.080	2.71/0.0230	3.48	90.9
4.12/0.080	2.64/0.0220	3.49	91.8
Experiment		3.56	91.5

[a] Calculations performed using MC BOSS [66] with the CHARMM combination rules. Partial atomic charges (C = −0.27 and H = 0.09) were identical for all three simulations.
[b] Lennard-Jones parameters are R_{min}/ε in angstroms and kilocalories per mole, respectively.
[c] Heat of vaporization in kilocalories per mole and molecule volume in cubic angstroms at −89°C [56].

parameters [12]. In practice, however, limitations in the ability of QM approaches to accurately treat dispersion interactions [69–71] make VDW parameters derived solely from QM data yield condensed phase properties in poor agreement with experiment [72,73]. Recent work has combined the reproduction of experimental properties with QM data to optimize VDW parameters while minimizing problems associated with parameter correlation. In that study QM data for helium and neon atoms interacting with alkanes were used to obtain the relative values of the VDW parameters while the reproduction of pure solvent properties was used to determine their absolute values, yielding good agreement for both pure solvent properties and free energies of aqueous solvation [63]. The reproduction of both experimental pure solvent and free energies of aqueous solvation has also been used to derive improved parameters [29]. From these studies it is evident that optimization of the VDW parameters is one of the most difficult aspects of force field optimization but also of significant importance for producing well-behaved force fields.

Development of models to treat electrostatic interactions between molecules represents one of the most central, and best studied, areas in force field development. For biological molecules, the computational limitations discussed above have led to the use of the Coulombic model included in Eq. (3). Despite its simplistic form, the volume of work done on the optimization of partial atomic charges, as well as the appropriate dielectric constant, has been huge. The present discussion is limited to currently applied approaches to the optimization of partial atomic charges. These approaches are all dominated by the reproduction of target data from QM calculations, although the target data can be supplemented with experimental data on interaction energies and orientations and molecular dipole moments when such data are available.

Method 1 is based on optimizing partial atomic charges to reproduce the electrostatic potential (ESP) around a molecule determined via QM calculations. Programs are available to perform this operation [74,75], and some of these methodologies have been incorporated into the GAUSSIAN suite of programs [76]. A variation of the method, in which the charges on atoms with minimal solvent accessibility are restrained, termed RESP [77,78], has been developed and is the basis for the partial atomic charges used in the 1995 AMBER force field. The goal of the ESP approach is to produce partial atomic charges that reproduce the electrostatic field created by the molecule. The limitation of this approach is that the polarization effect associated with the condensed phase environment is not explicitly included, although the tendency for the HF/6-31G* QM level of theory to overestimate dipole moments has been suggested to account for this deficiency. In addition, experimental dipole moments can be included in the charge-fitting procedure. An alternative method, used in the OPLS, MMFF, and CHARMM force fields, is to base the partial atomic charges on the reproduction of minimum interaction energies and distances between small-molecule dimers and small molecule–water interacting pairs determined from QM calculations [6,53]. In this approach a series of small molecule–water (monohydrate) complexes are subjected to QM calculations for different idealized interactions. The resulting minimum interaction energies and geometries, along with available dipole moments, are then used as the target data for the optimization of the partial atomic charges. Application of this approach in combination with pure solvent and aqueous solvent simulations has yielded offsets and scale factors that allow for the production of charges that yield reasonable condensed phase properties [67,79]. Advantages of this method are that the use of the monohydrates in the QM calculations allows for local electronic polarization to occur at the different interacting sites, and the use of the scale factors accounts for the multibody electronic polarization contributions that are not included explicitly in Eq. (3).

As for the dielectric constant, when explicit solvent molecules are included in the calculations, a value of 1, as in vacuum, should be used because the solvent molecules themselves will perform the charge screening. The omission of explicit solvent molecules can be partially accounted for by the use of an R-dependent dielectric, where the dielectric constant increases as the distance between the atoms, r_{ij}, increases (e.g., at a separation of 1 Å the dielectric constant equals 1; at a 3 Å separation the dielectric equals 3; and so on). Alternatives include sigmoidal dielectrics [80]; however, their use has not been widespread. In any case, it is important that the dielectric constant used for a computation correspond to that for which the force field being used was designed; use of alternative dielectric constants will lead to improper weighting of the different electrostatic interactions, which may lead to significant errors in the computations.

C. Explicit Solvent Models and the Importance of Balancing the External Interactions

Proper condensed phase simulations require that the non-bond interactions between different portions of the system under study be properly balanced. In biomolecular simulations this balance must occur between the solvent–solvent (e.g., water–water), solvent–solute (e.g., water–protein), and solute–solute (e.g., protein intramolecular) interactions [18,21]. Having such a balance is essential for proper partitioning of molecules or parts of molecules in different environments. For example, if the solvent–solute interaction of a glutamine side chain were overestimated, there would be a tendency for the side chain to move into and interact with the solvent. The first step in obtaining this balance is the treatment of the solvent–solvent interactions. The majority of biomolecular simulations are performed using the TIP3P [81] and SPC/E [82] water models.

The SPC/E water model is known to yield better pure solvent properties than the TIP3P model; however, this has been achieved by overestimating the water–dimer interaction energy (i.e., the solvent–solvent interactions are too favorable). Although this overestimation is justifiable considering the omission of explicit electronic polarizability from the force field, it will cause problems when trying to produce a balanced force field due to the need to overestimate the solute–solvent and solute–solute interaction energies in a compensatory fashion. Owing to this limitation, the TIP3P model is suggested to be a better choice for the development of a balanced force field. It is expected that water models that include electronic polarization will allow for better pure solvent properties while having the proper solvent–solvent interactions to allow for the development of balanced force fields. It is important when applying a force field to use the water model for which that particular force field was developed and tested. Furthermore, extensions of the selected force field must maintain compatibility with the originally selected water model.

D. Use of Quantum Mechanical Results as Target Data

Throughout this chapter and in Table 1 the inclusion of QM results as target data is evident, with the use of such data in the optimization of empirical forces fields leading to many improvements. Use of QM data alone, however, is insufficient for the optimization of parameters for condensed phase simulations. This is due to limitations in the ability to perform QM calculations at an adequate level combined with limitations in empirical force fields. As discussed above, QM data are insufficient for the treatment of dispersion interactions, disallowing their use alone for the optimization of Van der Waals parameters. The use of HF/6-31G*-calculated intermolecular interaction energies for the optimization of partial atomic charges has been successful because of extensive testing of the ability of optimized charges to reproduce experimentally determined condensed phase values, thereby allowing for the appropriate offsets and scaling factors to be determined (see below).

In many cases, results from QM calculations are the only data available for the determination of conformational energetics. However, there is a need for caution in using such data alone, as evidenced by recent work showing that the rigorous reproduction of QM energetic data for the alanine dipeptide leads to systematic variations in the conformation of the peptide backbone when applied to MD simulations of proteins [20]. Furthermore, QM data are typically obtained in the gas phase, and, as discussed above, significant changes in geometries, vibrations, and conformational energetics can occur in going from the gas phase to the condensed phase. Although the ideal potential energy function would properly model differences between the gas and condensed phases, this has yet to be realized. Thus, the use of QM results as target data for the optimization of force fields must include checks against experimentally accessible data whenever possible to ensure that parameters appropriate for the condensed phase are being produced.

E. Extension of Available Force Fields: Application to CHARMM

Selection of a force field is often based on the molecules of interest being treated by a particular force field. Although many of the force fields discussed above cover a wide range of functionalities, they may not be of the accuracy required for a particular study. For example, if a detailed atomistic picture or quantitative data are required on the binding of a series of structurally similar compounds to a protein, the use of a general force field may not be appropriate. In such cases it may be necessary to extend one of the force fields refined for biomolecular simulations to be able to treat the new molecules. When this is to be done, the optimization procedure must be the same as that used for the development of the original force field. In the remainder of this chapter a systematic procedure to obtain and optimize new force field parameters is presented. Due to my familiarity with the CHARMM force field, this procedure is consistent with those parameters. An outline of the parametrization procedure is presented in Figure 2. A similar protocol for the AMBER force field has been published [83] and can be supplemented with information from the AMBER web page.

1. Selection of Model Compounds

Step 1 of the parametrization process is the selection of the appropriate model compounds. In the case of small molecules, such as compounds of pharmaceutical interest, the model compound may be the desired molecule itself. In other cases it is desirable to select several small model compounds that can then be "connected" to create the final, desired molecule. Model compounds should be selected for which adequate experimental data exist, as listed in Table 1. Since in almost all cases QM data can be substituted when experimental data are absent (see comments on the use of QM data, above), the model compounds should be of a size that is accessible to QM calculations using a level of theory no lower than HF/6-31G*. This ensures that geometries, vibrational spectra, conformational energetics, and model compound–water interaction energies can all be performed at a level of theory such that the data obtained are of high enough quality to accurately replace and

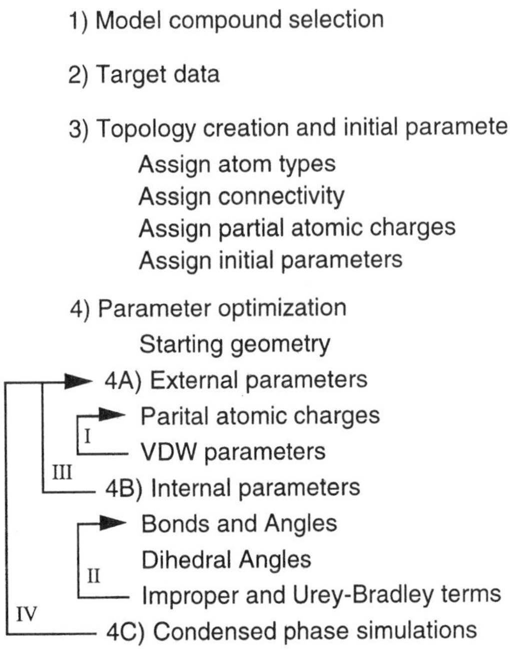

1) Model compound selection

2) Target data

3) Topology creation and initial parameter selection
      Assign atom types
      Assign connectivity
      Assign partial atomic charges
      Assign initial parameters

4) Parameter optimization
      Starting geometry
      4A) External parameters
         Parital atomic charges
         VDW parameters
      4B) Internal parameters
         Bonds and Angles
         Dihedral Angles
         Improper and Urey-Bradley terms
      4C) Condensed phase simulations

(Loop markers: I, II, III, IV)

Figure 2   Outline of the steps involved in the preparation of a force field for the inclusion of new molecules and optimization of the associated parameters. Iterative loops (I) over individual external terms, (II) over individual internal terms, (III) over the external and internal terms. In loop (IV) over the condensed phase simulations, both external terms and internal terms are included.

supplement the experimental data. Finally, the model compounds should be of such a size that when they are connected to create the final molecule, QM calculations of at least the HF/3-21G* level (though HF/6-31G* is preferable) can be performed to test the linkage.

For illustration of the parametrization concepts, methotrexate, the dihydrofolate reductase inhibitor, was selected as a model system. Its structure is shown in Figure 3a. Methotrexate itself is too large for QM calculations at a satisfactory level, requiring the use of smaller model compounds that represent the various parts of methotrexate. Examples of model compounds that could be used for the parametrization of methotrexate are included as compounds **1–3** in Figure 3a, which are, associated with the pteridine, benzene, and diacid moieties, respectively. It may be assumed that some experimental data would be available for the pteridine and diacid compounds and that information on the chemical connectivities internal to each compound could be obtained from a survey of the CSD [60]. Each of these compounds is of such a size that HF/6-31G* calculations are accessible, and at least HF/3-21G* calculations would be accessible to the dimers, as required to test the parameters connecting the individual model compounds. An alternative model compound would include the amino group with model **3**, yielding glutamic acid; however, that would require breaking the amide bond on compound **2**, which would cause the loss of some of the significant chemical characteristics of methotrexate. Of note is the use of capping methyl groups on compounds **1** and **2**. With **1** the methyl group will ensure that the partial atomic charges assigned to the pteridine ring accurately reflect the covalent bond to the remainder of the molecule. The same is true in the case of model compound **2**, although

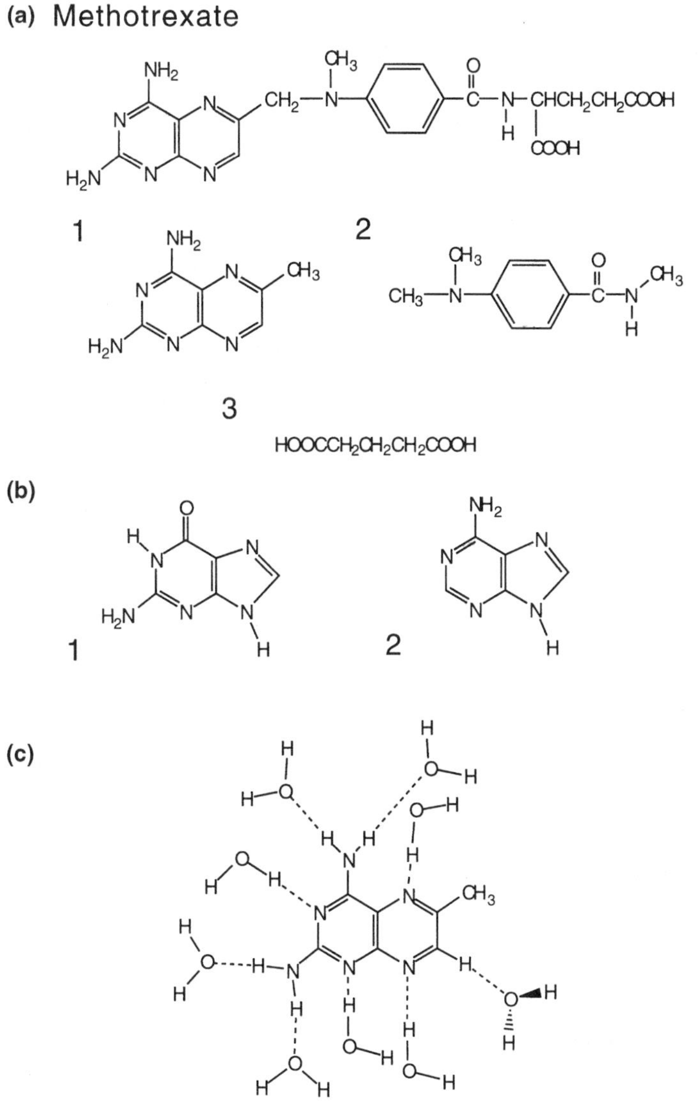

(a) Methotrexate

1   **2**

3

HOOCCH₂CH₂CH₂COOH

$HOOCCH_2CH_2CH_2COOH$

(b)

1   **2**

(c)

Figure 3 (a) Structure of methotrexate and the structures of three model compounds that could be used for parameter optimization of methotrexate. (b) The structures of (**1**) guanine and (**2**) adenine. (c) Interaction orientations between model compound of **1**(a) and water to be used in the optimization of the partial atomic charges. Note that in the optimization procedure the water–model compound dimers are treated individually (e.g., as monohydrates).

in this case the presence of the methyl groups is even more important; the properties of a primary amine, even in an amide, can be expected to differ significantly from those of the secondary amine present in methotrexate. Including the methyl cap ensures that the degree of substitution of the amine, or any other functional group, is the same in the model compound as in the final compound to be used in the calculations.

2. Target Data Identification

Simultaneous with the selection of the appropriate model compounds is the identification of the target data, because the availability of adequate target data in large part dictates the selection of the model compound. Included in Table 1 is a list of the various types of target data and their sources. Basically, the parameters for the new compounds will be optimized to reproduce the selected target data. Thus, the availability of more target data will allow the parameters to be optimized as accurately as possible while minimizing problems associated with parameter correlation, as discussed above. With respect to the types of target data, efforts should be made to identify as many experimental data as possible while at the same time being aware of possible limitations in those data (e.g., counterion contributions in IR spectra of ionic species). The experimental data can be supplemented and extended with QM data; however, the QM data themselves are limited due to the level of theory used in the calculations as well as the fact that they are typically restricted to the gas phase. As discussed above, target data associated with the condensed phase will greatly facilitate the optimization of a force field for condensed phase simulations.

3. Creation of Topology and Initial Parameter Selection

Once the model compounds are selected, the topology information (e.g., connectivity, atomic types, and preliminary partial atomic charges) must be input into the program and the necessary parameters supplied to perform the initial energy calculations. This is initiated by identifying molecules already present in the force field that closely mimic the model compound. In the case of model compound **1** in Figure 3a, the nucleic acid bases guanine and adenine, shown as compounds **1** and **2**, respectively, in Figure 3b, would be reasonable starting points. Although going from a 5,6 to a 6,6 fused ring system, the distribution of heteroatoms between the ring systems is similar and there are common amino substituents. The initial information for model compound **1** would be taken from guanine (e.g., assign atomic types and atomic connectivity). To this an additional aromatic carbon would be added to the five-membered ring and the atomic types on the two carbons in the new six-membered ring would have to be switched to those corresponding to six-membered rings. For the methyl group, atomic types found on thymine would be used. Atomic types for the second amino group on model compound **1**, which is a carbonyl in guanine, would be extracted from adenine. This would include information for both the second amino group and the unprotonated ring nitrogen. Completion of the topology for compound **1** in Figure 3a would involve the creation of reasonable partial atomic charges. In one approach, the charges would be derived on the basis of analogy to those in guanine and adenine; with the charges on the new aromatic carbon and covalently bound hydrogen set equivalent and of opposite sign, the now methylated aromatic carbon would be set to a charge of zero and the methyl group charges would be assigned a total charge of zero (e.g., $C = -0.27$, $H = 0.09$). Care must be taken at this stage that the total charge on the molecule is zero. Alternatively, charges from Mulliken population analysis of an HF/6-31G* [84] calculation could act as a starting point. Concerning the VDW parameters, assignment of the appropriate types of atoms to the model compound simultaneously assigns the VDW parameters.

At this point the information required by CHARMM to create the molecule is present, but the parameters necessary to perform energy calculations are not all available yet. In the case of CHARMM, the program is designed to report missing parameters when an energy calculation is requested. Taking advantage of this feature, missing parameters can be identified and added to the parameter file. The advantage of having the program identify the missing parameters is that only new parameters that are unique to your system will be added. It is these added parameters that will later be adjusted to improve the agreement between the empirical and target data properties for the model compound. Note that no parameters already present in the parameter file should be changed during the optimization procedure, because this would compromise the quality of the molecules that had previously been optimized. It is highly recommended that the use of wild cards to create the needed parameters be avoided, because it could compromise the ability to efficiently optimize the parameters.

4. Parameter Optimization

Empirical force field calculations during the optimization procedure should be performed in a fashion consistent with the final application of the force field. With recent developments in the Ewald method, particularly the particle mesh Ewald (PME) approach [85], it is possible to perform simulations of biological molecules in the condensed phase with effectively no cutoff of the non-bond interactions. Traditionally, to save computational resources, no atom–atom non-bond interactions beyond a specified distance are included in the calculation; the use of PME makes this simplification unnecessary (i.e., no distance-based truncation of non-bond interactions). Accordingly, all empirical calculations in the gas phase (e.g., water–model compound interactions, energy minimizations, torsional rotation surfaces) should be performed with no atom–atom truncation, and condensed phase calculations should be performed using PME. In addition, condensed phase calculations should also be used with a long-tail correction for the VDW interactions. Currently, such a correction is not present in CHARMM, although its implementation is in progress. Other considerations are the dielectric constant, which should be set to 1 for all calculations, and the 1,4 scaling factor (see legend of Fig. 1), which should also be set to 1.0 (no scaling).

Initiation of the parameter optimization procedure requires that an initial geometry of the model compound be obtained (see flow diagram in Fig. 2). The source of this can be an experimental, modeled, or QM-determined structure. What is important is that the geometry used represent the global energy minima and that it be reasonably close to the final empirical geometry that will be obtained from the parameter optimization procedure.

(a) External Parameters. The parameter optimization process is initiated with the external terms because of the significant influence those terms have on the final empirical geometries and conformational energetics. Since reasonable starting geometries can readily be assigned from an experimental or QM structure, the external parameters obtained from the initial round of parametrization can be expected to be close to the final values. Alternatively, starting the optimization procedures with the internal terms using very approximate external parameters could lead to extra iterations between the internal and external optimization procedures (loop III in Fig. 2) owing to possibly large changes in the geometries, vibrations, and conformational energetics when the external parameters were optimized during the first or second iteration. It must be emphasized that the external parameters are influenced by the internal terms such that iterations over the internal and external parameters are necessary (loop III in Fig. 2).

Partial Atomic Charges. Determination of the partial atomic charges requires minimum interaction energies and geometries for individual water molecules interacting with different sites on the model compounds. An example of the different interaction orientations is shown in Figure 3c for model compound **1**, Figure 3a. As may be seen, idealized interactions with all the hydrogen bonding sites as well as nonpolar sites are investigated. Note that the procedure is carried out only on the individual monohydrates (i.e., the model compound and one water molecule) and not on a system in which the model compound is interacting with multiple water molecules. Typically to obtain the QM target data the model compound is geometrically optimized at the HF/6-31G* level. Individual water molecules are then placed in idealized orientations and only the interaction distance and, in some cases, an additional angle are optimized at the HF/6-31G* level while the model compound and water intramolecular geometries are fixed; the water geometry is typically the TIP3P geometry. From this optimization the minimum interaction energy and distance are obtained. The interaction energy is determined on the basis of the difference between the total energy of the model compound–water supramolecular complex and the individual monomer energies. No correction for basis-set superposition error (BSSE) is required in the present approach. At this stage the QM interaction distances and energies are offset and scaled as follows to produce the final target data that will allow for the optimization of partial atomic charges that yield reasonable condensed phase properties. The offsets and scalings are performed as follows. The QM distances are decreased by 0.2 Å for polar–neutral interactions, by 0.1 Å for hydrogen bonds involving charged species, and not offset for interactions between water and nonpolar sites. Scaling of the interaction energies by 1.16 is performed for all interactions involving polar–neutral compounds and no scaling is performed for the charged compounds. The 1.16 scaling factor is based on the ratio of the TIP3P to HF/6-31G* water–dimer interaction energies. The overestimation of the TIP3P water interaction energy partially accounts for the omission of explicit polarizability in the force field and use of the same 1.16 scaling factor maintains the balance between the solvent–solvent and solute–solvent interactions. In addition to the water–model compound interactions it may also be useful to perform calculations to obtain model compound–model compound interaction information. Such data are most useful for small model compounds (e.g., methanol or NMA) and special cases, such as the Watson–Crick basepairs, where specific types of interactions dominate the properties of the system.

Once the target data are obtained, the partial atomic charges can then be optimized to reproduce the QM interaction offset distances and scaled energies. Along with reproduction of the QM interaction data the partial charge optimization can also include dipole moments (magnitude as well as direction), from either experiment or QM calculations, as target data. For polar–neutral compounds the empirical dipole moments are typically larger than the experimental or QM gas-phase values owing to the need to include electronic polarization effects implicitly in the charge distribution, though no well-defined scale factor is applied. When performing the charge optimization it is suggested that groups of atoms whose sum of partial atomic charges yields a unit charge be used. Typically unit charge groups of three to seven atoms are used; however, in the most recent version of the CHARMM nucleic acid parameters [86] the unit charges were summed over the entire bases. The use of unit charge groups allows for the use of the group truncation option in CHARMM [87] and simplifies the reorganization of the charge when combining model compounds into larger chemical entities (see below).

Aliphatic and aromatic partial atomic charges are a special case. In CHARMM all the aliphatic hydrogens are assigned a charge of 0.09, with the carbon charge adjusted to yield a unit charge. For example, in methane, with hydrogen charges of 0.09, the carbon charge is −0.36; and in methanol, with aliphatic hydrogen charges of 0.09 and charges of −0.66 and 0.43 on the hydroxyl oxygen and hydrogen atoms, the charge on the aliphatic carbon is set to −0.04 to yield a total charge of 0.0. The use of the same charge on all the aliphatic hydrogens, which was set to 0.09 based on the electrostatic contribution to the butane torsional surface [88], is justified by both the free energy of solvation of alkanes (S Fischer, M Karplus, personal communication) and the pure solvent properties of ethane being insensitive to the charge distribution. Concerning the aromatic carbons and hydrogens, charges of 0.115 and −0.115, based on the condensed phase properties of benzene [89], are used on all aromatic functional groups. The only exception are rings containing heteroatoms (e.g., pyridine), where the carbon–hydrogen pairs adjacent to the heteroatom may have different charges; these charges would be determined via interactions with water as presented above. At points of substitution of aromatic groups, the carbon charge is set to 0.0 except when substituted by a heteroatom (e.g., tyrosine), where a charge that yields good interaction energies of the substituent with water is assigned to the ring carbon (i.e., to the ζ atom in tyrosine).

Beyond its simplicity, the use of the same charges for alkanes and aromatic functional groups during the fitting procedure also allows charges to be systematically altered when small model compounds are connected. This is done by replacing an aliphatic or aromatic hydrogen with the new covalent bond required to connect the two model compounds and moving the charge previously on the hydrogen into the carbon to which it was attached. For example, to connect model compounds **1** and **2** in Figure 3a, a hydrogen from each of the terminal methyl groups would be removed, the charges on each of the methyl carbons increased by 0.09, and the carbon–carbon covalent bond created.

van der Waals Parameters. The van der Waals and Lennard-Jones (LJ) parameters are the most challenging of the parameters in Eqs. (2) and (3) to optimize. In the majority of cases, however, direct transfer of the VDW parameters based on analogy with atoms already in the force field will yield satisfactory results. This is particularly true in cases where ligands that interact with biological molecules are being parametrized. In the case of methotrexate (see Fig. 3a), all the atoms in the three model compounds are directly analogous to those in DNA (e.g., the pteridine ring) or in proteins (e.g., the benzene ring, the amide group, and the carboxylic acids). In cases where the condensed phase properties of the new molecules themselves are to be studied, formal optimization of the VDW parameters is required. This requires the identification of appropriate experimental condensed phase target data along with QM calculations on rare gas–model compound interactions (see Table 1). Detailed presentations of the methods to perform such optimizations have been published elsewhere [27,63]. Note that in cases where optimization of the VDW parameters is required, the majority of the VDW parameters can be transferred directly from the available parameters (e.g., alkane functional groups or polar hydrogens) and only the VDW parameters of one or two atoms (e.g., the sulfur and oxygen in dimethylsulfoxide) actually optimized. If the VDW parameters are included in the optimization procedure it is essential that the partial atomic charges be rechecked for agreement with the target data. If significant changes do occur, further optimization of the charges must be performed, followed by rechecking of the VDW parameters, and so on, in an iterative fashion (loop I in Fig. 2).

(b) Internal Parameters. Molecular geometries are dominated by the bond and angle equilibrium terms and the dihedral multiplicity and phase, whereas the vibrational spectra and conformational energetics are controlled primarily by the internal term force constants, as presented in Table 1. Typically, the vibrational spectra will be largely influenced by the bond and angle force constants while dihedral force constants control the conformational energetics. These relationships, however, are not absolute, requiring iteration over the different internal parameter optimization steps (see loop II in Fig. 2). Furthermore, the

external parameters can have a significant influence on the geometries, vibrations, and energetics. Thus, any change in those parameters requires that the internal parameters be rechecked with respect to the target data and additional optimization performed as required (see loop III in Fig. 2).

Bond, Angle, and Dihedral Terms. Adjustment of the bond and angle equilibrium values to reproduce the geometric target data is generally straightforward. It should be noted, however, that the minimized structure generally does not have bond lengths and angles that correspond directly to the equilibrium bond and angle parameters. This is due to the influence of the other terms in the potential energy function on the final geometry. In planar systems (e.g., aromatic rings) the angles should sum to 360° for the three angles around a single atom, to 540° for the five endocyclic angles in a five-membered ring, and to 720° for the six endocyclic angles in a six-membered ring to ensure that there is no angular strain at the planar minima. Initial optimization of the dihedral phase and multiplicity should be performed by assigning only one dihedral parameter to each torsion angle and assigning that dihedral the appropriate values for the type of covalent bond (e.g., an sp^3–sp^3 bond would have a multiplicity of 3 and a phase of 0.0). Note that for nonplanar systems incorrect selection of the dihedral phase and multiplicity will often lead to minima with the wrong conformation.

Force constant optimization is initially performed by reproducing experimental or QM vibrational spectra. Quantum mechanical frequencies generally have to be scaled to yield experimentally relevant values. This is best done by comparison with experimental data on the same molecule and calculation of the appropriate scale factor (see example with dimethylphosphate in Ref. 18); however, if no experimental data are available, then published scale factors associated with different QM levels of theory should be applied [90–92]. As mentioned above, the parameters should be optimized to reproduce the assignments of the frequencies as well as their numerical values. This is best performed by producing a potential energy distribution (PED), where the contributions of the different normal modes (e.g., symmetric methyl stretch, asymmetric stretch of water, and so on) are assigned to the individual frequencies [62,93]. The module MOLVIB [94] in CHARMM allows for calculation of PED for the empirically calculated vibrational spectra as well as for vibrational spectra calculated via QM calculations. Thus, the PED from the empirical vibrational spectra can be compared directly with assignments for experimental data based on isotopic substitution and/or QM-based assignments.

Final optimization of the dihedral parameters is performed on the basis of conformational energetics of the model compounds. Typically, only dihedrals containing all nonhydrogen atoms are used for adjustment of the conformational energies, with dihedrals that include terminal hydrogens optimized on the basis of the vibrational spectra. While experimentally determined conformational energetics are available for a wide variety of molecules, typically data from QM calculations at the HF/6-31G* level or higher are used. Use of QM data allows for calculation of entire energy surfaces and also yields the energies of all minima when several local minima are present. In addition, QM data include changes in geometry (i.e., bond lengths and angles) as a function of a torsional energy surface, allowing for an additional check of the bond and angle parameters.

Model compound **2** in Figure 3a offers a good example. In that molecule at least two torsional energy profiles should be calculated, one for the C_{methyl}—N—$C_{aromatic}$—$C_{aromatic}$ torsion and a second for the $C_{aromatic}$—$C_{aromatic}$—C=O torsion. Torsional surfaces for the methyl rotations and about the amide could also be investigated, although the strong analogy of these groups with previously parametrized functional groups indicates that the assigned parameters should yield reasonable results. Each torsional surface would be calculated at the HF/6-31G* level of theory or higher by fixing the selected torsion at a given value and relaxing the remainder of the molecule, followed by incrementing that dihedral by 15° or 30°, reoptimizing, and so on, until a complete surface is obtained (i.e., an adiabatic surface). Note that the energy from the fully optimized structure (global minima), with no dihedral constraints, should be used to offset the entire surface with respect to zero. Using these as target data, the corresponding empirical torsional energy surface is calculated and compared to the QM surface. The dihedral parameters are then optimized to maximize the agreement between the empirical and target surfaces. At this stage it is often helpful to add additional dihedrals with alternate multiplicities (i.e., create a Fourier series) to more accurately reproduce the target surface, although all the dihedral parameters (i.e., the force constant, phase, and multiplicity of a single dihedral term) contributing to a torsional surface should first be adjusted to reproduce the target surface. It should be noted that the empirical energy surface associated with rotation of a torsion angle will contain contributions from all the terms in the potential energy function, both external and internal, and not just from the dihedral term. This again emphasizes the necessity of iterating over all the individual optimization steps in order to obtain a final, consistent set of parameters.

Following initial optimization of all the bond, angle, and dihedral parameters, it is important to reemphasize that all the target data must be rechecked for convergence (see below) via an iterative approach (see loop II of Fig. 2). This is due to the parameter correlation problem. Even though excellent agreement may be achieved with the geometric target data initially, adjustment of the force constants often alters the minimized geometry such that the bond and angle equilibrium parameters must be readjusted; typically the dihedral multiplicity and phase are not significantly affected. Thus, an iterative approach must be applied for the different internal parameters to ensure that the target data are adequately reproduced (loop II in Fig. 2).

Improper and Urey–Bradley Terms.   When initially optimizing the internal parameters for a new molecule, only the bond, angle, and dihedral parameters should be included. If at the end of the iterative optimization of the internal parameters, agreement with the target data is still not satisfactory, then improper and Urey–Bradley terms can be added where needed. Improper terms are typically used to improve agreement with respect to out-of-plane distortions associated with planar groups (Wilson wags). For example, with model compound **2**, Figure 3a, improper terms could be added for the aromatic hydrogens as well as for the amine and amide substituents. Urey–Bradley terms are often helpful for the proper assignment of symmetric and asymmetric stretching modes in, for example, methyl groups. They can also be used for optimizing energy surfaces including changes in angles as a function of torsional surfaces. This approach has been used for proper treatment of the conformational energetics of dimethylphosphate and methylphosphate [18].

(c) Convergence Criteria.   As mentioned several times in the preceding section and as emphasized in Figure 2, the parameter optimization approach is an iterative procedure required to ensure satisfactory agreement with all available target data. Iterative cycles must be performed over the partial atomic charges and the VDW parameters (loop I), over the internal parameters (loop II), and over the external and internal parameter optimization protocols (loop III). Furthermore, in certain cases it may be necessary to introduce another iterative cycle based on additional condensed phase simulations (loop IV, see below), although this is typically not required for the optimization of small-molecule parameters.

With any iterative approach it is necessary to have convergence criteria in order to judge when to exit an iterative loop. In the case of the optimization of empirical force field parameters it is difficult to define rigorous criteria due to the often poorly defined and system-dependent nature of the target data; however, guidelines for such criteria are appropriate. In the case of the geometries, it is expected that bond lengths, angles, and torsion angles of the fully optimized model compound should all be within 0.02 Å, 1.0°, and 1.0° respectively, of the target data values. In cases where both condensed phase and gas-phase data are available, the condensed phase data should be weighted more than the gas-phase data, although in an ideal situation both should be accurately fit. It should be noted that, because of the harmonic nature of bonds and angles, values determined from energy minimization are generally equivalent to average values from MD simulations, simplifying the optimization procedure. This, however, is less true for torsional angles and for non-bond interactions, for which significant differences in minimized and dynamic average values can exist. With respect to vibrational data, generally a root-mean-square (RMS) difference of 10 cm^{-1} or an average difference of 5% between the empirical and target data should be considered satisfactory. Determination of these values, however, is generally difficult, owing to problems associated with unambiguous assignment of the normal modes to the individual frequencies. Typically the low frequency modes (below 500 cm^{-1}) associated with torsional deformations and out-of-plane wags and the high frequency modes (above 1500 cm^{-1}) associated with bond stretching are easy to assign, but significant mixing of different angle bending and ring deformation modes makes assignments in the intermediate range difficult. What should be considered when optimizing force constants to reproduce vibrational spectra is which modes will have the greatest impact on the final application for which the parameters are being developed. If that final application involves MD simulations, then the low frequency modes, which involve the largest spatial displacements, are the most important. Accordingly, efforts should be made to properly predict both the frequencies and assignments of these modes. For the 500–1500 cm^{-1} region, efforts should be made to ensure that frequencies dominated by specific normal modes are accurately predicted and that the general assignment patterns are similar between the empirical and target data. Finally, considering the simplicity of assigning stretching frequencies, the high frequency modes should be accurately assigned, although the common use of the SHAKE algorithm [95] to constrain covalent bonds during MD simulations, especially those involving hydrogens, leads to these modes often having no influence on the results of MD simulations.

With respect to conformational energetics, the most important task is to properly select the target data for the optimization. As discussed above, these data are typically from QM calculations, with the level of theory depending on the size of the model compound and available computational resources. Typically, the larger the basis set, the better, although the HF/6-31G* level has been shown to yield generally satisfactory results, especially for predicting the relative energies of local minima. Calculation of barrier heights, particularly in nonaromatic ring systems that include heteroatoms [96], often requires the use of MP2/6-31G* calculations. For molecules that contain multiple local minima that have similar energies, higher levels of theory are often required, and care must be taken that the actual order of the energies of the minima is correct [52]. Emphasis should also be placed on properly reproducing the energetics associated with barriers. This is especially true for barriers of 2 kcal/mol or less, which can be frequently crossed during room temperature MD simulations. It should be noted that many force field optimization and comparison studies omit information on energy barriers; such an omission can lead to poor energetic properties that could cause significant problems in MD simulations. Once a set of target energies have been selected, the parameters should be adjusted to reproduce these as accurately as possible, generally within 0.2 kcal/mol of the target data. The lower energy conformations should be weighted more than the high energy terms, because those conformations are sampled more frequently in MD simulations. If necessary, at a later time, higher level QM calculations can be performed on the model compound to obtain improved target data that can be used to reoptimize the dihedral parameters. Two final points concerning the dihedral parameters. First, in optimizing these terms to reproduce the conformational energetics, there is often a decrease in the agreement with the vibrational target data. In such cases, it is suggested that the agreement with the energetic data be maximized. The second point concerns the height of the empirical energy barriers compared with the target data. In the CHARMM force field, emphasis has been placed on making the energy barriers lower than that of the target data rather than higher when ideal agreement cannot be achieved. This will tend to make the molecule more flexible and therefore more sensitive to environmental conditions. Creation of artificially high energy barriers will make the molecule more rigid, possibly locking it in an undesirable conformation with respect to its surrounding environment.

The use of intermolecular minimum interaction energies and distances between the model compounds and water makes the assignment of convergence criteria for the partial atomic charges straightforward. Typically, the energetic average difference (average over all interaction energy pairs) should be less than 0.1 kcal/mol and the rms difference should be less than 0.5 kcal/mol. The small value of the average difference is important because it ensures that the overall solvation of the molecule will be reasonable, while the rms criterion of 0.5 kcal/mol ensures that no individual term is too far off the target data. Emphasis should be placed on accurately reproducing the more favorable interactions, which are expected to be the more important in MD or MC simulations, at the expense of the less favorable interactions. With distances, the rms difference should be less than 0.1 Å; note that the $1/r^{12}$ repulsive wall leads to the differences generally being larger than the QM values, especially in the analysis of data from MD simulations. For both energies and differences, the criteria presented above are with respect to the target data after they have been offset and scaled (see above). In the case of small-molecule dimers (e.g., Watson–Crick basepairs), the difficulty is again in selection of the appropriate target data, as with the conformational energetics discussed in the preceding paragraph, rather than the degree of convergence. Again, suitable experimental or QM data must be identified and then the empirical parameters must be optimized to reproduce both sets of data as closely as possible. If problems appear during application of the parameters, then the target data themselves must be reassessed and the parameters reoptimized as necessary.

Concerning the VDW parameters, the ability to directly apply previously optimized values makes convergence criteria unnecessary. If VDW parameter optimization is performed based on pure solvent or crystal simulations, then the heats of vaporization or sublimation should be within 2% of experimental values, and the calculated molecular or unit cell volumes should be also. If rare gas–model compound data are used, the references cited above should be referred to for a discussion of the convergence criteria.

(d) Condensed Phase Testing. In the majority of cases where parameters are being optimized for a small compound or series of compounds to use with an already available force field, final testing of the parameters via condensed phase simulations is often not necessary. Rigorous use of crystal data, especially survey data, in force field optimization is discussed elsewhere [86,97]. If, however, crystal structures for one or more of the compounds exist or a crystal structure of the compound bound to a macromolecule exists,

then additional testing is appropriate and can quickly identify any significant problems with the parameters. These tests are performed by creating the explicit crystal environment followed by energy minimization and MD simulations in that environment. The CRYSTAL module in CHARMM is useful for this purpose [98]. From the MD simulation, which generally converges within 100 ps or less [99], the averages of the unit cell parameters, internal geometries, and nonbond interaction distances can be determined and compared directly with the experimental values. Such comparisons are more valid for small-molecule crystal simulations than for small-molecule–macromolecular complexes owing to the higher resolution of the former, although both are useful. Note that since crystal structures are obtained at finite temperatures it is appropriate to compare data from MD simulations at the same temperature rather than data from energy minimizations, as has been previously shown [18,20,99]. If results from the crystal simulations are not satisfactory, then a careful analysis should be made to identify which parameters can be associated with the problem, and those parameters should be subjected to additional optimization as required. It should be noted that if discrepancies do occur, a careful examination of the experimental structure should also be made, as errors in the experimental structures are possible, especially for the lower resolution macromolecular structures.

V. FUTURE DIRECTIONS

Improvements in empirical force fields continue, as evidenced by further improvements in the AMBER and CHARMM force fields and the new BMS force field of nucleic acids published during 1998 and 1999 based on the same form of energy function as that shown in Eqs. (1), (2), and (3) [86,97,100]. Thus, future efforts will continue to involve additional optimization of empirical force fields and their extension to new molecules. In certain cases, however, force fields based on Eqs. (1)–(3) are inadequate. One such case involves bond-breaking and -making events, where QM/MM-based approaches are relevant (see Chapter 11). Proper application of QM/MM methods will require special parametrization to ensure that the QM and MM regions and interactions between them are properly balanced [101–103]. Other cases include biomolecules where local electronic or steric interactions dominate the biological phenomena being investigated. In these cases, extension of the force fields to include electronic polarizability or anisotropic (nonspherical) van der Waals surfaces will be helpful. Concerning anisotropic VDW parameter studies, examples include membrane bilayers that contain unsaturated lipids, for which the current spherical representations of atoms are poor models for the carbons involved in double bonds [27].

Efforts to introduce electronic polarizability [104–110] and anistropic VDW models [111] have been undertaken but have not yet been applied to the study of biomolecular systems. Although it is to be expected that these advances will soon be used in biomolecular simulations, it should be emphasized that they should not be applied in all situations. As part of selecting the appropriate force field for a particular study, one should consider whether the additional computational costs associated with the use of a force field that includes electronic polarizability, anisotropic VDW surfaces, or both is worth the gains associated with those terms for the particular phenomenon under study. Furthermore, it should be emphasized that additional effects associated with going from a gas-phase environment to a condensed phase environment, such as changes in conformational energetics, may not necessarily be accurately modeled by inclusion of these terms alone, suggesting that additional extension of the potential energy function may be required.

VI. CONCLUSION

Atomistic empirical force fields have been shown to be effective tools for the study of biomolecular systems, and it can be expected that their use will greatly expand in the future. However, these methods must be used with care, as the complexity of the systems involved and the number of simplifications employed in the mathematical treatment of these systems can yield results that may be misleading and consequently lead to the improper interpretation of data. To minimize the possibility of such an outcome, it is essential that the user of biomolecular force fields understand the assumptions implicit to them and the approaches used to derive and optimize them. That knowledge will allow users to better select the appropriate force field for their particular application as well as to judge if results from that application are significantly influenced by the simplifications used in the force fields. It is within this context that we can expect to gain the most from biomolecular simulations, thereby increasing our understanding of a wide variety of biophysical and biochemical processes.

REFERENCES

1. DM York, TS Lee, W Yang. J Am Chem Soc 118:10940–10941, 1996.
2. CL Brooks III, M Karplus, BM Pettitt. Proteins, A Theoretical Perspective: Dynamics, Structure, and Thermodynamics, Vol 71. New York: Wiley, 1988.
3. JA McCammon, SC Harvey. Dynamics of Proteins and Nucleic Acids. New York: Cambridge Univ Press, 1987.
4. U Burkert, NL Allinger. Molecular Mechanics. Washington, DC: American Chemical Society, 1982.
5. A Blondel, M Karplus. J Comput Chem 17:1132–1141, 1996.
6. WE Reiher III. Theoretical Studies of Hydrogen Bonding. Ph.D. Thesis, Harvard University, 1985.
7. T Lazaridis, M Karplus. Science 278:1928–1931, 1997.
8. BR Brooks, RE Bruccoleri, BD Olafson, DJ States, S Swaminathan, M Karplus. J Comput Chem 4:187–217, 1983.
9. AD MacKerell Jr. Protein force fields. In: PvR Schleyer, NL Allinger, T Clark, J Gasteiger, PA Kollman, HF Schaefer III, PR Schreiner, eds. The Encyclopedia of Computational Chemistry, Vol 3. Chichester, UK, Wiley 1998, pp 2191–2200.
10. DA Pearlman, DA Case, JW Caldwell, WS Ross, TE Cheatham, S Debolt, D Ferguson, G Seibel, P Kollman. Comput Phys Commun 91:1–41, 1995.
11. WF van Gunsteren, SR Billeter, AA Eising, PH Hünenberger, P Krüger, AE Mark, WRP Scott, IG Tironi. Biomolecular Simulation: The GROMOS96 Manual and User Guide. Zürich: BIOMOS, 1996.
12. AT Hagler, JR Maple, TS Thacher, GB Fitzgerald, U Dinur. In: WF van Gunsteren, PK Weiner, eds. Potential Energy Functions for Organic and Biomolecular Systems. Leiden: ESCOM, 1989, pp 149–167.
13. B Roux, M Karplus. J Comput Chem 16:690–704, 1995.
14. SL Mayo, BD Olafson, WA Goddard III. J Phys Chem 94:8897–8909, 1990.
15. S Lifson, AT Hagler, P Dauber. J Am Chem Soc 101:5111–5121, 1979.
16. NL Allinger, YH Yuh, J-L Lii. J Am Chem Soc 111:8551–8566, 1989.
17. TA Halgren. J Comput Chem 17:520–552, 1996.
18. AD MacKerell Jr, J Wiórkiewicz-Kuczera, M Karplus. J Am Chem Soc 117:11946–11975, 1995.

19. M Schlenkrich, J Brickmann, AD MacKerell Jr, M Karplus. In: KM Merz, B Roux, eds. Empirical Potential Energy Function for Phospholipids: Criteria for Parameter Optimization and Applications. Boston; Birkhaüser, 1996, pp 31–81.

20. AD MacKerell Jr, D Bashford, M Bellott, RL Dunbrack Jr, J Evanseck, MJ Field, S Fischer, J Gao, H Guo, S Ha, D Joseph, L Kuchnir, K Kuczera, FTK Lau, C Mattos, S Michnick, T Ngo, DT Nguyen, B Prodhom, WE Reiher III, B Roux, M Schlenkrich, J Smith, R Stote, J Straub, M Watanabe, J Wiorkiewicz-Kuczera, D Yin, M Karplus. J Phys Chem B 102: 3586–3616, 1998.

21. WD Cornell, P Cieplak, CI Bayly, IR Gould, KM Merz Jr, DM Ferguson, DC Spellmeyer, T Fox, JW Caldwell, PA Kollman. J Am Chem Soc 117:5179–5197, 1995.

22. SP Weiner, PA Kollman, DA Case, UC Singh, C Ghio, G Alagona, S Profeta, P Weiner. J Am Chem Soc 106:765–784, 1984.

23. E Neria, S Fischer, M Karplus. J Chem Phys 105:1902–1921, 1996.

24. M Ceccarelli, M Marchi. J Phys Chem B 101:2105–2108, 1997.

25. M Feig, BM Pettitt. Biophys J 75:134–149, 1998.

26. RM Venable, Y Zhang, BJ Hardy, RW Pastor. Science 262:223–226, 1993.

27. SE Feller, D Yin, RW Pastor, AD MacKerell Jr. Biophys J 73:2269–2279, 1997.

28. WL Jorgensen, DS Maxwell, J Tirado-Rives. J Am Chem Soc 118:11225–11236, 1996.

29. G Kaminski, EM Duffy, T Matsui, WL Jorgensen. J Phys Chem 98:13077–13082, 1994.

30. WL Jorgensen, J Tirado-Rives. J Am Chem Soc 110:1657–1666, 1988.

31. J Pranata, SG Wierschke, WL Jorgensen, J Am Chem Soc 113:2810–2819, 1991.

32. W Damm, A Frontera, J Tirado-Rives, WL Jorgensen. J Comput Chem 18:1955–1970, 1997.

33. J-H Lii, NL Allinger. J Comput Chem 19:1001–1016, 1998.

34. J-L Lii, NL Allinger. J Comput Chem 12:186–199, 1991.

35. JR Maple, M-J Hwang, KJ Jalkanen, TP Stockfisch, AT Halger. J Comput Chem 19:430–458, 1998.

36. H Sun. J Phys Chem B 102:7338–7364, 1998.

37. TA Halgren. J Comput Chem 17:490–519, 1996.

38. AK Rappé, CJ Colwell, WA Goddard III, WM Skiff. J Am Chem Soc 114:10024–10035, 1992.

39. FA Momany, R Rone. J Comput Chem 13:888, 1992.

40. G Némethy, KD Gibson, KA Palmer, CN Yoon, G Paterlini, A Zagari, S Rumsey, HA Scheraga. J Phys Chem 96:6472–6484, 1992.

41. A Liwo, S Oldziej, MR Pincus, RJ Wawak, S Rackovsky, HA Scheraga. J Comput Chem 18:849–876, 1997.

42. LA Mirny, EI Shakhnovich. J Mol Biol 264:1164–1179, 1996.

43. J Novotny, RE Bruccoleri, M Davis, KA Sharp. J Mol Biol 268:401–411, 1997.

44. D Eisenberg, AD McLachlan. Nature 319:199–203, 1986.

45. L Wesson, D Eisenberg. Protein Sci 1:227–235, 1992.

46. MK Gilson, ME Davis, BA Luty, JA McCammon. J Phys Chem 97:3591–3600, 1993.

47. M Schaefer, M Karplus. J Phys Chem 100:1578–1599, 1996.

48. M Nina, D Beglov, B Roux. J Phys Chem B 101:5239–5248, 1997.

49. D Sitkoff, KA Sharp, B Honig. J Phys Chem 98:1978–1988, 1994.

50. D Qui, PS Shenkin, FP Hollinger, WC Still. J Phys Chem A 101:3005–3014, 1997.

51. B Jayaram, D Sprous, DL Beveridge. J Phys Chem B 102:9571–9576, 1998.

52. TA Halgren. J Comput Chem 20:730–748, 1999.

53. WL Jorgensen. J Phys Chem 87:5304–5312, 1983.

54. JR Maple, U Dinur, AT Hagler. Proc Natl Acad Sci USA 85:5350–5354, 1988.

55. TS Thacher, AT Hagler, H Rabitz. J Am Chem Soc 113:2020–2033, 1991.

56. JD Cox, R Pilcher. Thermochemistry of Organic and Organometallic Compounds. London: Academic, 1970.

57. MA Spackman. Chem Rev 92:1769–1797, 1992.

58. FJ Millero. Chem Rev 71:147–176, 1971.
59. H Guo, M Karplus. J Phys Chem 96:7273–7287, 1992.
60. FH Allen, S Bellard, MD Brice, BA Cartwright, A Doubleday, H Higgs, T Hummelink, BG Hummelink-Peters, O Kennard, WDS Motherwell, JR Rodgers, DG Watson. Acta Cryst B35: 2331–2339, 1979.
61. V Barone, M Cossi, J Tomasi. J Comput Chem 19:404–417, 1998.
62. EB Wilson Jr, JC Decius, PC Cross. Molecular Vibrations. New York: McGraw-Hill, 1955.
63. D Yin, AD MacKerell Jr. J Comput Chem 19:334–348, 1998.
64. AD MacKerell Jr. J Chim Phys 94:1436–1447, 1997.
65. MD Beachy, D Chasman, RB Murphy, TA Halgren, RA Friesner. J Am Chem Soc 119: 5908–5920, 1997.
66. WL Jorgensen. BOSS. Version 3.5 of a computer program. New Haven, CT: Yale University, 1994.
67. AD MacKerell Jr, M Karplus. J Phys Chem 95:10559–10560, 1991.
68. A Warshel, S Lifson. J Chem Phys 53:582–594, 1970.
69. G Chalasinski, MM Szczesniak. Chem Rev 94:1723–1765, 1994.
70. B Jeziorski, R Moszynski, K Szalewicz. Chem Rev 94:1887–1930, 1994.
71. FB van Duijneveldt, JGCM van Duijneveldt-van de Rijdt, JH van Lenthe. Chem Rev 94: 1873–1885, 1994.
72. S Tsuzuki, T Uchimaru, K Tanabe, S Kuwajima. J Phys Chem 98:1830–1833, 1994.
73. J-R Hill. J Comput Chem 18:211–220, 1997.
74. LE Chirlian, MM Francl. J Comput Chem 8:894–905, 1987.
75. UC Singh, PA Kollman. J Comput Chem 5:129–145, 1984.
76. MJ Frisch, GW Trucks, HB Schlegel, PMW Gill, BG Johnson, MA Robb, JR Cheeseman, K Raghavachari, MA Al-Laham, VG Zakrzewski, JV Ortiz, JB Foresman, J Cioslowski, BB Stefanov, A Nanayakkara, M Challacombe, CY Peng, PY Ayala, W Chen, MW Wong, JL Andres, ES Replogle, R Gomperts, RL Martin, DJ Fox, JS Binkley, DJ Defrees, J Baker, JJP Stewart, M Head-Gordon, C Gonzalez, JA Pople. Gaussian 94. Ver C.3 of a computer program. Pittsburgh, PA: Gaussian, Inc, 1996.
77. CI Bayly, P Cieplak, WD Cornell, PA Kollman. J Phys Chem 97:10269–10280, 1993.
78. WD Cornell, P Cieplak, CE Bayly, PA Kollman. J Am Chem Soc 115:9620, 1993.
79. WL Jorgensen. J Phys Chem 90:1276–1284, 1986.
80. D Flatters, K Zakrzewska, R Lavery. J Comput Chem 18:1043–1055, 1997.
81. WL Jorgensen, J Chandrasekhar, JD Madura, RW Impey, ML Klein. J Chem Phys 79:926–935, 1983.
82. HJC Berendsen, JR Grigera, T Straatsma. J Phys Chem 91:6269–6271, 1987.
83. T Fox, PA Kollman. J Phys Chem B 102:8070–8079, 1998.
84. WJ Hehre, L Radom, P von R Schleyer, JA Pople. Ab Initio Molecular Orbital Theory. New York: J Wiley, 1986.
85. TA Darden, D York, LG Pedersen. J Chem Phys 98:10089–10092, 1993.
86. N Foloppe, AD MacKerell Jr. J Comp Chem 21:86–104, 2000.
87. RH Stote, DJ States, M Karplus. J Chim Phys 88:2419–2433, 1991.
88. JC Smith, M Karplus. J Am Chem Soc 114:801–812, 1992.
89. WL Jorgensen, DL Severance. J Am Chem Soc 112:4768–4774, 1990.
90. J Florián, BG Johnson. J Phys Chem 98:3681–3687, 1994.
91. AP Scott, L Radom. J Phys Chem 100:16502–16513, 1996.
92. J Baker, AA Jarzecki, P Pulay. J Phys Chem A 102:1412–1424, 1998.
93. P Pulay, G Fogarasi, F Pang, JE Boggs. J Am Chem Soc 101:2550–2560, 1979.
94. K Kuczera, JK Wiorkiewicz, M Karplus. MOLVIB: Program for the Analysis of Molecular Vibrations, CHARMM, Harvard University, 1993.
95. JP Ryckaert, G Ciccotti, HJC Berendsen. J Comput Phys 23:327–341, 1977.
96. N Foloppe, AD MacKerell Jr. J Phys Chem B 102:6669–6678, 1998.

97. DR Langley. J Biomol Struct Dynam 16:487–509, 1998.
98. MJ Field, M Karplus. CRYSTAL: Program for Crystal Calculations in CHARMM. Cambridge, MA: Harvard University, 1992.
99. DJ Tobias, K Tu, ML Klein. J Chim Phys 94:1482–1502, 1997.
100. TE Cheatham III, P Cieplak, PA Kollman. J Biomol Struct Dynam, 16:845–861, 1999.
101. LL Ho, AD MacKerell Jr, PA Bash. J Phys Chem 100:2588–2596, 1996.
102. PA Bash, LL Ho, AD MacKerell Jr, D Levine, P Hallstrom. Proc Natl Acad Sci USA 93: 3698–3703, 1996.
103. M Freindorf, J Gao. J Comput Chem 17:386–395, 1996.
104. LX Dang. J Phys Chem B 102:620–624, 1998.
105. SW Rick, BJ Berne. J Am Chem Soc 118:672–679, 1996.
106. P Linse, MA Carignano, G Karostrom. J Phys Chem B 101:1142–1147, 1997.
107. Y Sun, JW Caldwell, PA Kollman. J Phys Chem 99:10081–10085, 1995.
108. Y Ding, DN Bernardo, K Krogh-Jespersen, RM Levy. J Phys Chem 99:11575–11583, 1995.
109. J Gao, JJ Pavelites, D Habibollazadeh. J Phys Chem 100:2689–2697, 1996.
110. C-H Kiang, WA Goddard III. J Phys Chem 99:14334–14339, 1995.
111. MR Wilson, MP Allen, MA Warren, A Sauron, W Smith. J Comput Chem 18:478–488, 1997.

3
Dynamics Methods

Oren M. Becker
Tel Aviv University, Tel Aviv, Israel

Masakatsu Watanabe*
Moldyn, Inc., Cambridge, Massachusetts

I. INTRODUCTION

Molecular dynamics simulation, which provides the methodology for detailed microscopical modeling on the atomic scale, is a powerful and widely used tool in chemistry, physics, and materials science. This technique is a scheme for the study of the natural time evolution of the system that allows prediction of the static and dynamic properties of substances directly from the underlying interactions between the molecules.

Dynamical simulations monitor time-dependent processes in molecular systems in order to study their structural, dynamic, and thermodynamic properties by numerically solving an equation of motion, which is the formulation of the rules that govern the motion executed by the molecule. That is, molecular dynamics (MD) provides information about the time dependence and magnitude of fluctuations in both positions and velocities, whereas the Monte Carlo approach provides mainly positional information and gives only little information on time dependence.

Depending on the desired level of accuracy, the equation of motion to be numerically solved may be the classical equation of motion (Newton's), a stochastic equation of motion (Langevin's), a Brownian equation of motion, or even a combination of quantum and classical mechanics (QM/MM, see Chapter 11).

Good reviews of the application of dynamic simulation methods to biomolecules can be found in the books by Brooks et al. [1] and McCammon and Harvey [2]. Good short reviews on this topic can also be found in Refs. 3–5. More detailed discussions of dynamic simulation methodologies can be found in books by Allen and Tildesley [6], Frenkel and Smit [7], and Rapaport [8] and in the review by van Gunsteren [9].

---

* *Current affiliation*: Wavefunction, Inc., Irvine, California.

II. TYPES OF MOTIONS

Macromolecules in general, and proteins in particular, display a broad range of characteristic motions. These range from the fast and localized motions characteristic of atomic fluctuations to the slow large-scale motions involved in the folding transition. Many of these motions, on each and every time scale, have an important role in the biological function of proteins. For example, localized side-chain motion controls the diffusion of oxygen into and out of myoglobin and hemoglobin [1]. A more extensive ''medium-scale'' structural transition is involved, for example, in the hemoglobin R to T allosteric transition, which makes it such an efficient transport agent [1]. Finally, prion proteins exhibit a global structural transition of biological importance. These proteins undergo a global transition from an α-helical structure to a predominantly β-sheet structure, which is implicated in the onset of Creutzfeldt-Jacob disease (CJD) in humans and the ''mad cow'' disease in cattle (bovine spongiform encephalopathy; BSE) [10].

Table 1 gives a crude overview of the different types of motion executed by a protein and their characteristic time scales and amplitudes. These should be regarded as rough guidelines, because individual motions in specific systems may differ significantly from these estimates. Note that the motions executed by a protein span almost 20 orders of magnitude in characteristic time scales, from femtoseconds (10^{-15} s) to hours (10^4–10^5 s). They also cover a wide range of amplitudes (0.01–100 Å) and energies (0.1–100 kcal/mol).

An important characteristic of biomolecular motion is that the different types of motion are interdependent and coupled to one another. For example, a large-scale dynamic transition cannot occur without involving several medium-scale motions, such as helix rearrangements. Medium-scale motions cannot occur without involving small-scale motions, such as side-chain movement. Finally, even side-chain motions cannot occur without the presence of the very fast atomic fluctuations, which can be viewed as the ''lubricant'' that enables the whole molecular construction to move. From the point of view of dynamic

Table 1 An Overview of Characteristic Motions in Proteins

Type of motion	Functionality examples	Time and amplitude scales
Local motions Atomic fluctuation Side chain motion	Ligand docking flexibility Temporal diffusion pathways	Femtoseconds (fs) to picoseconds (ps) (10^{-15}–10^{-12} s); less than 1 Å
Medium-scale motions Loop motion Terminal-arm motion Rigid-body motion	Active site conformation adaptation Binding specificity	Nanoseconds (ns) to microseconds (μs) (10^{-9}–10^{-6} s); 1–5 Å
Large-scale motions Domain motion Subunit motion	Hinge-bending motion Allosteric transitions	Microseconds (μs) to milliseconds (ms) (10^{-6}–10^{-3} s); 5–10 Å
Global motions Helix-coil transition Folding/unfolding Subunit association	Hormone activation Protein functionality	Milleseconds (ms) to hours (10^{-3}–10^4 s); more than 10 Å

simulations, this has serious implications. It indicates that even in the study of slow large-scale motions (of biological importance) it is not possible to ignore the fast small-scale motions, which eventually are the ones that impose limitations on the simulation time step and length.

III. THE STATISTICAL MECHANICS BASIS OF MOLECULAR DYNAMICS

A classical system is described by a classical Hamiltonian, H, which is a function of both coordinates $\mathbf{r}$ and momenta $\mathbf{p}$. For regular molecular systems, where the potential energy function is independent of time and velocity, the Hamiltonian is equal to the total energy,

$$H = H(\mathbf{r}, \mathbf{p}) = K(\mathbf{p}) + U(\mathbf{r}) = \sum_i \frac{p_i}{2m_i} + U(\mathbf{r}) \tag{1}$$

where $K(\mathbf{p})$ is the kinetic energy, $U(\mathbf{r})$ is the potential energy, p_i is the momentum of particle i, and m_i the mass of particle i. A microscopic state of the system is therefore characterized by the set of values $\{\mathbf{r}, \mathbf{p}\}$, which corresponds to a point in the space defined by both coordinates $\mathbf{r}$ and momenta $\mathbf{p}$ (known as "phase space").

To obtain thermodynamic averages over a "canonical" ensemble, which is characterized by the macroscopic variables (N, V, T), it is necessary to know the probability of finding the system at each and every point (= state) in phase space. This probability distribution, $\rho(\mathbf{r}, \mathbf{p})$, is given by the Boltzmann distribution function,

$$\rho(\mathbf{r}, \mathbf{p}) = \frac{\exp\left[-H(\mathbf{r}, \mathbf{p})/k_B T\right]}{Z} \tag{2}$$

where the canonical partition function, Z, is an integral over all phase space of the Boltzmann factors $\exp\left[-H(\mathbf{r}, \mathbf{p})/k_B T\right]$, and k_B is the Boltzmann factor. Once this distribution function is known it can be used to calculate phase space averages of any dynamic variable $A(\mathbf{r}, \mathbf{p})$ of interest. Examples for dynamic variables are the position, the total energy, the kinetic energy, fluctuations, and any other function of $\mathbf{r}$ and/or $\mathbf{p}$. These averages,

$$\langle A(\mathbf{r}, \mathbf{p})\rangle_Z = \int_V d\mathbf{r} \int_{-\infty}^{\infty} d\mathbf{p} \, \rho(\mathbf{r}, \mathbf{p}) A(\mathbf{r}, \mathbf{p}) \tag{3}$$

are called "thermodynamic averages" or "ensemble averages" because they take into account every possible state of the system. However, in order to calculate these thermodynamic averages, it is necessary to simultaneously know the Boltzmann probability [Eq. (2)] for each and every state $\{\mathbf{r}, \mathbf{p}\}$, which is an extremely difficult computational task.

An alternative strategy for calculating systemwide averages is to follow the motion of a single point through phase space instead of averaging over the whole phase space all at once. That is, in this approach the motion of a single point (a single molecular state) through phase space is followed as a function of time, and the averages are calculated only over those points that were visited during the excursion. Averages calculated in this way are called "dynamic averages." The motion of a single point through phase space is obtained by integrating the system's equation of motion. Starting from a point $\{\mathbf{r}(0), \mathbf{p}(0)\}$, the integration procedure yields a trajectory that is the set of points $\{\mathbf{r}(t), \mathbf{p}(t)\}$

describing the state of the system at any successive time t. Dynamic averages of any dynamical variable $A(\mathbf{r}, \mathbf{p})$ can now be calculated along this trajectory as follows:

$$\langle A(\mathbf{r}, \mathbf{p})\rangle_\tau = \frac{1}{\tau} \int_0^\tau A(\mathbf{r}(t), \mathbf{p}(t))dt \tag{4}$$

where τ is the duration of the simulation. Compared to the previous approach, dynamic averaging is easier to perform. The two averaging strategies can be summarized as follows:

> *Thermodynamic average.* An average over *all* points in phase space at a *single* time.
> *Dynamic average.* An average over a *single* point in phase space at *all* times.

It is hoped that the point that is being dynamically followed will eventually cover all of phase space and that the dynamic average will converge to the desired thermodynamic average. A key concept that ties the two averaging strategies together is the *ergodic hypothesis*. This hypothesis states that for an infinitely long trajectory the thermodynamic ensemble average and the dynamic average become equivalent to each other,

$$\lim_{\tau \to \infty} \langle A(\mathbf{r}, \mathbf{p})\rangle_\tau = \langle A(\mathbf{r}, \mathbf{p})\rangle_Z \tag{5}$$

In other words, the ergodic hypothesis claims that when the trajectory becomes long enough, the point that generates it will eventually cover all of phase space, so the two averages become identical. For this hypothesis to hold, the system has to be at equilibrium (technically, at a stationary state). Also, there must not be any obstacle, such as a fragmented topology, that will prevent an infinitely long trajectory from covering all of phase space. A system that obeys these two conditions is said to be *ergodic*, and its hypothesis is the theoretical justification for using molecular dynamic simulations as a means for calculating thermodynamic averages of molecular systems. It is tacitly assumed that finite molecular dynamics trajectories are "long enough" in the ergodic sense.

IV. NEWTONIAN MOLECULAR DYNAMICS

A. Newton's Equation of Motion

The temporal behavior of molecules, which are quantum mechanical entities, is best described by the quantum mechanical equation of motion, i.e., the time-dependent Schrödinger equation. However, because this equation is extremely difficult to solve for large systems, a simpler classical mechanical description is often used to approximate the motion executed by the molecule's heavy atoms. Thus, in most computational studies of biomolecules, it is the classical mechanics Newtonian equation of motion that is being solved rather than the quantum mechanical equation.

In its most simplistic form, Newton's equation of motion (also known as Newton's second law of motion) states that

$$F_i = m_i a_i = m_i \ddot{r}_i \tag{6}$$

where F_i is the force acting on particle i, m_i is the mass of particle i, a_i is its acceleration, and $\ddot{r}_i$ is the second derivative of the particle position $\mathbf{r}$ with respect to time. The force F_i is determined by the gradient of the potential energy function, $U(\mathbf{r})$, discussed in Chapter 2, which is a function of all the atomic coordinates $\mathbf{r}$,

$$F_i = -\nabla_i U(\mathbf{r}) \tag{7}$$

Equation (7) is a second-order differential equation. A more general formulation of Newton's equation of motion is given in terms of the system's Hamiltonian, H [Eq. (1)]. Put in these terms, the classical equation of motion is written as a pair of coupled first-order differential equations:

$$\dot{r}_k = \frac{\partial H(\mathbf{r}, \mathbf{p})}{\partial p_k}; \qquad \dot{p}_k = -\frac{\partial H(\mathbf{r}, \mathbf{p})}{\partial r_k} \tag{8}$$

By substituting the definition of H [Eq. (1)] into Eq. (8), we regain Eq. (6). The first first-order differential equation in Eq. (8) becomes the standard definition of momentum, i.e., $p_i = m_i \dot{r}_i = m_i v_i$, while the second turns into Eq. (6). A set of two first-order differential equations is often easier to solve than a single second-order differential equation.

B. Properties of Newton's Equation of Motion

Newton's equation of motion has several characteristic properties, which will later serve as "handles" to ensure that the numerical solution is correct (Section V.C). These properties are

Conservation of energy. Assuming that U and H do not depend explicitly on time or velocity (so that $\partial H/\partial t = 0$), it is easy to show from Eq. (8) that the total derivative dH/dt is zero; i.e., the Hamiltonian is a constant of motion for Newton's equation. In other words, there is conservation of total energy under Newton's equation of motion.

Conservation of linear and angular momentum. If the potential function U depends only on particle separation (as is usual) and there is no external field applied, then Newton's equation of motion conserves the total linear momentum of the system, P,

$$P = \sum_i p_i \tag{9}$$

and the total angular momentum, L,

$$L = \sum_i r_i \times p_i = \sum_i m_i r_i \times \dot{r}_i \tag{10}$$

Time reversibility. The third property of Newton's equation of motion is that it is reversible in time. Changing the signs of all velocities (or momenta) will cause the molecule to retrace its trajectory. If the equations of motion are solved correctly, then the numerical trajectory should also have this property. Note, however, that in practice this time reversibility can be reproduced by numerical trajectories only over very short periods of time because of the chaotic nature of large molecular systems.

C. Molecular Dynamics: Computational Algorithms

Solving Newton's equation of motion requires a numerical procedure for integrating the differential equation. A standard method for solving ordinary differential equations, such as Newton's equation of motion, is the finite-difference approach. In this approach, the molecular coordinates and velocities at a time $t + \Delta t$ are obtained (to a sufficient degree of accuracy) from the molecular coordinates and velocities at an earlier time t. The equations are solved on a step-by-step basis. The choice of time interval Δt depends on the properties of the molecular system simulated, and Δt must be significantly smaller than the characteristic time of the motion studied (Section V.B).

A good starting point for understanding finite-difference methods is the Taylor expansion about time t of the position at time $t + \Delta t$,

$$\mathbf{r}(t + \Delta t) = \mathbf{r}(t) + \dot{\mathbf{r}}(t)\Delta t + \frac{1}{2}\ddot{\mathbf{r}}(t)\Delta t^2 + \cdots \tag{11}$$

Alternatively, this can be written as

$$\mathbf{r}(t + \Delta t) = \mathbf{r}(t) + \mathbf{v}(t)\Delta t + \frac{1}{2}\mathbf{a}(t)\Delta t^2 + \cdots \tag{12}$$

where $\mathbf{v}(t)$ is the velocity vector and $\mathbf{a}(t)$ is the acceleration. Because the integration proceeds in a stepwise fashion, and recalling Eq. (6), it is convenient to rewrite the above expansion in a discrete form. Using $\mathbf{r}_n$ to indicate the position at step n (at time t) and $\mathbf{r}_{n+1}$ to indicate the position at the next step, $n + 1$ (at time $t + \Delta t$), Eq. (12) can be written as

$$\mathbf{r}_{n+1} = \mathbf{r}_n + \mathbf{v}_n\Delta t + \frac{1}{2}\left(\frac{\mathbf{F}_n}{\mathbf{m}}\right)\Delta t^2 + O(\Delta t^3) \tag{13}$$

where $O(\Delta t^n)$ is the terms of order Δt^n or smaller. With this information the velocity $\mathbf{v}_{n+1}$ at time $n + 1$ can be crudely estimated, for example, as

$$\mathbf{v}_{n+1} = (\mathbf{r}_{n+1} - \mathbf{r}_n)/2 \tag{14}$$

Together, Eqs. (13) and (14) form an integration algorithm. Given the position $\mathbf{r}_n$, the velocity $\mathbf{v}_n$, and the force $\mathbf{F}_n$ at step n, these equations allow one to calculate (actually, estimate) the position $\mathbf{r}_{n+1}$ and velocity $\mathbf{v}_{n+1}$ at step $n + 1$. The formulation is highly trivial and results in a low quality integration algorithm (large errors). Other, more accurate, algorithms have been developed using the same kind of reasoning. In the following subsections we survey some of the more commonly used finite-difference integration algorithms, highlighting their advantages and disadvantages.

1. Verlet Integrator

The most common integration algorithm used in the study of biomolecules is due to Verlet [11]. The Verlet integrator is based on two Taylor expansions, a forward expansion ($t + \Delta t$) and a backward expansion ($t - \Delta t$),

$$\mathbf{r}_{n+1} = \mathbf{r}_n + \mathbf{v}_n\Delta t + \frac{1}{2}\left(\frac{\mathbf{F}_n}{\mathbf{m}}\right)\Delta t^2 + O(\Delta t^3) \tag{15a}$$

$$\mathbf{r}_{n-1} = \mathbf{r}_n + \mathbf{v}_n \Delta t + \frac{1}{2} \left(\frac{\mathbf{F}_n}{\mathbf{m}} \right) \Delta t^2 - O(\Delta t^3) \tag{15b}$$

The sum of the two expansions yields an algorithm for propagating the position,

$$\mathbf{r}_{n+1} = 2\mathbf{r}_n - \mathbf{r}_{n-1} + \frac{\mathbf{F}_n}{\mathbf{m}} \Delta t^2 + O(\Delta t^4) \tag{16}$$

Translated into a stream of commands, this algorithm is executed in two steps:

1. Use the current position $\mathbf{r}_n$ to calculate the current force $\mathbf{F}_n$.
2. Use the current and previous positions $\mathbf{r}_n$ and $\mathbf{r}_{n-1}$ together with the current force $\mathbf{F}_n$ (calculated in step 1) to calculate the position in the next step, $\mathbf{r}_{n+1}$, according to Eq. (16).

These two steps are repeated for every time step for each atom in the molecule. Subtracting Eq. (15b) from Eq (15a) yields a complementary algorithm for propagating the velocities,

$$\mathbf{v}_n = \frac{\mathbf{r}_{n+1} - \mathbf{r}_{n+1}}{2\Delta t} + O(\Delta t^2) \tag{17}$$

Figure 1a gives a graphical representation of the steps involved in a Verlet propagation. The algorithm embodied in Eqs. (16) and (17) provides a stable numerical method for solving Newton's equation of motion for systems ranging from simple fluids to biopolymers. Like any algorithm, the Verlet algorithm has advantages as well as disadvantages.

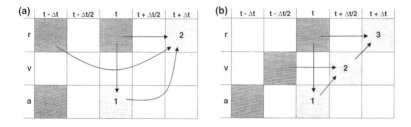

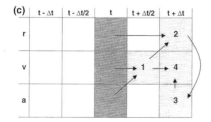

Figure 1 A stepwise view of the Verlet integration algorithm and its variants. (a) The basic Verlet method. (b) Leap-frog integration. (c) Velocity Verlet integration. At each algorithm dark and light gray cells indicate the initial and calculating variables, respectively. The numbers in the cells represent the orders in the calculation procedures. The arrows point from the data that are used in the calculation of the variable that is being calculated at each step.

Advantages of the Verlet algorithm are

1. The position integration is quite accurate [errors on the order of $O(\Delta t^4)$] and is independent of the velocity propagation, a fact that simplifies the position integration and reduces memory requirements.
2. The algorithm requires only a single force evaluation per integration cycle (computationally, force evaluations are the most ''expensive'' part of the simulation).
3. This formulation, which is based on forward and backward expansions, guarantees time reversibility (a property of the equation of motion).

Disadvantages of the Verlet algorithm are

1. The velocity propagation is subject to relatively large errors, on the order $O(\Delta t^2)$. Recall that an accurate estimate of the velocity is required for the kinetic energy evaluations. An added inconvenience is that v_n can be computed only if r_{n+1} is already known.
2. Further numerical errors are introduced when an $O(\Delta t^2)$ term is added to an $O(\Delta t^0)$ term.
3. The Verlet algorithm is not ''self-starting.'' A lower order Taylor expansion [e.g., Eq. (13)] is often used to initiate the propagation.
4. It must be modified to incorporate velocity-dependent forces or temperature scaling.

2. Leap-Frog Integrator

Modifications to the basic Verlet scheme have been proposed to tackle the above deficiencies, particularly to improve the velocity evaluation. One of these modifications is the leap-frog algorithm, so called for its half-step scheme: Velocities are evaluated at the midpoint of the position evaluation and vice versa [12,13]. The algorithm can be written as

$$\mathbf{r}_{n+1} = \mathbf{r}_n + \mathbf{v}_{n+1/2}\Delta t \tag{18a}$$

$$\mathbf{v}_{n+1/2} = \mathbf{v}_{n-1/2} + \frac{\mathbf{F}_n}{\mathbf{m}}\Delta t \tag{18b}$$

where $\mathbf{v}_{n\pm1/2}$ stands for the velocity at the mid-step time $[t \pm (1/2)\Delta t]$. Elimination of the velocities from these equations shows that the method is algebraically equivalent to the Verlet algorithm. Cast in the form of execution instructions, the leap-frog algorithm involves three steps:

1. Use the current position $\mathbf{r}_n$ to calculate the current force $\mathbf{F}_n$.
2. Use the current force $\mathbf{F}_n$ and previous mid-step velocity $\mathbf{v}_{n-1/2}$ to calculate the next mid-step velocity $\mathbf{v}_{n+1/2}$.
3. Use the current position $\mathbf{r}_n$ and the next mid-step velocity $\mathbf{v}_{n+1/2}$ (from step 2) to calculate the position in the next step, $\mathbf{r}_{n+1}$.

Figure 1b gives a graphical representation of the steps involved in the leap-frog propagation. The current velocity $\mathbf{v}_n$, which is necessary for calculating the kinetic energy, can be calculated as

$$\mathbf{v}_n = (\mathbf{v}_{n+1/2} + \mathbf{v}_{n-1/2})/2 \tag{19}$$

Advantages of the leap-frog algorithm are

1. It improves velocity evaluation.
2. The direct evaluation of velocities gives a useful handle for controlling the simulation temperature (via velocity scaling).
3. It reduces the numerical errors, since here $O(\Delta t^1)$ terms are added to $O(\Delta t^0)$ terms.

Disadvantages of the leap-frog algorithm are

1. It still does not handle the velocities in a completely satisfactory manner, because the velocities at time t are only approximated by Eq. (19).
2. This algorithm is computationally a little more expensive than the Verlet algorithm.

3. Velocity Verlet Integrator

An even better handling of the velocities is obtained by another variant of the basic Verlet integrator, known as the "velocity Verlet" algorithm. This is a Verlet-type algorithm that stores positions, velocities, and accelerations all at the same time t and minimizes round-off errors [14]. The velocity Verlet algorithm is written

$$\mathbf{r}_{n+1} = \mathbf{r}_n + \mathbf{v}_n \Delta t + \frac{1}{2} \left(\frac{\mathbf{F}_n}{\mathbf{m}} \right) \Delta t^2 \tag{20a}$$

$$\mathbf{v}_{n+1} = \mathbf{v}_n + \frac{1}{2} \left[\frac{\mathbf{F}_n}{\mathbf{m}} + \frac{\mathbf{F}_{n+1}}{\mathbf{m}} \right] \Delta t \tag{20b}$$

Again, elimination of the velocities from these equations recovers the Verlet algorithm. In practice, the velocity Verlet algorithm consists of the following steps:

1. Calculate the position $\mathbf{r}_{n+1}$ at time $t + \Delta t$ from Eq. (20a).
2. Calculate the velocity at mid-step $\mathbf{v}_{n+1/2}$ using the equation

$$\mathbf{v}_{n+1/2} = \mathbf{v}_n + \frac{1}{2} \left(\frac{\mathbf{F}_n}{\mathbf{m}} \right) \Delta t \tag{21}$$

3. Calculate the force $\mathbf{F}_{n+1}$ at time $t + \Delta t$.
4. Finally, complete the velocity move to $\mathbf{v}_n$ by using

$$\mathbf{v}_{n+1} = \mathbf{v}_{n+1/2} + \frac{1}{2} \left(\frac{\mathbf{F}_{n+1}}{\mathbf{m}} \right) \Delta t \tag{22}$$

At this point, the kinetic energy at time $t + \Delta t$ is available. Figure 1c gives a graphical representation of the steps involved in the velocity Verlet propagation.

Advantages of the velocity Verlet algorithm are

1. It is numerically very stable.
2. It is convenient and simple [the code of this method is a straightforward transcription of Eqs. (20)–(22)].
3. It provides an accurate evaluation of velocities and hence of the kinetic energy.

The main disadvantage of this algorithm is that it is computationally a little more expensive than the simpler Verlet or leap-frog algorithms (though the added accuracy often outweighs this slight overhead).

V. MOLECULAR DYNAMICS: SIMULATION PRACTICE

A. Assigning Initial Values

Newton's equation of motion is a second-order differential equation that requires two initial values for each degree of freedom in order to initiate the integration. These two initial values are typically a set of initial coordinates $\{r(0)\}$ and a set of initial velocities $\{v(0)\}$.

1. Initial Coordinates

The initial coordinates $\{r(0)\}$ are usually obtained from experimentally determined molecular structures, mainly from X-ray crystallography and NMR experiments. Alternatively, the initial coordinates can be based on computer models generated by a variety of modeling techniques (see Chapters 14 and 15). Note, however, that even the experimentally determined structures must often undergo some preparation steps before they can be used as initial structures in a dynamic simulation.

First, it is not possible to determine hydrogen atom positions by X-ray crystallography. Thus the coordinates for the many hydrogen atoms in the molecule are missing from X-ray coordinate files. These coordinates must be added to the initial structure before the simulation is started. Several algorithms are available for ensuring reasonable placement of hydrogens.

In some cases, whole parts of the protein are missing from the experimentally determined structure. At times, these omissions reflect flexible parts of the molecule that do not have a well-defined structure (such as loops). At other times, they reflect parts of the molecule (e.g., terminal sequences) that were intentionally removed to facilitate the crystallization process. In both cases, structural models may be used to fill in the gaps.

After all the coordinates are accounted for, it is good practice to refine the initial structure by submitting it to energy minimization (see Chapter 4). The role of this minimization is to relieve local stresses due to nonbonded overlaps, as well as to relax bond length and bond angle distortions in the experimental structure. The origin of these stresses is due both to the empirical nature of the energy function (Chapter 2) and to the average nature of the experimentally determined structures.

2. Initial Velocities

Unlike the initial coordinates, which can be obtained experimentally, the only relevant information available about atomic velocities is the system's temperature T, which determines the velocity distribution. In the absence of a better guideline, initial velocities (v_x, v_y, v_z) are usually randomly assigned from the standard Maxwellian velocity distribution at a temperature T,

$$P(v)dv = \left(\frac{m}{2\pi k_B T}\right)^{1/2} \exp\left[\frac{-mv^2}{2k_B T}\right] dv \tag{23}$$

This initial assignment is, of course, not at equilibrium. In particular, the expected velocity correlation between neighboring atoms is not guaranteed, and most likely it is nonexistent (i.e., in general, neighboring atoms, such as bonded pairs, are expected to move at similar velocities). Furthermore, the random assignment process may accidentally assign high velocities to a localized cluster of atoms, creating a "hot spot" that makes the simulation unstable. To overcome this problem, it is common practice to start a simulation with a "heat-up" phase. Velocities are initially assigned at a low temperature, which is then increased gradually allowing for dynamic relaxation. This slow heating continues until the simulation reaches the desired temperature.

In practice, heating is performed by increasing atomic velocities, either by reassigning new velocities from a Maxwellian distribution [Eq. (23)] at an elevated temperature or by scaling the velocities by a uniform factor. This heating process, as well as the dynamic simulation that follows, requires a measurable definition of the system's temperature T at time t. According to the equipartition theorem, the temperature, $T(t)$, at any given time t is defined in terms of the mean kinetic energy by

$$T(t) = \frac{1}{k_B N_{\text{dof}}} \sum_{i=1}^{N_{\text{dof}}} m_i |v_i|^2 \tag{24}$$

where N_{dof} is the number of unconstrained degrees of freedom in the system ($N_{\text{dof}} = 3N - n$, where N is the number of atoms and n is the number of constraints). It is clear from this expression that scaling the velocities by a factor of $[T_0/T(t)]^{1/2}$ will result in a mean kinetic energy corresponding to a desired temperature T_0.

Another problem related to the initial velocity assignment is the large total linear momentum P and total angular momentum L formed [Eqs. (9) and (10)]. These momenta cause a drift of the molecule's center of mass relative to the reference frame. Numerically, this drift hampers the computational efficiency of the simulation, overshadowing the smaller internal motions by the physically irrelevant translational drift and global rotation. Because Newton's equation of motion conserves linear and angular momenta, these momenta will not go away unless they are actively taken out. The ideal zero-drift situation is reached by periodically zeroing these momenta during the equilibration phase of the simulation.

B. Selecting the Integration Time Step

The size of the time step Δt is an important parameter that determines the magnitude of the error associated with each of the foregoing integration algorithms. On the one hand, a small time step means better integration quality. But on the other hand it also means that more integration steps are required for the same length of simulation. Thus, every simulation involves a trade-off between economy and accuracy. The time step in molecular dynamics simulations is one of the most important factors that balance this trade-off. In general, one would like to choose the largest possible time step that will still ensure an accurate simulation.

An appropriate time step should be small by comparison to the period of the fastest motion (highest frequency motion) in the system being simulated. If τ is the period of the fastest motion, a good rule of thumb for selecting Δt is

$$\tau/\Delta t \approx 20 \tag{25}$$

For biomolecules, such as proteins, the fastest motions are the stretching vibrations of the bonds connecting hydrogen atoms to heavy atoms (X—H stretching). The frequency of these motions is in the vicinity of 3000 cm^{-1}, which means periods of about 10 fs (1 $\times 10^{-14}$ s). Thus, an appropriate time step for simulating biomolecules would be $\Delta t \approx$

0.5 fs. This extremely small time step is the main reason that, with only a few exceptions, molecular dynamics simulations are limited today to the nanosecond (10^{-9} s) time scale.

Naturally, much effort is invested in developing advanced algorithms that allow for larger time steps that enable longer simulations. A basic rationale, common to many such approaches, is to remove the fast (high frequency) motions from the numerical integration and account for them in some other way. Because of the characteristic coupling between types of motion (Section III.A), this task is far from simple. A first step in this direction is to take the X—H stretching motions out of the numerical integration. If these stretching motions were accounted for in some other way, then the time step would be determined by the next fastest molecular motion, i.e., the X—X stretching modes with frequencies around 1500 cm^{-1}. According to relation (25), this elimination will increase the time step by a factor of 2 (to $\Delta t \approx 1.0$ fs), extending the length of the simulation by a similar factor at only a slight additional computational cost.

The algorithm that is usually employed to account for the hydrogen positions is SHAKE [15,16] (and its variant RATTLE [17]). Stated in a simplistic way, the SHAKE algorithm assumes that the length of the X—H bond can be considered constant. Because in a numerical simulation there are always fluctuations, this means that the deviation of the current length $d_k(t)$ of the kth bond from its ideal (constant) bond length d_k^0 must be smaller than some tolerance value ε,

$$s_k = [d_k(t)^2 - d_k^{0^2}]/d_k^{0^2} < \varepsilon \tag{26}$$

SHAKE is an iterative procedure that adjusts the atomic positions (of the hydrogen atoms in this case) after each integration step (of the heavy atoms) in order to simultaneously satisfy all the constraints. It iterates until s_k is smaller than ε for all values of k, (A more detailed description of the algorithm can be found in Refs. 14 and 16 and in appendix A1.4 of Ref. 2.) SHAKE can be applied not only to X—H type bonds but also to all bond-stretching motions in the system, allowing for time steps as large as 2 or 3 fs (depending on the details of the system). Although, SHAKE can in principle be applied to bending motions too, it was found that such constraints result in low quality simulations. This is due to the fact that the important coupling between bond angle motion and torsional motion is neglected [18].

More advanced methods for extending the length of molecular dynamics simulations are discussed in Section VIII.

C. Stability of Integration

An important issue in any numerical study is that of the accuracy and stability of the simulation. A simulation become unstable when it has picked up errors along the way and has become essentially meaningless. In general, it is unrealistic to expect that any approximate method of solution will follow the exact classical trajectory indefinitely. It is our goal, however, to maintain a stable simulation for at least the duration of interest for the specific study. Thus, the stability of the simulation must be gauged at all times. If one is lucky, an unstable simulation will also crash and not reach its designated termination. It may happen, however, that even though the unstable simulation reaches its designated termination, its content carries little merit.

The best gauges for the stability of any simulation are the constants of motion of the physical equation that is numerically solved, i.e., quantities that are expected to be conserved during the simulation. Since numerical fluctuations cannot be avoided, a dynamic variable $A(\mathbf{r},\mathbf{p})$ is considered numerically "conserved" if the ratio of its fluctuations to its value is below an acceptable tolerance ε, $\Delta A/A < \varepsilon$. The constants of motion for Newton's equation of motion were specified in Section IV.B.

> *Conservation of energy.* Newton's equation of motion conserves the total energy of the system, E (the Hamiltonian), which is the sum of potential and kinetic energies [Eq. (1)]. A fluctuation ratio that is considered adequate for a numerical solution of Newton's equation of motion is
>
> $$\frac{\Delta E}{E} < 10^{-4} \quad \text{or} \quad \log_{10}\left(\frac{\Delta E}{E}\right) < -4 \tag{29}$$
>
> *Conservation of linear and angular momenta.* After equilibrium is reached, the total linear momentum $\mathbf{P}$ [Eq. (9)] and total angular momentum $\mathbf{L}$ [Eq. (10)] also become constants of motion for Newton's equation and should be conserved. In advanced simulation schemes, where velocities are constantly manipulated, momentum conservation can no longer be used for gauging the stability of the simulation.
>
> *Time reversibility.* Newton's equation is reversible in time. For a numerical simulation to retain this property it should be able to retrace its path back to the initial configuration (when the sign of the time step Δt is changed to $-\Delta t$). However, because of chaos (which is part of most complex systems), even modest numerical errors make this backtracking possible only for short periods of time. Any two classical trajectories that are initially very close will eventually exponentially diverge from one another. In the same way, any small perturbation, even the tiny error associated with finite precision on the computer, will cause the computer trajectories to diverge from each other and from the exact classical trajectory (for examples, see pp. 76–77 in Ref. 6). Nonetheless, for short periods of time a stable integration should exhibit temporal reversibility.

D. Simulation Protocol and Some Tricks for Better Simulations

Every molecular dynamics simulation consists of several steps. Put together, these steps are called a "simulation protocol." The main steps common to most dynamic simulation protocols are the following.

1. *Preparation of the data.* Preparation of the initial coordinates (adding hydrogen atoms, minimization) and assignment of initial velocities.
2. *Heating up.* Gradual scaling of the velocities to the desired temperature, accompanied by short equilibration dynamics.
3. *Equilibration.* A relatively long dynamic simulation, whose goal is to ensure that the simulation is stable and free of erratic fluctuations. This step may take from tens of picoseconds to several hundred picoseconds.
4. *Production.* When the simulation is "equilibrated," the dynamic simulation is considered reliable. From this point on, the trajectory generated is stored for further analysis. Typical "production runs" take from several hundred picoseconds up to tens of nanoseconds (depending on the size of the system and the available computer power).

5. *Analysis.*   The resulting trajectory is submitted to careful analysis. (Refer to Section VI.)

We have presented a simple protocol to run MD simulations for systems of interest. There are, however, some "tricks" to improve the efficiency and accuracy of molecular dynamics simulations. Some of these techniques, which are discussed later in the book, are today considered standard practice. These methods address diverse issues ranging from efficient force field evaluation to simplified solvent representations.

One widely used trick is to apply bookkeeping to atom–atom interactions, commonly referred to as the neighbor list [11], which is illustrated in Figure 2. If we simulate a large *N*-particle system and use a cutoff that is smaller than the simulation box, then many particles do not contribute significantly to the energy of a particle *i*. It is advantageous, therefore, to exclude from the expensive energy calculation particle pairs that do not interact. This technique increases the efficiency of the simulations. Details of programming for this approach can be found in the books by Allen and Tildesley [6] and Frenkel and Smit [7].

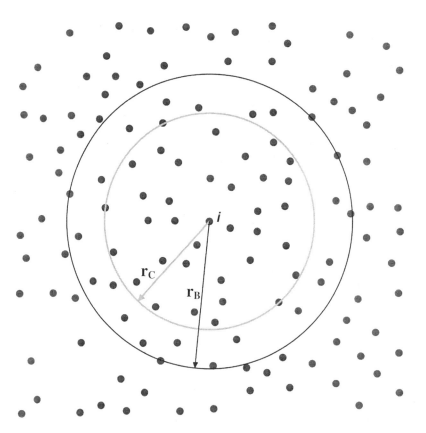

Figure 2   A particle *i* interacts mainly with particles that are within the cutoff radius r_C. The "neighbor list" contains only those particles that are within a sphere of radius $r_B > r_C$. Particles outside this sphere will not contribute to the force or energy affecting particle *i*. The use of a neighbor list that is periodically updated during the simulation reduces the computer time required in calculating pairwise interactions.

Other techniques can be found elsewhere in this volume. The following list includes pointers to several of these techniques:

1. Constant-temperature and constant-pressure simulations—Section VII.C, this chapter.
2. Constraint and multiple time step methods—Section VIII, this chapter.
3. Periodic boundary conditions—Chapter 5.
4. Long-range interactions and extended electrostatics—Chapter 5.
5. Solvation models—Chapter 7.

VI. ANALYSIS OF DYNAMIC TRAJECTORIES

An important issue, the significance of which is sometime underestimated, is the analysis of the resulting molecular dynamics trajectories. Clearly, the value of any computer simulation lies in the quality of the information extracted from it. In fact, it is good practice to plan the analysis procedure before starting the simulation, as the goals of the analysis will often determine the character of the simulation to be performed.

A. Time Series

The direct output of a dynamic simulation is a numerical trajectory, i.e., a series of system "snapshots" (coordinates and/or velocities) taken at equal time intervals $\Delta\tau$ from the full trajectory (the sampling interval $\Delta\tau$ is typically much larger than Δt). The size of the trajectory sampling interval, $\Delta\tau$, should be determined according to the time scale of the phenomenon that is being studied. For example, a 1 ps sampling interval may be a good choice for studying phenomena that take many tens of picoseconds but is clearly inappropriate for studying fast processes on the subpicosecond time scale.

Calculating any dynamic variable, $A(t)$, along the trajectory results in a "time series." Dynamic variables can be any function of coordinates and/or velocities. They may be relatively straightforward, such as total energy, specific bond lengths, or torsion angles of interest, or quite complex. Examples of the latter are the end-to-end distance in a protein (a quantity useful for studying protein folding), distances between a hydrogen bond donor and an acceptor, an angle formed between two helices, and so forth. The most straightforward analytical procedure is to plot the time series as a function of time. Such plots give a quick and easy overview of the simulation and are especially useful for picking up trends (e.g., drifts) or sudden transitions from one state to another.

Because a time series consists of instantaneous values taken at the trajectory sampling points, they tend to be very noisy. The level of noise can be reduced by simple smoothing procedures. A common smoothing procedure is to slide an "N-point window" along the data points and plot the "window averages" $\langle A(t)\rangle_N$ as a function of time [instead of plotting $A(t)$ itself]. The width of the window, N, and the suitability of the smoothing approach depend on the property that is being studied and on the ratio between the sampling interval $\Delta\tau$ and the characteristic time of the noise. For example, noise due to bond-stretching motions (time scale of 10–50 fs) can be smoothed by 0.5–1 ps windows. Alternatively, simply increasing the size of the sampling interval $\Delta\tau$ beyond the characteristic time scale of the "noise," to 0.5–1 ps in this case, will also reduce the effect of the noise.

B. Averages and Fluctuations

Time series plots give a useful overview of the processes studied. However, in order to compare different simulations to one another or to compare the simulation to experimental results it is necessary to calculate average values and measure fluctuations. The most common average is the root-mean-square (rms) average, which is given by the second moment of the distribution,

$$\langle A \rangle_{\text{rms}} = (\langle A^2 \rangle)^{1/2} = \left[\frac{1}{N_S} \sum_{i=1}^{N_S} (A_i)^2 \right]^{1/2} \tag{28}$$

where A is any dynamic variable and N_s is the number of "snapshots" in the trajectory. Root-mean-square fluctuations are calculated in the same way, with the fluctuation ΔA, which is described as a difference with respect to an average A, replacing the values A in Eq. (28).

Higher moments of the distribution are often of interest too, especially when nonisotropic or anharmonic processes are being studied. The third moment of the distribution reflects the skewness α_3 defined as

$$\alpha_3 = \langle A^3 \rangle / \langle A^2 \rangle^{3/2} \tag{29}$$

while the fourth moment reflects the excess kurtosis α_4 defined as

$$\alpha_4 = \langle A^4 \rangle / \langle A^2 \rangle^2 - 3 \tag{30}$$

Both α_3 and α_4 are zero for a Gaussian distribution.

C. Correlation Functions

A powerful analytical tool is the time correlation function. For any dynamic variable $A(t)$, such as bond lengths or dihedral angles, the time autocorrelation function $C_A(t)$ is defined as

$$C_A(t) = \langle A(t) \, A(0) \rangle \tag{31}$$

This function measures the correlation of the property $A(t)$ to itself at two different times separated by the time interval t, averaged over the whole trajectory. The auto-correlation function is *reversible* in time [i.e., $C_A(t) = C_A(-t)$], and it is *stationary* (i.e., $\langle A(t + t) \, A(t) \rangle = \langle A(t) \, A(0) \rangle$). In practical terms, the autocorrelation function is obtained by averaging the terms $\langle A(s + t) \, A(s) \rangle$ while sliding s along the trajectory.

A time "cross-correlation" function between dynamic variables $A(t)$ and $B(t)$ is defined in a similar way:

$$C_{AB}(t) = \langle A(t) \, B(0) \rangle \tag{32}$$

An important property of the time autocorrelation function $C_A(t)$ is that by taking its Fourier transform, $F\{C_A(t)\}_\omega$, one gets a spectral decomposition of all the frequencies that contribute to the motion. For example, consider the motion of a single particle in a harmonic potential (harmonic oscillator). The "time series" describing the position of the

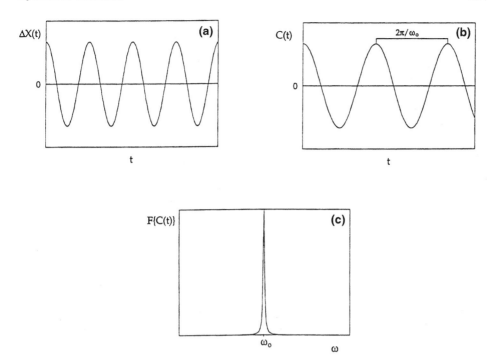

Figure 3 (a) A ''time series'' describing the position as a function of time of a particle moving in a one-dimensional harmonic well (harmonic oscillator). (b) The autocorrelation function of that motion; (c) its Fourier transform spectrum.

particle as a function of time is given by the $\cos(\omega_0 t)$ function (Fig. 3a). The autocorrelation function is given by a cosine function with a period $2\pi/\omega_0$ (Fig. 3b), and its Fourier transform gives a spectrum with a single sharp peak at ω_0 (Fig. 3c). The resulting frequency can be used to extract the (real or effective) local force constant $K_0 = m\omega_0^2$, where m is the mass of the system.

D. Potential of Mean Force

The potential of mean force is a useful analytical tool that results in an effective potential that reflects the average effect of all the other degrees of freedom on the dynamic variable of interest. Equation (2) indicates that given a potential function it is possible to calculate the probability for all states of the system (the Boltzmann relationship). The potential of mean force procedure works in the reverse direction. Given an observed distribution of values (from the trajectory), the corresponding effective potential function can be derived. The first step in this procedure is to organize the observed values of the dynamic variable, A, into a distribution function $\rho(A)$. From this distribution the ''effective potential'' or ''potential of mean force,'' $W(A)$, is calculated from the Boltzmann relation:

$$W(A) = -RT \ln [\rho(A)] \tag{33}$$

The advantage of a potential of mean force is that it reflects the effect of environmental factors on the behavior of the dynamic variable of interest. Such an effect may be the damping of the solvent or a correlated fluctuation, which reduces effective transition barriers.

E. Estimating Standard Errors

Computer simulation is an experimental science to the extent that calculated dynamic properties are subject to systematic and statistical errors. Sources of systematic error consist of size dependence, poor equilibration, non-bond interaction cutoff, etc. These should, of course, be estimated and eliminated where possible. It is also essential to obtain an estimate of the statistical significance of the results. Simulation averages are taken over runs of finite length, and this is the main cause of statistical imprecision in the mean values so obtained.

Statistical errors of dynamic properties could be expressed by breaking a simulation up into multiple blocks, taking the average from each block, and using those values for statistical analysis. In principle, a block analysis of dynamic properties could be carried out in much the same way as that applied to a static average. However, the block lengths would have to be substantial to make a reasonably accurate estimate of the errors. This approach is based on the assumption that each block is an independent sample.

Another approach is to run multiple MD simulations with different initial conditions. The recent work by Karplus and coworkers [19] observes that multiple trajectories with different initial conditions improve conformational sampling of proteins. They suggest that multiple trajectories should be used rather than a single long trajectory.

VII. OTHER MD SIMULATION APPROACHES

A. Stochastic Dynamics

There are cases in which one is interested in the motion of a biomolecule but wishes also to study the effect of different solvent environments on this motion. In other cases, one may be interested in studying the motion of one part of the protein (e.g., a side chain or a loop) as moving in a "solvent bath" provided by the remainder of the protein. One way to deal with these issues is, of course, to explicitly include all the additional components in the simulation (explicit water molecules, the whole protein, etc.). This solution is computationally very expensive, because much work is done on parts of the system that are of no direct interest to the study.

Another way is to reduce the magnitude of the problem by eliminating the explicit solvent degrees of freedom from the calculation and representing them in another way. Methods of this nature, which retain the framework of molecular dynamics but replace the solvent by a variety of simplified models, are discussed in Chapters 7 and 19 of this book. An alternative approach is to move away from Newtonian molecular dynamics toward stochastic dynamics.

The basic equation of motion for stochastic dynamics is the Langevin equation,

$$m_i \ddot{r}_i = -\nabla_i U(\mathbf{r}) - m_i \beta_i v_i(t) + R_i(t) \qquad (34)$$

which has three terms on the right-hand side. The first term accounts for molecular interactions and is the same as that used in Newtonian molecular dynamics [Eqs. (6) and (7). The other two terms are unique to stochastic dynamics and represent solvent effects. The second term is a dissipative drag force reflecting friction due to the solvent. This term is proportional to the velocity v_i and the friction coefficient β_i (which is related to the diffusion constant D_i). The third term is a random force $R_i(t)$ that represents stochastic collisions between solvent molecules and the solute. The stochastic force introduces energy into the system, and the friction force removes energy from the system (for further discussion see Ref. 20). The Langevin random force $R_i(t)$ on atom i is obtained from a random Gaussian distribution of zero mean and a variance related to the friction coefficient,

$$\langle R_i(t) \rangle = 0 \tag{35a}$$

$$\langle R_i(t)R_i(0) \rangle = 6m_i k_B T \delta(t) = 2D_i \, \delta(t) \tag{35b}$$

Sometimes an additional term, F_i^{mean}, is added to the right-hand side of Eq. (34). This ''mean force'' represents an average effect of degrees of freedom not explicitly treated in the simulation.

Because of the additional velocity-dependent forces (the dissipative forces), the straightforward finite-difference algorithms of the Verlet type cannot be used to integrate Langevin's equation. An algorithm may be derived that reduces to the Verlet algorithm in the limit of vanishing friction ($\beta_i \rightarrow 0$). This algorithm is obtained by adding the Langevin terms to the Verlet equation described in Eqs. (16) and (17). The resulting algorithm is of the order $O(\Delta t^3)$ and is valid for $\beta_i \Delta t < 1$.

B. Brownian Dynamics

Biomolecular motions that involve substantial displacements of the molecular surface, such as the motion of heavy particles in aqueous solution, are usually damped owing to the high viscosity of the surrounding solvent. In many such cases the damping effects are sufficiently great that internal forces are negligible and the motion has the character of a random walk. In other cases, such as the diffusion pairing of a substrate with its enzyme, the specific details of the internal dynamics are of little interest to the intermolecular motion. In such cases a further level of simplification can be introduced into the equation of motion. The relevant approximation applicable to such cases is the Brownian equation of motion, which is a diffusional analog of the molecular dynamics method. The Brownian equation can be easily derived from the stochastic Langevin equation presented in Eq. (34). If the inertial term on the left-hand side of the equation is small compared to the force terms on the right-hand side, it can be neglected, resulting in the diffusional Brownian equation of motion,

$$v_i(t) = \dot{r}_i = \frac{-\nabla V_i(\mathbf{r}) + F_i^{mean} + R_i(t)}{m_i \beta_i} \tag{36}$$

where the properties of the stochastic force $R_i(t)$ and its dependence on the friction coefficient D_i are given by Eqs. (35a) and (35b).

As with Newtonian molecular dynamics, a number of different algorithms have been developed to calculate the diffusional trajectories. An efficient algorithm for solving the Brownian equation of motion was introduced by Ermak and McCammon [21]. A detailed survey of this and other algorithms as well as their application can be found in Ref. 2.

C. Molecular Dynamics in Alternative Ensembles

The original molecular dynamics (MD) technique was used only to study the natural time evolution of a classical system of N particles in a volume V. In such simulations, the total energy, E, is a constant of motion, so that the time averages obtained by such a conventional MD simulation are equivalent to microcanonical ensemble averages. It is often scientifically more relevant to perform simulations in other ensembles, such as the canonical (*NVT*) that is associated with a Helmholtz free energy or the isothermal-isobaric (NPT) ensemble that is associated with a Gibbs free energy. Two rather different types of solutions to the problem of simulating alternative ensembles with the MD method have been proposed.

The first approach is based on introducing simple velocity or position rescaling into the standard Newtonian MD. The second approach has a dynamic origin and is based on a reformulation of the Lagrangian equations of motion for the system (so-called extended Lagrangian formulation.) In this section, we discuss several of the most widely used constant-temperature or constant-pressure schemes.

1. Constant-Temperature MD

The simplest method that keeps the temperature of a system constant during an MD simulation is to rescale the velocities at each time step by a factor of $(T_0/T)^{1/2}$, where T is the current instantaneous temperature [defined in Eq. (24)] and T_0 is the desired temperature. This method is commonly used in the equilibration phase of many MD simulations and has also been suggested as a means of performing "constant temperature molecular dynamics" [22]. A further refinement of the velocity-rescaling approach was proposed by Berendsen et al. [24], who used velocity rescaling to couple the system to a heat bath at a temperature T_0. Since heat coupling has a characteristic relaxation time, each velocity v is scaled by a factor λ, defined as

$$\lambda = \left[1 + \frac{\Delta t}{2\tau_T} \left(\frac{T_0}{T} - 1 \right) \right]^{1/2} \tag{37}$$

In this expression, Δt is the size of the integration time step, τ_T is a characteristic relaxation time, and T is the instantaneous temperature. In the simulation of water, they found a relaxation time of $\tau_T = 0.4$ ps to be appropriate. However, this method does not correspond exactly to the canonical ensemble.

An alternative method, proposed by Andersen [23], shows that the coupling to the heat bath is represented by stochastic impulsive forces that act occasionally on randomly selected particles. Between stochastic collisions, the system evolves at constant energy according to the normal Newtonian laws of motion. The stochastic collisions ensure that all accessible constant-energy shells are visited according to their Boltzmann weight and therefore yield a canonical ensemble.

To carry out this method, values are chosen for T_0, the desired temperature, and v, the mean frequency with which each particle experiences a stochastic collision. If successive collisions are uncorrected, then the distribution of time intervals between two successive stochastic collisions, $P(v, t)$, is of the Poisson form,

$$P(v, t) = v \exp(- vt) \tag{38}$$

A constant-temperature simulation now consists of the following steps:

1. Start with an initial set of positions and momenta and integrate the equation of motion.

2. The probability that any particular particle suffers a stochastic collision in a time interval of Δt is $v \, \Delta t$.
3. If particle i has been selected to undergo a collision, obtain its new velocity from a Maxwellian velocity distribution, defined in Eq. (23), corresponding to the desired temperature T_0. All other particles are unaffected by this collision.

Another popular approach to the isothermal (canonical) MD method was shown by Nosé [25]. This method for treating the dynamics of a system in contact with a thermal reservoir is to include a degree of freedom that represents that reservoir, so that one can perform deterministic MD at constant temperature by reformulating the Lagrangian equations of motion for this extended system. We can describe the Nosé approach as an illustration of an extended Lagrangian method. Energy is allowed to flow dynamically from the reservoir to the system and back; the reservoir has a certain thermal inertia associated with it. However, it is now more common to use the Nosé scheme in the implementation of Hoover [26].

To construct Nosé–Hoover constant-temperature molecular dynamics, an additional coordinate, s, and its conjugate momentum p_s are introduced. The Hamiltonian of the extended system of the N particles plus extended degrees of freedom can be expressed as

$$H_{\text{Nosé}} = \sum_{i=1}^{N} \frac{p_i^2}{2m_i s^2} + U(q) + \frac{p_s^2}{2Q} + \frac{g}{\beta} \ln s \tag{39}$$

where β is $1/k_B T$, Q is an effective "mass" associated with s, and g is a parameter related to the degrees of freedom in the system. The microcanonical distribution in the augmented set of variables in Eq. (39) is equivalent to a canonical distribution of the variables r_i and p_i/s. One of the disadvantages of the original Nosé approach, however, is that s can be interpreted as a scaling factor of the time step. This implies that the real time-step fluctuations occur during a simulation.

In a simulation it is not convenient to work with fluctuating time intervals. The real-variable formulation is therefore recommended. Hoover [26] showed that the equations derived by Nosé can be further simplified. He derived a slightly different set of equations that dispense with the time-scaling parameter s. To simplify the equations, we can introduce the thermodynamic friction coefficient, $\xi = p_s/Q$. The equations of motion then become

$$\dot{r}_i = \frac{p_i}{m_i} \tag{40a}$$

$$\dot{p}_i = -\frac{\partial U}{\partial r_i} - \xi p_i \tag{40b}$$

$$\dot{\xi} = \left(\sum_i \left(\frac{p_i^2}{m_i} \right) - \frac{g}{\beta} \right) Q^{-1} \tag{40c}$$

$$\frac{\dot{s}}{s} = \frac{d \ln s}{dt} = \xi \tag{40d}$$

Note that Eq. (40d), in fact, is redundant, because the other three equations form a closed set. Nonetheless, if we solve the equations of motion for s as well, we can use the following, $\overline{H}_{\text{Nosé}}$, as a diagnostic tool, because this quantity has to be conserved during the simulation even though $\overline{H}_{\text{Nosé}}$ is no longer a Hamiltonian:

$$\overline{H}_{\text{Nosé}} = \sum_{i=1}^{N} \frac{p_i^2}{2m_i} + U(q) + \frac{\xi^2 Q}{2} + \frac{gs}{\beta} \tag{41}$$

where $g = 3N$ in this real-variable formulation.

Theoretically, static quantities are independent of the value chosen for the parameter Q. In practice, however, we observe that quantities are Q-dependent because of the finite number of MD simulation steps. Too high a value of Q results in slow energy flow between the system and reservoir, and in the limit $Q \rightarrow \infty$ we regain conventional MD. On the other hand, if Q is too low, the energy undergoes long-lived, weakly damped oscillations, resulting in poor equilibration. Nosé suggested the choice of the adjustable parameter Q for the physical system. It may, however, be necessary to choose Q by trial and error in order to achieve satisfactory damping.

Both Andersen and Nosé–Hoover methods, indeed, have generated canonical distributions. However, sometimes the Nosé–Hoover thermostat runs into ergodicity problems and the desired distribution cannot be achieved. The Andersen method does not suffer from such problems, but its dynamics are less realistic than Nosé–Hoover. To alleviate the ergodicity problems, Martyna et al. [27] proposed a scheme in which the Nosé–Hoover thermostat is coupled to another thermostat or, if necessary, to a whole chain of thermostats. The coupling ensures that the thermostats are allowed to fluctuate. In the original Nosé–Hoover method, the thermostat variable does not fluctuate. This generalization of the original Nosé–Hoover method is also shown to generate a canonical distribution, but this approach no longer faces the ergodicity problems. Details of the programming for this approach may be obtained from the book by Frenkel and Smit [7].

2. Constant-Pressure MD

In a normal molecular dynamics simulation with repeating boundary conditions (i.e., periodic boundary condition), the volume is held fixed, whereas at constant pressure the volume of the system must fluctuate. In some simulation cases, such as simulations dealing with membranes, it is more advantageous to use the constant-pressure MD than the regular MD. Various schemes for prescribing the pressure of a molecular dynamics simulation have also been proposed and applied [23,24,28,29]. In all of these approaches it is inevitable that the system box must change its volume.

To include the volume as a dynamic variable, the equations of motion are determined in the analysis of a system in which the positions and momenta of all particles are scaled by a factor proportional to the cube root of the volume of the system. Andersen [23] originally proposed a method for constant-pressure MD that involves coupling the system to an external variable, V, the volume of the simulation box. This coupling mimics the action of a piston on a real system. The piston has a ''mass'' M_v [which has units of $(\text{mass})(\text{length})^{-4}$]. From the Lagrangian for this extended system, the equations of motion for the particles and the volume of the cube are

$$\dot{r}_i = \frac{P_i}{m_i} + \frac{1}{3}\left(\frac{\dot{V}}{V}\right) r_i \tag{42a}$$

$$\dot{p}_i = F_i - \frac{1}{3}\left(\frac{\dot{V}}{V}\right)p_i \tag{42b}$$

$$\ddot{V} = \frac{1}{M_v}[P(t) - P_0] \tag{42c}$$

where V is the volume, $P(t)$ is the instantaneous pressure, P_0 is the desired pressure, and r_i, p_i, m_i, and F_i are the position, momentum, mass, and force, respectively, for each particle i. Andersen proved that the solution to these equations produces trajectories in the isobaric-isoenthalpic (*NPH*) ensemble where the particle number, pressure, and enthalpy of the system are constant. Here the choice of piston mass determines the decay time of the volume fluctuations. It has been proven that equilibrium quantities are independent of M_v, but in practice M_v influences the dynamic properties in the simulations. Even though there is no precise formulation for choosing M_v, Andersen suggests that the piston mass may be chosen by trial and error to satisfy the length of time required for a sound wave to travel through the simulation cell.

An alternative procedure rescales the coordinates of each atom at each time step [24]. The atomic coordinate x and the characteristic distance for repeating boundary conditions, d, are rescaled to values μx and μd, respectively, where

$$\mu = \left[1 - \frac{\Delta t}{\tau_P}(P_0 - P)\right]^{1/3} \tag{43}$$

Here, Δt is the size of the time step, τ_p is a characteristic relaxation time, and P_0 is the pressure of the external constant-pressure bath. The instantaneous pressure can be calculated as follows:

$$P = \frac{2}{3V}\left[E_k + \frac{1}{2}\sum_{i<j} r_{ij} \cdot F_{ij}\right] \tag{44}$$

where V is the volume and E_k is the kinetic energy, r_{ij} is the vector from particle i to particle j, and F_{ij} is the force on particle j due to particle i. In simulations of water, values of $\tau_p = 0.01$–0.1 ps were found suitable. This method does not drastically alter the dynamic trajectories and is easy to program, but the appropriate ensemble has not been identified. Therefore, the meaning of fluctuations in any observed quantity cannot be determined.

An algorithm for performing a constant-pressure molecular dynamics simulation that resolves some unphysical observations in the extended system (Andersen's) method and Berendsen's methods was developed by Feller et al. [29]. This approach replaces the deterministic equations of motion with the piston degree of freedom added to the Langevin equations of motion. This eliminates the unphysical fluctuation of the volume associated with the piston mass. In addition, Klein and coworkers [30] present an advanced constant-pressure method to overcome an unphysical dependence of the choice of lattice in generated trajectories.

In the foregoing treatments of pressure feedback, the simulation volume retains its cubic form, so changes consist of uniform contractions and expansions. The method is readily extended to the case of a simulation region in which the lengths and directions of the edges are allowed to vary independently. Parrinello and Rahman [31] and Nosé and Klein [32] extended the Andersen method to the case of noncubic simulation cells and derived a new Lagrangian for the extended system. Though their equations of motion are different from Andersen's original equations, they produce an identical ensemble. This technique is particularly helpful in the study of crystal structures of proteins, because it allows for phase changes in the simulation, which may involve changes in the unit cell dimensions and angles.

These constant-pressure MD methods can be combined with a suitable temperature control mechanism, as discussed in the previous section, to produce a more useful method to control both pressure and temperature simultaneously during the simulation. There are several approaches. The simplest approach is to use the scaling factors. The details of the algorithm are given by Berendsen et al. [24]. Another approach [25,26] is to define the appropriate extended Lagrangian for simultaneously coupling pressure and temperature to produce the isothermal-isobaric (*NPT*) ensemble. Hoover [26] presents a detailed description of the Nosé–Hoover constant-temperature method with Andersen's constant-pressure method. Even though this extended system method is slightly more complicated to program, this is the best candidate for conducting *NPT* ensemble MD. Details on programming for this approach are available in the book by Rapaport [8]. In addition, the new Langevin piston method [29] for constant pressure can be easily extended to couple a Nosé–Hoover thermostat to obtain a constant-pressure and constant-temperature method.

VIII. ADVANCED SIMULATION TECHNIQUES

Computer simulations have become a valuable tool for the theoretical investigation of biomolecular systems. Unfortunately, these simulations are often computationally demanding tasks owing to the large number of particles as well as the complex nature of their associated interactions. A longstanding problem, however, is that molecular dynamics is typically limited to a time scale of 10^{-6} s (1 µs) or less. In an MD simulation, the most rapidly varying quantities, such as the bond lengths, limit the integration time step, while the more slowly varying molecular processes are of primary interest (see Table 1) and determine the simulation length required. This would make the simulation of molecular substances very expensive.

A variety of techniques have been introduced to increase the time step in molecular dynamics simulations in an attempt to surmount the strict time step limits in MD simulations so that long time scale simulations can be routinely undertaken. One such technique is to solve the equations of motion in the internal degree of freedom, so that bond stretching and angle bending can be treated as rigid. This technique is discussed in Chapter 6 of this book. Herein, a brief overview is presented of two approaches, constrained dynamics and multiple time step dynamics.

A. Constrained Dynamics

To avoid the situation in which high frequency motions, such as bond stretching and bond angle bending which limits the integration time step, it is customary to eliminate such degrees of freedom entirely by the simple expedient of replacing them with constraints. In general, the dynamics could satisfy many constraints simultaneously, e.g., many bond lengths and bond angles. Assuming that a total of n distance constraints are imposed on a particular molecule, the constraint σ_k for a fixed distance d_{ij} between atom i and j can be expressed as

$$\sigma_k = r_{ij}^2 - d_{ij}^2 = 0, \qquad k = 1, \ldots, n \tag{45}$$

The equations of motion follow directly from the Lagrangian formulation containing all constraints. The result is

$$m_i \ddot{r}_i = F_i + G_i \tag{46}$$

where F_i is the usual force term, m_i the mass of the ith atom, and the additional term G_i expresses the effect of the constraints on atom i, G_i can be written

$$G_i = - \sum_\alpha \lambda_\alpha \frac{\partial \sigma_\alpha}{\partial r_i} \tag{47}$$

Here α denotes the set of constraints that directly involve r_i and the $\{\lambda_\alpha\}$ are the Lagrange multipliers introduced into the problem.

There are various ways to obtain the solutions to this problem. The most straightforward method is to solve the full problem by first computing the Lagrange multipliers from the time-differentiated constraint equations and then using the values obtained to solve the equations of motion [7,8,37]. This method, however, is not computationally cheap because it requires a matrix inversion at every iteration. In practice, therefore, the problem is solved by a simple iterative scheme to satisfy the constraints. This scheme is called SHAKE [6,14] (see Section V.B). Note that the computational advantage has to be balanced against the additional work required to solve the constraint equations. This approach allows a modest increase in speed by a factor of 2 or 3 if all bonds are constrained.

Although constrained dynamics is usually discussed in the context of the geometrically constrained system described above, the same techniques can have many other applications. For instance, constant-pressure and constant-temperature dynamics can be imposed by using constraint methods [33,34]. Car and Parrinello [35] describe the use of the extended Lagrangian to maintain constraints in the context of their ab initio MD method. (For more details on the Car–Parrinello method, refer to the excellent review by Galli and Pasquarrello [36].)

B. Multiple Time Step Methods

According to the nature of the empirical potential energy function, described in Chapter 2, different motions can take place on different time scales, e.g., bond stretching and bond angle bending vs. dihedral angle librations and non-bond interactions. Multiple time step (MTS) methods [38–40,42] allow one to use different integration time steps in the same simulation so as to treat the time development of the slow and fast movements most effectively.

Tuckerman et al. [38] showed how to systematically derive time-reversible, area-preserving MD algorithms from the Liouville formulation of classical mechanics. Here, we briefly introduce the Liouville approach to the MTS method. The Liouville operator for a system of N degrees of freedom in Cartesian coordinates is defined as

$$iL = [\ldots, H] = \sum_{i=1}^N \left[\dot{r}_i \frac{\partial}{\partial r_i} + F_i(r) \frac{\partial}{\partial p_i} \right] = \dot{r} \frac{\partial}{\partial r} + F(r) \frac{\partial}{\partial p} \tag{48}$$

where $[\ldots, \ldots]$ is the Poisson bracket, H is the Hamiltonian of the system, r_i and p_i are the position and conjugate momentum at coordinate i, $\dot{r}_i$ is the time derivative of r_i, and F_i is the force acting on the ith degree of freedom. The state of the system, Γ, at time Δt is then given by

$$\Gamma[r(\Delta t), p(\Delta t)] = U(\Delta t) \cdot \Gamma[r(0), p(0)] \tag{49}$$

where $U(t)$ is composed of a classical time evolution operator, $e^{iL\Delta t}$, and Γ could be any arbitrary function that depends on all coordinates and momenta of the system.

We decompose the Liouville operator into two parts,

$$iL = iL_1 + iL_2 \tag{50}$$

Unfortunately, we cannot replace $e^{iL\Delta t}$ by $e^{iL_1\Delta t} e^{iL_2\Delta t}$, because iL_1 and iL_2 are noncommutative operators. Applying Trotter's theorem [41], however, we can decompose the propagator, $U(\Delta t)$:

$$U(\Delta t) = e^{iL\Delta t} = e^{iL_1\Delta t/2} e^{iL_2\Delta t} e^{iL_1\Delta t/2} \tag{51}$$

The idea is now to replace the formal solution of the Liouville equation by the discretized version. The middle term $e^{iL_2\Delta t}$ of the propagator in Eq. (51) can be further decomposed by an additional Trotter factorization to obtain

$$e^{iL_2\Delta t} = (e^{iL_2\Delta\tau})^n + O(n\Delta\tau^3) \tag{52}$$

where $\Delta t = n\,\Delta\tau$. Here the smaller time interval $\Delta\tau$ and the integer n determining the number of steps are chosen to guarantee stable dynamics for the system. Now Eq. (51) becomes

$$U(\Delta t) \approx e^{iL_1\Delta t/2} (e^{iL_2\Delta\tau})^n e^{iL_1\Delta t/2} \tag{53}$$

With the propagator written in this way, the equation of motion can be integrated by a multiple time step algorithm in Cartesian coordinates because Δt and $\Delta\tau$ are different integration time steps ($\Delta t > \Delta\tau$ when $n > 1$). As an example, the force terms are separated into two components

$$F(\mathbf{r}) = F_f(\mathbf{r}) + F_s(\mathbf{r}) \tag{54}$$

where F_f associates with "stiff" degrees of freedom or fast-varying forces, such as forces from bond-stretching, angle-bending motions, and F_s is associated with the rest of the contributions (i.e., slowly varying forces), such as forces from torsion motions and nonbond interaction. By introducing this decomposition into the Liouville operator we obtain

$$iL = \dot{r}\frac{\partial}{\partial r} + F_f(r)\frac{\partial}{\partial p} + F_s(r)\frac{\partial}{\partial p} \tag{55}$$

In this separation, the two Liouville operators, iL_1 and iL_2 of Eq. (50) can now be defined:

$$iL_2 = \dot{r}\frac{\partial}{\partial r} + F_f(r)\frac{\partial}{\partial p}; \qquad iL_1 = F_s(r)\frac{\partial}{\partial p} \tag{56}$$

The propagator $U(\Delta t)$ defined in Eq. (53) can now be implemented algorithmically as follows:

1. Starting with the initial state $[r(0), p(0)]$, generate the motion by using the propagator $e^{iL_1\Delta t/2}$.

2. Using the final state of step 1 as the initial state, generate the motion using the middle propagator $e^{iL_2\Delta\tau}$. Repeat this step n times.

3. Finally, starting with the state generated in step 2 as the initial state, generate the motion using the propagator $e^{iL_1\Delta t/2}$.

Corresponding implementations of the velocity Verlet operator can be easily derived for this Liouville propagator [38]. It should also be realized that the decomposition of iL into a sum of iL_1 and iL_2 is arbitrary. Other decompositions are possible and may lead to algorithms that are more convenient. One example is that in a typical MD simulation, a large portion of the computer processing time is spent in examining the non-bond pair interactions. These non-bond forces, therefore, can be divided into fast and slow parts based on distance by using a continuous switching function [42]. Applications of this MTS method to protein simulations have been shown to reduce the CPU time by a factor of 4–5 without altering dynamical properties [39,40,42]. In addition, this MTS approach shows significantly better performance enhancement in systems where the separation of fast and slow motions is pronounced [43].

C. Other Approaches and Future Direction

There are other approaches to obtaining efficiency in MD simulations. Examples include eigenvector-based schemes [44,45], implicit integration schemes [46], path optimization schemes [47], and a transition state theory approach [48]. Recently, a unique approach to overcome the time scale problems in MD was developed. Multibody order (N) dynamics [MBO(N)D] [49] is based on aggregating atoms of a macromolecule into groupings of interacting flexible and rigid bodies. Body flexibility is modeled by a truncated set of body-based normal modes. This approach allows for the unimportant high frequency modes of vibration, such as bond and angle motions, to be eliminated, leaving only the important lower frequency motions. This results in the use of a larger integration time step size, substantially reducing the computational time required for a given dynamic simulation. By coupling MBO(N)D with MTS described in the previous section, speed increases of up to 30-fold over conventional simulation methods have been realized in various MD simulations [49]. In addition to increasing computational efficiency, the approach also allows for a simplified analysis of dynamics simulations, as there are fewer degrees of freedom to consider.

Additionally, continuous developments of computer architectures, such as the clock speed of CPU chips and massive parallel computers, also help to carry out simulations of large biomolecules that require enormous computing power. In recent years, distributed memory parallel computers have been offering cost-effective computational power to researchers. This approach shows a great advantage in the size of the system (it is possible to run a million atoms in the system), although the simulation length is not scaled as well as the size because of the nature of solving equations of motion sequentially in time.

Finally, molecular modeling based on low resolution (coarse-grain) models has gained some attention in the field of biomolecular simulations [50]. This approach dramatically reduces the number of interaction sites by adapting a simple approach (e.g., a single site per residue) [51,52] or a multiple sites per residue approach (e.g., one backbone and one side chain interaction site per residue) [53,54]. These coarse-grain potentials are described by two categories: those based on statistical contact information derived from high resolution protein structures [51,52,54], and those base on established molecular mechanics force fields [53]. Coarse grain approaches are another way to gain a significant increase in speed and therefore begin to address large systems, such as protein–protein complexes, routinely.

Despite recent developments in algorithms and computer hardware, to bridge the gap between the time and size scales accessible by computer simulations and those required by experimental observations we still need to develop noble approaches.

REFERENCES

1. CL Brooks III, M Karplus, BM Pettitt, Proteins: A Theoretical Perspective of Dynamics, Structure and Thermodynamics. New York: Wiley, 1988.
2. JA McCammon, SC Harvey. Dynamics of Proteins and Nucleic Acids. Cambridge, UK: Cambridge Univ Press, 1987.
3. M Karplus, GA Petsko. Nature 347:631–639, 1990.
4. CL Brooks III, DA Case. Chem Rev 93:2487–2502, 1993.
5. WF van Gunsteren, HJC Berendsen. Angew Chem Int Ed Engl 29:992–1023, 1990.
6. MP Allen, DJ Tildesley. Computer Simulations of Liquids. New York: Oxford Univ Press, 1989.
7. D Frenkel, B Smit. Understanding Molecular Simulation from Algorithms to Applications. New York: Academic Press, 1996.
8. DC Rapaport. The Art of Molecular Dynamics Simulation. Cambridge, UK: Cambridge Univ Press, 1995.
9. WF van Gunsteren. Molecular dynamics and stochastic dynamics simulations: A primer. In: WF van Gunsteren, PK Weiner, AJ Wilkinson, eds. Computer Simulations of Biomolecular Systems. Leiden: ESCOM, 1993, pp 3–36.
10. SB Prusiner. Cell 93:337, 1998.
11. L Verlet. Phys Rev 159:98, 1967.
12. RW Hockney. Methods Comput Phys 9:136, 1970.
13. D Potter. Computational Physics. New York: Wiley, 1972, Chap 2.
14. W Swope, HC Andersen, H Berens, KR Wilson. J Chem Phys 76:637, 1992.
15. JP Ryckaert, G Ciccotti, HJC Berendsen. J Comput Phys 23:327, 1977.
16. WF van Gunsteren, HJC Berendsen. Mol Phys 34:1311, 1977.
17. HC Andersen. J Comput Phys 52:24, 1983.
18. WF van Gunsteren, M Karplus. Macromolecules 15:1528, 1982.
19. LSD Caves, JD Evanseck, M Karplus. Protein Sci 7:649, 1998.
20. F Reif. Fundamentals of Statistical and Thermal Physics. London: McGraw-Hill, 1965, Chap 15.5.
21. DL Ermak, JA McCammon. J Chem Phys 69:1352, 1978.
22. LV Woodcock. Chem Phys Lett 10:257, 1971.
23. HC Andersen. J Chem Phys 72:2384, 1980.
24. HJC Berendsen, JPM Postma, WF van Gunsteren, J Hermans. J Chem Phys 81:3684, 1984.
25. S Nosé. J Chem Phys 81:511, 1984.
26. WG Hoover. Phys Rev A 31:1695, 1985.
27. GJ Martyna, ML Klein, M Tuckerman. J Chem Phys 97:2635, 1992.
28. DJ Evans, GP Morriss. Comp Phys Repts 1:297, 1984.
29. SE Feller, Y Zhang, RW Pastor, BR Brooks. J Chem Phys 103:4613, 1995.
30. GJ Martyna, DJ Tobias, ML Klein. J Chem Phys 101:4177, 1994.
31. M Parrinello, A Rahman. Phys Rev Lett 45:1196, 1980.

32. S Nosé, ML Klein. Mol Phys 50:1055, 1983.
33. DJ Evans, GP Morriss. Phys Lett 98A:433, 1983.
34. DJ Evans, GP Morriss. Chem Phys 77:63, 1983.
35. R Car, M Parrinello. Phys Rev Lett 55:2471, 1985.
36. G Galli, A Pasquarello. First-principle molecular dynamics. In: MP Allen, DJ Tildesley, eds. Proceedings of the NATO ASI on Computer Simulation in Chemical Physics. Dordrecht: Kluwer, 1993, pp 261–313.
37. G Ciccotti, JP Ryckaert. Comp Phys Rep 4:345, 1986.
38. ME Tuckerman, BJ Berne, GJ Martyna. J Chem Phys 97:1990, 1992.
39. M Watanabe, M Karplus. J Chem Phys 99:8063, 1993, and references cited therein.
40. M Watanabe, M Karplus. J Phys Chem 99:5680, 1995.
41. HF Trotter. Proc Am Math Soc 10:545, 1959.
42. DD Humphreys, RA Friesner, BJ Berne. J Phys Chem 98:6885, 1994.
43. P Procacci, BJ Berne. J Chem Phys 101:2421, 1994.
44. A Amadei, ABM Linssen, BL deGroot, DMF vanAlten, HJC Berendsen. Biomol Struct Dynam 13:615, 1996.
45. BL deGroot, A Amadei, DMF vanAlten, HJC Berendsen. Biomol Struct Dynam 13:741, 1996.
46. CS Peskin, T Schlick. Commun Pure Appl Math 42:1001, 1989.
47. R Olender, R Elber. J Chem Phys 105:9299, 1996.
48. AF Voter. Phys Rev Lett 78:3908, 1997.
49. HM Chun, CE Padilla, DN Chin, M Watanabe, VI Karlov, HE Alper, K Soosaar, K Blair, O Becker, LSD Caves, R Nagle, DN Haney, BL Farmer. J Comput Chem 21:159, 2000.
50. T Haliloglu, I Bahar. Proteins 31:271, 1998.
51. M Hendlich, P Lackner, S Weitckus, H Floeckner, R Froschauer, K Gottsbacher, G Casari, MJ Sippl. J Mol Biol 216:167, 1990.
52. BA Reva, AV Finkelstein, MF Scanner, AJ Olson. Protein Eng 10:856, 1997.
53. T Head-Gordon, CL Brooks III. Biopolymers 31:77, 1991.
54. I Bahar, RL Jernian. J Mol Biol 266:195, 1997.

4

Conformational Analysis

Oren M. Becker
Tel Aviv University, Tel Aviv, Israel

I. BACKGROUND

The goal of conformational analysis is to shed light on conformational characteristics of flexible biomolecules and to gain insight into the relationship between their flexibility and their function. Because of the importance of this approach, conformational analysis plays a role in many computational projects ranging from computer-aided drug design to the analysis of molecular dynamics simulations and protein folding. In fact, most structure-based drug design projects today use conformational analysis techniques as part of their toolchest. As will be discussed in Chapter 16, in structure-based drug design a rational effort is applied to identifying potential drug molecules that bind favorably into a known three-dimensional (3D) binding site [1], the structure of which was determined through X-ray crystallography, NMR spectroscopy, or computer modeling. Because such an effort requires, among other things, structural compatibility between the drug candidate and the binding site, computational methods were developed to "dock" ligands into binding sites [2]. These docking calculations are used for screening large virtual molecular libraries, saving both time and money. However, although docking is fairly straightforward with rigid molecules, it becomes significantly more complicated when flexible molecules are considered. This is because flexible molecules can adopt many different conformations, each of which may, in principle, lead to successful docking.

Although there are a few "flexible docking" approaches that account for flexibility during the docking process itself, most docking applications rely on conformational analysis to deal with this problem (e.g., by generating a multitude of molecular conformations that are docked separately into the binding site). The importance of conformational analysis in the context of drug design extends beyond computational docking and screening. Conformational analysis is a major tool used to gain insight for future lead optimization. Furthermore, even when the 3D structure of the binding site is unknown, conformational analysis can yield insights into the structural characteristics of various drug candidates.

In a different context, conformational analysis is essential for the analysis of molecular dynamics simulations. As discussed in Chapter 3, the direct output of a molecular dynamics simulation is a set of conformations ("snapshots") that were saved along the trajectory. These conformations are subsequently analyzed in order to extract information about the system. However, if, during a long simulation, the molecule moves from one conformation class to another, averaging over the whole simulation is likely to be misleading. Conformational analysis allows one to first identify whether such drastic conformational transitions have occurred and then to focus the analysis on one group of conformations at a time.

In view of their importance it is not surprising that conformation sampling and analysis constitute a very active and innovative field of research that is relevant to biomolecules and inorganic molecular clusters alike. The following sections offer an introduction to the main methodologies that are used as part of a conformational analysis study. These are arranged according to the three main steps applied in such studies: (1) conformation sampling, (2) conformation optimization, and (3) conformational analysis.

II. CONFORMATION SAMPLING

Conformation sampling is a process used to generate the collection of molecular conformations that will later be analyzed. Ideally, all locally stable conformations of the molecule should be accounted for in order for the conformational analysis to be complete. However, owing to the complexity of proteins and even fairly small peptides it is impractical to perform such an enumeration (see Section II.D.1). The number of locally stable conformations increases so fast with the molecular size that the task of full enumeration becomes formidable. Even enumerating all possible $\{\phi, \psi\}$ conformations of a protein backbone rapidly becomes intractable. As a result, most conformational studies must rely on sampling techniques. The basic requirement from such sampling procedures is that the resulting conformational sample ("ensemble") will be representative of the system as a whole. This means that in most biomolecular studies a "canonical" ensemble, characterized by a constant temperature (see Chapter 3), is sought. Therefore sampling methods that were designed for canonical ensembles and that guarantee "detailed balance" are especially suitable for this task. Two such methods are high temperature molecular dynamics and Monte Carlo simulations. However, because of the complexity and volume of biomolecular conformational space, other sampling techniques, which do not adhere to the canonical ensemble constraint, are also often employed.

A. High Temperature Molecular Dynamics

Molecular dynamics simulations, which were discussed in Chapter 3, are among the most useful methods for sampling molecular conformational space. As the simulation proceeds, the classical trajectory that is traced is in fact a subset of the molecular conformations available to the molecule at that energy (for microcanonical simulations) or temperature (for canonical simulations). Assuming that the ergodic hypothesis holds (see Chapter 3), an infinitely long MD trajectory will cover all of conformational space. The problem with room temperature MD simulations is that a shorter finite-time trajectory is not likely to sample all of conformational space. Even a nanosecond MD trajectory will most likely be confined to limited regions of conformational space (Fig. 1a). The room temperature probability of crossing high energy barriers is often too small to be observed during a finite MD simulation.

A common solution that allows one to overcome the limited sampling by MD simulations at room temperature is simply to raise the temperature of the simulation. The additional kinetic energy available in a higher temperature simulation makes crossing high

(a)   (b)

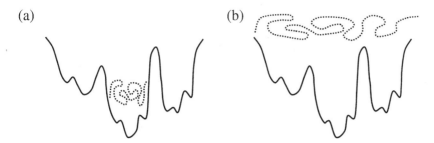

Figure 1   A schematic view of (a) a low temperature simulation that is confined by high energy barriers to a small region of the energy landscape and (b) a high temperature simulation that can overcome those barriers and sample a larger portion of conformational space.

energy barriers more likely and ensures a broad sampling of conformational space. In raising the simulation temperature to 1000 K or more, one takes advantage of the fact that chemical bonds cannot break in most biomolecular force fields (Chapter 2). Namely, the fact that bonds are modeled by a harmonic potential means that regardless of the simulation temperature these bonds can never spontaneously break, and the chemical integrity of the molecule remains intact. The effect of the unrealistically high temperatures employed is primarily to "shake" the system and allow the molecule to cross high energy barriers (Fig. 1b).

There is no definite rule regarding what temperature is "high temperature" in this context, as this depends on the character of the underlying energy landscape. Temperatures on the order of 1000 K are often used for sampling the conformations of peptides and proteins, because this temperature is below the temperature at which unwanted cis–trans transitions of the peptide bond frequently occur [3]. In other cases, such as for sampling the conformations of a ligand bound in a protein's active site, much lower temperatures must be used. Otherwise the ligand will dissociate and the simulation will sample the conformations of an unbound ligand rather than those of the bound ligand.

The main advantage of using MD for conformation sampling is that information of molecular forces is used to guide the search process into meaningful regions of the potential. A disadvantage associated with this sampling technique is the fact that high temperature simulations sample not only the regions of interest at room temperature but also regions that are inaccessible to the molecule at room temperature. To overcome this problem the sampled conformations have to be energy-minimized or preferably annealed before being considered as sampled conformations. These methods will be discussed in Section III.

B.   Monte Carlo Simulations

Monte Carlo search methods are stochastic techniques based on the use of random numbers and probability statistics to sample conformational space. The name "Monte Carlo" was originally coined by Metropolis and Ulam [4] during the Manhattan Project of World War II because of the similarity of this simulation technique to games of chance. Today a variety of Monte Carlo (MC) simulation methods are routinely used in diverse fields such as atmospheric studies, nuclear physics, traffic flow, and, of course, biochemistry and biophysics. In this section we focus on the application of the Monte Carlo method for conformational searching. More detailed in-depth accounts of these methods can be found in Refs. 5 and 6.

In performing a Monte Carlo sampling procedure we let the dice decide, again and again, how to proceed with the search process. In general, a Monte Carlo search consists of two steps: (1) generating a new "trial conformation" and (2) deciding whether the new conformation will be accepted or rejected.

Starting from any given conformation we "roll the dice," i.e., we let the computer choose random numbers, to decide what will be the next trial conformation. The precise details of how these moves are constructed vary from one study to another, but most share similar traits. For example, assuming that the search proceeds via polypeptide torsion moves, choosing a new trial conformation could include the following steps. First, roll the dice to randomly select an amino acid position along the polypeptide backbone. Second, randomly select which of the several rotatable bonds in that amino acid will be modified (e.g., the ϕ, ψ, or χ torsion angles). Finally, randomly select a new value for this torsion angle from a predefined set of values. In this example it took three separate random selections to generate a new trial conformation. Multiple torsion moves as well as Cartesian coordinate moves are among the many possible variations on this procedure.

Once a new "trial conformation" is created, it is necessary to determine whether this conformation will be accepted or rejected. If rejected, the above procedure will be repeated, randomly creating new trial conformations until one of them is accepted. If accepted, the new conformation becomes the "current" conformation, and the search process continues from it. The trial conformation is usually accepted or rejected according to a temperature-dependent probability of the Metropolis type,

$$p = \begin{cases} e^{-\beta\Delta U}, & e^{-\beta\Delta U} < 1 \\ 1, & e^{-\beta\Delta U} \geq 1 \end{cases} \quad \text{or} \quad p = \min[1, e^{-\beta\Delta U}] \tag{1}$$

where $\beta = 1/kT$ and ΔU is the change in the potential energy. This means that if the energy of the new trial conformation is lower than that of the current conformation, $\Delta U < 0$, it is always accepted. But even if the energy of the trial conformation is higher than the current energy, $\Delta U > 0$, there is a certain probability, proportional to the Boltzmann factor, that it will be accepted. To find out whether a higher energy trial conformation is accepted, a random number r in the range [0, 1] is selected and compared to the Metropolis probability defined in Eq. (1). If $r < p$, the conformation is accepted; otherwise it is rejected. This acceptance probability satisfies the principle of detailed balance, ensuring that if the process continues for a long enough time then a stationary solution will be achieved.

In Monte Carlo simulations, just as in MD simulations, temperature plays an important role. In general, MC simulations tend to move toward low energy states. However, at high temperatures (small β values) there is a significant probability of climbing up energy slopes, allowing the search process to cross high energy barriers. This probability becomes significantly smaller at low temperatures, and it vanishes altogether in the limit of $T \to 0$, where the method becomes equivalent to a minimization process. Thus, high temperature MC is often used to sample broad regions of conformational space.

As stated above, MC simulations are popular in many diverse fields. Their popularity is due mainly to their ease of use and their good convergence properties. Nonetheless, straightforward and application of MC methods to biomolecules is often problematic due to very low acceptance ratios, which significantly reduce the efficiency of the method. The reason for the low acceptance ratio, i.e., the ratio between accepted MC moves and total MC trial moves, is the compact character of most biomolecules. This means that many move attempts end up "bumping" into other parts of the molecule and are rejected because of the clash. This is true in particular of moves defined in Cartesian coordinates. To partially overcome this problem it is recommended that torsion move-sets be used. Advanced applications of MC methods for conformation sampling often involve various techniques for enhanced sampling [7,8].

A special comment must be added regarding the random number generator. Because the MC process is driven by random numbers, it is sensitive to the quality of the random number generator that is being used. The random number generator is supposed to generate uniformly distributed random numbers in the range [0, 1]. But in fact the computer does not generate random numbers at all. It uses a deterministic algorithm to generate a pseudo-random series of numbers with a finite periodicity. A high quality algorithm will have a long enough periodicity that the observed distribution does indeed appear random. However, there are many so-called random number generators on the market that are anything but random. It is good practice to check the random number generator prior to using it in an actual MC simulation. A simple histogram of 10,000 numbers or more will easily show whether or not the resulting distribution is uniform. Good random number generators are given in Ref. 9.

C. Genetic Algorithms

A genetic algorithm (GA) evolves a population of possible solutions through genetic operations, such as mutations and crossovers, to a final population of low energy conformations that minimize the energy function (the fitness function) [10,11]. For the purpose of conformational sampling, the translational, rotational, and internal degrees of freedom are encoded into "genes," which are represented by the real number values of those degrees of freedom [12]. Each individual conformation, named a "chromosome," consists of a collection of genes and is represented by the appropriate string of real numbers. A fitness value (energy) is assigned to each chromosome.

The GA evolutionary process iterates through the following two steps: (1) Generation of a children population from a parent population by means of genetic operators, and (2) a generation update step. The process starts with a random population of initial chromosomes. A new population is generated from the old one by the use of genetic operators. The two most fundamental operators are schematically depicted in Figure 2. The *mutation* operator (Fig. 2a) changes the value of a randomly selected gene by a random value, depending on the type of gene. The *crossover* operator (Fig. 2b) exchanges a set of genes between two parent chromosomes, creating two children with genes from both parents. Additional genetic operators, such as the *multiple simultaneous mutation* operator or a *migration* operator, which moves individual chromosomes from one subpopulation to another, may also be used.

Parents are selected for breeding based on their relative fitness and an external evolutionary "pressure," which directs the process to favor breeding by parents with higher fitness. Following each breeding cycle the chromosome population is updated. A common update scheme is to replace all but the most fit "elite" chromosomes of the original set. For example, a "generation update with an elitism of 2" indicates that all but the two most fit parent chromosomes are replaced [11]. The iteration through the two steps of the

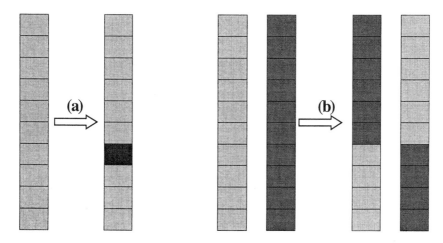

Figure 2 Genetic operators used to create a population of children chromosomes from a population of parent chromosomes. (a) Single-point mutation. A gene to be mutated is selected at random, and its value is modified. (b) One-point crossover. The crossover point is selected randomly, and the genes are exchanged between the two parents. Two children are created, each having genes from both parents.

evolutionary process continues until convergence or until the maximum number of steps is reached.

Application of GA to conformational searching typically uses the potential energy as the fitness function. The degrees of freedom encoded in the ''genes'' are usually only a subset of coordinates that are most significant for the conformational search. These may include backbone dihedral angles, side chain dihedral angles, or any other coordinate that may be considered useful. Application of GA to docking problems requires that the three center-of-mass coordinates and the three rotational Euler angles also be encoded in genes.

An advantage of GA for conformational analysis, besides its elegance, is that it is very easy to code and run. In addition, GA usually requires fewer iterations than either MD or MC to generate a large population of low energy conformations. On the other hand, because in every iteration whole populations are propagated, each iteration takes significantly more CPU time than in either of the other two methods. Furthermore, in many complex cases GA is known for its slow convergence.

A successful application of GA to conformation sampling is, for example, as a part of flexible docking [12–14]. It should be noted, however, that none of the three sampling methods discussed above, MD, MC, and GA, was shown to outperform the other two in any general way. In fact, a comparison of the three methods in the context of flexible docking showed similar efficiency for all three [12], although specific advantages are likely to exist for particular applications.

D. Other Search Methods

Many additional search methods have been devised in addition to the basic three just discussed. A few of them are outlined below. Note that whereas some methods, such as parallel tempering and J-walking, are improved or specialized versions of the basic techniques, other methods, such as systematic enumeration and distance geometry, follow different paths altogether.

1. Systematic Enumeration

Systematic sampling is, in principle, the most thorough method for searching molecular conformational space. The energy is sampled over the entire range of each degree of freedom (typically torsion angles) at regularly spaced intervals. Thus, the sampled conformations lie on an N-dimensional lattice (N being the number of degrees of freedom). Several problems make the systematic sampling procedure irrelevant to most biomolecular systems. A major problem associated with systematic searching is that the number of conformations generated rapidly becomes extremely large even for small molecules. For example, systematic search with six rotatable bonds sampled at 30° increments results in almost 3 million conformations. The number of samples can be reduced by limiting the range of rotation for a symmetrical substituent (e.g., 0–180° for a phenyl group) and/or increasing the rotation step size (all staggered conformations of a saturated $C—C$ bond can be sampled at 120° increments). Another problem arises from the fact that a systematic search is likely to generate a large number of unphysical conformations, in which one part of the molecule crosses over another part (e.g., a protein backbone that crosses over itself). As a result of the these limitations, systematic search is applied only to small molecules such as small peptides [15] or ligands in the context of flexible docking.

"Pruning" is a sophisticated way to reduce the computational requirements associated with systematic search. Pruning takes advantage of the fact that all conformations along a given branch of the search tree that are downstream from a high energy conformation will also have high energies and need not be calculated. For example, if an irrevocable clash such as the backbone folding over itself occurs after the fifth angle in an eight-angle systematic search, there is no need to continue sampling the three remaining angles. Pruning can be performed using geometrical constraints that must be satisfied for all conformations. An energy cutoff is a simplistic filter that can be applied at the end of the search to reduce the number of conformations stored. With this filter, conformations with energies above the cutoff are discarded. This filter, of course, does not reduce the number of calculations but does help in managing and analyzing the data.

2. Distance Geometry

Distance geometry is a general method for building conformational models of complex molecular systems based on a set of distance constraints. It is a purely geometric technique that generates structures that satisfy the given set of constraints without requiring a starting conformation or an energy function. The distance constraints, which are the input to the method, can be qualitative or approximate. As such they are defined by upper and lower bounds and often into a matrix form. The distance geometry approach converts, or "embeds," the uncertain distance constraints into 3D Cartesian coordinates.

In the basic metric matrix implementation of the distance constraint technique [16] one starts by generating a distance bounds matrix. This is an $N \times N$ square matrix (N the number of atoms) in which the upper bounds occupy the upper diagonal and the lower bounds are placed in the lower diagonal. The matrix is filled by information based on the bond structure, experimental data, or a hypothesis. After smoothing the distance bounds matrix, a new distance matrix is generated by random selection of distances between the bounds. The distance matrix is converted back into a 3D conformation after the distance matrix has been converted into a metric matrix and diagonalized. A new distance matrix randomly generated within the same bounds will result in another possible conformation that satisfies the model, and so forth. The tighter the bounds, the more restricted the search. Naturally, distance geometry is very useful in situations in which many distance constraints are known, in particular for suggesting models that agree with NMR data [17] (see Chapter 13). This method is also useful for pharmacophore modeling based on known bioactive molecules [18]. However, for general-purpose conformational searching this method is less appropriate unless the search space is limited by known constraints (such as active site data in a docking context). Computationally, distance geometry is limited to moderately sized systems (up to several thousand atoms) because it requires computationally expensive matrix manipulation. For a detailed review of distance geometry and its application to molecular problems, see Ref. 19.

3. Parallel Tempering and J-Walking

Energy barriers that confine the search to limited regions of conformational space restrict both molecular dynamics and Monte Carlo simulations, preventing them from reaching ergodicity. As discussed in Sections II.A and II.B, this limitation is often overcome by raising the simulation temperature. The higher temperature allows the simulation to overcome high energy barriers and extend the search space. However, it also causes the simulation to sample regions of conformational space that are irrelevant and inaccessible to room temperature molecules. The methods of parallel tempering and J-walking address the problem of how to raise the simulation temperature without wasting computational effort on inaccessible conformations.

The idea behind both parallel tempering and J-walking is to incorporate information obtained by ergodic high temperature simulations into low temperature simulations. By periodically passing information from high temperature simulations to low temperature simulations, these methods allow the low temperature system to overcome the barriers between separate regions. In practice, both methods, which can be applied in both MC and MD simulations, require propagation of at least two (and often more) parallel simulations—one at the desired low temperature and another at a high temperature. For simplicity we shall refer to the Monte Carlo implementation of these techniques; the molecular dynamics implementation is similar in nature.

In J-walking [20] the periodic MC trial probability for a simulation at temperature T is taken to be a Boltzmann distribution at a high temperature, T_J ($\beta_J = 1/kT_J$), The jumping temperature, T_J, is sufficiently high that the Metropolois walk can be assumed to be ergodic. This results in the acceptance probability,

$$p = \min[1, e^{-(\beta - \beta_J)\Delta U}] \tag{2}$$

where $\beta = 1/kT$, $\beta_J = 1/kT_J$, and ΔU is the change in the potential energy. In practice, at temperature T the trial moves are taken from the Metropolis distribution of Eq. (1) about 90% of the time, with jumps using Eq. (2) attempted about 10% of the time. The jumping conformations are generated with a Metropolis walk at temperature T_J. Several methods have been devised to overcome correlations between the low and high temperature walks.

In parallel tempering [21], the high and low temperature walkers exchange configurations, unlike in J-walking, in which only the high temperature walker feeds conformations to the low temperature walker. By exchanging conformations, parallel tempering satisfies detailed balance, assuming the two walks are long enough. Parallel tempering simulations can be performed for more than two temperatures [22]. It should be noted that the gaps between adjacent temperatures must be chosen so that the exchanges are accepted with sufficient frequency. This means that the distribution characteristics of the two temperatures must overlap so that a conformation generated at one temperature will have a significant probability of being accepted at the other temperature. If the temperature gap is too large, the likelihood of any conformation being accepted by the other simulation is very small.

III. CONFORMATION OPTIMIZATION

The structures generated by the various sampling methods often correspond to transient conformations. It is a desirable practice to bring these conformations to nearby local minima, which represent locally stable conformations, before further analysis is performed. The most widely used methods for this purpose are the various minimization techniques. These vary in accuracy and computational efficiency and are useful in many applications as well as conformational analysis. Minimization methods are used as a tool (often alongside other computational tools) in model building, preparation of the initial structure for molecular dynamics, preparation of structures for normal mode analysis, and more. Because of their importance we go into some detail in discussing the different minimization approaches. Another conformation optimization approach is simulated annealing. In the context of conformational analysis, simulated annealing is often used to optimize conformations generated by high temperature MD or MC simulations. It is also often used as an independent method for minimization and identification of the global minimum in a complex energy landscape.

A. Minimization

A drop of water that is placed on a hillside will roll down the slope, following the surface curvature, until it ends up in the valley at the bottom of the hill. This is a natural minimization process by which the drop minimizes its potential energy until it reaches a local minimum. Minimization algorithms are the analogous computational procedures that find minima for a given function. Because these procedures are ''downhill'' methods that are unable to cross energy barriers, they end up in local minima close to the point from which the minimization process started (Fig. 3a). It is very rare that a direct minimization method

(a)   (b)

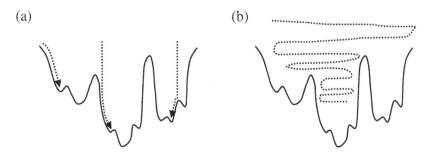

Figure 3   A schematic representation of two optimization schemes. (a) Minimization, which leads to the closest local minimum; (b) simulated annealing, which can overcome intermediate energy barriers along its path.

will find the global minimum of the function. This can be easily understood from our water drop analogy. The drop rolls down the slope until it reaches the smallest of hollows and stops there. It will not cross the nearest rise even if the valley on the other side of that rise is much deeper than the hollow it is in now.

In conformational analysis, minimization reduces the potential energy of a given conformation, typically by relieving local strains in the structure. However, the minimized structure will not stray far from the initial conformation. This is why the minimization process is considered an optimization technique and not a search method. Global optimization—namely, reaching the lowest point on the energy surface—cannot be addressed by straightforward minimization and is discussed in Chapter 17 (Protein Folding).

The goal of all minimization algorithms is to find a local minimum of a given function. They differ in how closely they try to mimic the way a drop of water or a small ball would roll down the slope, following the surface curvature, until it ends up at the ''bottom.'' Consider a Taylor expansion around a minimum point X_0 of the general one-dimensional function $F(X)$, which can be written as

$$F(X) = F(X_0) + (X - X_0)F'(X_0) + \frac{1}{2}(X - X_0)^2 F''(X_0) + \cdots \qquad (3)$$

where F' and F'' are the first and second derivatives of the function. The extension to a multidimensional function requires replacement of the variable X by the vector $\mathbf{X}$ and replacement of the derivatives by the appropriate gradients. Minimization algorithms are classified according to the amount of information about the function that is being used, represented by the highest derivative used by the algorithm. Algorithms that use only the value of the function and no derivative are called ''order 0'' algorithms. Algorithms that use the first derivative—the slope—of the function are denoted as ''order 1'' algorithms. Finally, ''order 2'' algorithms are algorithms that take advantage of the second derivative—the curvature—of the function. As in most computational techniques, higher order methods that use more information about the function are generally more accurate at the price of being computationally more expensive, taking more time and more resources than the lower order methods. This section offers a brief discussion of the most common minimization algorithms. Additional information can be found in Ref. 23.

1. Order 0 Algorithms

Order 0 minimization methods do not take the slope or the curvature properties of the energy surface into account. As a result, such methods are crude and can be used only with very simple energy surfaces, i.e., surfaces with a small number of local minima and monotonic behavior away from the minima. These methods are rarely used for macromolecular systems.

Grid searching is an example of an order 0 minimization algorithm. In this method a regular grid (e.g., a cubic grid) is placed over the energy surface and the value of the function at each node is calculated. The grid point with lowest energy is taken to represent the real minimum. The quality of a grid search and the computational effort associated with it depend, of course, on the density of the grid mesh. In some applications the convergence of the method is improved by gradually increasing the density of the grid in the vicinity of the best point of an earlier, coarser grid. This method is inefficient for large molecules and can easily converge to a false minimum. Grid searching as a minimization technique is rarely used in biomolecular studies.

Another order 0 method is the *downhill simplex method* [9]. A *simplex* is a geometrical element consisting, in N dimensions, of $N + 1$ points (vertices) and all their interconnecting line segments, polygonal faces, etc. In two dimensions, a simplex is a triangle; in three dimensions it is a tetrahedron. The downhill simplex method starts with $N + 1$ points, defining an initial simplex. The energy of each point is evaluated, and the method moves the point of the simplex from where the function is largest through the opposite face of the simplex to a lower point. Usually such ''reflections'' conserve the volume of the simplex, but when it can do so it will expand or contract the simplex to minimize the function. The minimization proceeds by taking the highest point in the new simplex and reflecting it through the opposite face and so forth. This process is schematically described in Figure 4. The simplex method is sometimes used for crude placement of starting conformations, for example in the context of docking. A detailed description of the simplex method can be found in Ref. 9.

2. Order 1 Algorithms

Order 1 algorithms, which represent a fair balance between accuracy and efficiency, are the most commonly used minimization methods in macromolecular simulations. As indicated by their name, these algorithms use the gradient of the function to direct the minimization process toward the nearest local minimum. They use information about the *slope* of the function but do not include information about its *curvature*. Thus, to compensate for the deficiency of curvature data, all order 1 minimization methods employ a stepwise iterative scheme. The iterations are used for recalculating the gradient and correcting the minimization approach pattern following changes in the direction of the slope.

In general, order 1 methods iterate over the following equation in order to perform the minimization until it converges or until it reaches a preset maximum number of steps:

$$\vec{\mathbf{r}}_k = \vec{\mathbf{r}}_{k-1} + \lambda_k \hat{\mathbf{S}}_k \tag{4}$$

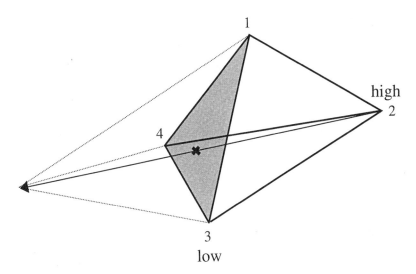

Figure 4 A representative step in the downhill simplex method. The original simplex, a tetrahedron in this case, is drawn with solid lines. The point with highest energy is reflected through the opposite triangular plane (shaded) to form a new simplex. The new vertex may represent symmetrical reflection, expansion, or contractions along the same direction.

where $\mathbf{r}_k$ is the new position at step k, $\mathbf{r}_{k-1}$ is the position at the previous step $k-1$, λ_k is the size of the step to be taken at step k, and $\mathbf{S}_k$ is the direction of that step. Various methods differ in the way they choose the step size and the step direction, i.e., in the way they try to compensate for the lack of curvature information. For example, the *steepest descent* method uses a variable step size for this purpose, whereas the *conjugated gradients* method takes advantage of its memory of previous steps. These two popular algorithms are described next. Information about other minimization methods can be found in Ref. 9.

(a) Steepest Descent.   Steepest descent (SD), which is the simplest minimization algorithm, follows the direction of the instantaneous gradients calculated at each iteration. This means that once the gradient of the energy function, $\mathbf{g}_k$, at the current position is calculated, the minimization step is taken in the direction opposite to it (i.e., in the direction of the force),

$$\mathbf{\hat{S}}_k = -\mathbf{\hat{g}}_k = -\nabla U(\mathbf{r}) \tag{5}$$

The step size, λ_k, is adjusted at each iteration to compensate for the lack of curvature information. If the energy of the new conformation is lower than that of the previous one, the algorithm assumes that the approach direction is correct and increases the step size λ_k by a small factor (often by the factor 1.2) to improve efficiency. However, if the energy of the new conformation turns out to higher, indicating that the real path has curved away from the current minimization direction, the step size λ_k is decreased (often by the factor 0.5) to allow the algorithm to correct the direction more efficiently. Because finite step sizes are used, the method does not flow smoothly down to the minimum but rather jitters around it. Furthermore, SD's imprecise approach to the minimum usually means that the method does not converge to one point and gets stuck in a limit cycle around the minimum. This means that SD often gets close to the minimum but rarely reaches it. Despite the relatively poor convergence of SD it is an efficient minimization procedure that is very useful for crude local optimization, such as relieving bad contacts in an initial structure before a dynamics simulation or as the first phase of a more complex minimization scheme.

(b) Conjugated Gradients.   To compensate for the missing curvature information, the conjugated gradients (CG) method makes use of its ''memory'' of gradients calculated in previous steps. The first step is taken in the direction of the force,

$$\mathbf{\hat{S}}_1 = -\mathbf{\hat{g}}_1 \tag{6}$$

but for all subsequent iterations, $k > 1$, the direction of the minimization step is determined as a weighted average of the current gradient and the direction taken in the previous iteration, i.e.,

$$\mathbf{\hat{S}}_k = -\mathbf{\hat{g}}_k + b_k \mathbf{\hat{S}}_{k-1} \tag{7}$$

The weight factor b_k is calculated as a ratio of the squares of the current and previous gradients.

$$b_k = |g_k|^2 / |g_{k-1}|^2 \tag{8}$$

For quadratic surfaces of dimension n, it can be shown that the method of conjugated gradients is very efficient, converging on the nth step. Nonetheless, even for nonquadratic surfaces such as molecular energy surfaces, the CG method converges much better than SD. A numerical disadvantage of the gradient memory employed in CG is that it accumulates numerical errors along the way. This problem is usually overcome by restarting the minimization process every so often; that is, at given intervals the gradient memory is wiped out by setting b_k equal to zero.

3. Order 2 Algorithms

Order 2 minimization algorithms, which use the second derivative (curvature) as well as the first derivative (slope) of the potential function, exhibit in many cases improved rate of convergence. For a molecule of N atoms these methods require calculating the $3N \times 3N$ Hessian matrix of second derivatives (for the coordinate set at step k)

$$[\mathbf{H}_k]_{ij} = \frac{\partial^2 U(\mathbf{r})}{\partial r_i \, \partial r_j} \tag{9}$$

in addition to the $3N$ vector of first derivatives $\mathbf{g}_k$ discussed above.

(a) Newton–Raphson. The Newton–Raphson minimization method is an order 2 method based on the assumption that, near the minimum, the energy can be approximated by a quadratic function. For a one-dimensional case, assuming that $F(X) = a + bX + cX^2$, the Newton–Raphson minimization can be formulated as

$$X^* = X - F'(X)/F''(X) \tag{10}$$

where X^* is the minimum, X is the current position, and F' and F'' are the first and second derivatives at the current position. Namely, for a quadratic function, this algorithm finds the minimum in a single step. This conclusion can be generalized to the multidimensional case:

$$\vec{\mathbf{S}}_k = -\mathbf{H}_k^{-1} \vec{\mathbf{g}}_k \tag{11}$$

For nonquadratic but monotonic surfaces, the Newton–Raphson minimization method can be applied near a minimum in an iterative way [24].

There are several reasons that Newton–Raphson minimization is rarely used in macromolecular studies. First, the highly nonquadratic macromolecular energy surface, which is characterized by a multitude of local minima, is unsuitable for the Newton–Raphson method. In such cases it is inefficient, at times even pathological, in behavior. It is, however, sometimes used to complete the minimization of a structure that was already minimized by another method. In such cases it is assumed that the starting point is close enough to the real minimum to justify the quadratic approximation. Second, the need to recalculate the Hessian matrix at every iteration makes this algorithm computationally expensive. Third, it is necessary to invert the second derivative matrix at every step, a difficult task for large systems.

(b) Adopted Basis Newton–Raphson. A derivative method that is suited for large systems such as proteins is the *adopted basis Newton–Raphson* (ABNR) algorithm [25]. Instead of calculating the full multidimensional curvature (i.e., the full Hessian matrix) at each minimization step, the ABNR method limits its calculation to a small subspace of dimension s in which the system has made the most progress in past moves. The idea is to add curvature information only in those directions where it is likely to contribute most. This way the system moves in the best direction in the restricted subspace. Because the dimensionality s of the subspace is taken to be between 4 and 10, ABNR is an efficient minimization method comparable in CPU time to the order 1 methods such as the conjugated gradient. ABNR is also comparable to CG in terms of convergence. Since the method

is not self-starting, the first steps are taken with an order 1 method, usually the steepest descent method.

4. Minimization Protocol

The foregoing discussion highlights the fact that the different minimization algorithms have relative strengths and weaknesses. As an example, Figure 5 compares the results of minimizing the same protein with the steepest descent (SD) and conjugated gradients (CG) methods. It is evident that although initially SD reduces the energy faster than CG, in the long run the latter outperforms the former. A similar result is obtained when CG is replaced by ABNR. A detailed comparison between the various minimization algorithms applied to a peptide and a protein is given in Ref. 25.

To optimize the minimization procedure it is usually best to combine several algorithms into a single minimization protocol, taking advantage of their relative strengths. A good minimization scheme will usually start with SD and then use CG or ANBR to finish the job. The number of steps to be used in each phase depends on the goal of the minimization and on the character of the system. When high quality minimization is required, the minimization can be completed with Newton–Raphson. The termination criterion is usually defined in terms of the gradient RMS (GRMS), which is defined as the root-mean-square of all $3N$ gradients (or forces).

B. Simulated Annealing

Simulated annealing is a popular method that is often used for global optimization, i.e., finding the global minimum of a potential energy surface. The method takes its name from the natural annealing process in which a glass or a metal is first heated and then slowly cooled into a stable low energy state. The key factor in this process is slow cooling. If the cooling is done too fast, the materials will end up in unstable brittle states. Alterna-

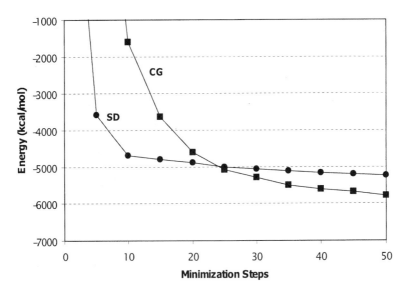

Figure 5 A comparison of steepest descent (SD) minimization and conjugated gradients (CG) minimization of the same protein.

tively phrased, heating up the system shakes and rattles the molecule around the energy landscape, infusing it with thermal energy, analogous to kT, enabling it to jump out of its initial local minimum. The gradual cooling that is subsequently applied decreases the amplitude of these shakes, bringing the details of the energy surface back into focus and causing the system to slowly settle down to a lower energy minimum. Figure 3b is a schematic representation of the simulated annealing process. In fact, we see that simulated annealing bridges between the high temperature conformational sampling simulations that are insensitive to the energy barriers (Fig. 1b) and the low temperature situations, which are sensitive to the details of the energy landscape (Fig. 1a).

Simulated annealing can be easily implemented in both molecular dynamics and Monte Carlo simulations. In molecular dynamics, the temperature is controlled through coupling to a heat bath (Chapter 3); with simulated annealing, the temperature of the bath is decreasing gradually. In Monte Carlo the trial move is accepted or rejected according to a temperature-dependent probability of the Metropolis type [Eq. (1)]. In simulated annealing MC, the temperature used in the acceptance probability is gradually decreased. It should be noted that it is not necessary to anneal all the way to 0 K, because once the kinetic energy kT gets below the characteristic barrier height, a significant change cannot occur. Thus, many simulated annealing protocols cool to room temperature (or somewhat below) and are followed by a local minimization algorithm to remove the excess energy. Specific implementations vary in cooling schedules, initial temperatures, the possibility of repeated heating ''spikes,'' etc. A detailed account of the method can be found in Ref. 26.

Although simulated annealing is often considered a global optimization method, this is not the case when biomolecules are concerned. It can be shown that in systems characterized by a broad distribution of energy scales, a simulated annealing trajectory (either MD or MC) will have to be extremely long before it is able to find the global minimum. Since the energy landscape of proteins is broadly distributed and rough, this means that simulated annealing is an inefficient and infeasible strategy for protein folding. Nonetheless, even with biomolecules, simulated annealing remains a very useful method for local optimization. Its advantage is that, unlike direct minimization, which takes the molecule only as far as the nearest local minimum, simulated annealing is able to locate lower local minima further away from the initial conformation. An example of the application of this method can be found in Ref. 27. In the context of conformational analysis, simulated annealing is often used in conjunction with a high temperature sampling simulation. Each of the molecular structures generated by the high temperature simulation is first annealed back to room temperature before it is included in the conformational sample and subjected to further analysis [28].

IV. CONFORMATIONAL ANALYSIS

To extract the conformational properties of the molecule that is being studied, the conformational ensemble that was sampled and optimized must be analyzed. The analysis may focus on global properties, attempting to characterize features such as overall flexibility or to identify common trends in the conformation set. Alternatively, it may be used to identify a smaller subset of characteristic low energy conformations, which may be used to direct future drug development efforts. It should be stressed that the different conformational analysis tools can be applied to any collection of molecular conformations. These may be generated by the above sampling techniques but can also have an experimental origin, such as NMR models or different X-ray structures of the same molecule (or analogous molecules).

A. Similarity Measures

A similarity measure is required for quantitative comparison of one structure with another, and as such it must be defined before the analysis can commence. Structural similarity is often measured by a root-mean-square distance (RMSD) between two conformations. In Cartesian coordinates the RMS distance d_{ij} between conformation i and conformation j of a given molecule is defined as the minimum of the functional

$$d_{ij} = \left[\frac{1}{N} \sum_{k=1}^{N} |\mathbf{r}_k^{(i)} - \mathbf{r}_k^{(j)}|^2 \right]^{1/2} \tag{12}$$

where N is the number of atoms in the summation, k is an index over these atoms, and $\mathbf{r}_k^{(i)}$, $\mathbf{r}_k^{(ij)}$ are the Cartesian coordinates of atom k in conformations i and j. The minimum value of Eq. (12) is obtained by an optimal superposition of the two structures. The resulting RMS distances are usually compiled into a distance matrix Δ, where the elements Δ_{ij} are the RMS distances between conformations i and j.

Since the summation in Eq. (12) may be on any subset of atoms, it can be fine-tuned to best suit the problem at hand. The summation may be over the whole molecule, but it is very common to calculate conformational distances based only on non-hydrogen "heavy" atoms or, in the case of proteins, even based on only the backbone C_α atoms. Alternatively, in a study related to drug design one may consider, for example, focusing only on atoms that make up the pharmacophore region or that are otherwise known to be functionally important.

The conformational distance does not have to be defined in Cartesian coordinates. For comparing polypeptide chains it is likely that similarity in dihedral angle space is more important than similarity in Cartesian space. Two conformations of a linear molecule separated by a single low barrier dihedral torsion in the middle of the molecule would still be considered similar on the basis of dihedral space distance but will probably be considered very different on the basis of their distance in Cartesian space. The RMS distance is dihedral angle space differs from Eq. (12) because it has to take into account the 2π periodicity of the torsion angle,

$$d_{ij} = \left\{ \frac{1}{N} \sum_{k=1}^{N} \min[(\theta_k^{(j)} - \theta_k^{(j)})^2, (2\pi - \theta_k^{(i)} + \theta_k^{(j)})^2] \right\}^{1/2} \tag{13}$$

where N is the number of dihedral angles in the summation and $\theta_k^{(i)}$, $\theta_k^{(ij)}$ are the values of the dihedral angle θ_k in the two structures. As with the Cartesian distance, any appropriate subset of dihedral angles may be used, ranging from only the backbone ϕ, ψ angles to a full set that includes all the side chain χ angles.

It is up to the researcher to decide whether to use a Cartesian similarity measure or a dihedral measure and what elements to include in the summation [29]. It should be stressed that while the RMS distances perform well and are often used, there are no restrictions against other similarity measures. For example, similarity measures that emphasize chemical interactions, hydrophobicity, or the relative orientation of large molecular domains rather than local geometry may serve well if appropriately used.

B. Cluster Analysis

The distance matrix Δ, which holds the relative distances (by whatever similarity measure) between the individual conformations, is rarely informative by itself. For example, when sampling along a molecular dynamics trajectory, the Δ matrix can have a block diagonal form, indicating that the trajectory has moved from one conformational basin to another. Nonetheless, even in this case, the matrix in itself does not give reliable information about the size and shape of the respective basins. In general, the distance matrix requires further processing.

Cluster analysis is a common analytical technique used to group conformations. This approach highlights structural similarity, as defined by the distance measure being used, within a conformational sample. Starting from one selected conformation (often that of the lowest energy), all conformations that are within a given cutoff distance from this structure are grouped together into the first cluster C_1. Next, one of the conformations that were not grouped into the first cluster is selected, and a new cluster is formed around it. The process continues until all the conformations in the sample are assigned to a cluster C_i. This process often generates overlapping clusters, that is, clusters with nonzero intersection $C_i \cap C_j \neq 0$, The overlapping clusters are typically treated in one of two ways: (1) Group together the overlapping clusters C_i and C_j to form a single large cluster that is their union $C_i \cup C_j$ (Fig. 6a) or (2) make the overlapping clusters disjoint by removing their intersection $C_i \cap C_j$ from one of the clusters, typically the one that started with a higher energy conformation (Fig. 6b). Since the optimal cutoff distance by which to cluster the conformations is not a priori known, cluster analysis is usually performed hierarchically. Starting with a short cutoff the analysis is repeated again and again, each time with a larger cutoff distance. The results are often represented as a dendogram. More information about the various clustering algorithms can be found in Ref. 30.

In many conformational studies, cluster analysis is used as a way to focus future effort on a small set of characteristic conformations. One conformation, typically the lowest energy one, is picked from each of the highly populated conformational clusters. The resulting small number of distinctly different conformations are then used as starting points for further computational analysis (such as free energy simulations) or as a basis for generating a pharmacological hypothesis used for directing future drug development [18]. It should be noted, however, that conformational clusters generated by the above procedure do not necessarily represent the correct basin structure of the underlying energy landscape.

(a)   (b)

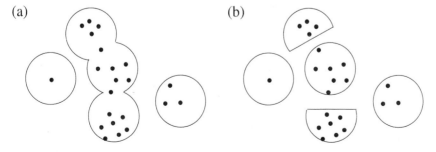

Figure 6 A schematic representation of two clustering methods, in which each point represents a single molecular conformation and the circles are the similarity cutoff distances used to define the clusters. (a) Three clusters are defined when overlapping clusters are grouped together. (b) Five clusters are defined when the overlaps are removed from one of the overlapping clusters.

It has been shown that similar conformations that belong to adjacent energy basins separated by high energy barriers are incorrectly grouped together by the straightforward cluster analysis [29].

C. Principal Component Analysis

An inherent problem associated with conformational analysis is the high dimensionality of the molecular conformational spaces. An N-atom molecule has $3N$ degrees of freedom, and its corresponding conformational space is $(3N - 6)$-dimensional. As a result, even relatively small molecules have very large conformational spaces, making them difficult to analyze. For example, a small heptapeptide may have a 100-dimensional or even 150-dimensional conformational space, depending on its precise amino acid composition. Principal component analysis (PCA) is a computational tool that reduces the effective dimensionality of molecular conformational spaces while retaining an accurate representation of the interconformational distances. This task is accomplished by projecting the original multidimensional data onto an optimal low-dimensional subspace, allowing visual inspection of conformational spaces and of dynamic trajectories that traverse these spaces. Principal component analysis (PCA) was introduced to protein simulations under the name quasi-harmonic analysis by Ichiye and Karplus [31] and is becoming widely used for a variety of applications involving sampling and visualization of conformational spaces [32–35]. A review of principal component analysis can be found in Ref. 36.

How does principal component analysis work? Consider, for example, the two-dimensional distribution of points shown in Figure 7a. This distribution clearly has a strong linear component and is closer to a one-dimensional distribution than to a full two-dimensional distribution. However, from the one-dimensional projections of this distribution on the two orthogonal axes X and Y you would not know that. In fact, you would probably conclude, based only on these projections, that the data points are homogeneously distributed in two dimensions. A simple axes rotation is all it takes to reveal that the data points

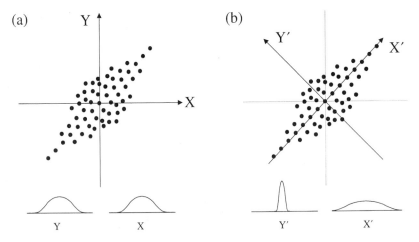

Figure 7 (a) A two-dimensional distribution of points and their one-dimensional projections on the original axes. Judging just from the 1D projections one would probably conclude that the original distribution is homogeneously distributed in two dimensions. (b) The same distribution of points and their 1D projections on the new axes set obtained by PCA by using a similarity transformation. The new 1D projections highlight the strong 1D character of the distribution.

are preferentially spread along one dimension, as reflected by the broad distribution along one of the new axes and a narrow distribution along the other in Figure 7b. The above procedure is what is done by PCA. Starting from a multidimensional distribution of data points (in our case, of molecular conformations), PCA performs a similarity transformation on the original axes to find a new set of axes that best fits the data. The first new axis is selected such that the variance of the distribution along it is the largest possible. The second axis is placed orthogonal to the first in the direction of the second largest variance of the distribution, and so forth. In this new axes set it is usually possible to identify a low-dimensional subspace that captures most of the relative distances between individual conformations.

In general, two related techniques may be used: principal component analysis (PCA) and principal coordinate analysis (PCoorA). Both methods start from the $n \times m$ data matrix $\mathbf{M}$, which holds the m coordinates defining n conformations in an m-dimensional space. That is, each matrix element M_{ij} is equal to q_{ij}, the jth coordinate of the ith conformation. From this starting point PCA and PCoorA follow different routes.

Principal component analysis (PCA) takes the m-coordinate vectors $\mathbf{q}$ associated with the conformation sample and calculates the square $m \times m$ $\mathbf{M}^T\mathbf{M}$ matrix, reflecting the relationships between the *coordinates*. This matrix, also known as the covariance matrix $\mathbf{C}$, is defined as

$$\mathbf{C} = \langle (\mathbf{q} - \langle \mathbf{q} \rangle)(\mathbf{q} - \langle \mathbf{q} \rangle)^T \rangle \tag{14}$$

where the averaging is over the conformation sample (in Cartesian space $m = 3N$ for an N-atom molecule). The covariance matrix $\mathbf{C}$ is diagonalized to obtain the eigenvectors that capture most of the variation in atomic position fluctuations.

Principal coordinate analysis (PCoorA) [37], on the other hand, operates on the square $n \times n$ $\mathbf{M}\mathbf{M}^T$ matrix, reflecting the relationships between the *conformations*. The elements of this matrix, also known as the distance matrix $\mathbf{\Delta}$, are distances d_{ij} between two conformations i and j [such as those defined in Eqs. (12) and (13)]. Since the distances d_{ij} can also be obtained from the $n \times n$ matrix $\mathbf{A}$ of latent roots (eigenvectors), one can use this matrix for the projection, defining $A_{ij} = -1/2d_{ij}^2$ and $A_{ii} = 0$ (for $i, j = 1, 2, \ldots, n$). To guarantee that the matrix $\mathbf{A}$ has a zero root (and thus guarantee that it corresponds to a real configuration) it is "centered," so that the sum of every row and the sum of every column of $\mathbf{A}$ is zero. This centering, which does not alter the distances d_{ij}, is defined as

$$A_{ij}^* = A_{ij} - \langle A_{ij} \rangle_i - \langle A_{ij} \rangle_j + 2\langle A_{ij} \rangle_{ij} \tag{15}$$

where $\langle \cdot \rangle_k$ is the mean over all specific indices $k = i, j, ij$. The centered matrix $\mathbf{A}^*$ is diagonalized using standard matrix algebra to obtain the latent eigenvectors and the diagonal matrix of eigenvalues. The resulting eigenvalues (normalized) give the percentage of the projection of the original distribution on the new coordinate set, and the eigenvectors (scaled by their corresponding eigenvalues) give the new coordinates of the original points in the new axes set. For a more detailed description of this method, see Refs. 29 and 37.

It should be stressed that PCA and PCoorA are dual methods that give the same analytical results. Using one or the other is simply a matter of convenience, whether one prefers to work with the covariance matrix $\mathbf{C}$ or with the distance matrix Δ.

As stated earlier, the main motivation for using either PCA or PCA is to construct a low-dimensional representation of the original high-dimensional data. The notion behind this approach is that the effective (or essential, as some call it [33]) dimensionality of a molecular conformational space is significantly smaller than its full dimensionality ($3N$-6 degrees of freedom for an N-atom molecule). Following the PCA procedure, each new

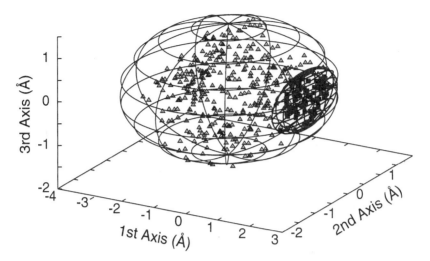

Figure 8 A joint principal coordinate projection of the occupied regions in the conformational spaces of linear (Ala)$_6$ (triangles) and its conformational constraint counterpart, cyclic-(Ala)$_6$ (squares), onto the optimal 3D principal axes. The symbols indicate the projected conformations, and the ellipsoids engulf the volume occupied by the projected points. This projection shows that the conformational volume accessible to the cyclic analog is only a small subset of the conformational volume accessible to the linear peptide, (Adapted from Ref. 41.)

axis k is associated with a normalized eigenvalue λ_k that reflects the relative weight of that axis in reproducing the original data. An axis with a high λ_k value is significant for the projection, whereas axes with small λ_k values are insignificant. By sorting the new axes according to their λ_k weight it is possible to select a small subset of effective coordinates that capture most of the conformational relationships of the original high-dimensional space. The quality of such a projection can be estimated by the average difference between conformation distances reconstructed in the low s-dimensional subspace $d_{ij}^{(s)}$ and the original distance d_{ij}. The reconstructed distances are defined as

$$d_{ij}^{(s)2} = \sum_{k=1}^{s} (Q_{ik} - Q_{jk})^2 \tag{16}$$

where Q_{ik} is the coordinate of the ith conformation along the kth new (principal) axis. It can be shown that the average deviation of the distances in s dimensions from the exact distances is given by the sum of the first s eigenvalues,

$$\langle d_{ij}^2 - d_{ij}^{(s)2} \rangle_{ij} = 1 - \sum_{k=1}^{s} \lambda_k \tag{17}$$

Thus by summing the normalized eigenvalues of the first s dimensions one can judge the quality of a projection onto that subspace.

Fortunately, it was found that in polypeptide systems the effective dimensionality of conformational spaces is significantly smaller than the dimensionality of the full space, with only a few principal axes contributing to the projection [38–41]. In fact, in many cases a projection quality of 70–90% can be achieved in as few as three dimensions [42], opening the way for real 3D visualization of molecular conformational space. Figure 8 shows a 3D visualization of the conformational spaces of two hexapeptides, $(Ala)_6$ and cyclic-$(Ala)_6$, jointly projected on the same principal coordinate set. The comparison shows that the conformation volume occupied by the linear peptide is about 10 times larger than the conformation volume occupied by its conformationally constrained counterpart. Quantifying relative flexibility of analogous peptides through joint principal projections was shown to be useful, for example, in predicting their relative bioactivity [41].

V. CONCLUSION

In this chapter we surveyed a variety of computational methods that contribute to the "conformational analysis" of complex molecules. These include methods for searching and sampling the molecular conformation space, methods for local optimization of the sampled conformations, and basic analytical techniques. In practice, many variations of the basic methodologies are reported as researchers continuously try to improve and enhance these procedures. The different methods are often used in conjunction to form a complete conformational analysis study of a bimolecular system of interest. However, each of the different procedures can also be used separately as part of computational studies with other goals. The need for these analytical techniques is to a large extent brought about by the continuous increase is simulation times, which generates more data than ever before, requiring systematic ways to interpret and represent them.

REFERENCES

1. G Jolles, KRH Wooldridge. Drug Design: Fact or Fantasy? London: Academic Press, 1984.
2. DA Gschwend, AC Good, ID Kuntz. J Mol Recogn 9:175, 1996.
3. RE Bruccoleri, M Karplus. Conformational sampling using high-temperature molecular dynamics. Biopolymers 29:1847–1862, 1990.
4. N Metropolis, S Ulam. The Monte Carlo method. J Am Stat Assoc 44:335–341, 1949.
5. MP Allen, DJ Tildesley. Computer Simulations of Liquids. Oxford: Oxford Univ Press, 1989.
6. D Frenkel, B Smit. Understanding Molecular Simulation: From Algorithms to Applications. San Diego: Academic Press, 1996.
7. J Cao, BJ Berne. Monte Carlo methods for accelerating barrier crossing: Anti-force-bias and variable step algorithms. J Chem Phys 92:1980–1985, 1990.
8. I Andricioaei, JE Straub. On Monte Carlo and molecular dynamics methods inspired by Tsallis statistics: Methodology, optimization, and application to atomic clusters. J Chem Phys 107:9117–9124, 1997.
9. WH Press, BP Flannery, SA Teukolsky, WT Vetterling. Numerical Recipes: The Art of Scientific Computing. Cambridge, UK: Cambridge Univ Press, 1989.
10. DE Goldberg. Genetic Algorithms in Search, Optimization and Machine Learning. Reading, MA: Addison-Wesley, 1989.
11. LD Davis. Handbook of Genetic Algorithms. New York: Van Nostrand Reinhold, 1991.
12. M Vieth, JD Hirst, BN Dominy, H Daigler, CL Brooks III. Assessing search strategies for flexible docking. J Comput Chem 19:1623–1631, 1998.
13. RS Judson, EP Jaeger, AM Treasurywala. A genetic algorithm based method for docking flexible molecules. THEOCHEM 114:191–206, 1994.
14. CM Oshiro, ID Kuntz, JS Dixson. Flexible ligand docking using a genetic algorithm. J Comput-Aided Mol Des 9:113–130, 1995.
15. R Bruccoleri, M Karplus. Prediction of the folding of short polypeptide segments by uniform conformational sampling. Biopolymers 26:137–168, 1987.

16. GM Crippen, TF Havel. Stable calculations of coordinates from distance information. Acta Cryst A 34:282, 1978.

17. TF Havel. An evaluation of computational strategies for use in the determination of protein structure from distance constraints obtained by nuclear magnetic resonance. Prog Biophys Mol Biol 56:43, 1991.

18. RP Sheridan, R Nilakatan, JS Dixson, R Venkataraghavan. The ensemble approach to distance geometry: Application to the nicotinic pharmacophore. J Med Chem 29:899–906, 1986.

19. JM Blaney, JS Dixon. Distance geometry in molecular modeling. In: KB Lipkowitz, DB Boyd, eds. Reviews in Computational Chemistry, Vol 5. New York: VCH, pp 299–335.

20. DD Frantz, DL Freeman, JD Doll. Reducing quasi-ergodic behavior in Monte Carlo simulations by J-walking: Applications to atomic clusters. J Chem Phys 93:2769-2784, 1990.

21. E Marinari, G Parisi. Europhys Lett 19:451, 1992.

22. M Falcioni, MW Deem. A biased Monte Carlo scheme for zeolite structure solution. J Chem Phys 110:1754–1766, 1999.

23. JA McCammon, SC Harvey. Dynamics of Proteins and Nucleic Acids. Cambridge, UK: Cambridge Univ Press, 1987.

24. RH Boyd. J Chem Phys 49:2574, 1968.

25. BR Brooks, RE Bruccoleri, BD Olafson, DJ States, S Swaminathan, M Karplus. CHARMM: A program for macromolecular energy, minimization, and dynamics calculations. J Comput Chem 4:187–217, 1983.

26. E Aarts, J Korst. Simulated Annealing and Boltzmann Machines. New York: Wiley, 1990.

27. C Wilson, S Doniach. Proteins: Struct Funct Genet 6:193, 1989.

28. Y Levy, OM Becker. Effect of conformational constraints on the topography of complex potential energy surfaces. Phys Rev Lett 81:1126–1129, 1998.

29. OM Becker. Geometrical versus topological clustering: An insight into conformation mapping. Proteins 27: 213–226, 1997.

30. H Spath. Cluster-Analysis Algorithms for Data Reduction and Classification of Objects. Chichester: Ellis Horwood, 1980.

31. T Ichiye, M Karplus. Collective motions in proteins: A covariance analysis of atomic fluctuations in molecular dynamics and normal mode simulations. Proteins: Struct Funct Genet 11: 205–217, 1991.

32. ANE Garcia. Large-amplitude nonlinear motions in proteins. Phys Rev Lett 68:2696–2699, 1992.

33. A Amadei, ABM Linssen, HJC Berendsen. Essential dynamics of proteins. Proteins 17:412–425, 1993.

34. S Hayward, A Kitao, N Go. Harmonic and anharmonic aspects in the dynamics of BPTI: A normal mode analysis and principal component analysis. Protein Sci 3:936–943, 1994.

35. OM Becker. Quantitative visualization of a macromolecular potential energy funnel'. J Mol Struct (THEOCHEM) 398–399:507–516, 1997.

36. DA Case. Curr Opin Struct Biol 4:285–290, 1994.

37. JC Gower. Some distance properties of latent root and vector methods used in multivariant analysis. Biometrika 53:325–338, 1966.

38. R Abagyan, P Argos. Optimal protocol and trajectory visualization for conformational searches of peptides and proteins. J Mol Biol 225:519–532, 1992.

39. JM Troyer, FE Cohen. Protein conformational landscapes: Energy minimization and clustering of a long molecular dynamics trajectory. Proteins 23:97–110, 1995.

40. LSD Caves, JD Evanseck, M Karplus. Locally accessible conformations of proteins: Multiple molecular dynamics simulations of crambin. Protein Sci 7:649–666, 1998.

41. OM Becker, Y Levy, O Ravitz. Flexibility, conformation spaces, and bioactivity. J Phys Chem B 104:2123–2135, 2000.

42. OM Becker. Principal coordinate maps of molecular potential energy surfaces. J Comput Chem 19:1255–1267, 1998.

5

Treatment of Long-Range Forces and Potential

Thomas A. Darden

National Institute of Environmental Health Sciences, National Institutes of Health, Research Triangle Park, North Carolina

I. INTRODUCTION: BASIC ELECTROSTATICS

The process of performing a molecular simulation can be divided into two main tasks: (1) the generation of molecular conformations and (2) the evaluation of the potential energy for each of these conformations. The systems we are interested in simulating will typically consist of a protein, DNA, or other biomolecule of interest (possibly a complex of one or more of these) together with some representation of its environment (the solvent plus salt and other small molecules). By a molecular conformation we mean a particular arrangement of the atoms comprising the system of interest. This arrangement is typically described using Cartesian coordinates for the atoms, but internal coordinates and various reduced or coarse-grained descriptions of the system are also popular. Chapter 3 discusses these latter representations. In this chapter we assume a Cartesian coordinate description, although much of the discussion will carry over to the other representations. In addition, other chapters in this book discuss methods for generating conformations and force fields for evaluating the conformational energies.

Our focus in this chapter is on long-range forces and methods for efficiently evaluating them. By long-range forces we mean forces originating in electrostatics. Since they are long-range, the definition of the solvent environment through boundary conditions becomes an integral part of the problem of evaluating the electrostatic interactions, so we also need to discuss boundary conditions.

Our problem involves electrostatic interactions, so perhaps it is best to begin by reviewing Coulomb's law and some of its consequences. By the end of the eighteenth century, physicists had established the basic facts of electrostatics. Charges come in two types called positive and negative, and the total charge on an isolated system is conserved during physical or chemical processes. Like charges repel, and unlike charges attract. Coulomb quantified this latter observation: The force between two charged bodies at rest is proportional to the product of their charges and inversely proportional to the square of the distance between them. Imagine a pair of charged particles having charges q_1 and q_2. For simplicity, the first charge q_1 is placed at the origin of our Cartesian coordinate system, and the other charge q_2 is at position $\mathbf{r} = (x, y, z)$, a distance r away [$r = (x^2 + y^2 + z^2)^{1/2}$]. Then the force $\mathbf{F}$ (a vector) on charge q_2 due to charge q_1 is given by

$$\mathbf{F} = k\,\frac{q_1 q_2}{r^2}\,\hat{\mathbf{r}} \tag{1}$$

whereas the force on q_1 due to q_2 is $-\mathbf{F}$.

In Eq. (1), $\hat{\mathbf{r}} = \mathbf{r}/r$ is the unit vector in the direction from charge q_1 to q_2, and k is the Coulomb force constant, which depends on the units being used. In many texts k is written as $1/4\pi\varepsilon_0$, where ε_0 is the permittivity of a vacuum. In Gaussian or cgs units the constant k is 1. If q_1 and q_2 are in units of elementary charge, so that for example the charge of a sodium ion is 1, distance is measured in angstroms, and energy is measured in kilocalories per mole (kcal/mol), k is approximately 332. For convenience k will be taken to be equal to 1 in the rest of this chapter.

The electrostatic energy of the pair of charges is the work that is required to move the second charge q_2 against the electrostatic force, from infinity (where the pair experiences no interaction) along a path to the point $\mathbf{r}$. This work is the integral of the dot product $\mathbf{F} \cdot d\mathbf{r}$ (we are now dealing with vectors) of the force $\mathbf{F}$ with the infinitesmal displacement $d\mathbf{r}$ along the path from infinity to $\mathbf{r}$, and is given by

$$U = \frac{q_1 q_2}{r} \tag{2}$$

Note that we have not defined which path we take from infinity to $\mathbf{r}$, but it turns out not to matter. This is because $\mathbf{F}$ is *curl-free* away from the origin.

The force should be minus the gradient of the energy, $\mathbf{F} = -\nabla U$, where the gradient, written ∇U, is a vector whose direction is that of the maximal increase in U and whose length is equal to its rate of increase. In Cartesian coordinates the gradient can be calculated by taking partial derivatives with respect to components: $\nabla U = (\partial U/\partial x, \partial U/\partial y, \partial U/\partial z)$. Applying this to U given by Eq. (2), we see that indeed $\mathbf{F} = -\nabla U$, where $\mathbf{F}$ is given by Coulomb's law in Eq. (1).

The electric field $\mathbf{E}$ at charge q_2 due to charge q_1 is defined by

$$\mathbf{E} = \frac{q_1}{r^2}\,\hat{\mathbf{r}} \tag{3}$$

and can be thought of as the force on a unit test charge at the position of charge q_2 due to charge q_1 (i.e., set $q_2 = 1$). Note that the force on q_2 is now given by $\mathbf{F} = q_2\mathbf{E}$. The electric field due to q_1 can also be defined by Eq. (3) at arbitrary points $\mathbf{r} \neq 0$, i.e., there need not be a charge present at the point $\mathbf{r}$.

The electrostatic potential $\phi(\mathbf{r})$ at the position $\mathbf{r}$ due to the charge q_1 is the energy of the pair when the second charge is a unit test charge:

$$\phi(\mathbf{r}) = \frac{q_1}{r} \tag{4}$$

Note that $\phi(\mathbf{r})$ is the work required to move a unit test particle from infinity to $\mathbf{r}$ when q_1 is at the origin. Following the above argument, this work is obtained by taking the integral of the field $\mathbf{E}$ dotted with displacement, and thus, as above, the field $\mathbf{E}$ is the negative of the gradient of the potential ϕ, or $\mathbf{E} = -\nabla\phi$. As with the field, the potential $\phi(\mathbf{r})$ can be defined at a point $\mathbf{r}$ even when there is no charge at the point.

If the charge q_1 is at the point $\mathbf{r}_1$ instead of the origin, the above results are all slightly modified. Relative to charge q_1, the charge q_2 is now at position $\mathbf{r} - \mathbf{r}_1$, which has components $(x - x_1, y - y_1, z - z_1)$ and length $|\mathbf{r} - \mathbf{r}_1| = [(x - x_1)^2 + (y - y_1)^2 + (z - z_1)^2]^{1/2}$. Thus, for example, the electrostatic potential $\phi(\mathbf{r})$ at position $\mathbf{r}$ due to the charge q_1 is now given by

$$\phi(\mathbf{r}) = \frac{q_1}{|\mathbf{r} - \mathbf{r}_1|} \tag{5}$$

Another important fact from elementary electrostatics is the *superposition principle*: The electrostatic interaction between any two charged particles in a system is unaffected by the presence of the other charges. From Coulomb's law and the superposition principle we can derive the electrostatic energy of a system of charged particles, which is the energy (due to electrostatic repulsion or attraction) required to assemble the particles in their current configuration. Imagine a system of N charged particles. For example, in most current force fields the atoms of the macromolecular simulation system are parametrized using partial charges at the atomic nuclei to simply represent the charge distribution in the system. The charges $q_1, q_2, \ldots, q_N$ are at positions $\mathbf{r}_1, \mathbf{r}_2, \ldots, \mathbf{r}_N$, where, e.g., $\mathbf{r}_i = (x_i, y_i, z_i)$, and the distance between charges i and j is the Cartesian distance $|\mathbf{r}_i - \mathbf{r}_j|$ between $\mathbf{r}_i$ and $\mathbf{r}_j$, which is the length, as defined above, of the vector $\mathbf{r}_i - \mathbf{r}_j$, and which we denote by r_{ij}. Then, due to the superposition principle and Eq. (2), the electrostatic energy due to the whole system of charges is given by

$$U = \sum_{i=1}^{N-1} \sum_{j=i+1}^{N} \frac{q_i q_j}{r_{ij}} = \frac{1}{2} \sum_{i=1}^{N} \sum_{j \neq i} \frac{q_i q_j}{r_{ij}} = \frac{1}{2} \sum_{i=1}^{N} q_i \phi(\mathbf{r}_i) \tag{6}$$

where the electrostatic potential $\phi(\mathbf{r}_i)$ at charge i due to the other charges, which is obtained from Eq. (4) and the superposition principle, is given by

$$\phi(\mathbf{r}_i) = \sum_{j \neq i} \frac{q_j}{r_{ij}} \tag{7}$$

These early results of Coulomb and his contemporaries led to the full development of classical electrostatics and electrodynamics in the nineteenth century, culminating with Maxwell's equations. We do not consider electrodynamics at all in this chapter, and our discussion of electrostatics is necessarily brief. However, we need to introduce Gauss' law and Poisson's equation, which are consequences of Coulomb's law.

The surface integral of the field $\mathbf{E}$ over some surface S is the sum of the infinitesmal quantity $(\mathbf{E} \cdot \mathbf{n})da$ over all points of S, where $\mathbf{n}$ is the unit vector normal to the surface at a point and da is the infinitesimal element of surface area there. Equation (3) gives the electric field $\mathbf{E}$ at a point $\mathbf{r}$ due to a charge at the origin. The direction of the field is parallel to the vector from the origin to $\mathbf{r}$, and its strength is given by the charge q_1 divided by the square of distance r. At any other point on the surface of the sphere of radius r about the origin, the electric field has the same strength and is also perpendicular to the surface of the sphere, or parallel to the unit normal vector $\mathbf{n}$. Thus the surface integral of $\mathbf{E}$ over the sphere can be simply calculated. It is the constant field strength q_1/r^2 times the sum of da over the sphere, or q_1/r^2 times the total surface area $4\pi r^2$, which is $4\pi q_1$. Gauss' law generalizes this result. If q_1 is generalized to an arbitrary collection of charges having total charge Q, enclosed in a volume V with closed surface ∂V, the surface integral over ∂V of the field due to the collection of charges is given by $4\pi Q$.

If instead of discrete charges the charge is described by a smooth charge density $\rho(\mathbf{r})$, the total charge Q contained within a volume V is given by the integral of ρ. Noting that the electric field is the negative of the gradient of ϕ, we can write *Gauss' law* as

$$\int_{\partial V} \nabla\phi(\mathbf{r}) \cdot \mathbf{n} \, da = -4\pi Q = -4\pi \int_V \rho(\mathbf{r}) \, d\mathbf{r} \tag{8}$$

Invoking the divergence theorem from calculus we can rewrite this as

$$\int_V \nabla \cdot \nabla\phi(\mathbf{r}) \, d\mathbf{r} = -4\pi \int_V \rho(\mathbf{r}) \, d\mathbf{r} \tag{9}$$

Poisson's equation is the pointwise form of this latter equation:

$$\nabla \cdot \nabla\phi(\mathbf{r}) = -4\pi\rho(\mathbf{r}) \tag{10}$$

If we are considering electrostatics in dielectric media instead of in vacuo, which is necessary for a continuum treatment of solvation, Gauss' law and Poisson's equation must be generalized. In dielectric media, Gauss' law is expressed in terms of the electric displacement $\mathbf{D}$ in place of the electric field $\mathbf{E}$. In a linear isotropic dielectric medium, $\mathbf{D} = \varepsilon\mathbf{E}$, where ε is the permittivity of the medium. Gauss' law in such a medium states that the surface integral of $\mathbf{D}$ over any closed surface is given by 4π times the total charge contained within.

As an example of the application of Gauss' law in a dielectric medium, let us consider a simple continuum model of an ion in water. The ion is modeled as a point charge q at the center of a sphere of radius a that is immersed in a dielectric continuum. The interior of the sphere has dielectric constant 1, whereas the continuum has a larger dielectric constant ε. In the dielectric continuum, by spherical symmetry and Gauss' law, the electric field at any point $\mathbf{r}$ distant $r > a$ from the center of the ion is given by $\mathbf{E}(\mathbf{r}) = (q/\varepsilon r^2)\,\hat{\mathbf{r}}$. Thus the electrostatic potential there is given by $\phi(\mathbf{r}) = q/\varepsilon r$. Inside the sphere the electric field is given for $r \neq 0$ by $\mathbf{E}(\mathbf{r}) = (q/r^2)\hat{\mathbf{r}}$. The electrostatic potential at $\mathbf{r}$ is given by its value on the surface of the ionic sphere plus the work to move the charge inside against the field, or $\phi(\mathbf{r}) = q/\varepsilon a + q/r - q/a$. If we remove the work required to move the test charge in from infinity to $\mathbf{r}$ against the unscreened vacuum electrostatic field due to the ion, which is given by q/r, we have the work performed by the dielectric medium in moving the test charge, or $\phi_{\text{diel}}(\mathbf{r}) = -(1 - 1/\varepsilon)q/a$. This result can be used to calculate the work of charging the ion in the dielectric medium, arriving at the Born solvation free energy,

$$\Delta G = -\left(1 - \frac{1}{\varepsilon}\right)\frac{q^2}{2a} \tag{11}$$

Although the continuum model of the ion could be analyzed by Gauss' law together with spherical symmetry, in order to treat more general continuum models of electrostatics such as solvated proteins we need to consider media that have a position-specific permittivity $\varepsilon(\mathbf{r})$. For these a more general variant of Poisson's equation holds:

$$\nabla \cdot \varepsilon(\mathbf{r})\nabla\phi(\mathbf{r}) = -4\pi\rho_{\text{free}}(\mathbf{r}) \tag{12}$$

which is derived in the same way as Eq. (10). In this equation ρ_{free} refers to the free charge density (for example, the charge density of the protein in a continuum treatment), as opposed to the charge density induced in the dielectric continuum or at dielectric boundaries.

Early in the twentieth century physicists established that molecules are composed of positively charged nuclei and negatively charged electrons. Given their tiny size and nonclassical behavior, exemplified by the Heisenberg uncertainty principle, it is remarkable (at least to me) that Eq. (1) can be considered *exact* as a description of the electrostatic forces acting between the atomic nuclei and electrons making up molecules and molecular systems. For those readers who are skeptical, and perhaps you should be skeptical of such a claim, I recommend the very readable introduction to Jackson's electrodynamics book [1].

Thus if electrons as well as nuclei could be treated classically, molecular simulation would be much simpler. Since the nonelectrostatic forces operating on the system (strong nuclear force, weak force, etc.) are negligibly weak over molecular scales, the evaluation of the potential energy of a conformation would be a straightforward application of Eq. (6), where the charged particles would now be atomic nuclei or electrons, whose charges are known precisely. There would be no need for empirical parameters, and our only problem would be to generate a sufficient number of conformations of nuclei and electrons to average over. Unfortunately, electrons cannot be described classically in terms of configurations like the nuclear configurations; electron density is the relevant physical observable. However, it can be said that all molecular interactions can be derived from Coulomb's law and the principles of quantum mechanics, together expressed in the Schrödinger equation.

The only problem with the foregoing approach to molecular interactions is that the accurate solution of Schrödinger's equation is possible only for very small systems, due to the limitations in current algorithms and computer power. For systems of biological interest, molecular interactions must be approximated by the use of empirical force fields made up of parametrized terms, most of which bear no recognizable relation to Coulomb's law. Nonetheless the force fields in use today all include terms describing electrostatic interactions. This is due at least in part to the following facts.

1. At long range, complex interactions having their origin in quantum mechanics are negligible, and molecular interactions can be described by classical electrostatics of nuclei and electron density.
2. At long range, the charge density need not be known precisely, and the interactions are well approximated by using a simplified representation, such as partial charges at the nuclei.
3. Although electrostatic interactions between pairs of molecules may be weak in many cases, a consequence of their long-range nature is that in large systems the energy due to electrostatic interactions must be calculated between all the pairs of the system and thus will dominate all other interactions. It is essential to include them.

As motivation for the rest of the chapter, a few further observations can be made. First, the calculation of full electrostatics is expensive. A typical molecular mechanics potential function is of the form

$$E = E_b + E_\theta + E_\phi + E_{vdw} + E_{es} \tag{13}$$

where E is the total molecular mechanics energy, E_b and E_θ are harmonic terms describing bond and angle vibrations, E_ϕ is a torsion term to describe the energetics of rotations about bonds, and E_{vdw} and E_{es} are non-bond terms to describe interactions between atom pairs that are not part of a common bond, valence angle, or torsion angle. The van der Waals interactions in E_{vdw} are described by dispersion and exchange repulsion terms originating in quantum mechanics, whereas E_{es} are the Coulombic interactions given by Eq. (6).

In large systems the computer time required to calculate the potential energy of a particular molecular conformation is dominated by the cost of calculating the non-bond interactions. This is due to the fact that the number of non-bond pairs is so much larger than the number of terms involved in the bond, angle, and torsion interactions. For example, in a system of 10^4 atoms, which is typical in current biomolecular simulations, there are about 10^4 bond terms and roughly the same number of angle and torsion terms (actually there are much fewer torsions if the system is mainly water, which is often the case). In contrast, there are $N(N - 1)/2$ or about 5×10^7 non-bond pairs to be calculated. If these were calculated in a straightforward way the simulation would be very expensive.

The non-bond dispersion and repulsion terms decay rapidly, and thus the calculation of E_{vdw} can be approximated by restricting the sum over all non-bond pairs to that over neighboring pairs (i.e., interactions are typically cut off at 8–10 Å), reducing the number of interaction pairs by more than a factor of 10 in our example. However, electrostatic interactions do not decay rapidly, and thus the above cutoff approximation is quite inaccurate for E_{es}. Until recently, in order to speed the calculation, cutoffs were applied to electrostatic interactions as well as to the other more rapidly decaying non-bond interactions. Although this approximation may not always cause severe artifacts in the simulation, it is to be avoided whenever possible, especially if the simulation system contains mobile charged entities. A simple example of the problems that can occur was given by Auffinger and Beveridge [2]. They simulated a mix of sodium and chloride ions in water, under periodic boundary conditions. When cutoffs were applied to the electrostatic interactions they observed a strong tendency for ions of the same charge to be separated by distances close to the cutoff distance even when large cutoffs were used.

Ions of the same charge repel each other unless they are further apart than the cutoff distance, in which case they do not interact at all. Many workers had assumed that the dielectric screening of water would diminish artifacts due to the use of cutoffs. However, normal dielectric screening *does not occur* when cutoffs are applied. Bader and Chandler [3] observed a strong attractive well for the potential of mean force for the Fe^{2+}–Fe^{3+} ion pair when cutoffs were applied, whereas the Ewald summation (see below) led to normal dielectric screening of the repulsion. Thus cutoffs are not a viable option for accurate simulations, and therefore efficient calculation of full electrostatics is the focus of intense effort by a number of groups. Describing the currently popular methods is one goal of this chapter.

A second observation is that owing to the slow decay of Coulombic interactions with distance r, it is not straightforward to calculate the actual value of the Coulombic energy or potential in certain circumstances. This observation is by now well known for lattice sums in periodic boundary conditions, which we treat later in this chapter. However, it is also true for simple situations in nonperiodic boundary conditions. A simple example is the problem of calculating the electrostatic potential ϕ of a neutral atom, e.g., neon, at the center of an isotropic bath of water molecules. For simplicity, assume that the neon atom is at the origin of our coordinate system and the water molecules surrounding it are modeled with partial charges at the oxygen and hydrogen positions. We thus have a collection of charges, and we wish to calculate the electrostatic potential due to them at the origin. The electrostatic potential due to one of the charges q_i at distance r_i from the origin is simply q_i/r_i. If the water bath is infinite, we cannot immediately sum these individual contributions; instead, we must devise a limiting process that involves a sequence of finite calculations. For example, for $0 < r < \infty$, let $S_1(r)$ denote the sum of q_i/r_i over all water atoms with $r_i < r$, and let $S_2(r)$ denote the sum of q_i/r_i over all water atoms belonging to water molecules whose oxygen is closer than r to the origin. Then consider the limits of $S_1(r)$ and $S_2(r)$ as $r \rightarrow \infty$. These two converge rapidly, but to different limits! In fact, the limit of $S_1(r)$ is positive, whereas that of $S_2(r)$ is negative [4]. Thus although the electrostatic potential at the neon atom is clearly a physically reasonable quantity, it is not immediately clear how to calculate it. Hummer et al. [5] and Ashbaugh and Wood [6] have argued that the atom-based summation $S_1(r)$ leads to the correct result, whereas Aqvist and Hansson [7] have argued that a molecule-based summation such as $S_2(r)$ is correct. Let us consider this dilemma carefully.

The difficulty in calculating the potential at the center of the water bath is due to the slow decay of the Coulombic interactions, leading to the conditional convergence of infinite Coulombic sums. To explain the notion of *conditional* versus *absolute* convergence, consider the simple infinite series $1 - 1/2 + 1/3 - 1/4 + 1/5 - \cdots$. An infinite series converges absolutely if the sum of the absolute values of its terms converges. In this case that would mean that the sum $1 + 1/2 + 1/3 + 1/4 + 1/5 + \cdots$ converges, which is not true. Thus this series is not absolutely convergent. The original alternating series, however, does converge, i.e., the sequence of its successive partial sums converges to a real number. However, if the series is rearranged as $1 + 1/3 - 1/2 + 1/5 + 1/7 - 1/4 + \cdots$, that is, two positive terms followed by a negative term, the partial sums now converge to a different limit [8]. The series is said to be *conditionally convergent*, meaning that the result is conditional on the method of summation. To summarize so far, it is possible for an infinite series to converge but only conditionally, meaning that the answer you obtain depends on the algorithm for calculating it. This strange behavior is not possible if a series converges absolutely.

If, in our example of the water bath, the individual terms q_i/r_i were replaced by their absolute values $|q_i|/r_i$, the resulting sum would be infinite. Thus the Coulombic sum is not absolutely convergent, and it is no surprise that $S_2(r)$ and $S_1(r)$ converge to different limits. If Coulombic interactions decayed faster than the inverse third power of distance, as do Lennard-Jones non-bond terms, their sums would converge absolutely and calculating them would be straightforward. However, since Coulombic interactions are long-range, care is needed in their calculation in order to arrive at a physically correct result.

Because of the spherical symmetry of the system, we can use Gauss' law to deduce the correct value of the potential of a neon atom at the center of an infinite water bath. The electrostatic potential of the neon atom is the work required to bring a unit test charge from infinity, against the field **E** produced by the water molecules, to the origin. For $0 < r < \infty$, let $Q(r)$ be the sum of q_i over all water atoms with $r_i < r$. By Gauss' law the surface integral of **E** over the sphere of radius r is $4\pi Q(r)$ (note that we are performing the calculation in vacuo instead of in a dielectric medium, i.e., the water molecules will provide the intrinsic dielectric screening), and by spherical symmetry the field at any point on the surface of the sphere is $Q(r)/r^2$. Using this result and simple calculus we can calculate the potential ϕ at the origin. The result agrees with the limit of $S_1(r)$, and the dilemma is resolved. The details and application to ionic charging free energies are found in Ref. 9.

Realistic models of proteins or other macromolecules in solution must include some description of the bulk solvent environment. Ideally this would be an infinite bath of water including the appropriate salt concentration. Unfortunately, current simulations are limited to 10^6 atoms or less, which is not sufficient to model bulk behavior. Thus the connection to bulk solvation is implemented through boundary conditions. As shown above, the long-range nature of Coulombic interactions could lead to confusion even if we had ideal bulk solvent present in the simulation, and thus it is plausible that the choice of boundary conditions can have a nonnegligible effect on the results of a simulation. A second goal of this chapter is to introduce the common choices of boundary conditions and to discuss what is currently known about the nature and size of artifacts due to long-range electrostatics under various boundary conditions.

II. CONTINUUM BOUNDARY CONDITIONS

In continuum boundary conditions the protein or other macromolecule is treated as a macroscopic body surrounded by a featureless continuum representing the solvent. The internal forces of the protein are described by using the standard force field including the Coulombic interactions in Eq. (6), whereas the forces due to the presence of the continuum solvent are described by solvation terms derived from macroscopic electrostatics and fluid dynamics.

Due to limitations in computer power, early protein and DNA simulations [10,11] used a particularly simple variant of this approach. The effect of the missing solvent is approximated by using an effective dielectric function: The electrostatic energy $U = q_1q_2/r$ between two charges q_1 and q_2 is replaced by $U = q_1q_2/[\varepsilon(r)r]$, and the electrostatic forces are obtained by differentiating the energy. The earliest implementation used the simple choice $\varepsilon(r) = r$, whereas later variants [12,13] employed a sigmoidal form in which $\varepsilon(r)$ approached 1 as $r \to 0$ whereas $\varepsilon(r) \approx 78$, the dielectric constant of bulk water for $r > 20$ Å. The use of an effective dielectric causes interactions between distant charges to be screened heavily while neighboring charge pairs experience nearly the full Coulombic interaction, thus approximating the dielectric screening of charge interactions in water.

Although the use of effective dielectrics could account approximately for the dielectric screening of charge pairs in water, it failed to account for the tendency of charged and polar residues to hydrogen bond with water molecules. This led to excessive hydrogen bonding between charged and polar groups on the surface of proteins. One important consequence was that the relative free energies between conformations of a peptide were found to be artifactually affected [14]. A more dramatic consequence was that this methodology together with existing force fields failed to distinguish between correctly and incorrectly folded proteins [15]. Eisenberg and McLachlan [16] and later Ooi et al. [17] proposed surface area based self-energy terms to model the tendency of a charged or polar group to be exposed to solvent. Later Still et al. [18] proposed the generalized Born method, a computationally tractable electrostatic model that simultaneously accounts for dielectric screening of charge pairs as well as for the self-energy of charged and polar groups in the presence of a dielectric. These developments are described in Chapter 7.

Within the continuum approximation, a rigorous approach to the electrostatic free energy of proteins and other biomolecules in solution is provided by the Poisson equation, Eq. (12). In this approach a protein in water is modeled as a low dielectric region carrying a fixed charge distribution and surrounded by a high dielectric region. The boundary between the two regions is defined by a molecular surface analogous to the solvent-accessible surface of the protein. The atomic radii used to define the boundary are similar to the initial Born radii used in the generalized Born approach and likewise are fit to reproduce experimental solvation energies. Although analytic solutions to Eq. (12) exist for certain special cases, in general the equation must be solved numerically. Because solutions to Eq. (12) are needed for later discussion on the influence of boundary conditions in simulations, we briefly discuss this topic.

A variety of algorithms, including finite difference, boundary element, finite element, and multigrid, have been implemented for solving Eq. (12) for biomolecules. For example, the boundary element methods are based on the application of Gauss' law to dielectric boundaries. The charges in a solute molecule induce a surface charge distribution at the dielectric boundary. From Gauss' law one can show that the surface charge density σ_{pol} at the boundary between the solute and solvent is given by

$$\sigma_{pol}(\mathbf{r}) = + \frac{1}{4\pi} \left(1 - \frac{1}{\varepsilon} \right) \mathbf{E}(\mathbf{r}) \cdot \mathbf{n}(\mathbf{r}) \tag{14}$$

where $\mathbf{r}$ is a point on the boundary, $\mathbf{E}(\mathbf{r})$ the electric field at $\mathbf{r}$, and $\mathbf{n}$ the unit normal vector to the surface there. The electric field depends on the charges of the solute as well as on the other induced surface charge elements (including a self term). Thus this equation must be solved iteratively. The precision of the solution depends on the accuracy of the surface representation. The boundary element approach generalizes easily to more complex descriptions of the solute charge density such as polarizable dipoles or quantum chemical charge densities [19]. Also, because at each iteration the electric field due to a collection of point charges (fixed and induced) is needed, it is straightforward to adapt algorithms for rapid calculation of electrostatic sums to improve the performance of the boundary element method. For example, Bharadwaj et al. [20] combine the fast multipole method with their implementation of the boundary element method to arrive at an algorithm that should be optimal for large systems.

The finite difference approach has proved to be the most popular algorithm for solving the Poisson equation. In this approach the molecule is placed inside a cell representing the solvent bath. The cube is then divided into a fine grid (usually the grid size must be 0.5 Å or finer for precision). The Poisson partial differential equation, Eq. (12), is discretized on the grid. The atomic charges are distributed over the neighboring grid points as a sampled charge density ρ, and the dielectric constant is interpolated near the dielectric boundary between solute and solvent. Finally, the values of the potential ϕ at the boundary of the containing cell must be assigned. These boundary values are usually not known a priori, but the boundary is assumed to be sufficiently distant that the details of potential assignment there do not affect the resulting potential at the molecule. One choice for a boundary potential is to use a sum of screened electrostatic potentials due to the solute charges. A modification is to use a succession of calculations, where a first-pass calculation of the potential on the grid is performed using some estimate for the potential at the boundary of the cell. A second, smaller cell, still containing the molecule of interest, is then regridded. The potential obtained at the boundary of this smaller cell from the first-pass solution is taken as the boundary condition for a second-pass solution. This can be iterated to a third pass, and so on, but in practice the potential in the interior of the cell is found to converge rapidly. The discretization of the differential equation (12) results in a linear system of equations that must be solved iteratively. Efficient iterative schemes have been developed [21].

The continuum treatment of electrostatics can also model salt effects by generalizing the Poisson equation (12) to the Poisson–Boltzmann equation. The finite difference approach to solving Eq. (12) extends naturally to treating the Poisson–Boltzmann equation [21], and the boundary element method can be extended as well [19].

III. FINITE BOUNDARY CONDITIONS

In finite boundary conditions the solute molecule is surrounded by a finite layer of explicit solvent. The missing bulk solvent is modeled by some form of boundary potential at the vacuum/solvent interface. A host of such potentials have been proposed, from the simple spherical half-harmonic potential, which models a hydrophobic container [22], to stochastic boundary conditions [23], which surround the finite system with shells of particles obeying simplified dynamics, and finally to the Beglov and Roux spherical solvent boundary potential [24], which approximates the exact potential of mean force due to the bulk solvent by a superposition of physically motivated terms.

The electrostatic effect of the missing bulk is usually approximated by dielectric continuum theory. The finite system including the layer of explicit solvent is treated as a low dielectric region embedded in a high dielectric continuum. The electrostatic potential at an atom is given by the solution of the Poisson equation, Eq. (12). Although this equation can be solved numerically as discussed above, for simulations a more efficient treatment is necessary. If the finite system is spherical, the reaction potential due to the continuum can be expanded in a series involving the total charge, dipole, quadrupole, and higher order multipoles of the system. The reaction potential is approximated by keeping a finite number of terms [24]. Another approximation is the image approximation [25], in which the multipole series is rearranged and the leading term in the rearranged series is identified as the potential due to image charges whose positions are defined in terms of the original charge positions. Just as in the continuum treatment of solvents, the reaction potential is sensitive to the distance between the system charges and the dielectric boundary. Thus it is important to ensure that charges do not approach this boundary during the simulation.

Although the number of atoms in a macromolecular simulation under finite boundary conditions is less than under periodic boundary conditions, straightforward evaluation of the Coulomb sum, Eq. (6), will still prove prohibitively expensive for large proteins in solution. One simple approach to reduce the cost is the "twin-range" approach [26]. In this method the more distant interactions are simply calculated less often (e.g., every M steps), and their effect is stored in memory to be applied as a constant force (or preferably all at once as an impulse [27]). The reasoning is that the step-to-step variations in the positions of distant charges have only a small relative effect on the potential due to them. Although this approach alleviates the problem for modest system sizes, it does not eliminate the basic order N^2 nature of the Coulomb sum and thus will not work for large systems.

Another approach to reducing the cost of Coulombic interactions is to treat neighboring interactions explicitly while approximating distant interactions by a multipole expansion. In Figure 1a the group of charges $q^{(1)}, q^{(2)}, \ldots, q^{(K)}$ at positions $\mathbf{r}^{(k)} = (r_1^{(k)}, r_2^{(k)}, r_3^{(k)})$, $k = 1, \ldots, K$, are all close to the point $\mathbf{b} = (b_1, b_2, b_3)$, so that their distances $|\mathbf{r}^{(k)} - \mathbf{b}|$, $k = 1, \ldots, K$, are all small compared to $|\mathbf{b} - \mathbf{r}|$. Then the electrostatic potential due to $q^{(1)}, q^{(2)}, \ldots, q^{(K)}$ evaluated at the point $\mathbf{r} = (r_1, r_2, r_3)$ can be approximated by a multipole expansion about the point $\mathbf{b}$. For example, the potential due to charge $q^{(1)}$

(a)

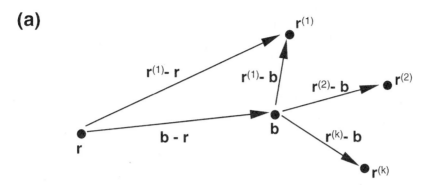

(b)

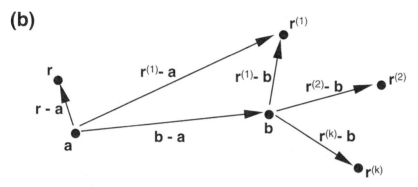

Figure 1 (a) Multipole expansion of the potential at **r** due to charges near **b**. Assume that $|\mathbf{r}^{(j)} - \mathbf{b}| < |\mathbf{b} - \mathbf{r}|$ for $j = 1, \ldots, k$. (b) Taylor expansion of the electrostatic potential at **r** about **a**. Assume $|\mathbf{r} - \mathbf{a}| + |\mathbf{r}^{(j)} - \mathbf{b}| < |\mathbf{b} - \mathbf{r}|$ for $j = 1, \ldots, k$.

evaluated at **r** is $q^{(1)}/|\mathbf{r}^{(1)} - \mathbf{r}| = q^{(1)}/|\mathbf{r}^{(1)} - \mathbf{b} + \mathbf{b} - \mathbf{r}|$. Since $|\mathbf{r}^{(1)} - \mathbf{b}|$ is small compared to $|\mathbf{b} - \mathbf{r}|$, the potential due to $q^{(1)}$ can be expanded as a Taylor series. A first-order (dipole) approximation about **b** would be

$$\frac{q^{(1)}}{|\mathbf{r}^{(1)} - \mathbf{r}|} \approx \frac{q^{(1)}}{|\mathbf{b} - \mathbf{r}|} - \sum_{i=1}^{3} \frac{q^{(1)}\,(r_i^{(1)} - b_i)(b_i - r_i)}{|\mathbf{b} - \mathbf{r}|^3} \tag{15}$$

whereas a second-order (quadrupole) approximation would be

$$\begin{aligned}
\frac{q^{(1)}}{|\mathbf{r}^{(1)} - \mathbf{r}|} &\approx \frac{q^{(1)}}{|\mathbf{b} - \mathbf{r}|} - \sum_{i=1}^{3} \frac{q^{(1)}\,(r_i^{(1)} - b_i)(b_i - r_i)}{|\mathbf{b} - \mathbf{r}|^3} \\[2mm]
&\quad + \frac{3}{2} \sum_{i=1}^{3} \sum_{j=1}^{3} \frac{q^{(1)}\,(r_i^{(1)} - b_i)(r_j^{(1)} - b_j)(b_i - r_i)(b_j - r_j)}{|\mathbf{b} - \mathbf{r}|^5} \\[2mm]
&\quad - \frac{1}{2} \sum_{i=1}^{3} \frac{q^{(1)}\,(r_i^{(1)} - b_i)^2}{|\mathbf{b} - \mathbf{r}|^3}
\end{aligned} \tag{16}$$

Third-order (octupole) or higher order multipole approximations can be employed for more accuracy.

Using the superposition principle, the second-order (quadrupole) approximation to the potential due to $q^{(1)}, q^{(2)}, \ldots, q^{(K)}$ evaluated at $\mathbf{r}$ is obtained by simply summing terms over the charges $q^{(k)}, k = 1, \ldots, K$. The result is

$$\sum_{k=1}^{K} \frac{q^{(k)}}{|\mathbf{r}^{(k)} - \mathbf{r}|} \approx \frac{Q}{|\mathbf{b} - \mathbf{r}|} - \sum_{i=1}^{3} \frac{d_i(b_i - r_i)}{|\mathbf{b} - \mathbf{r}|^3}$$

$$+ \frac{3}{2} \sum_{i=1}^{3} \sum_{j=1}^{3} \frac{\Theta_{ij}(b_i - r_i)(b_j - r_j)}{|\mathbf{b} - \mathbf{r}|^5} - \frac{1}{2} \sum_{i=1}^{3} \frac{\Theta_{ii}}{|\mathbf{b} - \mathbf{r}|^3} \qquad (17)$$

where $Q = \sum_{k=1}^{K} q^{(k)}$ is the total charge of the group, $\mathbf{d}$ its dipole moment, with $d_i = \sum_{k=1}^{K} q^{(k)} (r_i^{(k)} - b_i)$, and Θ its quadrupole moment, with $\Theta_{ij} = \sum_{k=1}^{K} q^{(k)} (r_i^{(k)} - b_i) (r_j^{(k)} - b_j)$.

A straightforward application of this multiple approximation is as follows. The simulation cell containing the molecule plus solvent is divided into subcells of similar size. At the start of the electrostatic calculation, the total charge, dipole moment, and quadrupole moment of each subcell are calculated. Each subcell is surrounded by up to 26 adjacent subcells. The electrostatic potential at position $\mathbf{r}$ is approximated by first calculating exactly the potential at $\mathbf{r}$ due to charges in the same or adjacent subcells. The potential due to charges in a nonadjacent subcell is approximated using Eq. (17), and these approximate potentials are summed over the nonadjacent subcells $\mathbf{b}$. Because the number of subcells is proportional to the number of charges, this method to obtain the potential at $\mathbf{r}$ is still an order N algorithm, and thus the calculation of the energy by summing charge times potential as in Eq. (6) is of order N^2. However, it can be made fast in comparison with the direct sum over all charge pairs by tuning the division into subcells so that the number of calculations on the right-hand side of Eq. (17) is substantially less than that on the left.

A further improvement can be seen for the situation depicted in Figure 1b. Let ϕ_b $(\mathbf{r})$ denote the potential due to the charges in the cell about point $\mathbf{b}$, evaluated at the point $\mathbf{r}$. Let $\mathbf{a}$ be the center of the subcell containing q. Then $\phi_b (\mathbf{r})$ can be approximated by a second-order Taylor expansion about $\mathbf{a}$:

$$\phi_b(\mathbf{r}) \approx \phi_b(\mathbf{a}) + \sum_{i=1}^{3} \frac{\partial \phi_b}{\partial r_i}(\mathbf{a})(r_i - a_i) \qquad (18)$$

$$+ \frac{1}{2} \sum_{i=1}^{3} \sum_{j=1}^{3} \frac{\partial^2 \phi_b}{\partial r_i \partial r_j}(\mathbf{a})(r_i - a_i)(r_j - a_j)$$

The electrostatic potential at $\mathbf{r}$ due to nonadjacent cells can be approximated by summing the second-order Taylor expansion, Eq. (18), for $\phi_b (\mathbf{r})$ over all nonadjacent cells $\mathbf{b}$. Thus $\phi_b (\mathbf{a})$, as well as $\partial \phi_b / \partial r_i(\mathbf{a})$ and $\partial^2 \phi_b / \partial r_i \partial r_i(\mathbf{a})$ for $i, j = 1, 2, 3$, are summed over all $\mathbf{b}$ to get the coefficients for the Taylor expansion, which is then used to approximate the potential at all the points in the cell centered at $\mathbf{a}$. This offers improved speed at the cost of a further approximation.

The algorithm outlined above is a level 1 cell multipole or Cartesian multipole algorithm [28]. A number of modifications are possible. Accuracy can be raised by using higher order expansions, which unfortunately are more expensive. The cost can be alleviated by using a "twin-range" approach as above, calculating the Taylor coefficients only every M steps and approximating them as constants in Eq. (18) during intermediate steps. Instead of Cartesian multipole expansions, an expansion in spherical harmonics can be performed [29]. This is more difficult to program but becomes more efficient with high-order approximations, because fewer terms are needed in the approximation at a given order. Another strategy to improve accuracy [30] is to use the exact potential due to nonadjacent cells and its derivatives evaluated at **a** in the Taylor series approximation for charges in sub-cell **a**.

All of the above algorithms are of order N^2, i.e., their cost grows with the square of the system size, and hence they are inefficient for large systems. The key to further improvements is to realize that more distant charges can be grouped into larger subcells, because, referring to Figure 1a, the important parameter in the approximation (16) is the ratio $|\mathbf{r}^{(1)} - \mathbf{b}|/|\mathbf{b} - \mathbf{r}|$. This insight inspired the tree codes, culminating in the fast multipole algorithm (FMA) [29]. In these the initial cell is divided into eight child cells, called level 1 cells. These in turn are divided into eight child cells each, yielding 64 level 2 cells, and so on, down to the Lth level containing 8^L subcells. The algorithm is described by a two-pass procedure. In the upward pass, multipole expansions out to some order are calculated for each subcell at each level, beginning with the lowest level L and proceeding upward.

The multipole expansion of a subcell can be calculated by using the expansions of its eight child cells using translation operators. The second or downward pass begins at level 2. For each of the 64 level 2 cells, the Taylor expansion due to nonadjacent level 2 cells is calculated as above at its center. For the level 3 cells, the Taylor expansion due to nonadjacent level 3 cells can be calculated more efficiently. First the Taylor expansion of the cell's level 2 parent cell is translated to its center using a translation operator. This accounts for the potential due to all nonadjacent level 3 cells except for those cells whose level 2 parents are adjacent to the level 2 parent of the cell in question. The Taylor expansion for these level 3 cells is calculated as above and added to the translated level 2 Taylor expansion. This process continues down to the lowest level L. After this downward pass, the Taylor expansion at each level L cell is available. The potential at a charge is then approximated as above, a sum of direct interactions due to charges in the same or adjacent level L cell, plus a Taylor expansion to approximate the potential due to charges in nonadjacent level L cells. A clear description of the algorithm, for more advanced readers, is given by White and Head-Gordon [31].

The FMA, using the above tree recursion, is a very general approach to reducing the cost of electrostatic sums. Using it, the cost of calculating the energy and forces for a system of N charges is an order N operation, i.e., its cost grows linearly with system size. The cost depends on the order of the multipole expansion as well as the level of the tree. The greater the tree depth L, the smaller the ultimate subcells are, which in turn lowers the number of interaction pairs to be calculated explicitly. However, as L increases, more work is performed in calculating the Taylor expansion of the long-range interactions. The accuracy depends mainly on the order of the expansion.

Since empirical force fields do not accurately estimate the true interatomic forces, it is difficult a priori to say how accurate the fast multipole approximation to the exact Coulomb potential and forces (exact in terms of the sum over partial charges) should be. Probably a good rule is to make sure that at each atom the approximate electrostatic force is within a few percent relative error of the true electrostatic force, obtained by explicitly summing over all atom pairs, i.e., $|\tilde{\mathbf{F}}_i - \mathbf{F}_i| < 0.05|\mathbf{F}_i|$, for all atoms i, where $\mathbf{F}_i$ is the exact and $\tilde{\mathbf{F}}_i$ the approximate atomic force at atom i. Another commonly used measure of accuracy is the root-mean-square force error, given by $(\sum_i |\tilde{\mathbf{F}}_i - \mathbf{F}_i|^2 / \sum_i |\mathbf{F}_i|^2)^{1/2}$. For systems of 10^4–10^5 atoms, which are typical for proteins in solvent, the RMS force error should be less than 0.1% to ensure that all atomic forces are accurate to within a few percent. Another measure of accuracy is the extent of energy conservation in molecular dynamics. Unfortunately, the multipole expansion algorithms are not particularly good at energy conservation [32]. The reason is that as atoms move during the simulation, some leave the subcell they are in and enter a new subcell. As they do, the multipole expansions for the new and old subcells are changed discontinuously. Furthermore, some interactions involving those atoms, which were calculated exactly, are now calculated approximately. Thus the energy and forces are not continuous functions of the particle positions, and atomic movement causes small discontinuous jumps in the energy, degrading energy conservation. If the multipole expansion were exact, the energy and forces would be continuous and energy would be conserved. The more accurate the approximation, the better the energy conservation. On this basis, Bishop et al. [32] recommend RMS force errors of $\approx 10^{-6}$. Furthermore, the expansions must be calculated frequently, e.g., every few femtoseconds, which further increases the expense.

IV. PERIODIC BOUNDARY CONDITIONS

Early simulations treated systems with only short-range interactions, e.g., liquid argon, using a Lennard-Jones potential to model exchange repulsion and dispersion interactions. System sizes were severely limited by the available computer power, and thus careful consideration was given to the choice of boundary conditions. The early simulations all employed periodic boundary conditions to minimize surface effects. Periodic boundary conditions are depicted in two dimensions in Figure 2. The system of atoms is contained in the central cell, which is then replicated infinitely in both dimensions, although only the neighboring replica cells are shown in the figure. In three dimensions the atoms of the simulation system are contained within a unit cell analogous to the unit cell familiar to X-ray crystallographers. For simplicity let us assume that the unit cell is a cube of side L, centered on the origin. However, note that all of the following discussion generalizes to arbitrary unit cells. This unit cell is then replicated infinitely in all directions, meaning that each atom i having position $\mathbf{r}_i = (x_i, y_i, z_i)$ in the central cell has an infinite number of image atoms with positions $\mathbf{r}_i + \mathbf{n}L = (x_i + n_1 L, y_i + n_2 L, z_i + n_3 L)$ for all possible integer triples $\mathbf{n} = (n_1, n_2, n_3)$. Under periodic boundary conditions, the distance r_{ij} between atoms i and j is taken to be the minimum image distance, $r_{ij} = \min_{\mathbf{n}} |\mathbf{r}_j + \mathbf{n}L - \mathbf{r}_i|$, i.e., the distance between atom i and the closest image of atom j.

If the interactions between atoms were short-ranged, such as Lennard-Jones interactions, the interactions could be restricted to neighboring pairs. For example, for the atom depicted as solid in the central cell in Figure 2, we could restrict consideration to the shaded atoms within the cutoff circle. However, as noted above we cannot truncate electrostatic interactions in this way. Historically a number of modified truncation approaches have been developed. Truncation based on charge groups was used because the potential due to a neutral group decays faster and abrupt atom-based truncation of the Coulomb interaction leads to instabilities when standard molecular dynamics schemes are used. Energy conservation is also poor under abrupt group-based truncation schemes. Steinbach and Brooks [33] showed that this was due to free rotation of polar groups just outside the

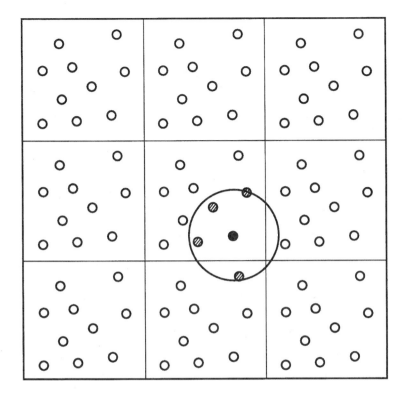

Figure 2 Periodic boundary conditions in two dimensions. The central simulation cell is replicated infinitely in both directions.

cutoff distance, leading to discontinuous change in energy when the groups reentered the cutoff distance. A variety of shifting or switching functions that modify the Coulomb potential to smoothly truncate it have been proposed. An optimal atom-based force-shifted cutoff was proposed by Steinbach and Brooks [33]. This leads to stable dynamics, even for highly polar or charged systems. However, the relationship of these modified potentials to basic electrostatics is not clear. Furthermore, it was shown later by Feller et al. [34] that these methods lead to significant artifacts in some simulations. The reaction field methods [35] attempt to correct for interactions between an atom such as the solid atom in Figure 2 and atoms outside the cutoff by replacing the latter with a dielectric continuum. The atoms inside the cutoff induce a reaction field in the continuum that is added to the explicit Coulomb interactions between the central atoms and those within the cutoff. Although this method has been very successful for simulations of liquids, it is inappropriate for macro-molecules, because the atoms outside a typical cutoff include a mix of solute and solvent for which different dielectric constants are appropriate.

Ewald summation has been applied successfully for many years to liquid simulations [35] and is now becoming a standard for macromolecular simulations [36]. For this reason we focus on Ewald summation for the remainder of this section.

Under periodic boundary conditions, the electrostatic potential of a point $\mathbf{r}$ in the central cell, which does not coincide with any atomic position $\mathbf{r}_i$, $i = 1, \ldots, N$, is given by summing the direct Coulomb potential over all atoms and all their images:

$$\phi(\mathbf{r}) = \sum_{\mathbf{n}} \sum_{i=1}^{N} \frac{q_i}{|\mathbf{r}_i + \mathbf{n}L - \mathbf{r}|} \tag{19}$$

where the outer sum is over all integer triples $\mathbf{n} = (n_1, n_2, n_3)$ and $\mathbf{n}L = (n_1 L, n_2 L, n_3 L)$.

This potential $\phi(\mathbf{r})$ is infinite if the central cell is not neutral, i.e., the sum of q_i is not zero, and otherwise is an example of a conditionally convergent infinite series, as discussed above, so a careful treatment is necessary. The potential depends on the order of summation, that is, the order in which partial sums over $\mathbf{n}$ are computed. For example, for positive integers K, define $\phi_K(\mathbf{r})$ as

$$\phi_K(\mathbf{r}) = \sum_{|\mathbf{n}| \leq K} \sum_{i=1}^{N} \frac{q_i}{|\mathbf{r}_i + \mathbf{n}L - \mathbf{r}|} \tag{20}$$

and consider the limit of $\phi_K(\mathbf{r})$ as $K \to \infty$. This limiting process is depicted in Figure 3. For large K, $\phi_K(\mathbf{r})$ is the potential at a point $\mathbf{r}$ in the central cell of a spherical macroscopic crystal. DeLeeuw et al. [37] discuss the limit of this potential as well as the potential for more general macroscopic shapes. In addition they discuss the case where the macroscopic crystal is immersed in a dielectric continuum. As discussed in Section II, the crystal induces a reaction field in the continuum that in turn affects the potential at points $\mathbf{r}$ in the central cell. DeLeeuw et al. show that when the continuum has an infinite dielectric constant (conducting boundary conditions) the potential is independent of the macroscopic shape of the crystal and is equal to the well-known Ewald potential. Otherwise, the potential is given by the Ewald potential plus a term that depends on the dielectric constant of

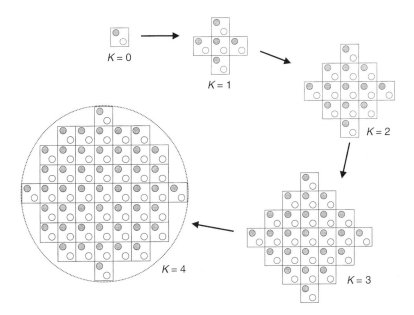

Figure 3 Periodic boundary conditions realized as the limit of finite clusters of replicated simulation cells. The limit depends in general on the asymptotic shape of the clusters; here it is spherical. Cations are presented as shaded circles; anions as open circles.

the continuum, the macroscopic shape of the crystal, and the dipole moment of the replicated unit cell. For some shapes, such as the spherical crystal depicted in Figure 3, they obtain a simple expression for this added term.

Although the work of DeLeeuw et al. is too complex to discuss further here, we can outline the derivation of the Ewald potential. The key is to imagine that each charge in the crystal is surrounded by a pair of continuous charge densities. One is a Gaussian-shaped density centered at $\mathbf{r}_i$ and having total charge $-q_i$ and is called the counterion density. The other density, called the co-ion density, is also centered at $\mathbf{r}_i$ and has identical shape but opposite charge to the counterion density. The potential at $\mathbf{r}$ due to q_i must be equal to the potential due to q_i together with the two canceling Gaussian densities. This is true for all atoms i and all their periodic images. We split the total potential $\phi(\mathbf{r})$ into the potential $\phi_1(\mathbf{r})$ due to the density ρ_1 given by the charges together with their counterion densities, depicted in Figure 4a, and the potential $\phi_2(\mathbf{r})$ due to the density ρ_2 given by the co-ion densities about the charges, depicted in Figure 4b. If we express Poisson's equation, Eq. (10), in spherical polar coordinates, we can solve for the potential due to q_i and its counterion density, obtaining $q_i\,\mathrm{erfc}\,(\beta|\mathbf{r}_i - \mathbf{r}|)/|\mathbf{r}_i - \mathbf{r}|$, where erfc $(\cdot)$ is the complementary error function and β is a parameter characterizing the width of the Gaussian density. This potential is summed over all i and all $\mathbf{n}$ to obtain $\phi_1(\mathbf{r})$.

To obtain the potential $\phi_2(\mathbf{r})$, we note that $\phi_2(\mathbf{r} + \mathbf{m}L) = \phi_2(\mathbf{r})$ for any integer triple $\mathbf{m}$, because this is simply a shifted sum over images. Thus $\phi_2(\mathbf{r})$ is a periodic function of $\mathbf{r}$. Similarly, the source density ρ_2, given by the sum of the co-ion densities over atoms

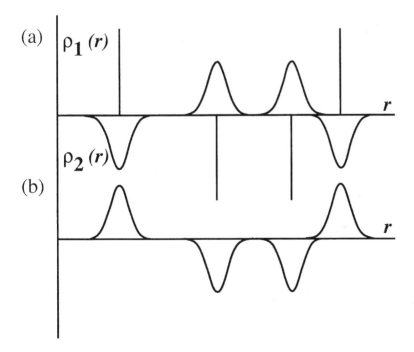

Figure 4 (a) Density ρ_1, the sum of point charges represented by vertical lines, plus Gaussian counterion density. This gives rise to short-range direct sum potential. (b) Density ρ_2, the Gaussian co-ion density. This gives rise to long-range reciprocal sum potential.

and their periodic images, is a periodic function. Thus we can expand ϕ_2 and ρ_2 as three-dimensional Fourier series. The Fourier series expansion of a periodic function g on a cube of side L is given by

$$g(\mathbf{r}) = \sum_{\mathbf{m}} \hat{g}(\mathbf{m}) \exp\left(2\pi i \frac{\mathbf{m} \cdot \mathbf{r}}{L}\right) \tag{21}$$

where the sum is over all integer triples $\mathbf{m}$ and the Fourier coefficient $\hat{g}(\mathbf{m})$ is given by

$$\hat{g}(\mathbf{m}) = \frac{1}{L^3} \int \int \int g(\mathbf{r}) \exp\left[-2\pi i \frac{\mathbf{m} \cdot \mathbf{r}}{L}\right] d\mathbf{r} \tag{22}$$

where the integral is over a cube of side L.

Now let $g(\mathbf{r})$ be given by the Gaussian co-ion density at a charge q_i, added to the co-ion density of all its periodic images. In this case the Fourier coefficient $\hat{g}(\mathbf{m})$ can be obtained in closed form by a clever trick. In Eq. (22), the integral over the cube of the complex exponential times the infinite sum of shifted Gaussians can be reexpressed as a single infinite integral that is the Fourier transform of the original co-ion density. The Fourier transform of a Gaussian density can be obtained in closed form in terms of another Gaussian. This result is added to that for the co-ion densities about the other charges and all their images to get $\hat{\rho}_2(\mathbf{m})$.

The Laplacian operator $\nabla \cdot \nabla$, or ∇^2, has a simple form in the Fourier domain: $\widehat{\nabla^2 g}(\mathbf{m}) = -4\pi^2 m^2 \hat{g}(\mathbf{m})$. This can be seen by differentiating the Fourier expansion, Eq. (21), and matching up coefficients. Thus, expanding ϕ_2 and ρ_2 in Fourier series, from Poisson's equation, Eq. (10), we get $-4\pi^2 m^2 \hat{\phi}_2(\mathbf{m}) = -4\pi \hat{\rho}_2(\mathbf{m})$, or $\hat{\phi}_2(\mathbf{m}) = \hat{\rho}_2(\mathbf{m})/\pi m^2$, for $m^2 \neq 0$. We set $\hat{\phi}_2(0, 0, 0) = 0$. This last choice requires that $\hat{\rho}_2(0, 0, 0) = 0$ (neutral unit cell) and involves the treatment of the limit process discussed by DeLeeuw at al. [37].

Putting these observations together, we can solve for the potential ϕ_2, obtaining

$$\phi_2(\mathbf{r}) = \frac{1}{\pi L} \sum_{\mathbf{m} \neq 0} \frac{\exp[-\pi^2 m^2 / \beta^2 L^2]}{m^2} \sum_{i=1}^{N} q_i \exp\left[2\pi i \frac{\mathbf{m} \cdot (\mathbf{r} - \mathbf{r}_i)}{L}\right]$$

$$= \frac{1}{\pi L} \sum_{\mathbf{m} \neq 0} \frac{\exp[-\pi^2 m^2 / \beta^2 L^2]}{m^2} S(\mathbf{m}) \exp\left[2\pi i \frac{\mathbf{m} \cdot \mathbf{r}}{L}\right] \tag{23}$$

where the structure factor $S(\mathbf{m})$ is the sum of $q_i \exp(-2\pi \mathbf{m} \cdot \mathbf{r}_i/L)$ over the N charges in the cube. For a fleshed-out version of this derivation of the Ewald potential, see Kittel [38].

Thus, for a point $\mathbf{r}$ in the central cell that does not coincide with any atomic position $\mathbf{r}_i$, $i = 1, \ldots, N$, the electrostatic potential $\phi(\mathbf{r})$ in Eq. (19) can be rewritten in the Ewald formulation as $\phi_1(\mathbf{r}) + \phi_2(\mathbf{r})$. The electrostatic potential at atom i is the potential due to all other atoms j together with their images as well as all nontrivial periodic images of atom i itself. This is like the potential $\phi(\mathbf{r})$ except that the (infinite) potential due to i itself is missing. Thus the potential at i can be obtained by removing the potential $q_i/|\mathbf{r}_i - \mathbf{r}|$

from $\phi_1(\mathbf{r}) + \phi_2(\mathbf{r})$ and then taking the limit as $\mathbf{r} \to \mathbf{r}_i$. The result is the electrostatic potential $\phi_{Ew}(\mathbf{r}_i)$ in the Ewald formulation and is given by

$$\phi_{Ew}(\mathbf{r}_i) = \sum_{\mathbf{n}}{}' \sum_j q_j \frac{\text{erfc}(\beta|\mathbf{r}_j - \mathbf{r}_i + \mathbf{n}|)}{|\mathbf{r}_j - \mathbf{r}_i + \mathbf{n}|} - q_i \frac{2\beta}{\sqrt{\pi}}$$
$$+ \frac{1}{\pi L} \sum_{\mathbf{m} \neq 0} \frac{\exp[-\pi^2 \, m^2/\beta^2 L^2]}{m^2} S(\mathbf{m}) \exp\left[2\pi i \frac{\mathbf{m} \cdot \mathbf{r}_i}{L}\right] \tag{24}$$

where the prime over the first sum means that $j \neq i$ when $\mathbf{n} = 0$. The energy U_{Ew} of the central unit cell is obtained by summing $(1/2)q_i\phi_{Ew}(\mathbf{r}_i)$ over the atoms:

$$U_{Ew} = \sum_{\mathbf{n}}{}' \sum_i \sum_j q_i q_j \frac{\text{erfc}(\beta|\mathbf{r}_j - \mathbf{r}_i|)}{|\mathbf{r}_j - \mathbf{r}_i|}$$
$$+ \frac{1}{2\pi L} \sum_{\mathbf{m} \neq 0} \frac{\exp[-\pi^2 \, m^2/\beta^2 L^2]}{m^2} |S(\mathbf{m})|^2 - \sum_{i=1}^{N} q_i^2 \frac{\beta}{\sqrt{\pi}} \tag{25}$$

where, as above, the prime over the first sum means that $j \neq i$ when $\mathbf{n} = 0$. The forces on atoms can be obtained by differentiating U_{Ew} with respect to their positions. In expression [25], the first term on the right-hand side is referred to as the direct sum, the second term as the reciprocal sum, and the third as the self-term.

Before discussing the computational cost of Ewald summation and the recent fast methods for calculating it, we discuss a variation on the above expressions, where the energy U_{Ew} is given in terms of the Ewald pair potential $\psi_{Ew}(\mathbf{r})$ together with the Wigner self-energy. At the same time we discuss the standard method for dealing with unit cells that have a net charge. Imagine that the unit cell contains a single unit charge at its center, the origin. Because the unit cell is not neutral, in addition to the Gaussian counterion and co-ion densities we add a uniform density over the cell, with total charge -1. These densities are then replicated along with the charge in all the image cells as above. To derive the potential in this case, we approximate the uniform density by gridding the cube with a $K \times K \times K$ grid. At each grid point we place a charge of $-1/K^3$. Because the unit cell is now neutral, the potential at a point $\mathbf{r}$ not at the origin and not on the grid is given by $\phi_1(\mathbf{r}) + \phi_2(\mathbf{r})$. We obtain $\psi_{Ew}(\mathbf{r})$ for $\mathbf{r}$ not at the origin by passing to the limit as $K \to \infty$. The result is

$$\psi_{Ew}(\mathbf{r}) = \sum_{\mathbf{n}} \frac{\text{erfc}(\beta|\mathbf{r} + \mathbf{n}|)}{|\mathbf{r} + \mathbf{n}|}$$
$$+ \frac{1}{\pi L} \sum_{\mathbf{m} \neq 0} \frac{\exp[-\pi^2 m^2/\beta^2 L^2]}{m^2} \exp\left[2\pi i \frac{\mathbf{m} \cdot \mathbf{r}}{L}\right] - \frac{\pi}{L^3 \beta^2} \tag{26}$$

The Wigner self-energy ζ is given by the limit of $\psi_{Ew}(\mathbf{r}) - 1/r$ as $r \to 0$:

$$\zeta = \sum_{\mathbf{n}=0} \frac{\text{erfc}(\beta|\mathbf{n}|)}{|\mathbf{n}|} + \frac{1}{\pi L} \sum_{\mathbf{m} \neq 0} \frac{\exp[-\pi^2 m^2/\beta^2 L^2]}{m^2} - \frac{\pi}{L^3 \beta^2} - \frac{2\beta}{\sqrt{\pi}} \tag{27}$$

For cubes of size L, $\zeta \approx 2.837297/L$. Using the Ewald pair potential we get another expression sometimes used for the energy of the unit cell,

$$U_{Ew} = \frac{1}{2} \sum_{i \neq j} q_i q_j \psi_{Ew}(\mathbf{r}_i - \mathbf{r}_j) + \frac{\zeta}{2} \sum_{i=1}^{N} q_i^2 \qquad (28)$$

This expression agrees with the previous expression Eq. (25), for neutral unit cells. If the unit cell is not neutral, the energy expression in Eq. (25) should be modified by adding the term $- \pi Q/2L^3\beta^2$ to the right-hand side, where Q is the total charge in the system. The energy expressions (25) and (28) will then agree.

The energy of the unit cell as well as the atomic forces obtained by differentiation are invariant to the choice of the parameter β. This parameter can then be chosen for computational convenience. For example, since erfc $(\beta r)/r$ decays exponentially, β can be chosen so that erfc$(\beta r)/r$ becomes negligible for $r > L/2$, and the direct sum can be restricted to minimum image pairs. In this case the number of structure factors $S(\mathbf{m})$ needed to converge the reciprocal sum does not grow with system size, and since each structure factor is an order N calculation, the reciprocal sum is also of order N. However, the computational cost of the direct sum will increase as N^2, because all charge pairs are calculated. On the other hand, β can be chosen such that the direct sum can be truncated at a fixed cutoff, e.g., 10 Å. In this case the direct sum becomes an order N calculation. Unfortunately, however, the number of structure factors needed to converge the reciprocal sum increases linearly with system size, making the reciprocal sum an order N^2 calculation. An optimal choice of β with system size makes both the direct and reciprocal sums order $N^{3/2}$ calculations [39].

Since the fast multipole algorithms (FMAs) outlined above are order N algorithms, extensions of them to periodic boundary conditions should be faster than Ewald summation for large systems. One interesting approach to a periodic FMA is that of Lambert et al. [40]. They numerically implement the limit process of DeLeeuw et al. [37] depicted in Figure 3. The unit cell is replicated a number of times to form a finite crystal. The FMA is applied to this crystal in such a way that the computational cost is only proportional to the size of the original cell. In this method the correction term of DeLeeuw et al. must be removed to arrive at the Ewald unit cell energy. Another approach to extending FMAs to periodic boundary conditions is that of Figuerido et al. [41]. They derive Ewald sums for the multipole expansion of the unit cell charges. This expression is then used to calculate the interactions due to nonadjacent cells in the infinite periodic array of unit cells. The rest of the fast multipole algorithm proceeds as before.

An entirely different approach to fast Ewald sums is based on the following observation. If the charges of the unit cell happened to be laid out on a regular $K \times K \times K$ grid, then the structure factors $S(\mathbf{m})$ for reciprocal vectors within the $K \times K \times K$ array could be calculated very quickly using the fast Fourier transform (FFT) algorithm. In fact, in the case that β is chosen such that the direct sum can be truncated at a regular cutoff, the order N number of structure factors necessary to converge the reciprocal sum can be calculated in order $N \log N$ using the FFT, making the whole Ewald sum an order $N \log N$ calculation. The particle mesh approaches to Ewald summation are based on reducing the usual case of irregularly positioned charges to that of regularly gridded charges.

The original particle mesh (P3M) approach of Hockney and Eastwood [42] treats the reciprocal space problem from the standpoint of numerically solving the Poisson equation under periodic boundary conditions with the Gaussian co-ion densities as the source density ρ on the right-hand side of Eq. (10). Although a straightforward approach is to sample the Gaussians on the grid, this is too expensive. Grid densities are typically about 1 Å, and the Gaussian is significant over a sphere of radius equal to the cutoff, e.g. 9 Å, so the number of grid points sampled is more than 1000 in this case, which is too expensive. They propose a narrower sampling density that is smooth and needs be sampled over only a much smaller number of grid points, together with a compensation factor in reciprocal space by which the original factor $1/m^2$ in the reciprocal Ewald sum is multiplied.

The particle mesh Ewald (PME) algorithm [43,44] instead approximates the structure factors $S(\mathbf{m})$ by interpolating the complex exponentials appearing in Eq. (23). Thus $\exp(2\pi\mathbf{m} \cdot \mathbf{r}_i/L)$ is rewritten as a linear combination of its values at nearby grid points, and so q_i times the complex exponential is rewritten as a sum of weighting factors times q_i times the complex exponential at the grid points. Thus, to approximate $S(\mathbf{m})$, first grid the charges using the weighting factors and then apply the FFT. Note that implicitly we are assuming that the weighting factors are independent of $\mathbf{m}$. In the original PME [43], which used a local Lagrangian interpolation, this was true. In the smooth PME [44], which uses B-spline interpolation, there is an additional constant $\lambda(\mathbf{m})$ that multiplies the FFT-based structure factors. A later modification [45] uses least squares B-spline approximation, which slightly modifies the constant. In the smooth PME, reciprocal space forces are obtained by analytically differentiating the approximate reciprocal space energy, using the smoothness of the B-splines. Alternatively, one could compute the Ewald reciprocal sum forces by differentiating the energy equation, Eq. (23), and approximating the forces using the above interpolation. This force-interpolated PME is more accurate than the smooth PME and rigorously conserves momentum but is more expensive, requiring a total of four FFT evaluations compared to two for the smooth PME. A fuller description of the mechanics of the PME algorithm together with a review of applications is given by Darden et al. [36].

The issue of which approach to Ewald sums is most efficient for a given system size has been plagued by controversy. Probably the best comparison is that by Pollock and Glosli [46]. They implement optimized versions of Ewald summation, FMA and P3M. They conclude that for system sizes of any conceivable interest, the P3M algorithm is most efficient. Interestingly, they also show that P3M can be used to efficiently calculate energies and forces for finite boundary conditions, using a box containing the cluster and a clever filter function in reciprocal space. The particle-mesh-based algorithms are excellent at energy conservation, which is an additional advantage. On the other hand, the FMA may scale better in highly parallel implementations because of the high communication needs of the FFT. In addition, since the expensive part of the FMA is due to long-range interactions, the FMA may be more appropriate for multiple time step implementations [41]. The algorithms for P3M and the force-interpolated PME are essentially identical, differing only in the form of the modification to the reciprocal space weighting factors $\exp[-\pi^2 m^2/\beta^2 L^2]/m^2$. The sampling density for the P3M turns out to be a shifted B-spline, so the weighting factors are very similar. Thus for the same grid density and order of interpolation, the computational costs of the P3M and force-interpolated PME are the same. In the case that contributions to the Ewald sum from high frequency reciprocal vectors $\mathbf{m}$ outside the $K \times K \times K$ array can be neglected, the expressions for P3M and force-interpolated PME become equivalent, and the accuracy and efficiency are thus equivalent [45]. Under all reasonable simulation parameters it was found that the errors due to neglect of high frequency reciprocal vectors were small compared to remaining errors, so the above two algorithms are equivalent for practical purposes. For typical simulation parameters (9 Å cutoff, RMS force error $\approx 10^{-4}$) the smooth PME is more efficient than either P3M or force-interpolated PME, because its accuracy is only marginally less than that of P3M and it requires fewer FFTs. However, for small cutoffs or higher accuracy the latter are more accurate and the smooth PME must compensate with higher order interpolation or denser grids, and it becomes less clear which is more efficient.

Regardless of which algorithm is used for fast calculation of Ewald sums, the computational cost is now competitive with the cost of cutoff calculations, and there is no longer a need to employ cutoffs for purposes of efficiency. Since Ewald summation is the natural expression of Coulomb's law in periodic boundary conditions, it is the recommended approach if periodic boundary conditions are to be used in a simulation.

V. STRENGTHS AND WEAKNESSES OF VARIOUS APPROACHES

In this final section, we recapitulate the relationship between long-range electrostatics and boundary conditions while attempting to assess the strengths and weaknesses of the three choices we have outlined.

As noted above, the continuum methods are treated fully elsewhere in the book, so we restrict ourselves to a few comments. The main advantage of the continuum approach over the other choices is its greater efficiency. The effect of solvent on the electrostatic free energy of the solute, which is considerable, is calculated by a model theory rather than through explicit time averaging over the many degrees of freedom involved. The main disadvantage of continuum methods is that explicit atomic level interactions of interest between solute and solvent such as the ''spine of hydration'' of water molecules and/ or sodium ions in the minor groove of B-DNA [47] cannot be treated. For our purposes in this chapter, however, we are interested in the qualitative insights afforded by continuum methods, particularly with respect to long-range electrostatic effects.

The continuum treatment is widely believed to give the correct long-range behavior for electrostatic interactions in solution. For this reason it can be used as a guide to the discussion of long-range effects in explicit solvent simulations. Recently Poisson's equation, Eq. (12), was used to estimate the effect of finite system size on solvation free energies calculated under periodic boundary conditions [48]. For example, in the finite difference approach to solving the Poisson equation (12), one could implement periodic boundary conditions in place of the usual estimate of the electrostatic potential at the boundary of the cell [49]. The difference between the continuum solvation free energy calculated using periodic boundary conditions and that calculated using the standard method will depend in general on the conformation of the solute. This conformational free energy difference, which will vanish in the limit of large cell size, can be used as an estimate of the system-size-dependent artifacts due to the use of periodic boundary conditions.

In principle, finite boundary conditions share much of the advantages of continuum models while adding explicit atomic level solvent–solute interactions. The amount of solvent used is typically smaller than in systems simulated under periodic boundary conditions, leading to greater efficiency. In addition, the long-range effects due to solvent can be modeled correctly using Poisson's equation, whereas, as mentioned above, there are finite system size artifacts due to long-range electrostatics under periodic boundary conditions. On the other hand, there are artifacts in calculated free energies due to the presence of the vacuum/solvent boundary [9]. There is a cost to transport charge across this interface, called the surface potential of water. This potential is substantial in current water models, although it seems to be small in real water. For spherically symmetrical systems the surface potential is easy to calculate and free energies are easily adjusted. It is not clear yet whether the artifact can be fully fixed in nonspherical systems, although it seems likely.

There are other inherent limitations to the finite boundary method. Due to the finite size of the system, water molecules cannot undergo natural unrestricted diffusion, and examination of water dynamics must be restricted to interior positions. Other transport properties may be affected as well by the size constraint. Existing boundary potentials cannot fully prevent artifactual structure and dynamics of the water near the vacuum/ solvent interface, and thus the layer of solvent must be large enough to allow sufficient bulklike solvent about the solute molecule, reducing the cost advantage over periodic boundary conditions. However, these potentials will probably continue to improve, leading to more accurate simulations with minimal solvation.

Periodic boundary conditions have long been a popular choice for simulations using explicit solvent. Using such a simulation, the density and pressure of the system can be easily specified. Furthermore, since the energy of the system is periodic, there are no preferred positions within the simulation cell. Regardless of position within the central cell, an atom is surrounded on all sides by atoms or their images. Unlike finite boundary conditions, there is no vacuum interface to be "fixed" via boundary potentials. Thus it is believed that periodic boundary conditions best model the environment of bulk matter, at least when long-range electrostatics are not involved.

Wood and Parker [50] discuss the possible artifacts due to periodic boundary conditions, in the context of short-range interactions. They suggest the still sensible approach of varying the unit cell size and shape to empirically test for the size of artifacts in finite systems. The implicit assumption is that these artifacts should vanish in the limit of large unit cell sizes. These early simulations established that artifacts due to periodicity in typical simulations (e.g., no phase transitions occurring) are negligible for systems that have only short-range interactions. When electrostatic interactions are involved, there are finite size periodicity artifacts, and the assumption that these artifacts vanish in the limit of large unit cell sizes is not immediately evident, due to the subtleties of conditional convergence discussed above. However, it seems clear from recent work [9,51] that the correct limiting behavior is reached for large unit cell sizes, and thus the question becomes, At what rate do the artifacts vanish?

Smith and Pettitt [52] studied the free energy of rotation of a dipole in an ideal dielectric at various temperatures. They conclude that finite size artifacts are not large in high dielectric solvents such as water at room temperature. Problems may arise in low dielectric solvents. Similar conclusions about the effect of solvent dielectric were reached in studies of charging free energies by Hummer et al. [51]. In the case of simple point ions they were able to correct these artifacts. The consensus at this point is that finite size artifacts due to periodicity are manageable in simulations of molecules in water at room temperature. The artifacts may be much larger in simulations of charged or polar molecules in low dielectric solvents [49]. In any case the suggestion of Wood and Parker [50] that the results be examined as a function of unit cell size is still pertinent.

REFERENCES

1. JD Jackson. Classical Electrodynamics. New York: Wiley, 1975.
2. P Auffinger, DL Beveridge. Chem Phys Lett 234:413, 1995.
3. JS Bader, D Chandler. J Phys Chem 96:6423, 1992.
4. G Hummer, LR Pratt, AE Garcia, BJ Berne, SW Rick. J Phys Chem B 101:3017, 1997.

5. G Hummer, LR Pratt, AE Garcia, S Garde, B Berne, SW Rick. J Phys Chem B 102:3841, 1998.
6. H Ashbaugh, R Wood. J Phys Chem B 102:3844, 1998.
7. J Aqvist, T Hansson. J Phys Chem B 102:3837, 1998.
8. GH Hardy. A Course of Pure Mathematics. Cambridge, UK: Cambridge Univ Press, 1975.
9. T Darden, D Pearlman, L Pedersen. J Chem Phys 109:10921, 1998.
10. JA McCammon, PG Wolynes, M Karplus. Biochemistry 18:927, 1979.
11. M Levitt. Cold Spring Harbor Symp Quant Biol 47:251, 1982.
12. R Lavery, H Sklenar, K Zakrzewski, B Pullman. J Biomol Struct Dynam 3:989, 1986.
13. BE Hingerty, RH Ritchie, TL Ferrel, JE Turner. Biopolymers 24:427, 1985.
14. BM Pettitt, M Karplus. Chem Phys Lett 121:194, 1985.
15. J Novotny, R Bruccoleri, M Karplus. J Mol Biol 177:787, 1984.
16. D Eisenberg, AD, McLachlan. Nature 319:199, 1986.
17. T Ooi, M Oobatake, G Nemethy, HA, Scheraga. Proc Natl Acad Sci USA 84:3086, 1987.
18. WC Still, A Tempczyk, RC Hawley, T Hendrickson. J Am Chem Soc 112:6127, 1990.
19. AA Rashin. J Phys Chem 94:1725, 1990.
20. R Bharadwaj, A Windemuth, S Sridharan, B Honig, A Nicholls. J Comput Chem 16:898, 1995.
21. A Nicholls, B Honig. J Comput Chem 12:435, 1991.
22. AC Belch, ML Berkowitz. Chem Phys Lett 113:278, 1985.
23. ML Berkowitz, JA McCammon. Chem Phys Lett 90:215, 1982.
24. D Beglov, B Roux. J Chem Phys 100:9050, 1994.
25. H Friedman. Mol Phys 29:1533, 1975.
26. HJC Berendsen. In: J Hermans, ed. Molecular Dynamics and Protein Structure. Western Springs, IL: Polycrystal Book Service, 1985.
27. JJ Biesiadecki, RD Skeel. J Comput Phys 109:318, 1993.
28. HQ Ding, N Karasawa, WA Goddard. J Chem Phys 97:4309, 1992.
29. L Greengard, V Rokhlin. J Comput Phys 73:325, 1987.
30. FS Lee, A Warshel. J Chem Phys 97:3100, 1992.
31. CA White, M Head-Gordon. J Chem Phys 101:6593, 1994.
32. T Bishop, R Skeel, K Schulten. J Comput Chem 18:1785, 1997.
33. PJ Steinbach, BR Brooks. J Comput Chem 15:667, 1994.
34. SE Feller, RW Pastor, A Rojnuckarin, S Bogusz, BR Brooks. J Phys Chem 100:17011, 1996.
35. MP Allen, DJ Tildesley. Computer Simulation of Liquids. Oxford, UK: Clarendon Press, 1987.
36. T Darden, L Perera, L Li, L Pedersen. Struct Fold Des 7:R55, 1999.
37. SW DeLeeuw, JW Perram, ER Smith. Proc Roy Soc Lond A373:27, 1980.
38. C Kittel. Introduction to Solid State Physics. New York: Wiley, 1986.
39. JW Perram, HG Petersen, SW DeLeeuw. Mol Phys 65:875, 1988.
40. CG Lambert, TA Darden, JA Board. J Comput Phys 126:274, 1996.
41. F Figuerido, R Levy, R Zhou, B Berne. J Chem Phys 106:9835, 1997.
42. RW Hockney, JW Eastwood. Computer Simulation Using Particles. New York: McGraw-Hill, 1981.
43. TA Darden, DM York, LG Pedersen. J Chem Phys 98:10089, 1993.
44. U Essmann, L Perera, ML Berkowitz, T Darden, H Lee, LG Pedersen. J Chem Phys 103: 8577, 1995.
45. T Darden, A Toukmaji, L Pedersen. J Chim Phys 94:1346, 1997.
46. E Pollock, J Glosli. Comput Phys Commun 95:93, 1996.
47. X Shui, L McFail-Isom, G Hu, LD Williams. Biochemistry 37:8341, 1998.
48. G Hummer, L Pratt, A Garcia. J Phys Chem A 102:7885, 1998.
49. PH Hunenberger, JA McCammon. J Chem Phys 110:1856, 1999.
50. WW Wood, FR Parker. In: G Ciccotti, D Frenkel, IR McDonald, eds. Simulations of Liquids and Solids. Amsterdam: North-Holland, 1987.
51. G Hummer, LR Pratt, AE Garcia. J Phys Chem 100:1206, 1996.
52. PE Smith, BM Pettitt. J Chem Phys 105:4289, 1996.

6

Internal Coordinate Simulation Method

Alexey K. Mazur

Institut de Biologie Physico-Chimique, CNRS, Paris, France

I. INTRODUCTION

In this chapter I outline the general principles of modeling biomacromolecules with internal coordinates as independent variables. This approach was generally preferred in the early period of computer conformational analysis when hardware computer resources were strongly limited [1]. In the last two decades, mainly because of the growing interest in molecular dynamics (MD), Cartesian coordinate approaches gradually became predominant, and one readily sees that just by looking into the index of this book. Nevertheless, internal coordinates continue to be employed, notably, in conformational searches based on energy minimization and Monte Carlo (MC) [2] and in normal mode analysis [3]. My main objective is to give a consistent exposition of the basic algorithms of this methodology and its underlying philosophy, with special emphasis on recent advances in the internal coordinate molecular dynamics (ICMD) techniques.

More traditional applications of internal coordinates, notably normal mode analysis and MC calculations, are considered elsewhere in this book. In the recent literature there are excellent discussions of specific applications of internal coordinates, notably in studies of protein folding [4] and energy minimization of nucleic acids [5].

II. INTERNAL AND CARTESIAN COORDINATES

The term ''internal coordinates'' usually refers to bond lengths, valence angles and dihedrals. They completely define relative atomic positions thus giving an alternative to the Cartesian coordinate description of molecular structures. Dihedrals corresponding to rotations around single bonds are most important because all other internal coordinates are usually considered fixed at their standard values, and the representation thus obtained is referred to as the standard geometry approximation [6]. For both proteins and nucleic acids the standard geometry approximation reduces the number of degrees of freedom from $3N$ to approximately $0.4N$, where N is the total number of atoms. Freezing of 'unimportant' variables accelerates minimization of the potential energy as well as equilibration in Monte Carlo calculations just because the space dimension is the principal parameter that determines the theoretical rate of convergence of iterative algorithms. It is important also that higher order minimizers that require much computer memory to store the Hessian matrix remain affordable even for very large systems. It should be noted, however, that because of the non-linear relationship between internal and Cartesian coordinates the distinction between them is not reduced to the foregoing simple arithmetic. To begin with, let us consider the following instructive example.

Figure 1 compares the courses of energy minimization with different choices of coordinates. A standard geometry initial conformation was minimized in three modes: (1) with all degrees of freedom and Cartesian coordinates as variables, (2) with all degrees of freedom but internal coordinates as variables, and (3) with fixed standard geometry. All computations were made with the same program code employing a conjugate gradient minimizer with analytical gradients. Figure 1a demonstrates that, as expected, the minimum is most rapidly found with the standard geometry approximation. With all degrees

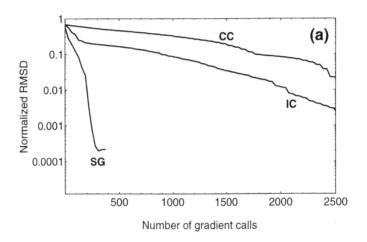

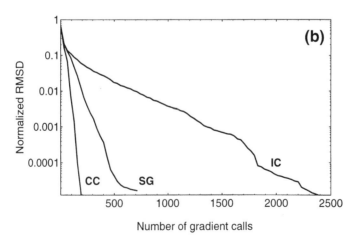

Figure 1 The course of energy minimization of a DNA duplex with different choices of coordinates. The rate of convergence is monitored by the decrease of the RMSD from the final local minimum structure, which was very similar in all three cases, with the number of gradient calls. The RMSD was normalized by its initial value. CC, IC, and SG stand for Cartesian coordinates, $3N$ internal coordinates, and standard geometry, respectively.

of freedom the structure changes much more slowly, but we note that the rate of convergence is noticeably higher when internal rather than Cartesian coordinates are used, even though the space dimension is $3N$ in both cases. The internal coordinate minimization goes faster because internal coordinates better correspond to the local potential energy landscape. The energy gradient is an invariant vector that does not depend on the choice of coordinates, and so is the direction computed by the minimizer. Once it is chosen, however, the minimizer moves the structure along a straight line in the corresponding space. In Cartesian coordinates the profiles of the potential energy are very complex, and any straight path quickly goes to a wall. In contrast, curved atomic trajectories corresponding to straight lines in the internal coordinate space make possible much longer moves.

A clear manifestation of the foregoing effect is exhibited in Figure 1b. This graph shows results of a similar minimization test but with additional harmonic restraints that pulled atomic Cartesian coordinates to the final minimum energy values. Now the potential energy landscape in Cartesian coordinate space is greatly simplified, giving a dramatic acceleration of convergence compared to internal coordinates. As a result, convergence appears even faster than with the standard geometry approximation in spite of the difference in the number of variables. In practice, regardless of the number of variables and the type of minimizer, internal coordinates are always preferable in unconstrained minimization. In contrast, for example, in crystallographic root-mean-square refinement with a high weight of experimental restraints Cartesian coordinates should give faster convergence and lower final R factors.

The local energy minimization is arguably the clearest domain in molecular modeling, but we see that even here the difference between the two coordinate sets is far from trivial. It becomes much more complicated, however, when the specific features of macromolecular systems are considered. One feature is the multiple minima problem often discussed in connection with protein folding [2]. It is usually tackled with hybrid MC and MD techniques such as simulated annealing or MC minimization. Common examples are the protein folding by global minimization of some target function (not necessarily energy) and structure determination based on experimental data. In these calculations, called conformational searches, one looks for the structures that satisfy certain conditions and does not care how well the intermediate steps correspond to the physical reality. The standard geometry approximation offers a whole list of specific advantages for such studies.

First, larger MC steps are possible due to the same effect as in the foregoing minimization example. Second, larger MD steps are possible because freezing of bond length and bond angles eliminates the fastest motions. Third, molecular models can tolerate strong stimulation, such as by elevated temperature and strong stochastic forces, and still maintain a correct geometry of chemical groups. In addition, freezing of bond length and bond angles removes the small-scale ''roughness'' from the energy landscape of a macromolecule, thus vastly reducing the density of insignificant local minima. Exact evaluation of such density is a difficult task, but nevertheless this intuitive suggestion agrees with many practical observations. For example, in terms of root-mean-square distance (RMSD) of atomic coordinates, the standard geometry approximation results in a significantly larger radius of convergence for energy minimization from random states [7]. A similar effect has been reported for simulated annealing of protein conformations in crystallographic refinement [8].

At present, conformational searches provide for the most important application of computer molecular modeling in biology. In contrast, in statistical physics, from which MC and MD methods were originally borrowed, they are primarily used for studying physical phenomena connected with thermal molecular motions. In such investigations exhaustive sampling is indispensable. In simple words this means that if an event is considered, it must occur many times in MC or MD trajectories, and if a parameter is measured, every state that contributes a distinct individual value to the average must be visited many times. Unfortunately, with the presently available computer power, hardly any biologically important event and hardly any system can be both correctly and accurately modeled in such a sense. Nevertheless, this line of research has many long-term prospects in molecular biophysics, and in the remaining part of this section I will briefly comment on the problems connected with the application of internal coordinates in such studies.

In "true simulations" physical realism is the goal, and the question arises, What part of such realism is sacrificed with the elimination of "unimportant" degrees of freedom? This issue appears to be rather complicated. It has been debated many times in the literature, but no consensus seems to have been reached [6,9–16]. Without going into details, I briefly summarize here the two opposite lines of argumentation, denoting them (A) and (B).

(A1) Freezing of bonds and angles deforms the phase space of the molecule and perturbs the time averages. The MD results, therefore, require a complicated correction with the so-called metric tensor, which undermines any gain in efficiency due to elimination of variables [10,17–20].

(B1) The metrics effect is very significant in special theoretical examples, like a freely joined chain. In simulations of polymer solutions of alkanes, however, it only slightly affects the static ensemble properties even at high temperatures [21]. Its possible role in common biological applications of MD has not yet been studied. With the recently developed fast recursive algorithms for computing the metric tensor [22], such corrections became affordable, and comparative calculations will probably appear in the near future.

(B2) With their frequencies beyond 1000 cm^{-1}, the bond length and bond angle oscillations occupy the ground state at room temperature. The classical harmonic treatment makes them "too flexible."

(A2) In spite of the high individual frequencies, bond length and bond angle vibrations participate in quasi-classical low frequency collective normal modes. Bond angle bending is necessary for the flexibility of five-membered rings, which plays a key role in the polymorphism of nucleic acids.

(B2) Usually, the role of these vibrations is not crucial, and with bond lengths and bond angles fixed the corresponding collective modes are only modified, not eliminated. Significant variations of valence angles in strained structures, as in furanose rings of nucleic acids, can be treated with special algorithms.

(A3) Bond lengths and bond angles vary in protein crystal structures.

(B3) These variations are related to the refinement procedures much more than to the experimental data [23] and are generally larger than in high resolution structures of small molecules. In MD calculations with harmonic bond lengths and bond angles they are still higher.

(A4) Bond angle bending makes a nonnegligible contribution to conformational entropy and can affect computed equilibrium populations [11].

(B4) The corresponding estimates are valid only in harmonic approximation; therefore, they are inapplicable to normal temperature conditions. The harmonic approximation tends to underestimate the relative weight of torsion fluctuations. In general, because of the sampling limitations, conformational entropies cannot be accurately evaluated, and it remains unclear how large their relative contribution is to the conformational stability of biopolymers, let alone the specific bond angle part.

(A5) In calculations, freezing of bond angles noticeably reduces the rate of torsion transitions [12,21], which sometimes is also attributed to the bond angle contribution to the transition state entropy.

(B5) Generally, in molecular modeling, such discrepancies should be corrected by adjusting force field parameters. One should note, however, that this specific effect has more than one interpretation. The bond angle motion contributes entropy both to saddle points and to minimum energy states, and, in principle, it can affect reaction rates both ways. It is known, on the other hand, that the rate of barrier crossings in classical MD is always overestimated because of the so-called zero point energy problem [24,25]. The "zero point energy" is the ground-level energy of molecular vibrations. In reality it is unexchangeable, but in classical dynamics it serves as a vast, easily accessible energy reservoir for barrier-crossing fluctuations. The imposition of constraints is, in fact, one of the ways to correct for this deficiency.

The foregoing polemic is of significant theoretical interest, but, in my opinion, none of these arguments can justify or disqualify this or that specific technique. For the latter purpose, computed results must be compared with experimental data, and this cannot be replaced by calculations with and without constraints. Because of the severe sampling limitations, the experimental basis for such comparisons is extremely narrow, and suitable experimental data can instead be used for fitting force field parameters. At present, and into the foreseeable future, the sampling power of MC and MD methods remains the key limiting factor in all biological applications, especially in "true simulations." Compared to other means commonly used to improve the sampling, such as high temperature, the standard geometry constraints propose a relatively harmless remedy anyway.

III. PRINCIPLES OF MODELING WITH INTERNAL COORDINATES

Every phase of internal coordinate modeling admits many methodological variations, and I do not attempt to review them all. I outline only the standard problems encountered in any particular domain of application and the common practical solutions.

A. Selection of Variables

First of all, one needs to choose the local coordinate frame of a molecule and position it in space. Figure 2a shows the global coordinate frame xyz and the local frame $x'y'z'$ bound with the molecule. The origin of the local frame coincides with the first atom. Its three Cartesian coordinates are included in the whole set and are varied directly by integrators and minimizers, like any other independent variable. The angular orientation of the local frame is determined by a quaternion. The principles of application of quaternions in mechanics are beyond this book; they are explained in detail in well-known standard texts

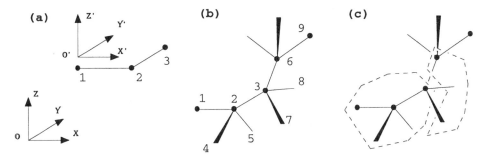

Figure 2 Illustration of the definition of internal coordinates. Main chain atoms are shown as filled circles.

[26,27]. We note only that a rigid-body quaternion involves four components bound by a normalization condition, which can be viewed as polar coordinates on a four-dimensional sphere isomorphous to the group of rigid-body rotations. Physically, the rotational motion involves three degrees of freedom and is completely determined by three Cartesian components of the angular velocity. The derivatives of the four quaternion components are related to the angular velocity by standard differential relations called kinematics equations. The latter are used both in dynamics and in energy minimization to compute variations of quaternions corresponding to infinitesimal rigid-body rotations.

The second atom always rests on axis $O'x'$, and the third atom always rests in plane $x'O'y'$. The second atom moves along the $O'x'$ axis if bond length 1-2 is not fixed. Similarly, the third atom moves in the $x'O'y'$ plane if bond 2-3 and/or bond angle 1-2-3 are variable. With these conventions, the six rigid-body degrees of freedom of frame $x'y'z'$ always complement the internal degrees of freedom in the molecule to the full set required. The fourth and the following atoms are added one by one, following a certain predefined order. At each step, one bond length, one planar angle, and one dihedral angle are used to position an atom. Note that the number of bonds available equals the number of atoms only for tree topologies; the number of bond angles and dihedrals is much larger. There is, therefore, considerable freedom in both atom ordering and the selection of bonds and angles employed as independent variables. In Figure 2b atoms at branches are assigned consecutive numbers starting from the main chain direction. For example, atom 6 is positioned with dihedral 1-2-3-6, angle 2-3-6, and bond 3-6. The corresponding triples for atoms 7 and 8 are (3-6-2-7, 6-3-7, 3-7) and (3-7-6-8, 7-3-8, 3-8). Only these bonds and angles are included in the set of independent internal coordinates.

In Figure 2b the main chain branch is positioned with true torsion, while the other branches employ so-called improper torsions. The utility of this seemingly strange choice becomes clear when the standard geometry approximation is required. In this case bonds, valence angles, and ''improper torsions'' are fixed, and the molecule is divided into rigid bodies. With the above definition, atoms can be sorted by using the following simple algorithm. Atoms are checked one by one according to their ordering, and if one of the internal coordinates of a given atom is not fixed it forms a new rigid body. If that is not the case, the atom is assigned to the rigid body of the lower branching point or to that of the previous atom, depending on whether it belongs to the main chain or to a secondary branch. This rule is easy to program and gives the standard geometry representation automatically regardless of the chemical nature of the molecule. In addition, it can produce reasonable models with some bond angles free, which is necessary for nucleic acids, for instance. Figure 2c shows the standard geometry rigid bodies for this specific example.

B. Energy Gradients

As we see, internal coordinates are always defined together with the transformation rules that tell how atomic Cartesian coordinates must be computed. This construction determines also how the structure changes in response to variation of internal coordinates, and it is easy to see that the infinitesimal displacements of atoms correspond to the motion of a tree. Namely, the first three atoms form the base, internal coordinates close to the base move almost the whole molecule, whereas variables in a specific branch move only the upper part of that branch. For example, when torsion 1-2-3-6 in Figure 2b is varied then, by definition, the first three atoms are not affected, whereas the fourth and the following atoms move together. Let us introduce the term ''articulated body'' to refer to the set of atoms that move when variable q_i is varied and denote it D_i. If q_i is a translational variable, like a bond length or a Cartesian coordinate of the first atom, the articulated body is translated along a fixed direction, and it is an easy matter to show that the partial derivative of the potential energy, $\partial U/\partial q_i$, equals the component of the total force applied to D_i along this direction. Therefore it can be computed as

$$-\frac{\partial U}{\partial q_i} = \mathbf{e}_i \sum_{\alpha \in D_i} \mathbf{f}_\alpha + f_i \qquad (1)$$

where $\mathbf{e}_i$ is the unit vector of the translation, $\mathbf{f}_\alpha$ are forces applied to atoms, and f_i is the possible additional force due to the bond stretching potential. Similarly, when q_i is a rotational variable, such as a torsion, a valence angle or one of the rigid body rotations of the base, the articulated body rotates around a certain fixed axis and $\partial U/\partial q_i$ is equal to the projection upon it of the total torque applied to D_i. It is computed as

$$-\frac{\partial U}{\partial q_i} = \mathbf{e}_i \sum_{\alpha \in D_i} \mathbf{r}_\alpha \times \mathbf{f}_\alpha - \mathbf{e}_i \mathbf{r}_i \times \sum_{\alpha \in D_i} \mathbf{f}_\alpha + t_i \qquad (2)$$

where $\mathbf{e}_i$ is the unit vector of the rotation axis, $\mathbf{r}_i$ is its position vector, $\mathbf{r}_\alpha$ are atom position vectors, and t_i is the possible additional torque due to the torsion or bond angle bending potential.

When the energy gradient is evaluated by a computer the calculation of atomic forces $\mathbf{f}_\alpha$ produced by bonded and nonbonded interactions takes a vast majority of time. Summations involved in Eqs. (1) and (2) are rapid because the articulated bodies fit one into another like in a Russian doll. In an unbranched chain $D_1 \supset D_2 \supset \cdots \supset D_{n-1} \supset D_n$, and therefore the sums in Eqs. (1) and (2) can be computed starting from the tip of the chain, moving to the base, and at the ith variable adding only the contribution from subset D_i/D_{i+1}. Extension to trees is straightforward. The idea of such recurrent calculations belongs to Gō and collaborators [28], and it appears in many other applications. As a result, in terms of computer time, energy gradients with respect to internal coordinates are no more expensive than atomic forces in Cartesian coordinate calculations. In fact, even a small savings can be achieved, because the last terms in Eqs. (1) and (2) can be evaluated directly.

IV. INTERNAL COORDINATE MOLECULAR DYNAMICS

A. Main Problems and Historical Perspective

Molecular dynamics has emerged as an application to molecules of the general method of point particles, with Cartesian coordinates and Newton's equations, and it was first applied to flexible polymeric molecules more than 20 years ago [29,26,30]. It was already clear at that time that harmonic potentials that keep bond lengths and bond angles close to their standard values severely limited the time step and that it would be desirable to get rid of these "uninteresting" degrees of freedom. No wonder, therefore, that early attempts to apply internal coordinates with the standard geometry approximation in MD were made at the same time [19,31]. This way, however, appeared too complicated and has been abandoned in favor of an alternative approach proposed by Ryckaert et al. [32], which consists in imposing holonomic distance constraints upon a system of point particles governed by Newton's equations. Their method, now called constraint dynamics, is reviewed elsewhere in this book. However, although it seemed initially that not only bond lengths but also bond angles, dihedral angles, and larger rigid groups could be fixed by using triangulation, this was found to be true only for very small molecules. In large complex polymers such as proteins, even bond angles cannot be fixed in this way [12].

The intrinsic limitation of constraint dynamics can be qualitatively understood from the underlying physical model. Imposing a distance constraint implies that a reaction force is introduced that is applied along a line joining two atoms. Reactions must be calculated at each time step to balance all other forces in the system, which for large branched molecules presents a difficult problem because they are all coupled and form a system of algebraic equations solved by iterations. With only bond lengths to hydrogens fixed, reactions are coupled in small groups, and iterations converge rapidly. With all bond lengths fixed, reactions are coupled globally, and looping becomes possible. In general, in this case the convergence is not guaranteed [33], but in practice it remains acceptable if reactions at the same atom are applied at obtuse angles. The last condition breaks down when bond angles are triangulated, and as a result the convergence becomes too slow.

These difficulties have led to a revival of work on internal coordinate approaches, and to date several such techniques have been reported based on methods of rigid-body dynamics [8,19,34–37] and the Lagrange–Hamilton formalism [38–42]. These methods often have little in common in their analytical formulations, but they all may be reasonably referred to as internal coordinate molecular dynamics (ICMD) to underline their main distinction from conventional MD: They all consider molecular motion in the space of generalized internal coordinates rather than in the usual Cartesian coordinate space. Their main goal is to compute long-duration macromolecular trajectories with acceptable accuracy but at a lower cost than Cartesian coordinate MD with bond length constraints. This task turned out to be more complicated than it seemed initially.

Two problems inherent in ICMD were clear from the beginning. The first is the derivation of equations of motion for large molecules, which was the main subject of the initial reports [19,34,38,39]. The second problem is the cost of additional calculations, which must be low enough to be compensated for by an increase in the step size. The need to invert the mass matrix to obtain generalized accelerations was the main obstacle. Generally, it is a full positive definite matrix; therefore, a direct inversion scales as $O(n^3)$ with the number of degrees of freedom and quickly becomes impractical when n exceeds 100. It turned out, fortunately, that this problem had been earlier solved in robot mechanics [43–46]. There are several recursive $O(n)$ algorithms that can rapidly compute exact generalized accelerations if the system can be treated as a tree of articulated rigid bodies. They are directly applicable to molecular models considered in the previous section, and when these algorithms were first implemented [8,36] it appeared that, as in Newtonian dynamics, the cost of the time step could be made close to that of the evaluation of forces.

Yet another difficulty was encountered in the numerical integration of dynamics equations. The general structure of the internal coordinate equations precludes the use of familiar Verlet or leapfrog algorithms, and that is why, at first, general-purpose predictor-corrector and Runge–Kutta integrators were used [8,36,39,40]. The results, however, clearly indicated that the quality of trajectories is much inferior to the conventional MD, even though the possibility of a considerable increase in time step length was demonstrated on some examples [36,39,40]. This difficulty could not be anticipated, because only recently was it realized that the exceptional stability of the integrators of the Störmer–Verlet–leapfrog group is bound to their symplectic property, which in turn is due to the fact that the Newtonian equations are essentially Hamiltonian. A very recent approach [42] seems to overcome this difficulty, and it has been demonstrated that ICMD is able to give a net gain in terms of computations per picosecond of dynamics.

The last problem to be mentioned concerns the physical factors that limit time steps in ICMD. All biopolymers have hierarchical spatial organization, and this structural hierarchy is naturally mapped onto the spectrum of their motions. That is, the fast motions involve individual atoms and chemical groups, whereas the slow ones correspond to displacements of secondary structures, domains, etc. Every such movement considered separately can be characterized by a certain maximum time step, and in this sense one can say that there exists a hierarchy of fast motions and, accordingly, step size limits. The lowest limit is determined by bond-stretching vibrations of hydrogens. It was always assumed that stretching of bonds between non-hydrogen atoms follows next [47], but in fact, until very recently, other fast motions did not attract much attention because only bond length constraints were technically possible anyway. With the development of ICMD, this issue acquired practical importance, and simultaneously it became possible to try different sorts of constraints and study the hierarchy of fast motions in larger detail. It was found [48] that, in proteins, this hierarchy does not always agree with the common intuitive suggestions. For instance, very fast collective vibrations in which hydrogen bonding plays a major role are rather common. On the other hand, nonbonded interatom interactions impose ubiquitous anharmonic limitations starting from rather small step sizes. This type of limitation is most important for ICMD, but, unfortunately, it is also the most difficult to reveal and overcome.

The last two problems have been realized only recently, and additional progress in these research directions may be expected in the near future. At present it is clear that with the standard geometry approximation all time step limitations below 10 fs can be overcome rather easily. This time step increase gives a substantial net increase in performance compared to conventional MD. The possibility of larger step sizes now looks problematic, although it has been demonstrated for small molecules. Larger steps should be possible, however, with constraints beyond the standard geometry approximation.

B. Dynamics of Molecular Trees

For the model of free point particles the Newtonian equations present by far the simplest and most efficient analytical formalism. In contrast, for chains of rigid bodies, there are several different, but equally applicable, analytical methods in mechanics, with their specific advantages and disadvantages. Recent studies in this field made clear, however, that the analytical difficulties connected with the size and chemical complexity of biological macromolecules can probably be overcome with any such formalism, and the main question is whether a given method is numerically efficient. Until now, only a few approaches have been able to treat large molecules, and they all take advantage of tree topologies in fast recurrent algorithms similar to the one outlined in Section III.B. The best performance has been achieved by combining the fast mass matrix inversion technique resulting from the Newton–Euler analysis of rigid-body dynamics [35] with equations of motion in canonical variables [42], which make possible symplectic numerical integration.

The symplectic property is a key feature of an integrator in the calculation of long-time trajectories of classical mechanics [49]. The term ''symplectic'' means that the discrete mapping corresponding to one time step must conserve the set of symplectic invariants of mechanics, one of which is the phase volume [50]. This condition looks very complex, but in practice it just means that one iteration of the integrator can be represented as a sequence of moves, each of which is an exact mechanical trajectory corresponding to some Hamiltonian. For instance, leapfrog or Verlet steps can be represented as several moves with only the kinetic or potential part of the full Hamiltonian used. However, such steps are possible only in the space of canonical variables, and that is why the generalized velocities and accelerations corresponding to internal coordinates are not appropriate as dynamic variables.

By definition, the vector of conjugate momenta corresponding to the vector of generalized internal coordinates $\mathbf{q}$ is

$$\mathbf{p} = \frac{\partial L(\mathbf{q}, \dot{\mathbf{q}})}{\partial \dot{\mathbf{q}}} = \mathbf{M}(\mathbf{q})\dot{\mathbf{q}} \tag{3}$$

where L is the Lagrangian and $\mathbf{M}(\mathbf{q})$ is the mass matrix. It can be shown [42] that the conjugate momentum of a translational variable is given by the projection of the total Cartesian momentum of the articulated body onto the direction of translation. The conjugate momentum of a rotational variable is the projection of the angular momentum of the articulated body onto the rotation axis. Neither of entities is convenient as an independent variable, but one avoids the difficulties by using equations of motion of the form

$$\dot{\mathbf{p}} = -\frac{\partial U}{\partial \mathbf{q}} + \mathbf{w}(\mathbf{q}, \dot{\mathbf{q}}) \tag{4a}$$

and

$$\dot{\mathbf{q}} = \mathbf{M}^{-1}\mathbf{p} \tag{4b}$$

where $\mathbf{w}(\mathbf{q}, \dot{\mathbf{q}})$ is an inertial term. For translational and rotational variables, respectively, its components read

$$w_i = \dot{\mathbf{e}}_i \mathbf{P}_i \tag{5a}$$

and

$$w_i = \dot{\mathbf{e}}_i \mathbf{Q}_i - \mathbf{P}_i (\mathbf{e}_i \times \dot{\mathbf{r}}_i) \tag{5b}$$

where $\mathbf{P}_i$ and $\mathbf{Q}_i$ are the translational and angular momenta, respectively, of the articulated body D_i. Similarly to forces and torques in Eqs. (1) and (2), these can be rapidly computed by a recurrent summation. Similar summation techniques are employed in computing the product $\mathbf{M}^{-1}\mathbf{p}$ in Eq. (4b). The corresponding algorithms originate from robot mechanics and are based on a special factorization of the mass matrix. A very clear unified presentation of these methods is given by Jain [51].

Equations (4) are called quasi-Hamiltonian because, even though they employ generalized velocities, they describe the motion in the space of canonical variables. Accordingly, numerical trajectories computed with appropriate integrators will conserve the symplectic structure. For example, an implicit leapfrog integrator can be expressed as

$$\mathbf{f}_n = \mathbf{f}(\mathbf{q}_n) \tag{6a}$$

$$\circ \quad \mathbf{q}_{n+1/2} = \mathbf{q}_n + \dot{\mathbf{q}}_{n+1/2}\frac{h}{2} \tag{6b}$$

$$\circ \quad \mathbf{p}_{n+1/2} = \mathbf{p}_{n-1/2} + \mathbf{f}_n h + (\mathbf{w}_{n-1/2} + \mathbf{w}_{n+1/2})\frac{h}{2} \tag{6c}$$

$$\circ \quad \mathbf{q}_{n+1/2} = \mathbf{M}^{-1}_{n+1/2}\,\mathbf{p}_{n+1/2} \tag{6d}$$

$$\mathbf{q}_{n+1} = \mathbf{q}_n + \dot{\mathbf{q}}_{n+1/2}h \tag{6e}$$

where the conventional notation is used for denoting on-step and half-step values. The lines marked by circles ($\circ$) are iterated until the convergence of Eqs. (6b) and (6c). When the mass matrix does not depend on coordinates, $\mathbf{w}(\mathbf{q}, \dot{\mathbf{q}})$ vanishes, and this integrator is reduced to the standard leapfrog. It is symplectic in the same sense as leapfrog, namely, the symplectic structure is conserved for pairs $(\mathbf{p}_{n-1/2}, \mathbf{q}_n)$ and $(\mathbf{p}_{n+1/2}, \mathbf{q}_n)$.

C. Simulation of Flexible Rings

Treatment of flexible rings is a special and inherently difficult task for algorithms that use specific advantages of tree topologies. If such a topology is imposed on a ring, it will be broken once all internal coordinates start to change independently. Several possible ways out of this can be considered. The simplest consists in applying harmonic restraints to the broken ring bonds. In this case, in dynamics, the time step may be limited by the frequencies introduced by these restraints. The rigorous but complex way is to treat some of the internal coordinates as dependent variables and exclude them from equations of motion [52]. However, this involves mass matrix transformations that would be incompatible with the fast inversion algorithms. The third way is to impose ring closure constraints explicitly, similarly to the method of constraints in Cartesian MD. The last possibility has been recently checked, and it gives an acceptable solution [53,54]. This difficulty is most critical for simulations of nucleic acids in which the bases are connected with the sugar–phosphate backbone via five-membered rings, and we now consider this specific example.

Figure 3 shows how the tree is constructed for a sugar ring in a nucleic acid. Ring atoms are numbered 1, . . . , 5 corresponding to C4′, C3′, C2′, C1′, O4′. The base is placed at the 5′ end; the main chain goes along the backbone with branching for bases at C3′ atoms. The ring conformation is determined by five valence and dihedral angles $q_1, \ldots, q_5$ indicated by arrows. The bond C4′ $\cdots$ O4′ shown by the broken line is excluded from the tree and replaced by the distance constraint

$$C = |\mathbf{r}_5 - \mathbf{r}_1| - l_{15} = 0. \tag{7}$$

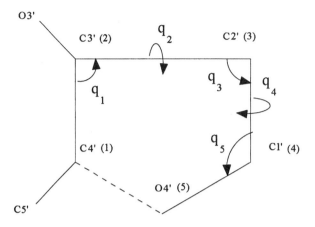

Figure 3 The underlying tree of a furanose ring in nucleic acids. Atoms are numbered 1, . . . , 5 corresponding to the natural tree ordering. All bond lengths are fixed. Arrows illustrate five internal coordinates that determine the ring conformation.

Here and below, $\mathbf{r}_i$, $\mathbf{l}_{ij}$, and $\mathbf{e}_{ij}$, $i, j = 1, . . . , 5$, denote atomic position vectors, atom–atom distances, and the corresponding unit vectors, respectively. In order to construct a correctly closed conformation, variables $q_1, . . . , q_4$ are considered independent, and the last valence angle q_5 is computed from Eq. (7) as follows. Variables $q_1, . . . , q_4$ determine the orientation of the plane of q_5 specified by vector $\mathbf{e}_{34}$ and an in-plane unit vector $\mathbf{e}_{345}$ orthogonal to it. In the basis of these two vectors, condition (7) results in

$$x(\mathbf{e}_{14}\mathbf{e}_{34}) + y(\mathbf{e}_{14}\mathbf{e}_{345}) = \frac{l_{15}^2 - l_{45}^2 - l_{14}^2}{2l_{14}l_{45}} \tag{8a}$$

$$x^2 + y^2 = 1 \tag{8b}$$

where x and y are local coordinates of vector $\mathbf{e}_{45}$. This system is reduced to a square equation and gives a single $x > 0$ solution, which solves the problem.

When equations of motion are integrated, all five generalized coordinates shown in Figure 3 are considered independent. The constraint condition of Eq. (7) means, however, that there is an additional reaction force applied between atoms C4′ and O4′. Such forces in all sugar rings result in a generalized reaction force $\mathbf{f}^\perp$ that has to be added to other forces in the system. Reactions depend upon both coordinates and velocities, but it appears, fortunately, that their explicit calculation is unnecessary. It is sufficient that the components of velocities along constrained bonds be canceled, which is achieved by projecting the vector of generalized velocities, predicted with constraints ignored, upon a certain multidimensional plane. Integrator (6) is modified as follows:

$$\mathbf{f}_n = \mathbf{f}(\mathbf{q}_n) \tag{9a}$$

$$\circ \quad \mathbf{q}_{n+1/2} = \mathbf{q}_n + \dot{\mathbf{q}}_{n+1/2}\frac{h}{2} \tag{9b}$$

$$\circ \quad \tilde{\mathbf{p}}_{n+1/2} = \mathbf{p}_{n-1/2} + \mathbf{f}_n h + (\mathbf{w}_{n-1/2} \, \mathbf{w}_{n+1/2})\frac{h}{2} + \mathbf{f}_{n-1/2}^\perp \frac{h}{2} \tag{9c}$$

$$\circ \quad \dot{\mathbf{q}}_{n+1/2} = \mathbf{T}_{n+1/2}\mathbf{M}_{n+1/2}^{-1}\tilde{\mathbf{p}}_{n+1/2} \tag{9d}$$

$$\mathbf{p}_{n+1/2} = \mathbf{M}_{n+1/2}\dot{\mathbf{q}}_{n+1/2} \tag{9e}$$

$$\mathbf{f}_{n+1/2}^{\perp}\frac{h}{2} = \mathbf{p}_{n+1/2} - \tilde{\mathbf{p}}_{n+1/2} \tag{9f}$$

$$\mathbf{q}_{n+1} = \mathbf{q}_n + \dot{\mathbf{q}}_{n+1/2}\,h \tag{9g}$$

where $\mathbf{T}$ is the matrix of the corresponding projection operator. It is computed for a half-step conformation with correctly closed rings. These additional computations only slightly reduce the net performance because it is still largely determined by the cost of evaluation of atom–atom forces [54].

V. PRACTICAL EXAMPLES

A. Time Step Limitations

Testing time step limitations plays an important role in ICMD because, in many cases, several alternative models of the same system can be constructed, with different spectra of fast motions. In general, in MD the step-size maximum depends on the system being studied, and for a given algorithm it is determined by its ability to conserve the total energy in microcanonical ensemble conditions [26,30]. For the leapfrog discretization the most appropriate method consists in checking the variation of the average total energy computed with different time steps [55]. The same test trajectory is computed starting from the same constant-energy hypersurface, and the average total energy is compared with the "ideal" value, i.e., its low time step limit. With growing time steps, the average total energy normally deviates upward, and a deviation of $0.2D[U]$, where U is the potential energy and $D[\cdot]$ denotes the operator of time variance, is taken as an upper acceptable level. The step-size maximum thus determined is denoted as h_c and is said to be "characteristic."

Figure 4 shows the results of two such time step tests for a hydrated B-DNA dodecamer duplex [54]. All bases were rigid except for rotation of thymine methyls. Bond lengths were fixed as well as all bond angles except those centered at sugar ring atoms. It is always interesting to check whether the time step is limited by harmonic or anharmonic motions. To distinguish them, virtually harmonic conditions are simulated by reducing the temperature to very low values so that the time step becomes limited by the highest frequency normal mode. In Figure 4a, for instance, the two traces corresponding to low and normal temperatures leave the band of acceptable deviation with a similar time step of around 4 fs, which indicates the harmonic nature of the limitation. The expected fastest harmonic mode in this case is the scissors H—C—H vibration with a frequency around 1600 cm^{-1}, which according to theory [55] should limit h_c to approximately 3.6 fs. In order to raise h_c to the level of 10 fs, inertias of hydrogen-only rigid bodies, as well as rigid bodies in flexible rings, are increased by different empirically adjusted increments. In the case of a scissors hydrogen, for example, an inertia $\mathbf{I}_{ij} = \mu\delta_{ij}$ is added at the position of the carbon atom, where δ_{ij} is the Kronecker delta and $\mu = 9$ amu $\cdot$ Å^2. This means that the hydrogen is no longer considered as a point mass but as a rigid body of the same, but redistributed, mass, which helps to scale down the scissors frequency by a factor of 3.

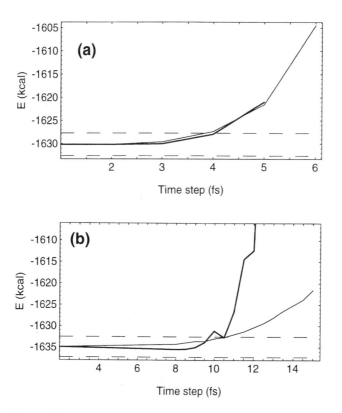

Figure 4 Time step dependence of the average total energy for two models of a partially hydrated dodecamer DNA duplex. Thinner traces show results for virtually harmonic conditions when temperature was lowered to 1 K. The DNA molecule has fixed bond lengths, rigid bases, and fixed valence angles except for the intra- and extracyclic bond angles in sugars. (a) No modifications of inertia; (b) inertia modified as explained in the text. (From Ref. 54.)

Testing of the resulting model system is shown in Figure 4b. We see that both the low and room temperature h_c values have increased to the desired 10 fs level.

Modification of inertia of hydrogen-only rigid bodies is a simple and safe way to balance different frequencies in the system, and it usually allows one to raise h_c to 10 fs. Unfortunately, the further increase appears problematic because of various anharmonic effects produced by collisions between non-hydrogen atoms [48].

B. Standard Geometry Versus Unconstrained Simulations

In our last example we return to the issue of the possible damaging effects of the standard geometry constraints. Two long trajectories have been computed for a partially hydrated dodecamer DNA duplex of the previous example, first by using ICMD and second with Cartesian coordinate molecular dynamics without constraints [54]. Both trajectories started from the same initial conformation with RMSD of 2.6 Å from the canonical B-DNA form. Figure 5 shows the time evolution of RMSD from the canonical A and B conformations. Each point in the figure corresponds to a 15 ps interval and shows an average RMSD value. We see that both trajectories approach the canonical B-DNA, while the RMSD

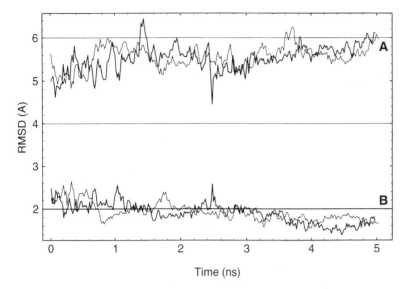

Figure 5 Time dependence of RMSD of atomic coordinates from canonical A- and B-DNA forms in two trajectories of a partially hydrated dodecamer duplex. The A and B (A and B correspond to A and B forms) trajectories started from the same state and were computed with internal and Cartesian coordinates as independent variables, respectively. (From Ref. 54.)

from A-DNA increases and reaches the level corresponding to the difference between the canonical A and B forms. The RMSD from B-DNA falls below the 2 Å level, and in both cases the final RMSD from the crystallographic conformation is around 1.3 Å. The RMSD between the two final computed states is around 1.1 Å, which is within the range of short time scale fluctuations in dynamics, while the overall drift from the initial state goes beyond 2.5 Å.

These two duplex models have 646 and 2264 internal degrees of freedom, respectively. In spite of this large difference they show very similar behavior in terms of atomic position fluctuations as well as in terms of fluctuations of torsions, sugar pseudorotation, and DNA helical parameters [54]. Apparently, the standard geometry model, which is allowed to move only along narrow paths in the full unconstrained configurational space, still keeps enough low energy tracks to sample from the main areas defined by a given temperature of 300 K. This example shows that the differences between the trajectories computed by ICMD and Cartesian MD at least are not readily seen, and, probably, for many applications they are not essential. It should be noted at the same time that the Cartesian coordinate trajectory was computed with a lower time step of 2 fs and took nearly five times as much computer time.

VI. CONCLUDING REMARKS

Internal coordinate molecular modeling is an efficient instrument with specific advantages that make it an indispensable complement to other existing approaches. It is best suited for simulation and analysis of large-scale structural transformations in biomacro-molecules, and at present ICMD is generally considered the most powerful tool in conformational searches, notably in NMR-based structural refinement [56]. Its application to physical problems involves certain controversial and unclear aspects that hold significant theoretical interest for future studies. The slow but steady progress in the development of these methods in recent years suggests that their performance and scope of application will continue to grow.

REFERENCES

1. HA Scheraga. Chem Rev 71:195–217, 1971.
2. KD Gibson, HA Scheraga. In: RH Sarma, MH Sarma, eds. Structure and Expression, Vol 1, From Proteins to Ribosomes. New York: Adenine Press, 1988, pp 67–94.
3. N Gō, T Noguti, T Nishikawa. Proc Natl Acad Sci USA 80:3696–3700, 1983.
4. RA Abagyan. In: WF van Gunsteren, PK Weiner, AJ Wilkinson, eds. Computer Simulation of Biomolecular Systems. Dordrecht: Kluwer, 1998, pp 363–394.
5. R Lavery. Adv Comput Biol 1:69–145, 1994.
6. N Gō, HA Scheraga. J Chem Phys 51:4751–4767, 1969.
7. R Abagyan, M Totrov, D Kuznetsov. J Comput Chem 15:488–506, 1994.
8. LM Rice, AT Brünger. Proteins: Struct Funct Genet 19:277–290, 1994.
9. N Gō, HA Scheraga. Macromolecules 9:535–542, 1976.
10. WF van Gunsteren. Mol Phys 40:1015–1019, 1980.
11. M Karplus, JN Kushick. Macromolecules 14:325–332, 1981.
12. WF van Gunsteren, M Karplus. Macromolecules 15:1528–1544, 1982.
13. IK Roterman, KD Gibson, HA Scheraga. J Biomol Struct Dyn 7:391–419, 1989.
14. IK Roterman, MH Lambert, KD Gibson, HA Scheraga. J Biomol Struct Dyn 7:421–453, 1989.
15. PA Kollman, KK Dill. J Biomol Struct Dyn 8:1103–1107, 1991.
16. KD Gibson, HA Scheraga. J Biomol Struct Dyn 8:1108–1111, 1991.
17. M Fixman. J Chem Phys 69:1527–1537, 1978.
18. E Helfand. J Chem Phys 71:5000–5007, 1979.
19. MR Pear, JH Weiner. J Chem Phys 71:212–224, 1979.
20. NG van Kampen. Phys Rep 124:69–160, 1985.
21. S Toxvaerd. J Chem Phys 87:6140–6143, 1987.
22. A Jain. J Comput Phys 136:289–297, 1997.
23. RA Laskowski, DS Moss, JM Thornton. J Mol Biol 231:1049–1067, 1993.
24. RA Marcus. Ber Bunsenges Phys Chem 136:190–197, 1977.
25. M Ben-Nun, RD Levine. J Chem Phys 105:8136–8141, 1996.
26. MP Allen, DJ Tildesley. Computer Simulation of Liquids. Oxford, UK: Clarendon Press, 1987.
27. H Goldstein. Classical Mechanics. Reading, MA: Addison-Wesley, 1980.
28. H Abe, W Braun, T Noguti, N Gō. Comput Chem 8:239–247, 1984.
29. RW Hockney, JW Eastwood. Computer Simulation Using Particles. New York: McGraw-Hill, 1981.
30. JM Haile. Molecular Dynamics Simulations: Elementary Methods. New York: Wiley-Interscience, 1992.
31. JP Ryckaert, A Bellemans. Chem Phys Lett 30:123–125, 1975.
32. JP Ryckaert, G Ciccotti, HJC Berendsen. J Comput Phys 23:327–341, 1977.
33. E Barth, K Kuczera, B Leimkuhler, RD Skeel. J Comput Chem 16:1192–1209, 1995.
34. JW Perram, HG Petersen. Mol Phys 65:861–874, 1988.
35. A Jain, N Vaidehi, G Rodriguez. J Comput Phys 106:258–268, 1993.
36. AM Mathiowetz, A Jain, N Karasawa, WA Goddard III. Proteins: Struct Funct Genet 20:227–247, 1994.
37. GR Kneller, K Hinsen. Phys Rev E 50:1559–1564, 1994.

38. AK Mazur, RA Abagyan. J Biomol Struct Dyn 6:815–832, 1989.
39. AK Mazur, VE Dorofeyev, RA Abagyan. J Comput Phys 92:261–272, 1991.
40. VE Dorofeyev, AK Mazur. J Comput Phys 107:359–366, 1993.
41. KD Gibson, HA Scheraga. J Comput Chem 11:468–486, 1990.
42. AK Mazur. J Comput Chem 18:1354–1364, 1997.
43. AF Vereshchagin. Eng Cybernet 6:1343–1346, 1974.
44. R Featherstone. Robot Dynamics Algorithms. Boston: Kluwer, 1987.
45. G Rodriguez. IEEE J Robot Automat RA-3:624–639, 1987.
46. G Rodriguez, K Kreutz-Delgado. IEEE Trans Robot Automat 8:65–75, 1992.
47. WF van Gunsteren, HJC Berendsen. Angew Chem 29:992–1023, 1990.
48. AK Mazur. J Phys Chem B 102:473–479, 1998.
49. JM Sanz-Serna, MP Calvo. Numerical Hamiltonian Problems. London: Chapman and Hall, 1994.
50. VI Arnold. Mathematical Methods of Classical Mechanics. New York: Springer-Verlag, 1978.
51. A Jain. J Guidance, Control Dyn 14:531–542, 1991.
52. RA Abagyan, AK Mazur. J Biomol Struct Dyn 6:833–845, 1989.
53. AK Mazur. J Chem Phys 111:1407–1414, 1999.
54. AK Mazur. J Am Chem Soc 120:10928–10937, 1998.
55. AK Mazur. J Comput Phys 136:354–365, 1997.
56. P Güntert. Quart Rev Biophys 31:145–237, 1998.

7
Implicit Solvent Models

Benoît Roux
Weill Medical College of Cornell University, New York, New York

I. INTRODUCTION

An understanding of a wide variety of phenomena concerning conformational stabilities and molecule–molecule association (protein–protein, protein–ligand, and protein–nucleic acid) requires consideration of solvation effects. In particular, a quantitative assessment of the relative contribution of hydrophobic and electrostatic interactions in macromolecular recognition is a problem of central importance in biology.

There is no doubt that molecular dynamics simulations in which a large number of solvent molecules are treated explicitly represent one of the most detailed approaches to the study of the influence of solvation on complex biomolecules [1]. The approach, which is illustrated schematically in Figure 1, consists in constructing detailed atomic models of the solvated macromolecular system and, having described the microscopic forces with a potential function, applying Newton's classical equation $F = ma$ to literally ''simulate'' the dynamic motions of all the atoms as a function of time [1,2]. The calculated classical trajectory, though an approximation to the real world, provides ultimate detailed information about the time course of the atomic motions, which is difficult to access experimentally. However, statistical convergence is an important issue because the net influence of solvation results from an averaging over a large number of configurations. In addition, a large number of solvent molecules are required to realistically model a dense system. Thus, in practical situations a significant fraction of the computer time is used to calculate the detailed trajectory of the solvent molecules even though it is often the solute that is of interest.

An alternative approach, illustrated schematically in Figure 2, consists in incorporating the influence of the solvent implicitly. Such approximate schemes can provide useful quantitative estimates of solvation free energies while remaining computationally tractable. Implicit solvent approaches avoid the statistical errors associated with averages extracted from simulations with a large number of solvent molecules. Furthermore, implicit solvent models are sometimes better suited for particularly complex situations. For example, an explicit representation of the cellular membrane potential would require prohibitively large atomic simulation systems and is currently impractical. Finally, implicit solvent representations can be very useful conceptual tools for analyzing the results of

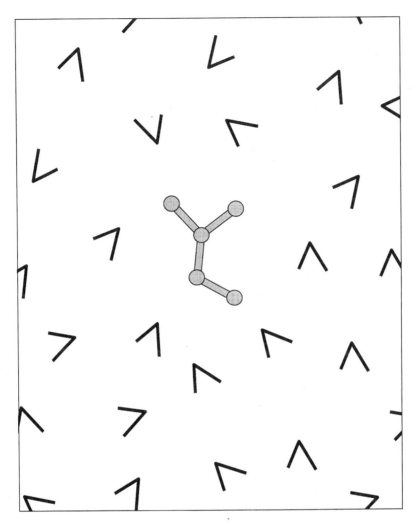

Figure 1 Schematic representation of an atomic model of a biomolecular solute surrounded by explicit water molecules.

simulations generated with explicit solvent molecules and to better understand the nature of solvation phenomena in general. The complexity of the environment in which biomolecules must perform their functions is such that information extracted from simple theoretical models may be helpful to further our understanding of these systems.

In this chapter we provide an introductory overview of the implicit solvent models commonly used in biomolecular simulations. A number of questions concerning the formulation and development of implicit solvent models are addressed. In Section II, we begin by providing a rigorous formulation of implicit solvent from statistical mechanics. In addition, the fundamental concept of the potential of mean force (PMF) is introduced. In Section III, a decomposition of the PMF in terms of nonpolar and electrostatic contributions is elaborated. Owing to its importance in biophysics, Section IV is devoted entirely to classical continuum electrostatics. For the sake of completeness, other computational

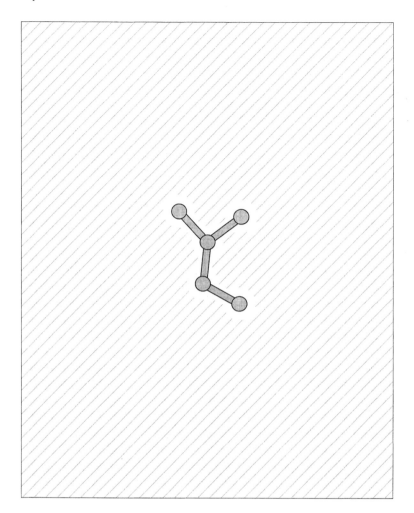

Figure 2   Schematic representation of a biomolecular solute in a solvent environment that is taken into account implicitly.

schemes such as statistical mechanical integral equations, implicit/explicit solvent boundary potential, solvent-accessible surface area (SASA), and knowledge-based potentials are briefly reviewed in Section V. Finally, the chapter is concluded in Section VI with a short summary of the principal ideas.

II.   BASIC FORMULATION OF IMPLICIT SOLVENT

A.   The Potential of Mean Force

As a first step, it is important to establish implicit solvent models on fundamental principles. For the sake of concreteness, let us consider a solute **u** immersed in a bulk solution **v**. The configuration of the solute is represented by the vector $\mathbf{X} \equiv \{\mathbf{x}_1, \mathbf{x}_2, \ldots\}$. All other degrees of freedom of the bulk solution surrounding the solute, which may include solvent molecules as well as mobile counterions, are represented by the vector $\mathbf{Y}$. It is expected that the system is fluctuating over a large number of configurations. It is therefore necessary to consider the problem from a statistical point of view. For a system in equilibrium with a thermal bath at temperature T, the probability of a given configuration $(\mathbf{X}, \mathbf{Y})$ is given by the function $P(\mathbf{X}, \mathbf{Y})$ [3],

$$P(\mathbf{X}, \mathbf{Y}) = \frac{\exp\{-U(\mathbf{X},\mathbf{Y})/k_BT\}}{\int d\mathbf{X}\, d\mathbf{Y}\, \exp\{-U(\mathbf{X},\mathbf{Y})/k_BT\}} \tag{1}$$

where $U(\mathbf{X}, \mathbf{Y})$ is the total potential energy of the system. For the sake of simplicity, we neglect nonadditive interactions and assume that the total potential energy can be written as

$$U(\mathbf{X}, \mathbf{Y}) = U_u(\mathbf{X}) + U_{vv}(\mathbf{Y}) + U_{uv}(\mathbf{X}, \mathbf{Y}) \tag{2}$$

where $U_u(\mathbf{X})$ is the intramolecular potential of the solute, $U_{vv}(\mathbf{Y})$ is the solvent–solvent potential, and $U_{uv}(\mathbf{X}, \mathbf{Y})$ is the solute–solvent potential. All observable properties of the system are fundamentally related to averages weighted by the probability function $P(\mathbf{X}, \mathbf{Y})$. For example, the average of any quantity $Q(\mathbf{X})$ depending on the solute configuration is given by

$$\langle Q \rangle = \int d\mathbf{X}\, d\mathbf{Y}\, Q(\mathbf{X})P(\mathbf{X}, \mathbf{Y}) \tag{3}$$

An important question is whether one can rigorously express such an average without referring explicitly to the solvent degrees of freedom. In other words, Is it possible to avoid explicit reference to the solvent in the mathematical description of the molecular system and still obtain rigorously correct properties? The answer to this question is yes. A reduced probability distribution $\bar{P}(\mathbf{X})$ that depends only on the solute configuration can be defined as

$$\bar{P}(\mathbf{X}) = \int d\mathbf{Y}\, P(\mathbf{X}, \mathbf{Y}) \tag{4}$$

The reduced probability distribution does not depend explicitly on the solvent coordinates $\mathbf{Y}$, although it incorporates the average influence of the solvent on the solute. The operation symbolized by Eq. (4) is commonly described by saying that the solvent coordinates $\mathbf{Y}$ have been "integrated out." In a system at temperature T, the reduced probability has the form

$$\bar{P}(\mathbf{X}) = \frac{\int d\mathbf{Y}\, \exp\{-[U_u(\mathbf{X}) + U_{vv}(\mathbf{Y}) + U_{uv}(\mathbf{X}, \mathbf{Y})]/k_BT\}}{\int d\mathbf{X}\, d\mathbf{Y}\, \exp\{-[U_u(\mathbf{X}) + U_{vv}(\mathbf{Y}) + U_{uv}(\mathbf{X}, \mathbf{Y})]/k_BT\}} \tag{5}$$

$$= \frac{\exp\{-W(\mathbf{X})/k_BT\}}{\int d\mathbf{X}\, \exp\{-W(\mathbf{X})/k_BT\}}$$

The function $W(\mathbf{X})$ is called the potential of mean force (PMF). The fundamental concept of the PMF was first introduced by Kirkwood [4] to describe the average structure of liquids. It is a simple matter to show that the gradient of $W(\mathbf{X})$ in Cartesian coordinates is related to the average force,

$$\frac{\partial W(\mathbf{X})}{\partial \mathbf{x}_i} = \left\langle \frac{\partial U}{\partial \mathbf{x}_i} \right\rangle_{(\mathbf{X})} \equiv -\langle \mathbf{F}_i \rangle_{(\mathbf{X})} \tag{6}$$

where the symbol $\langle \cdots \rangle_{(X)}$ represents an average over all coordinates of the solvent, with the solute in the fixed configuration specified by $\mathbf{X}$. All solvent effects are included in $W(\mathbf{X})$ and consequently in the reduced distribution function $\overline{P}(\mathbf{X})$. The PMF is an effective configuration-dependent free energy potential $W(\mathbf{X})$ that makes no explicit reference to the solvent degrees of freedom, such that no information about the influence of solvent on equilibrium properties is lost.

As long as the normalization condition given by Eq. (5) is satisfied, an arbitrary offset constant may be added to $W(\mathbf{X})$ without affecting averages in Eq. (3). The absolute value of the PMF is thus unimportant. For convenience, it is possible to choose the value of the free energy $W(\mathbf{X})$ relative to a reference system from which the solute–solvent interactions are absent. The free energy $W(\mathbf{X})$ may thus be expressed as

$$\exp\{-W(\mathbf{X})/k_\mathrm{B}T\} = \frac{\int d\mathbf{Y}\,\exp\{-[U_\mathrm{u}(\mathbf{X}) + U_\mathrm{vv}(\mathbf{Y}) + U_\mathrm{uv}(\mathbf{X},\,\mathbf{Y})]/k_\mathrm{B}T\}}{\int d\mathbf{Y}\,\exp\{-U_\mathrm{vv}(\mathbf{Y})/k_\mathrm{B}T\}} \tag{7}$$

It is customary to write $W(\mathbf{X}) = U_\mathrm{u}(\mathbf{X}) + \Delta W(\mathbf{X})$, where $U_\mathrm{u}(\mathbf{X})$ is the intramolecular solute potential and $\Delta W(\mathbf{X})$ is the solvent-induced influence. In practice, ΔW depends on $\mathbf{X}$, the configuration of the solute, as well as on thermodynamic variables such as the temperature T and the pressure p.

B. Relative and Absolute Values: Reversible Work

As shown by Eq. (6) the PMF is the reversible work done by the average force. It is possible to express relative values of the PMF between different solute configurations $\mathbf{X}_1$ and $\mathbf{X}_2$ using Eq. (6) and the reversible work theorem [4]:

$$W(\mathbf{X}_2) = W(\mathbf{X}_1) + \int_{\mathbf{X}_1}^{\mathbf{X}_2} \sum_i d\mathbf{x}_i \cdot \frac{\partial W(\mathbf{X})}{\partial \mathbf{x}_i} \tag{8}$$

$$= W(\mathbf{X}_1) - \int_{\mathbf{X}_1}^{\mathbf{X}_2} \sum_i d\mathbf{x}_i \cdot \langle \mathbf{F}_i \rangle_{(X)}$$

This relationship makes it clear that the PMF is not equal to an average potential energy because one needs to compute a reversible work against an average force to get $W(\mathbf{X})$. It is also possible to express the free energy in terms of a thermodynamic integral. Introducing the thermodynamic solute–solvent coupling parameter λ [4], we write the potential energy as

$$U(\mathbf{X},\,\mathbf{Y};\,\lambda) = U_\mathrm{u}(\mathbf{X}) + U_\mathrm{vv}(\mathbf{Y}) + U_\mathrm{uv}(\mathbf{X},\,\mathbf{Y};\,\lambda) \tag{9}$$

constructed such that $\lambda = 0$ corresponds to a noninteracting reference system with $U_\mathrm{uv}(\mathbf{X},\,\mathbf{Y};\,0) = 0$ and $\lambda = 1$ corresponds to the fully interacting system. As long as the end points are respected, any form of thermodynamic coupling is correct. Therefore, we have

$$\Delta W(\mathbf{X}) = \int_0^1 d\lambda \left\langle \frac{\partial U_\mathrm{uv}}{\partial \lambda} \right\rangle_{(X,\lambda)} \tag{10}$$

where the symbol $\langle \cdots \rangle_{(X,\lambda)}$ represents an average over all coordinates of the solvent for a solute in the fixed configuration $\mathbf{X}$ with thermodynamic coupling λ. It may be noted that $\partial U_{uv}/\partial\lambda$ in Eq. (10) plays the role of a generalized thermodynamic force similar to that of $\partial U/\partial\mathbf{x}_i$ in Eq. (8).

III. DECOMPOSITION OF THE FREE ENERGY

Intermolecular forces are dominated by short-range harsh repulsive interactions arising from Pauli's exclusion principle, van der Waals attractive forces arising from quantum dispersion, and long-range electrostatic interactions arising from the nonuniform charge distribution. It is convenient to express the potential energy $U_{uv}(\mathbf{X}, \mathbf{Y})$ as a sum of electrostatic contributions and the remaining nonpolar (nonelectrostatic) contributions,

$$U_{uv}(\mathbf{X}, \mathbf{Y}) = U_{uv}^{(np)}(\mathbf{X}, \mathbf{Y}) + U_{uv}^{(elec)}(\mathbf{X}, \mathbf{Y}) \tag{11}$$

Although such a representation of the microscopic non-bonded interactions does not follow directly from a quantum mechanical description of the Born–Oppenheimer energy surface, it is commonly used in most force fields for computer simulations of biomolecules (e.g., AMBER [5], CHARMM [6], OPLS [7]). The separation of the solute–solvent interactions in Eq. (11) is useful for decomposing the reversible work that defines the function $W(\mathbf{X})$. The total free energy of a solute in a fixed configuration $\mathbf{X}$ may be expressed rigorously as the reversible thermodynamic work needed to construct the system in a step-by-step process. In a first step, the nonpolar solute–solvent interactions are switched ''on'' in the absence of any solute–solvent electrostatic interactions; in a second step, the solute–solvent electrostatic interactions are switched ''on'' in the presence of the solute–solvent nonpolar interactions. The solute is kept in the fixed configuration $\mathbf{X}$ throughout the whole process, and the intramolecular potential energy does not vary during this process. By construction, the total PMF is

$$W(\mathbf{X}) = U_{u}(\mathbf{X}) + \Delta W^{(np)}(\mathbf{X}) + \Delta W^{(elec)}(\mathbf{X}) \tag{12}$$

where the nonpolar solvation contribution is

$$\exp\{-\Delta W^{(np)}(\mathbf{X})/k_BT\} = \frac{\int d\mathbf{Y}\, \exp\{-[U_{vv}(\mathbf{Y}) + U_{uv}^{(np)}(\mathbf{X}, \mathbf{Y})]/k_BT\}}{\int d\mathbf{Y}\, \exp\{-U_{vv}(\mathbf{Y})/k_BT\}} \tag{13}$$

and the electrostatic solvation contribution is

$$\exp\{-\Delta W^{(elec)}(\mathbf{X})/k_BT\} = \frac{\int d\mathbf{Y}\, \exp\{-[U_{vv}(\mathbf{Y}) + U_{uv}^{(np)}(\mathbf{X}, \mathbf{Y}) + U_{uv}^{(elec)}(\mathbf{X}, \mathbf{Y})]/k_BT\}}{\int d\mathbf{Y}\, \exp\{-[U_{vv}(\mathbf{Y}) + U_{uv}^{(np)}(\mathbf{X}, \mathbf{Y})]/k_BT\}} \tag{14}$$

Combining Eqs. (12)–(14) yields Eq. (7) directly. Although such a free energy decomposition is path-dependent [8], it provides a useful and rigorous framework for understanding the nature of solvation and for constructing suitable approximations to the nonpolar and electrostatic free energy contributions.

In the following sections, we describe an implicit solvent model based on this free energy decomposition that is widely used in biophysics. It consists in representing the nonpolar free energy contributions on the basis of the solvent-accessible surface area

(SASA), a concept introduced by Lee and Richards [9], and the electrostatic free energy contribution on the basis of the Poisson–Boltzmann (PB) equation of macroscopic electrostatics, an idea that goes back to Born [10], Debye and Hückel [11], Kirkwood [12], and Onsager [13]. The combination of these two approximations forms the SASA/PB implicit solvent model. In the next section we analyze the microscopic significance of the nonpolar and electrostatic free energy contributions and describe the SASA/PB implicit solvent model.

A. Nonpolar Free Energy Contribution

To clarify the significance of $\Delta W^{(np)}$, let us first consider the special case of a nonpolar molecule solvated in liquid water. We assume that the electrostatic free energy contribution is negligible. Typically, the solute–solvent van der Waals dispersion interactions are relatively weak and the nonpolar free energy contribution is dominated by the reversible work needed to displace the solvent molecules to accommodate the short-range harsh repulsive solute–solvent interaction. For this reason, $\Delta W^{(np)}$ is often refered to as the ''free energy of cavity formation.'' The reversible thermodynamic work corresponding to this process is positive and unfavorable. It gives rise to two aspects of the hydrophobic effect: hydrophobic solvation and hydrophobic interaction [14]. The former phenomenon is responsible for the poor solubility of nonpolar molecules in water; the latter accounts for the propensity of nonpolar molecules to cluster and form aggregates in water.

Modern understanding of the hydrophobic effect attributes it primarily to a decrease in the number of hydrogen bonds that can be achieved by the water molecules when they are near a nonpolar surface. This view is confirmed by computer simulations of nonpolar solutes in water [15]. To a first approximation, the magnitude of the free energy associated with the nonpolar contribution can thus be considered to be proportional to the number of solvent molecules in the first solvation shell. This idea leads to a convenient and attractive approximation that is used extensively in biophysical applications [9,16–18]. It consists in assuming that the nonpolar free energy contribution is directly related to the SASA [9],

$$\Delta W^{(np)}(\mathbf{X}) = \gamma_v \mathcal{A}_{tot}(\mathbf{X}) \tag{15}$$

where γ_v has the dimension of a surface tension and $\mathcal{A}_{tot}(\mathbf{X})$ is the configuration-dependent SASA (note that both polar and nonpolar chemical groups must be included in the SASA for a correct estimate of $\Delta W^{(np)}$). As pointed out by Tanford [19], there should be a close relationship between the macroscopic oil–water surface tension, interfacial free energies, and the magnitude of the hydrophobic effect. However, in practical applications, the surface tension γ_v is usually adjusted empirically to reproduce the solvation free energy of alkane molecules in water [18]. Its value is typically around 20–30 cal/(mol · $\overset{\circ}{A}^2$), whereas the macroscopic oil–water surface tension is around 70 cal/(mol · $\overset{\circ}{A}^2$) [19]. The difference between the optimal parameter γ_v for alkanes and the true macroscopic surface tension for oil/water interfaces reflects the influence of the microscopic length scale and the crudeness of the SASA model. A simple statistical mechanical approach describing the free energy of inserting hard spheres in water, called scaled particle theory (SPT), provides an important conceptual basis for understanding some of the limitations of SASA models [20–22]. It is clear that the SASA does not provide an ultimate representation of the nonpolar contribution to the solvation free energy. Other theories based on cavity distributions in liquid water [23,24] and long-range perturbation of water structure near large obstacles [25] are currently being explored. A quantitative description of the hydrophobic effect remains a central problem in theoretical chemical physics and biophysics.

B. Electrostatic Free Energy Contribution

The electrostatic free energy contribution in Eq. (14) may be expressed as a thermodynamic integration corresponding to a reversible process between two states of the system: no solute–solvent electrostatic interactions ($\lambda = 0$) and full electrostatic solute–solvent interactions ($\lambda = 1$). The electrostatic free energy has a particularly simple form if the thermodynamic parameter λ corresponds to a scaling of the solute charges, i.e., $U_{uv}^{(elec)}$ ($\mathbf{X}$, $\mathbf{Y}; \lambda$) = $\lambda U_{uv}^{(elec)}$ ($\mathbf{X}, \mathbf{Y}$), and the coupling is linear,

$$\Delta W^{elec}(\mathbf{X}) = \int_0^1 d\lambda \langle U_{uv}^{(elec)} \rangle_{(\lambda)} \tag{16}$$

For this reason, the quantity $\Delta W^{(elec)}(\mathbf{X})$ is often called the "charging free energy." If one assumes that the solvent responds linearly to the charge of the solute, then $\langle U_{uv}^{(elec)} \rangle_{(\lambda)}$ is proportional to λ and the charging free energy can be written as

$$\Delta W^{elec}(\mathbf{X}) = \int_0^1 d\lambda \sum_i q_i \Phi_{rf}(\mathbf{x}_i; \lambda) \approx \frac{1}{2} \sum_i q_i \Phi_{rf}(\mathbf{x}_i; \lambda = 1) \tag{17}$$

where $\Phi_{rf}(\mathbf{x}_i; \lambda = 1)$ is the solvent field acting on the ith solute atomic charge located at position $\mathbf{x}_i$ in reaction to the presence of all the solute charges (in the following, the coupling parameter λ will be omitted for the sake of simplicity). The "reaction field" is thus the electrostatic potential exerted on the solute by the solvent that it has polarized. The assumption of linear response implies that $\Delta W^{elec} = (1/2)\langle U_{uv}^{(elec)} \rangle$, a relationship, that is often observed in calculations based on simulations with explicit solvent [26,27]. The factor 1/2 is a characteristic signature of linear solvent response.

The dominant effects giving rise to the charging free energy are often modeled on the basis of classical continuum electrostatics. This approximation, in which the polar solvent is represented as a structureless continuum dielectric medium, was originally pioneered by Born in 1920 to calculate the hydration free energy of spherical ions [10]. It was later extended by Kirkwood [12] and Onsager [13] for the treatment of arbitrary charge distributions inside a spherical cavity. Nowadays, the treatment of solutes of arbitrary shape is possible with the use of powerful computers and numerical methods. In many cases, this is an excellent approximation. The classical electrostatics approach is remarkably successful in reproducing the electrostatic contribution to the solvation free energy of small solutes [26,28] or amino acids [27], as shown by comparisons to free energy simulations with explicit solvent. Applications to biophysical systems are reviewed in Refs. 29 and 30. Because of its importance the next section is devoted completely to this topic.

IV. CLASSICAL CONTINUUM ELECTROSTATICS

A. The Poisson Equation for Macroscopic Media

The continuum electrostatic approximation is based on the assumption that the solvent polarization density of the solvent at a position $\mathbf{r}$ in space is linearly related to the total local electric field at that position. The Poisson equation for macroscopic continuum media follows from those assumptions about the local and linear electrostatic response of the solvent [31]:

$$\nabla \cdot [\varepsilon(\mathbf{r})\nabla\phi(\mathbf{r})] = -4\pi\rho_u(\mathbf{r}) \tag{18}$$

where $\phi(\mathbf{r})$, $\rho_u(\mathbf{r})$, and $\varepsilon(\mathbf{r})$ are the electrostatic potential, the charge density of the solute, and the position-dependent dielectric constant at the point $\mathbf{r}$, respectively. The Poisson equation (18) can be solved numerically by mapping the system onto a discrete grid and using a finite-difference relaxation algorithm [32,33]. Several programs are available for computing the electrostatic potential using this approach, e.g., DelPhi [33,34], UHBD [35], and the PBEQ module [27,36] incorporated in the simulation program CHARMM [37]. Alternatively, one can use an approach based on finite elements distributed at the dielectric boundary (the boundary element method) [38]. Significant improvements can be obtained with this approach by using efficient algorithms for generating the mesh at the dielectric boundaries [39,41]. FAMBE is one program that is available to compute the electrostatic potential using this method [42]. Finally, a different (but physically equivalent) approach to incorporate the influence of a polar solvent, in which the solvent is modeled by a discrete lattice of dipoles that reorient under the influence of applied electric fields, has been proposed and developed by Warshel and coworkers [43].

It is generally assumed that the dielectric constant is uniform everywhere except in the vicinity of the solute/solvent boundary. If all the solute degrees of freedom are treated explicitly and the influence of induced electronic polarization is neglected, the position-dependent dielectric constant $\varepsilon(\mathbf{r})$ varies sharply from 1, in the interior of the solute, to ε_v in the bulk solvent region outside the solute. Such a form for $\varepsilon(\mathbf{r})$ follows rigorously from an analysis based on a statistical mechanical integral equation under the assumption that there are only short-range direct correlations in the solvent [44]. To estimate the electrostatic contribution to the solvation free energy, the reaction field Φ_{rf} used in Eq. (17) is obtained as the electrostatic potential calculated from Eq. (18) with the nonuniform dielectric constant $\varepsilon(\mathbf{r})$, minus the electrostatic potential calculated with a uniform dielectric constant of 1.

Results obtained using macroscopic continuum electrostatics for biomolecular solutes depend sensitively on atomic partial charges assigned to the nuclei and the location of the dielectric boundary between the solute and the solvent. The dielectric boundary can be constructed on the basis of the molecular surface [34,35] or the solvent-accessible surface (constructed as a surface formed by overlapping spheres) [27]. The parametrization of an accurate continuum electrostatic model thus requires the development of optimal sets of atomic radii for the solutes of interest. Various parametrization schemes aimed at reproducing the solvation free energy of a collection of molecules have been suggested [27,28]. From a fundamental point of view, the dielectric boundary is closely related to the nearest density peak in the solute–solvent distribution function [45]. As a consequence, the optimal radius of an atom is not a property of that atom alone but is an effective empirical parameter that depends on its charge, on its neighbors in the solute, and also on the nature of the molecules forming the bulk solvent. In contrast to the radii, the partial charges of the solute are generally taken from one of the standard biomolecular force fields without modification and are not considered as free parameters.

Continuum electrostatic approaches based on the Poisson equation have been used to address a wide variety of problems in biology. One particularly useful application is in the determination of the protonation state of titratable groups in proteins [46]. For further details readers are referred to the reviews of Honig and Nicholls [29] and Sharp and Honig [30].

B. Electrostatic Forces and Analytic Gradients

In most practical applications of continuum electrostatics, the solute is considered to be in a fixed conformation. However, this procedure has obvious limitations, because it ignores the importance of conformational flexibility. To proceed further requires knowledge of the "electrostatic solvation forces" associated with the continuum electrostatics description of the solvent, i.e., the analytic first derivative of the solvation free energy with respect to the atomic coordinates of the solute. The computation of analytic gradients of the free energy of solvation with respect to nuclear coordinates is important for efficient geometric optimization based on energy minization, conformational searches, and dynamics. Analytic gradients for finite-difference solutions to the Poisson equation have been presented by Gilson et al. [47] and Im et al. [36]. Boundary element methods can also be used very effectively for computing analytic gradients [48]. Nonetheless, when repeated evaluation of the solvation energy is requested, the solution to the classical electrostatic problem and the calculation of analytic gradients may be too expensive computationally. For this purpose, approximations to the exact continuum electrostatics based on semianalytical functions have been developed. This is possible, in principle, because the free energy can be expressed as a superposition of pairwise additive terms (which depends on the geometry of the solute/solvent dielectric interface) in virtue of the linearity of continuum electrostatics. The general strategy of semianalytical approaches is to design a suitable closed-form pairwise deshielding function for the charge–charge coupling. One of the most popular approximation is the generalized Born (GB) [49], although alternative formulations such as the field integrated electrostatic approach (FIESTA) [50], the inducible multipole solvation model (IMS) [51], the analytical continuum electrostatics approach (ACE) [52], and the solvation models (SMx) of Cramer and Truhlar [53] are also based on this general idea. Semianalytical approximations such as GB represent a very promising approach for implicitly incorporating the influence of the solvent in biomolecular simulations. Extensions and improvements to the original form of the GB deshielding function have been proposed and parametrized [54–58]. The results have been compared with those from numerical continuum electrostatic calculations [57–59] and explicit solvent simulations [60,61]. The GB approximation has been applied to various problems, e.g., protein and nucleic acid stability [62–64], conformational searches [65,66], macromolecular association [67], and ligand binding [68].

C. Treatment of Ionic Strength

The concentration of salt in physiological systems is on the order of 150 mM, which corresponds to approximately 350 water molecules for each cation–anion pair. For this reason, investigations of salt effects in biological systems using detailed atomic models and molecular dynamic simulations become rapidly prohibitive, and mean-field treatments based on continuum electrostatics are advantageous. Such approximations, which were pioneered by Debye and Hückel [11], are valid at moderately low ionic concentration when core–core interactions between the mobile ions can be neglected. Briefly, the spatial density throughout the solvent is assumed to depend only on the local electrostatic potential $\rho_i(\mathbf{r}) = \bar{\rho}_i \exp\{-q_i\phi(\mathbf{r})/k_B T\}$, where i refers to a specific ion type (e.g., counterion or co-ion) and $\bar{\rho}_i$ is the number density in the bulk solution. The total ion charge density (summed over the different ion types) is then inserted explicitly in the Poisson equation with the solute charge $\rho_u(\mathbf{r})$, resulting in the nonlinear form of the Poisson–Boltzmann (PB) equation. Linearization with respect to the potential ϕ yields the familiar Debye–Hückel approximation [11,69],

$$\nabla \cdot [\varepsilon(\mathbf{r})\nabla\phi(\mathbf{r})] - \bar{\kappa}^2(\mathbf{r})\phi(\mathbf{r}) = -4\pi\rho^{(u)}(\mathbf{r}) \tag{19}$$

where $\bar{\kappa}^2(\mathbf{r})$ is the space-dependent screening factor, which varies from zero in the interior of the solute to $4\pi\sum_i q_i^2 \bar{\rho}_i/k_B T$ in the bulk solvent. The spatial dependence of $\bar{\kappa}^2(\mathbf{r})$ is often assumed to be similar to that of $\varepsilon(\mathbf{r})$, though that is not necessary. The PB equation (linear and nonlinear) is a particularly simple and powerful approach to address questions about the influence of salt on complex biological systems. In particular, it has been used to examine the salt dependence of the conformational stability of nucleic acids [70,71] and protein–DNA association [72].

D. Treatment of a Transmembrane Potential

The electrostatic free energy of a macromolecule embedded in a membrane in the presence of a membrane potential V can be expressed as the sum of three separate terms involving the capacitance C of the system, the reaction field $\Phi_{rf}(\mathbf{r})$, and the membrane potential field $\Phi_{mp}(\mathbf{r})$ [73],

$$\Delta W^{elec} = \frac{1}{2}CV^2 + \frac{1}{2}\sum_i q_i\Phi_{rf}(\mathbf{x}_i) + \left[\sum_i q_i\Phi_{mp}(\mathbf{x}_i)\right]V \tag{20}$$

where q_i and $\mathbf{x}_i$ are the charge and position, respectively, of solute i. Generally, the capacitive energy contribution is negligible. The function $\Phi_{mp}(\mathbf{x}_i)$ corresponds to the fraction of the electrostatic transmembrane potential interacting with a charge of the solute. It is calculated by solving a modified version of the linear PB equation,

$$\nabla \cdot [\varepsilon(\mathbf{r})\nabla\Phi_{mp}(\mathbf{r})] - \bar{\kappa}^2(\mathbf{r})[\Phi_{mp}(\mathbf{r}) - \Theta(\mathbf{r})] = 0 \tag{21}$$

where the function $\Theta(\mathbf{r})$ is equal to 1 on the side of the membrane that is contact with the bulk solution set to the reference potential V, and zero otherwise. The Θ function in Eq. (21) ensures that the mobile ions are in equilibrium with the bath with which they are in contact. In the case of a perfectly planar system, the electric field across the membrane is constant and $\Phi_{mp}(\mathbf{x})$ is a linear function corresponding roughly to a fraction of the membrane thickness (for this reason, it is often referred to as the "electric distance" [74,75]). If the shape of the protein/solution interface is irregular, the interaction of the solute charges with the membrane potential is more complicated than the simple linear field.

Simple considerations show that the membrane potential cannot be treated with computer simulations, and continuum electrostatic methods may constitute the only practical approach to address such questions. The capacitance of a typical lipid membrane is on the order of 1 μF/cm^2, which corresponds to a thickness of approximately 25 Å and a dielectric constant of 2 for the hydrophobic core of a bilayer. In the presence of a membrane potential the bulk solution remains electrically neutral and a small charge imbalance is distributed in the neighborhood of the interfaces. The membrane potential arises from a strikingly small accumulation of net charge relative to the bulk ion density. Typical physiological conditions correspond to a membrane potential on the order of 100 mV and a salt concentration of 150 mM. In this situation, the net charge per unit area is $CV = 10^{-7}$ C/cm^2, which corresponds to only one atomic unit charge per (130 Å)^2 of surface. For molecular dynamics simulations, a minimal salt solution at a concentration of 150 mM with a membrane system of cross-sectional area (130 Å)^2 containing about 100 ion pairs would require nearly 50,000 water molecules and 500 phospholipid molecules, for a total of more than 200,000 atoms, which is computationally prohibitive. At the present time, the modified PB, Eq. (21), with membrane potential may provide the only practical way to address questions about the membrane potential and its influence on the configurational free energy of intrinsic protein. The approach has been implemented in the PBEQ module [27,36,73] of the biomolecular simulation program CHARMM [37] and has been used to calculate the influence of the transmembrane potential on the insertion of an α-helix into a membrane.

V. MISCELLANEOUS APPROACHES

A. Statistical Mechanical Integral Equations

The average solvent structure caused by granularity, packing, and hydrogen bonding gives rise to important effects that are ignored by continuum electrostatic approaches. Statistical mechanical theories based on distribution functions and integral equations are sophisticated approaches that can provide a rigorous framework for incorporating such effects into a description of solvation [3,76]. A complete review of integral equations would be beyond the scope of this chapter; therefore we provide only a brief overview of this vast field.

One important class of integral equation theories is based on the reference interaction site model (RISM) proposed by Chandler [77]. These RISM theories have been used to study the conformation of small peptides in liquid water [78–80]. However, the approach is not appropriate for large molecular solutes such as proteins and nucleic acids. Because RISM is based on a reduction to site–site, solute–solvent radially symmetrical distribution functions, there is a loss of information about the three-dimensional spatial organization of the solvent density around a macromolecular solute of irregular shape. To circumvent this limitation, extensions of RISM-like theories for three-dimensional space (3d-RISM) have been proposed [81,82],

$$c_\alpha(\mathbf{r}) = \exp[-U_\alpha(\mathbf{r})/k_\mathrm{B}T + h_\alpha(\mathbf{r}) - c_\alpha(\mathbf{r})] - h_\alpha(\mathbf{r}) + c_\alpha(\mathbf{r}) - 1 \qquad (22)$$

and

$$\bar{\rho}h_\alpha(\mathbf{r}) = \int d\mathbf{r}' \sum_\gamma c_\gamma(\mathbf{r}')\chi_{\gamma\alpha}(\mathbf{r} - \mathbf{r}') \qquad (23)$$

where $U_\alpha(\mathbf{r})$ is the solute–solvent interaction on the solvent site α, $h_\alpha(\mathbf{r})$ is the solute–solvent site correlation function $h_\alpha(\mathbf{r}) \equiv [\rho_\alpha(\mathbf{r})/\bar{\rho} - 1]$, $c_\alpha(\mathbf{r})$ is the solute–solvent site direct correlation function, and $\chi_{\gamma\alpha}(\mathbf{r} - \mathbf{r}')$ is the density susceptibility of the uniform unperturbed liquid. The solvent susceptibility (an input in this approach) is related to the equilibrium site–site density susceptibility of the uniform unperturbed liquid. Numerical solutions of the 3d-RISM equation indicate that this approach is able to incorporate important features of hydration such as hydrogen bonding and packing of the solvent molecules in the first solvation shell [81,82]. Recent advances allow the accurate estimate of the solvation free energy for nonpolar as well as polar biomolecules [83].

Other statistical mechanical theories are also currently being explored. An extension to the mean spherical approximation integral equation in three dimensions (3d-MSA) describing the distribution function of a liquid of spherical molecules with an embedded dipole around a polar solute [44] as well as an integral equation describing the structure of water molecules in terms of sticky interaction points [84,85] were formulated and solved numerically. A theory based on an expansion in terms of two- and three-body correlation functions has been proposed to describe the hydration structure around nucleic acids [86] and proteins [87]. A theory for inhomogeneous fluids in the neighborhood of large nonpolar solutes was proposed to describe the hydrophobic effect [25].

B. Solvent Boundary Potentials and Implicit/Explicit Mixed Schemes

A description in which all atomic and structural details of the solvent molecules are ignored may not always be desirable. In some cases, it may be advantageous to use a mixed scheme that combines an implicit solvent model with a limited number of explicit solvent molecules. An intermediate approach, illustrated schematically in Figure 3, consists in including a small number of explicit solvent molecules in the vicinity of the solute while representing the influence of the remaining bulk with an effective solvent boundary potential [88–95]. The first to design such a simulation method appropriate for liquids were Berkowitz and McCammon [88]. In their method, the many-body system was divided into three main spherical regions: a central reaction region, a buffer region, and a surrounding static reservoir region. The forces arising from the reservoir region were calculated from fixed atomic centers. Instead of using explicit fixed atomic centers in the bath region, Brooks and Karplus introduced a mean force field approximation (MFFA) to calculate a soft boundary potential representing the average influence of the reservoir region on the reaction region [89]. In the MFFA treatment, the boundary potential was calculated by integrating all contributions to the average force arising from the reservoir region. The MFFA approach was extended by Brunger et al. [90] for the simulation of bulk water. A similar potential for water droplets of TIP4P was developed by Essex and Jorgensen [91]. The average electrostatic reaction field was taken into account in the surface constrained all-atom solvent (SCAAS) treatment of King and Warshel [93], and in the reaction field with exclusion (RFE) of Rullmann and van Duijnen [94].

The problem was reformulated on the basis of a separation of the multidimensional solute–solvent configurational integral in terms of ''inner'' solvent molecules nearest to the solute, and the remaining ''outer'' bulk solvent molecules [95]. Following this formulation, the solvent boundary potential was identified as the solvation free energy of an effective cluster comprising the solute and inner explicit solvent molecules embedded in a large hard sphere. The hard sphere corresponds to a configurational restriction on the outer bulk solvent molecules; its radius is variable, such that it includes the most distant inner solvent molecule. An approximate spherical solvent boundary potential (SSBP) based on this formulation has been implemented in the biomolecular simulation program CHARMM [37]. Using computer simulations it was shown that SSBP yields solvation free energies that do not depend sensitively on the number of explicit water molecules [95].

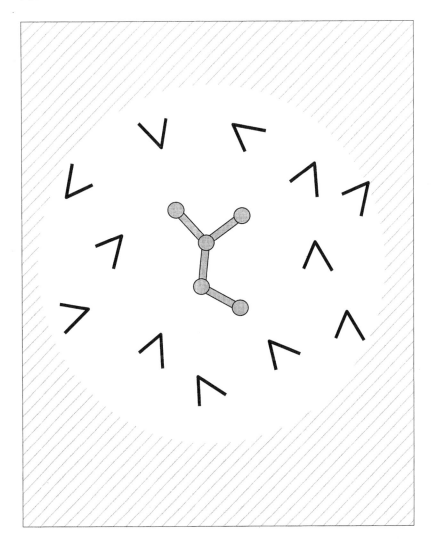

Figure 3 Schematic representation of a mixed explicit–implicit solvent treatment. A small number of water molecules are included explicitly in the vicinity of the solute while the influence of the remaining bulk is taken into account implicitly.

C. Solvent-Accessible Surface Area Models

In Section III we described an approximation to the nonpolar free energy contribution based on the concept of the solvent-accessible surface area (SASA) [see Eq. (15)]. In the SASA/PB implicit solvent model, the nonpolar free energy contribution is complemented by a macroscopic continuum electrostatic calculation based on the PB equation, thus yielding an approximation to the total free energy, $\Delta W = \Delta W^{(\mathrm{np})} + \Delta W^{(\mathrm{elec})}$. A different implicit solvent model, which also makes use of the concept of SASA, is based on the assumption that the entire solvation free energy of a solute can be expressed in terms of a linear sum of atomic contributions weighted by partial exposed surface area,

$$\Delta W(\mathbf{X}) = \sum_i \gamma_i \mathcal{A}_i(\mathbf{X}) \tag{24}$$

Here, $\mathcal{A}_i(\mathbf{X})$ is the partial SASA of atom i (which depends on the solute configuration $\mathbf{X}$), and γ_i is an atomic free energy per unit area associated with atom i. We refer to those models as "full SASA." Because it is so simple, this approach is widely used in computations on biomolecules [96–98]. Variations of the solvent-exposed area models are the shell model of Scheraga [99,100], the excluded-volume model of Colonna-Cesari and Sander [101,102], and the Gaussian model of Lazaridis and Karplus [103]. Full SASA models have been used for investigating the thermal denaturation of proteins [103] and to examine protein–protein association [104].

One important limitation of full SASA models is the difficulty of taking into account the dielectric shielding of electrostatic interactions between charged particles in a physically realistic way. The SASA model incorporates, in an average way, the free energy cost of taking a charged particle and burying it in the interior of the protein. In the continuum electrostatic description that corresponds to the self-interaction energy, i.e., the interaction of a charge with its own reaction field. However, as two charged particles move from the solvent to the nonpolar core of the protein, their electrostatic interaction should also vary progressively from a fully to an incompletely shielded form. Thus, full SASA approximations require further assumptions about the treatment of electrostatic interactions and dielectric shielding in practical applications. For example, in full SASA models residues carrying a net charge are usually neutralized and a distance-dependent dielectric function is introduced to shield the Coulomb potential at large distances [98,103].

D. Knowledge-Based Potentials

One of the greatest problems in predicting the three-dimensional fold of a protein is the need to search over a large number of possible configurations to find the global free energy minimum. For extensive configurational searches, it is necessary to use a free energy function $W(\mathbf{X})$ that is as simple and inexpensive as possible. Knowledge-based potentials are the simplest free energy functions that can be designed for this purpose. Such potentials are constructed empirically from statistical analyses of known protein structures taken from structural databases [105]. The general idea is that the number of residue pairs at a certain distance observed in the database follows the statistics of a thermal ensemble, in other words a Boltzmann principle [106]. Equivalently, it is assumed that the observed probability of finding a pair of residues at a distance R in a protein structure is related to the Boltzmann factor of an effective distance-dependent free energy. The simplest potentials distinguish only two types of residues: nonpolar and polar [105]. Usually no attempts are made to establish a realistic description of the microscopic interactions at the atomic level, though some comparisons have been made with explicit solvent simulations [60]. For example, one of the simplest potentials, designed by Sippl [105], is attractive for pairs of nonpolar residues and repulsive for pairs of polar residues. Nevertheless, the resulting structures that are obtained via conformational searches, usually with an additional restraint on the protein radius of gyration, are reasonable: The nonpolar residues tend to form a hydrophobic core in the center of the structure, whereas the polar residues tend to be located at the protein surface. A growing number of potentials are constructed on the basis of similar ideas [107–111]. In 1996, Mirny and Shakhnovich [112] reexamined the methods for deriving knowledge-based potentials for protein folding. Their potential is obtained by a global optimization procedure that simultaneously maximizes thermodynamic stability for all proteins in the database. This field is in rapid expansion, and it is beyond the scope of the present review to cover all possible developments. For more information, see Refs. 113–115 and references therein.

VI. SUMMARY

A statistical mechanical formulation of implicit solvent representations provides a robust theoretical framework for understanding the influence of solvation biomolecular systems. A decomposition of the free energy in terms of nonpolar and electrostatic contributions, $\Delta W = \Delta W^{(np)} + \Delta W^{(elec)}$, is central to many approximate treatments. An attractive and widely used treatment consists in representing the nonpolar contribution $\Delta W^{(np)}$ by a SASA surface tension term with Eq. (15) and the electrostatic contribution $\Delta W^{(elec)}$ by using the finite-difference PB, Eq. (19). These two approximations constitute the SASA/PB implicit solvent model. Although SASA/PB does not incorporate solvation effects with all atomic details, it nevertheless relies on a physically consistent picture of solvation. A relationship with first principles and statistical mechanics can be established, and the significance of the approximations at the microscopic level can be clarified. The results can be compared with computer simulations including explicit solvent molecules [15,26–28]. Implicit solvent models based on the SASA/PB approximation have been used to address a wide range of questions concerning biomolecular systems, e.g., to discriminate misfolded proteins [116], assess the conformational stability of nucleic acids [71], and examine protein–ligand [117], protein–DNA [72], and protein–membrane association [73,118].

It is possible to go beyond the SASA/PB approximation and develop better approximations to current implicit solvent representations with sophisticated statistical mechanical models based on distribution functions or integral equations (see Section V.A). An alternative intermediate approach consists in including a small number of explicit solvent molecules near the solute while the influence of the remain bulk solvent molecules is taken into account implicitly (see Section V.B). On the other hand, in some cases it is necessary to use a treatment that is markedly simpler than SASA/PB to carry out extensive conformational searches. In such situations, it possible to use empirical models that describe the entire solvation free energy on the basis of the SASA (see Section V.C). An even simpler class of approximations consists in using information-based potentials constructed to mimic and reproduce the statistical trends observed in macromolecular structures (see Section V.D). Although the microscopic basis of these approximations is not yet formally linked to a statistical mechanical formulation of implicit solvent, full SASA models and empirical information-based potentials may be very effective for particular problems.

REFERENCES

1. CL Brooks III, M Karplus, BM Pettitt. Proteins. A Theoretical Perspective of Dynamics, Structure and Thermodynamics. Adv Chem Physics, Vol 71. New York: Wiley, 1988.
2. MP Allen, DJ Tildesley. Computer Simulation of Liquids. Oxford, UK: Clarendon Press, 1989.
3. DA McQuarrie. Statistical Mechanics. New York: Harper and Row, 1976.

4. JG Kirkwood. J Chem Phys 3:300, 1935.
5. WD Cornell, P Cieplak, CI Bayly, IR Gould, KM Merz Jr, DM Ferguson, DC Spellmeyer, T Fox, JW Caldwell, PA Kollman. J Am Chem Soc 117:5179–5197, 1995.
6. AD MacKerell Jr, D Bashford, M Bellot, RL Dunbrack, JD Evanseck, MJ Field, S Fischer, J Gao, H Guo, S Ha, D Joseph-McCarthy, L Kuchnir, K Kuczera, FTK Lau, C Mattos, S Michnick, T Ngo, DT Nguyen, B Prodhom, WE Reiher III, B Roux, B Schlenkrich, J Smith, R Stote, J Straub, M Watanabe, J Wiorkiewicz-Kuczera, M Karplus. J Phys Chem B 102: 3586–3616, 1998.
7. WL Jorgensen, DS Maxwell, J Tirado-Rives. J Am Chem Soc 118:11225–11236, 1996.
8. S Boresch, G Archontis, M Karplus. Proteins 20:25, 1994.
9. B Lee, FM Richards. J Mol Biol 55:379, 1971.
10. M Born. Z Phys 1:45, 1920.
11. P Debye, E Hückel. Phys Z 24:305–325, 1923.
12. JG Kirkwood. J Chem Phys 2:351, 1934.
13. L Onsager. J Am Chem Soc 58:1468–1493, 1936.
14. K Lum, D Chandler, JD Weeks. Adv Protein Chem 14:1, 1959.
15. JP Postma, HC Berendsen, JR Haak. Faraday Symp Chem Soc 17:55, 1982.
16. K Sharp, A Nicholls, R Fine, B Honig. Science 252:106–109, 1991.
17. K Sharp, A Nicholls, R Friedman, B Honig. Biochemistry 30:9696–9697, 1991.
18. T Simonson, AT Brünger. J Phys Chem 98:4683–4694, 1994.
19. C Tanford. Proc Natl Acad Sci USA 76:4175–4176, 1979.
20. H Reiss. Adv Chem Phys 9:1–84, 1965.
21. F Stillinger. J Solut Chem 2:141–158, 1973.
22. RA Pierotti. Chem Rev 76:717–726, 1976.
23. G Hummer, S Garde, AE Garcia, A Pohorille, LR Pratt. Proc Natl Acad Sci USA 93:8951, 1996.
24. G Hummer, LR Pratt, AE Garcia. J Chem Phys 107:9275, 1997.
25. K Lum, D Chandler, JD Weeks. J Phys Chem 1999.
26. A Jean-Charles, A Nicholls, K Sharp, B Honig, A Tempczyk, T Hendrickson, WC Still. J Am Chem Soc 113:1454–1455, 1991.
27. M Nina, D Beglov, B Roux. J Phys Chem 101:5239–5248, 1997.
28. D Sitkoff, KA Sharp, B Honig. J Phys Chem 98:1978, 1994.
29. B Honig, A Nicholls. Science 268:1144, 1995.
30. KA Sharp, B Honig. Annu Rev Biophys Biophys Chem 19:301–332, 1990.
31. JD Jackson. Classical Electrodynamics. New York: Wiley, 1962.
32. J Warwicker, HC Watson. J Mol Biol 157:671–679, 1982.
33. I Klapper, R Hagstrom, R Fine, K Sharp, B Honig. Proteins 1:47, 1986.
34. MK Gilson, KA Sharp, BH Honig. J Comput Chem 9:327–335, 1987.
35. ME Davis, JD Madura, B Luty, JA McCammon. Comput Phys Commun 62:187–197, 1991.
36. W Im, D Beglov, B Roux. Comput Phys Commun 111:59–75, 1998.
37. BR Brooks, RE Bruccoleri, BD Olafson, DJ States, S Swaminathan, M Karplus. J Comput Chem 4:187–217, 1983.
38. R Zauhar, R Morgan. J Mol Biol 186:815–820, 1985.
39. J Liang, S Subramaniam. Biophys J 73:1830–1841, 1997.
40. CM Cortis, RA Friesner. J Comput Chem 18:1570–1590, 1997.
41. CM Cortis, RA Friesner. J Comput Chem 18:1591–1608, 1997.
42. YN Vorobjev, HA Scheraga. J Comput Chem 18:569–583, 1997.
43. A Warshel, J Åqvist. Annu Rev Biophys Biophys Chem 20:267–298, 1991.
44. D Beglov, B Roux. J Chem Phys 104:8678–8689, 1996.
45. B Roux, H-A Yu, M Karplus. J Phys Chem 94:4683–4688, 1990.
46. J Antosiewicz, JA McCammon, MK Gilson. Biochemistry 18:7819–7833, 1996.
47. MK Gilson, JA McCammon, JD Madura. J Comput Chem 16:1081, 1995.

48. M Friedrichs, RH Zhou, SR Edinger, RA Friesner. J Phys Chem B 103:3057–3061, 1999.
49. WC Still, A Tempczyk, RC Hawley, T Henderickson. J Am Chem Soc 112:6127–6129, 1990.
50. H Sklenar, F Eisenhaber, M Poncin, R Lavery. In: DL Beveridge, R Lavery, eds. Theoretical Biochemistry and Molecular Biophysics. New York: Adenine Press, 1990.
51. ME Davis. J Chem Phys 100:5149–5159, 1994.
52. M Schaefer, M Karplus. J Phys Chem 100:1578–1599, 1996.
53. CJ Cramer, DG Truhlar. In: KB Lipkowitz, ed. Reviews in Computational Chemistry. New York: VCH, 1995.
54. SR Edinger, C Cortis, PS Shenkin, RA Friesner. J Phys Chem B 101:1190, 1997.
55. B Jayaram, D Sprous, DL Beveridge. J Phys Chem B 102:9571–9576, 1998.
56. B Jayaram, Y Liu, DL Beveridge. J Chem Phys 109:1465–1471, 1998.
57. A Ghosh, CS Rapp, RA Friesner. J Phys Chem B 102:10983–10990, 1998.
58. BN Dominy, CL Brooks. J Phys Chem B 103:3765–3773, 1999.
59. SB Dixit, B Jayaram. J Biomol Struct Dyn 16:237–242, 1998.
60. D Mohanty, BN Dominy, A Kolinski, CL Brooks, J Skolnick. Proteins 35:447–452, 1999.
61. M Scarsi, J Apostolakis, A Caflisch. J Phys Chem B 102:3637–3641, 1998.
62. J Srinivasan, MW Trevathan, P Beroza, DA Case. Theor Chem Acc 101:426–434, 1999.
63. B Jayaram, KJ McConnell, SB Dixit, DL Beveridge. J Comput Phys 151:333–357, 1999.
64. R Luo, L David, H Hung, J Devaney, MK Gilson. J Phys Chem B 103:727–736, 1999.
65. J Srinivasan, J Miller, PA Kollman, DA Case. J Biomol Struct Dyn 16:671, 1998.
66. CS Rapp, RA Friesner. Proteins 35:173–183, 1999.
67. R Luo, MS Head, JA Given, MK Gilson. Biophys Chem 78:183–193, 1999.
68. J Bostrom, PO Norrby, T Liljefors. J Computer-Aided Mol Des 12:383–396, 1998.
69. RH Fowler, EA Guggenheim. Statistical Thermodynamics. Oxford, UK: Cambridge Univ Press, 1939.
70. KA Sharp, B Honig. Curr Opin Struct Biol 5:323–328, 1995.
71. VK Misra, B Honig. Biochemistry 35:1115–1124, 1996.
72. VK Misra, JL Hecht, AS Yang, B Honig. Biophys J 75:2262–2273, 1998.
73. B Roux. Biophys J 73:2980–2989, 1997.
74. B Hille. Ionic Channels of Excitable Membranes. 2nd ed. Sunderland, MA: Sinauer, 1992.
75. FJ Sigworth. Quart Rev Biophys 27:1–40, 1993.
76. JP Hansen, IR McDonald. Theory of Simple Liquids, 2nd ed. London: Academic Press, 1986.
77. D Chandler. The equilibrium theory of polyatomic fluids. In: EW Montroll, JL Lebowitz, eds. The Liquid State of Matter: Fluids, Simple and Complex, Vol. 8. Amsterdam: North-Holland, 1982.
78. MB Pettitt, M Karplus. Chem Phys Lett 121:194–201, 1985.
79. MB Pettitt, M Karplus. Chem Phys Lett 136:383–386, 1987.
80. WF Lau, BM Pettitt. Biopolymer 26:1817–1831, 1987.
81. D Beglov, B Roux. J Phys Chem 101:7821–7826, 1997.
82. CM Cortis, PJ Rossky, RA Friesner. J Chem Phys 107:6400–6414, 1997.
83. Q Du, D Beglov, B Roux. J Phys Chem B 104:796–805, 2000.
84. Y Liu, T Ichiye. Chem Phys Lett 231:380–386, 1994.
85. J-K Hyun, CS Babu, T Ichiye. J Chem Phys 99:5187–5185, 1995.
86. G Hummer, DM Soumpasis. Phys Rev E 50:5085–5095, 1994.
87. AE García, G Hummer, DM Soumpasis. Biophys J 72(2):A104, 1997.
88. M Berkowitz, JA McCammon. Chem Phys Lett 90:215, 1982.
89. CL Brooks III, M Karplus. J Chem Phys 79:6312–6325, 1983.
90. A Brunger, CL Brooks III, M Karplus. Chem Phys Lett 105:495–500, 1984.
91. JW Essex, WL Jorgensen. J Comput Chem 16:951–972, 1995.
92. A Warshel, G King. Chem Phys Lett 121:124, 1985.
93. G King, A Warshel. J Chem Phys 91:3647, 1989.

94. JA Rullmann, PTh van Duijnen. Mol Phys 61:293, 1987.
95. D Beglov, B Roux. J Chem Phys 100:9050–9063, 1994.
96. D Eisenberg, A McClachlan. Nature 319:199–203, 1986.
97. L Wesson, D Eisenberg. Protein Sci 1:227–235, 1992.
98. F Fraternali, WF van Gunsteren. J Mol Biol 256:939–948, 1996.
99. HA Scheraga. Acc Chem Res 12:7–14, 1979.
100. YK Kang, KD Gibson, G Nemethy, H Scheraga. J Phys Chem 92:4739–4742, 1988.
101. F Colonna-Cesari, C Sander. Biophys J 57:1103–1107, 1990.
102. P Stouten, C Frömmel, H Nakamura, C Sander. Mol Simul 10:97–120, 1993.
103. T Lazaridis, M Karplus. Science 278:1928–1931, 1997.
104. MD Cummings, TN Hart, RJ Read. Protein Sci 4:2087–2089, 1995.
105. MJ Sippl. J Mol Biol 213:859–883, 1990.
106. MJ Sippl. J Computer-Aided Mol Des 7:473–501, 1993.
107. K Yue, KA Dill. Protein Sci 5:254–261, 1996.
108. I Bahar, RL Jernigan. J Mol Biol 266:195–214, 1997.
109. SH Bryant, CE Lawrence. Proteins 16:92–112, 1993.
110. SE DeBolt, J Skolnick. Protein Eng 9:637–655, 1996.
111. KK Koretke, Z Luthey-Schulten, PG Wolynes. Protein Sci 5:1043–1059, 1996.
112. LA Mirny, EI Shakhnovich. J Mol Biol 264:1164–1179, 1996.
113. S Vajda, M Sippl, J Novotny. Curr Opin Struct Biol 7:222–228, 1997.
114. DT Jones, JM Thornton. Curr Opin Struct Biol 6:210–216, 1996.
115. RL Jernigan, I Bahar. Curr Opin Struct Biol 6:195–209, 1996.
116. YN Vorobjev, JC Almagro, J Hermans. Protein Struct Funct Genet 32:399–413, 1998.
117. KA Sharp. Protein 1:39–48, 1998.
118. N Ben-Tal, A Ben-Shaul, A Nicholls, BH Honig. Biophys J 70:1803–1812, 1996.

8

Normal Mode Analysis of Biological Molecules

Steven Hayward
University of East Anglia, Norwich, England

I. INTRODUCTION

Normal mode analysis exists as one of the two main simulation techniques used to probe the large-scale internal dynamics of biological molecules. It has a direct connection to the experimental techniques of infrared and Raman spectroscopy, and the process of comparing these experimental results with the results of normal mode analysis continues. However, these experimental techniques are not yet able to access directly the lowest frequency modes of motion that are thought to relate to the functional motions in proteins or other large biological molecules. It is these modes, with frequencies of the order of 1 cm^{-1}, that mainly concern this chapter.

Normal mode analysis was first applied to proteins in the early 1980s [1–3]. Much of the literature on normal mode analysis of biological molecules concerns the prediction of functionally relevant motions. In these studies it is always assumed that the soft normal modes, i.e., those with the lowest frequencies and largest fluctuations, are the ones that are functionally relevant. The ultimate justification for this assumption must come from comparisons to experimental data. Several studies have been made in which the predictions of a normal mode analysis have been compared to functional transitions derived from two X-ray conformers [4–7]. These studies do indeed suggest that the low frequency normal modes are functionally relevant, but in no case has it been found that the lowest frequency normal mode corresponds exactly to a functional mode. Indeed, one would not expect this to be the case.

Normal mode analysis is a harmonic analysis that assumes that, over the range of thermal fluctuations, the conformational energy surface can be characterized by the parabolic approximation to a single energy minimum. However, there exists abundant evidence, both experimental [8] and computational [9], that the harmonic approximation breaks down spectacularly for proteins at physiological temperatures, where, far from performing harmonic motion in a single energy minimum, the state point visits multiple minima, crossing energy barriers of various heights. Even if the motion within a single energy minimum is representative of the motion within all energy minima, as appears to be the case [10,11], barrier-crossing events would be expected to have an even greater influence on the overall motion of the molecule, with no obvious relation to the motion within individual minima. Given the level of the approximation, then, the relative success of the normal mode analysis is surprising and intriguing.

In the following, the method itself is introduced, as are the various techniques used to perform normal mode analysis on large molecules. The method of normal mode refinement is described, as is the place of normal mode analysis in efforts to characterize the nature of a protein's conformational energy surface.

II. NORMAL MODE ANALYSIS IN CARTESIAN COORDINATE SPACE

This section describes the basic methodology of normal mode analysis. Owing to its long history it has been described in detail in the context of many different fields. However, to aid in understanding subsequent sections of this chapter, it is described here in some detail.

To do a normal mode analysis one needs a set of coordinates, a force field describing the interactions between constituent atoms, and software to perform the required calculations. A normal mode analysis requires three main calculations. The first is the minimization of the conformational potential energy as a function of the atomic Cartesian coordinates. The various energy minimization techniques have been described previously in this volume. To find a true minimum, line search algorithms are often used in the later stages of minimization. At a minimum, the potential energy can be expanded in a Taylor series in terms of mass-weighted coordinates, $q_i = \sqrt{m_j}\,\Delta x_j$, $q_{i+1} = \sqrt{m_j}\,\Delta y_j$, $q_{i+2} = \sqrt{m_j}\,\Delta z_j$, where j labels the N atoms, and i the $3N$ Cartesian coordinates, to give

$$V = \frac{1}{2} \sum_{i,j=1}^{3N} \left.\frac{\partial^2 V}{\partial q_i \partial q_j}\right|_0 q_i q_j + \cdots \tag{1}$$

The first term in the expansion, the value of the energy at the minimum, has been set to zero. The linear terms are also zero, as the first derivatives of the energy (the force) at a minimum are also zero. In normal mode analysis the higher order terms are neglected, and the second derivatives calculated at the minimum are assumed to characterize the energy surface over fluctuations that are far from the minimum. These second derivatives are the elements of the symmetric matrix, $\mathbf{F}$, that is often called the Hessian. The calculation of the Hessian is the second major calculation in a normal mode analysis. The Langrangian, the kinetic energy minus the potential energy, can be written as

$$L = \frac{1}{2} \sum_{i=1}^{3N} \dot{q}_i^2 - \frac{1}{2} \sum_{i,j=1}^{3N} \left.\frac{\partial^2 V}{\partial q_i \partial q_j}\right|_0 q_i q_j \tag{2a}$$

or in vector-matrix form as

$$L = \frac{1}{2} \dot{\mathbf{q}}^t \dot{\mathbf{q}} - \frac{1}{2} \mathbf{q}^t \mathbf{F} \mathbf{q} \tag{2b}$$

where the superscript t denotes the transpose. This determines the dynamics of the system.

As $\mathbf{F}$ is a symmetric matrix, there exists an orthogonal transformation that diagonalizes $\mathbf{F}$:

$$\mathbf{W}^t\mathbf{F}\mathbf{W} = \Omega \tag{3}$$

where Ω is the diagonal matrix, $\mathbf{W}$ is a matrix of order $3N$, and $\mathbf{W}^t\mathbf{W} = \mathbf{I}$ or $\mathbf{W}\mathbf{W}^t = \mathbf{I}$, where $\mathbf{I}$ is the identity matrix. Simple rearrangement of Eq. (3) shows that it can also be regarded as an eigenvalue equation, where each element of Ω is an eigenvalue associated with a specific column of $\mathbf{W}$, the eigenvector. Using $\mathbf{W}$, a new coordinate set can be defined:

$$\mathbf{Q} = \mathbf{W}^t\mathbf{q} \tag{4}$$

or

$$Q_i = \sum_{k=1}^{3N} W_{ki}q_k \tag{5}$$

where $i = 1, \ldots, 3N$. Each Q_i is a collective variable termed a normal mode coordinate. Substitution of Eq. (4) into Eq. (2b) leads to $3N$ independent equations of motion for each normal mode coordinate, each with a solution $Q_i = A_i \cos(\omega_i t + \varepsilon_i)$, showing that each normal mode coordinate oscillates sinusoidally with an angular frequency ω_i. Each normal mode coordinate specifies a set of atomic displacements through Eq. (4):

$$q_k = W_{ki}A_i \cos(\omega_i t + \varepsilon_i) \tag{6}$$

The pattern of motion, or normal mode, is fully specified by the W_{ki}'s, i.e., the eigenvector associated with ω_i. Therefore, normal modes and their frequencies of oscillation are determined by the eigenvectors and eigenvalues of $\mathbf{F}$. The diagonalization of $\mathbf{F}$, then, is the final major computational challenge in performing a normal mode analysis. Six normal modes describing the rigid-body motion of the whole molecule have frequencies of zero and are usually eliminated from any further analysis. The general solution to the equation of motion involves a sum of the terms of Eq. (6) (over the normal mode index i), each with a different amplitude and phase. The precise amplitude and phase of each normal mode is determined by the initial conditions.

Application of the equipartition law shows that for a molecule in thermal equilibrium,

$$\langle Q_i^2 \rangle = \frac{k_B T}{\omega_i^2} \tag{7a}$$

or for the mass-weighted atomic coordinates,

$$\langle q_i^2 \rangle = k_B T \sum_{k=1}^{3N-6} \left(\frac{W_{ik}}{\omega_k}\right)^2 \tag{7b}$$

where k_B is Boltzmann's constant and T the absolute temperature. Equation (7b) is often used to calculate the mean-square fluctuations (MSFs) of atoms for comparison to those derived from the atomic B-factors in X-ray crystallography. From Eq. (7b) it is easy to show that

$$\left\langle \sum_{i=1}^{3N-6} q_i^2 \right\rangle = \sum_{i=1}^{3N-6} \langle q_i^2 \rangle = k_B T \sum_{k=1}^{3N-6} \frac{1}{\omega_k^2} \tag{8}$$

Equation (8) shows that it is the fluctuations of the lowest frequency modes that contribute most to the overall fluctuation of the molecule. For example, in the case of lysozyme, the lowest frequency normal mode (out of a total of 6057) accounts for 13% of the total mass-weighted MSF. It is for this reason that it is common to analyze just the lowest frequency modes for the large-scale functional motions.

Covariances between the Cartesian coordinates can also be calculated using

$$\langle q_i q_j \rangle = k_B T \sum_{k=1}^{3N-6} W_{ik} W_{jk} \frac{1}{\omega_k^2} \tag{9a}$$

or, in matrix form,

$$\mathbf{U} = \langle \mathbf{q}\mathbf{q}^t \rangle = k_B T \mathbf{W} \boldsymbol{\Omega}^{-1} \mathbf{W}^t \tag{9b}$$

where the zero frequency eigenvalues and eigenvectors are eliminated. Equation (9b) shows how one can perform a quasi-harmonic analysis by diagonalizing the variance–covariance matrix of atomic fluctuations determined from a molecular dynamics simulation to attain a set of quasi-harmonic modes and effective frequencies. Note that a quasi-harmonic analysis necessarily includes any anharmonic effects in the molecular dynamics simulation.

Many thermodynamic quantities can be calculated from the set of normal mode frequencies. In calculating these quantities, one must always be aware that the harmonic approximation may not provide an adequate physical model of a biological molecule under physiological conditions.

III. NORMAL MODE ANALYSIS OF LARGE BIOLOGICAL MOLECULES

On the face of it, a normal mode analysis of many biological molecules is a daunting task. A normal mode analysis of the protein citrate synthase, a homodimer of 874 residues, would involve the diagonalization of a matrix whose order is in the tens of thousands. Out of the proteins whose structures have been solved, this is by no means outstandingly large, but a full-scale diagonalization of the Hessian of citrate synthase is still not feasible. However, methods exist that can reduce the size of the calculation considerably without sacrificing accuracy. Indeed, the diagonalization of a large symmetric matrix is a ubiquitous task for which many different numerical techniques have been developed [12]. As an alternative, reduced basis sets can be used [13] that, if chosen carefully, can considerably reduce the size of the Hessian without compromising accuracy. By far the most popular among these is the dihedral angle space normal mode analysis [1].

Further reductions can be achieved by taking symmetry into account, an approach that holds promise for the analysis of large oligomeric proteins such as virus capsids [14].

Combining all these techniques suggests that, provided minimization has been achieved, a number of the lowest frequency normal modes of any protein can be accurately determined.

A. Determination of the Eigenvalues and Eigenvectors of a Large Hessian

The determination of some of the eigenvalues and eigenvectors of a large real symmetric matrix has a long history in numerical science. Of particular interest in the normal mode analysis of large biomolecules, where only some of the lowest frequency normal modes are required, are iterative techniques such as the Lanczos algorithm [12], which is an efficient algorithm suited to the computation of a few outer eigenvalues and eigenvectors. In its simplest form the Lanczos algorithm starts with an initial vector, which could be a guess of the lowest frequency normal mode, and iteratively increases the dimension of the space one dimension at a time by application of an operator that could be the Hessian itself. It can be shown that the subsequent projection of the Hessian into this space and diagonalization of the resulting reduced matrix will produce better and better approximations to the outer eigenvalues and eigenvectors as the dimension of the space increases. Sophisticated techniques are available that make the algorithm very efficient. This approach has been used with an operator that should lead to faster convergence and is particularly suited to a diagonally dominant matrix [13,15]. Although the Hessian is not a naturally sparse matrix, due primarily to the long range Coulomb interaction, cutoff methods have been used to create a sparse matrix that can then be transformed to a diagonally dominant form.

A block Lanczos algorithm (where one starts with more than one vector) has been used to calculate the first 120 normal modes of citrate synthase [4]. In this calculation no apparent use was made of symmetry, but it appears that to save memory a short cutoff of 7.5 Å was used to create a sparse matrix. The results suggested some overlap between the low frequency normal modes and functional modes determined from the two X-ray conformers.

Although the Lanczos is a fast efficient algorithm, it does not necessarily give savings in memory. To save memory a number of techniques divide the molecule into smaller parts that correspond to subspaces within which the Hessian can be expressed as a matrix of much lower order. These smaller matrices are then diagonalized. The methods described below show how one then proceeds to achieve good approximations to the true low frequency modes by combining results from subspaces of lower dimension.

Mouawad and Perahia [16] developed a technique, which they call "diagonalization in a mixed basis," whereby the Hessian is projected into a subspace spanned by a union of the space defined by combining low frequency eigenvectors of the parts, with a space defined by a selected set of Cartesian coordinate vectors. An iterative process has been devised whereby diagonalization in this subspace and other subspaces, created by selecting new sets of Cartesian coordinate vectors, produces new low frequency eigenvectors with which one can repeat the whole process. It converges to yield good approximations to the true low frequency modes. This technique has been applied to some very large proteins such as hemoglobin [7] (600 residues), for which the first 203 lowest frequency modes were estimated, and aspartate transcarbamylase [6], a dodecamer of some 2760 residues for which the first 53 modes were calculated.

A related method is the component synthesis method [17], which uses a so-called static condition to model the interactions between parts of a molecule whose corresponding diagonal blocks in the Hessian are first diagonalized. It has been combined with a residue clustering algorithm that provides a hierarchy of parts, which at the lowest level provides small enough matrices for efficient diagonalization [18]. It has been applied to double-helical DNA [17] and the protein crambin [18].

In another promising method, based on the effective Hamiltonian theory used in quantum chemistry [19], the protein is divided into "blocks" that comprise one or more residues. The Hessian is then projected into the subspace defined by the rigid-body motions of these blocks. The resulting low frequency modes are then perturbed by the higher frequency modes that result from projecting the Hessian into the space defined by intrablock motions. It has been shown that the method scales with N^2, where N is the total number of degrees of freedom.

Finally, any symmetry that a molecule possesses can be exploited by the methods of group theory to reduce the Hessian to a number of independent submatrices [20]. If the symmetry group to which the molecule belongs has been determined, then the procedures of group theory show how one can construct a basis (the basis for the irreducible representations) in which the Hessian comprises a number of smaller independent submatrices lined along the diagonal. These independent submatrices can be individually diagonalized and the eigenvectors and eigenvalues of the whole system reconstructed. This process gives a considerable saving on the diagonalization of the whole Hessian. For example, if the molecule has a tenfold rotational symmetry (i.e., it belongs to the group C_{10}), the order of the blocks is one tenth that of the whole Hessian. Group theory has been applied to normal mode analyses of the gramicidin A dimer [21] and one layer of the tobacco mosaic virus protein disk [22]. The whole problem of performing normal mode analysis on symmetrical protein assemblies in dihedral angle space has been tackled by Gibrat et al. [14]. This promises to make viable the normal mode analyses of very large molecular assemblies such as those of virus capsids and crystals.

B.   Normal Mode Analysis in Dihedral Angle Space

For most large biological molecules, most of the flexibility arises from torsional or dihedral angle rotations. In comparison, bond lengths and bond angles are comparatively rigid. This automatically leads one to the idea of performing a normal mode analysis in dihedral angle space. In such an analysis the bond lengths and angles are kept fixed at their minimum energy values. At least for bond length fluctuations, the fact that they correspond to a permanent ground quantum oscillator state at physiological temperatures suggests that constraining them is not unphysical. For proteins the ratio of the number of rotatable dihedral angles to Cartesian coordinates is about $1:8$ in proteins and about $1:11$ for nucleic acids [23]. This provides a considerable saving, but there is the practical disadvantage that a normal mode analysis in dihedral angle space is technically more involved.

The dihedral angle space procedure is complicated by the fact that the kinetic energy term cannot be expressed as a simple function of the dihedral angle variables as it can in the case of the mass-weighted Cartesian coordinates. There are six degrees of freedom for the overall rigid-body motion of the molecule, and formulating the kinetic energy in a way that ensures that changes in dihedral angles do not result in the movement of the center of mass of the molecule or an overall rotation (by use of the Eckart conditions) means that the equations of motion can be solved for the internal motion alone. This is usually done by calculating the so-called **K** matrix, which is the linear term in a Taylor expansion,

$$K_{ij} = \frac{\partial q_i}{\partial \phi_j} \tag{10}$$

where q_i is the ith mass-weighted Cartesian coordinate, and ϕ_j the jth dihedral angle. The derivative is calculated at the energy minimum, subject to the six constraints [24]. The kinetic energy is then expressed by way of the mass tensor, often denoted **H**, as

$$T = \frac{1}{2} \dot{\phi}' \mathbf{H} \dot{\phi} \tag{11}$$

where $\dot{\phi}$ is the column vector whose elements are the time derivatives of the dihedral angles, and $\mathbf{H} = \mathbf{K}'\mathbf{K}$. The solution of the equations of motion proceeds in the same way as for the Cartesian coordinates, except that one must solve a generalized eigenvalue problem whereby the eigenvector matrix $\mathbf{V}$ now simultaneously diagonalizes $\mathbf{H}$ to the identity matrix $\mathbf{I}$ and the Hessian to the eigenvalue matrix Ω. To convert between the normal mode coordinates, which now describe collective variations in dihedral angles, and the mass-weighted Cartesian coordinates, one can use the matrix

$$\mathbf{W} = \mathbf{KV} \tag{12}$$

where $\mathbf{WW}' = \mathbf{I}$.

As a dihedral angle space analysis is equivalent to a Cartesian coordinate space analysis where the bond angles and lengths are kept fixed, one can compare the results from a conventional Cartesian coordinate space analysis to those from a dihedral angle space analysis to access the effects of bond angle and bond length variations [25,26]. Although low frequency modes can be represented solely by dihedral angle variations, allowing bond length and bond angle variations has the indirect effect of making their amplitudes of fluctuation larger. A study of the protein bovine pancreatic trypsin inhibitor (BPTI) showed that the important subspaces (the subspaces within which most of the fluctuation occurs) overlap considerably, implying that dihedral angle space normal mode analysis is a viable alternative to Cartesian coordinate space analysis [26]. In fact, it has been argued that if one includes second-order terms in the conversion from dihedral angle fluctuations to Cartesian coordinate fluctuations, then the results from a dihedral angle space analysis are, in fact, better than those from a Cartesian coordinate space analysis because the harmonic approximation is valid over a wider range in dihedral angle space than in Cartesian coordinate space [27]. This approach has been used to calculate nuclear magnetic resonance (NMR) order parameters in proteins [28].

Dihedral angle space normal mode analyses of a number of proteins (not all of them large by current standards) have been made, including BPTI [1,2,26,27,29–31], lysozyme [30,32–35], G-actin [36], myoglobin [37–39], epidermal growth factor [40], bacterio-phage 434 Cro and 434 repressor [41], and subtilisin–Eglin c complex [42]; nucleic acids, including tRNA [43]; and a double-stranded DNA dodecamer [23]. In the latter, flexibility arising from the pseudorotation of the furanose rings was included [44].

C. Approximate Methods

The methods described so far are, within the framework of the analysis, accurate, in the sense that they are implemented by using basically the same detailed force fields for the bonded and non-bonded interactions that are used in molecular dynamics simulations. Tirion [45] showed that a "single parameter model" can give results similar to those of these detailed analyses. The model replaces the Hessian by a matrix whose elements are zero for any pair of atoms separated by a distance greater than a cutoff distance (equal to the sum of the van der Waals radii plus a distance parameter R_c that models the decay of the interaction) and have values according to a simple Hookean pairwise potential with the same force constant C for all pairs of atoms within the cutoff distance of each other. By adjusting C and R_c, the predictions from conventional normal mode analyses and this simplified model are found to be astoundingly similar. Interestingly, for good fits to the conventional normal mode analysis data, the product CR_c^2 has a constant value, indicating a universal "bond strength" of 3 kJ/mol. As distances between atoms are calculated directly from their X-ray-determined coordinates, costly energy minimization is avoided.

Bahar et al. [46] have used this kind of approach to predict the B-factors of 12 X-ray structures. Elements in the "Hessian" corresponding to atom pairs separated by a distance of less than 7 Å are set to zero, and the remainder have the same value dependent on a single adjustable parameter. Generally B-factor predictions for the α-carbons compare very well with the B-factors measured by X-ray crystallography. Figure 1 shows the result for the subunit A of endodeoxyribonuclease I complexed with actin.

Given that one does not need to perform an energy minimization and that the "Hessian" is very sparse, it is not surprising that the computation time is reported to be at least one order of magnitude less than for a conventional normal mode analysis.

IV. NORMAL MODE REFINEMENT

The normal mode refinement method is based on the idea of the normal mode important subspace. That is, there exists a subspace of considerably lower dimension than $3N$, within which most of the fluctuation of the molecule undergoing the experiment occurs, and a number of the low frequency normal mode eigenvectors span this same subspace. In its application to X-ray diffraction data, it was developed by Kidera et al. [33] and Kidera and Go [47,48] and independently by Diamond [49]. Brueschweiler and Case [50] applied it to NMR data.

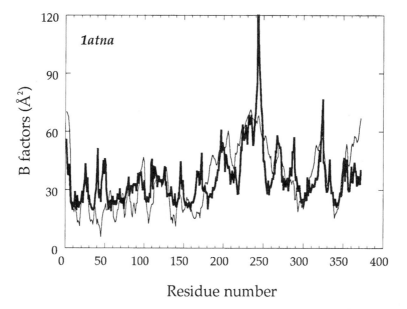

Figure 1 Experimental B-factors of α-carbon atoms (thin curve) compared with those predicted using their single-parameter model for subunit A of endodeoxyribonuclease I complexed with actin (thick curve). Calculations were performed using both subunits A and D comprising 633 residues. Their parameter is adjusted such that the area under the predicted curve equals the area under the experimental curve. (From Ref. 46.)

A. Normal Mode X-Ray Refinement

In the procedure of X-ray refinement, the positions of the atoms and their fluctuations appear as parameters in the structure factor. These parameters are varied to match the experimentally determined structure factor. The term pertaining to the fluctuations is the Debye–Waller factor in which the atomic fluctuations are represented by the atomic distribution tensor:

$$\mathbf{U}_j = \begin{pmatrix} \langle \Delta x_j\, \Delta x_j \rangle & \langle \Delta x_j\, \Delta y_j \rangle & \langle \Delta x_j\, \Delta z_j \rangle \\ \langle \Delta y_j\, \Delta x_j \rangle & \langle \Delta y_j\, \Delta y_j \rangle & \langle \Delta y_j\, \Delta z_j \rangle \\ \langle \Delta z_j\, \Delta x_j \rangle & \langle \Delta z_j\, \Delta y_j \rangle & \langle \Delta z_j\, \Delta z_j \rangle \end{pmatrix} \tag{13}$$

where j is the atomic index. In a common method of refinement it is often assumed that atomic fluctuations are isotropic and independent. In such a case the atomic distribution tensor is a scalar quantity, the B-factor. It is the B-factors that are varied in the refinement process. The normal mode refinement method requires two refinement procedures. The first uses the isotropic B-factors to refine a structure well enough for a normal mode analysis to be performed. Then the process of normal mode refinement begins by expressing the atomic distribution tensor in terms of the fluctuations of a selected number, M, of the lowest frequency normal modes through

$$\mathbf{U}_j = \mathbf{W}_j \Lambda^{\text{ref}} \mathbf{W}_j^t \tag{14}$$

This equation is a variant of Eq. (9b), but here $\mathbf{W}_j$ is the $3 \times M$ portion of $\mathbf{W}$ corresponding to atom j and the first M normal mode eigenvectors, and Λ^{ref} is the $M \times M$ variance–covariance matrix of the M lowest frequency normal mode coordinates. Note that Λ^{ref} would equal the corresponding portion of the diagonal matrix $k_B T \Omega^{-1}$ from Eq. (9b) if the dynamics of the real molecule in the experiment were perfectly described by the normal mode analysis. In the real situation this is not the case, so the elements of Λ^{ref} are the parameters to be varied in the refinement process. If M is small enough that $M \times M$ is less than N, the number of isotropic B-factors, there can be a significant saving in the number of parameters and, as has been reported, an improvement in the R-factor [47] and R_{free}-factor [51]. If the six modes describing external motion are included, then simulation showed that it could distinguish between internal and external fluctuations [48]. Perhaps the most significant advantage is that, unlike the normal atomic B-factor refinement methods, correlations between atoms can be determined. Kidera and Go, together with Inaka and Matsushima, tested it on real diffraction data from human lysozyme [33]. Figure 2 shows the result of a comparison of the internal root-mean-square fluctuations (RMSFs) of human lysozyme residues determined from the normal mode refinement procedure in comparison with the results from a normal mode analysis on that molecule. The correspondence between the two is remarkable and unexpected. Diamond [49] presented a similar analysis on the protein BPTI.

B. Normal Mode NMR Refinement

The normal mode NMR refinement method of Brueschweiler and Case [50] can be applied to experimentally measurable quantities such as order parameters or nuclear Overhauser spectroscopy (NOSEY) intensities. Unlike the X-ray case, the expression of these quanti-

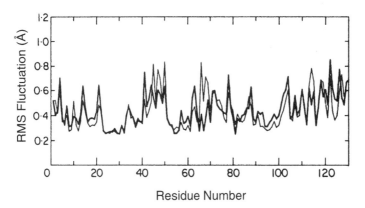

Figure 2 Internal RMSF of residues (average over heavy atoms) determined for human lysozyme by the X-ray normal mode refinement method applied to real X-ray data (heavy curve), in comparison with results from a normal mode analysis on a single isolated lysozyme molecule (lightweight curve). (From Ref. 33.)

ties in terms of the normal mode eigenvectors is not so natural. However, they still can be approximately expressed in terms of the variance and covariances of internuclear vectors of spin pairs, which are in turn expressed in terms of the lowest frequency normal modes. The method then proceeds in much the same way as in the X-ray case. The method was demonstrated on a solvated 25-residue zinc finger peptide on which a molecular dynamics simulation was performed. Low frequency normal modes were used to fit the order parameters for backbone C—H, N—H spin pairs calculated directly from the simulation. The predictive power of the method was tested by calculating the order parameters of spin pairs not included in the refinement. The results showed general agreement with the directly calculated quantities, although some large discrepancies were seen in individual cases. Applying the method to NOESY intensities allows one also to refine the predicted averaged structure and gain information on correlated motion in analogy to the X-ray refinement method.

C. Validity of the Concept of a Normal Mode Important Subspace

As already mentioned in Section I, normal mode analysis is based on a physical model that is quite far from reality for a biological molecule under physiological conditions. Although some studies found partial overlap between some of the lowest frequency modes and the functional mode determined from two X-ray conformers, in general it would be fanciful to expect anything more than a moderate correspondence to individual normal modes. However, the expectation that there exists a larger subspace spanned by the first M lowest frequency normal modes (where M may be between 10% and 20% of $3N$) that is also spanned by the modes with the largest fluctuation in the real molecule is more realistic. If this were the case, then the low frequency normal modes would contain information on the modes of largest fluctuation in the real molecule, which would also be largely determined by barrier-crossing motions. The success of the normal mode refinement method itself is the ultimate test of the validity of this concept.

V. NORMAL MODE ANALYSIS AND REALITY

A number of studies have compared normal mode analysis predictions with results from more realistic simulation techniques or experiments. These studies shed light on the nature of the conformational energy surface and the effect of solvent.

A. The Solvent Effect

Normal mode analyses are usually performed in a single minimum using vacuum force fields. Molecular dynamics simulations of proteins in vacuum, however, reveal not one minimum but multiple minima in the energy surface. One effect of adding solvent is to increase the density of local minima over that found in vacuum [52,53]. Extended RISM calculations have shown for the protein melittin that adding the solvation free energy to its conformational potential energy results in two minima [54] along the direction of the first normal mode, rather than one. In addition to the change in the conformational energy surface, the solvent causes damping and other hydrodynamic effects. One can attempt to incorporate these into a normal mode analysis by using a variant of the normal mode analysis developed by Lamm and Szabo [55], called the Langevin mode analysis. The resulting generalized eigenvalue problem involves the simultaneous diagonalization of the Hessian and the friction matrix. The resulting modes display the extra feature of damping. Time correlation functions and spectral densities can be calculated directly from the Langevin modes. This approach was applied to the protein crambin and a DNA duplex to reveal that a number of modes had overdamped motions [56]. One obstacle to performing a Langevin mode analysis is the accurate determination of the off-diagonal hydrodynamic terms in the friction matrix. The spectral density determined from inelastic neutron scattering experiments for BPTI shows a shallower rise from zero frequency than normal mode calculations predict [57,58]. It has been shown that this is directly due to frictional damping of the low frequency modes and can be reproduced by the Langevin mode analysis [52,53].

B. Anharmonicity and Normal Mode Analysis

The emerging model for protein dynamics is one that incorporates the dual aspects of motion within minima, combined with transitions between minima. The normal mode analysis can be seen as addressing directly only one of these two features. Ironically, however, one variant of normal mode analysis can be used to help address the other feature, namely the transition of energy barriers. To determine barrier heights for transitions occurring in a molecular dynamics simulation, instantaneous normal modes can be determined by diagonalizing instantaneous force matrices at selected configurations along the trajectory. Negative eigenvalues indicate local negative curvature possibly arising from energy barriers in the multiple-minima surface. Simulations performed at different temperatures can give information on the distribution of barrier heights [59,60].

The physical model of protein dynamics indicated above would be greatly simplified if all minima were identical. Janezic et al. [10] performed 201 normal mode analyses starting minimizations from frames along a 1 ns vacuum molecular dynamics simulation of BPTI. Comparing normal modes by taking inner products revealed that in general the normal modes remained stable, indicating similar minima. Using this assumption, a normal mode analysis performed in one single minimum can be used to analyze anharmonic effects in the molecular dynamics simulation. By performing a quasi-harmonic analysis of a 200 ps vacuum molecular dynamics simulation on BPTI and taking inner products between the normal mode and the quasi-harmonic modes or principal modes, fluctuations along the quasi-harmonic modes could be analyzed for their harmonic and anharmonic contributions [11]. The "anharmonicity factor" for the ith quasi-harmonic mode was defined as

$$\alpha_i^2 = \frac{\langle q_i^2 \rangle}{\langle q_i^2 \rangle^{\mathrm{har}}} \tag{15}$$

where

$$\langle q_i^2 \rangle^{\mathrm{har}} = k_B T \sum_{k=1}^{3N-6} \frac{g_{ik}^2}{\omega_k^2} \tag{16}$$

and g_{ik} is the inner product value between the kth normal mode and the ith quasi-harmonic mode. A value of 1.0 would indicate that the fluctuation in the mode can be predicted by the normal mode analysis. A value greater than 1.0 would indicate fluctuation beyond what could be predicted by the normal mode analysis. Modes with anharmonicity factors greater than 1.0 are termed "anharmonic modes"; those with values equal to 1.0, "harmonic modes." Figure 3 shows a plot of the anharmonicity factor versus quasi-

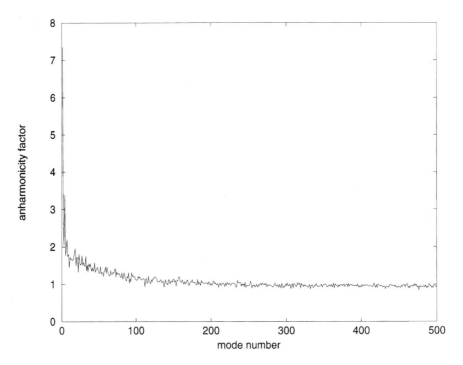

Figure 3 Anharmonicity factor versus quasi-harmonic mode number from a 200 ps vacuum simulation of BPTI. It can be seen that beyond about the 200th mode the anharmonicity factors are about 1.0, indicating harmonicity. Those below mode number 200 show progressively greater anharmonicity factors, indicating that they span a space within which energy barriers are crossed. A similar picture was found for a 1 ns simulation of human lysozyme in water [61]. (Adapted from Ref. 11.)

harmonic mode number for BPTI. Only 12% of the modes were found to be anharmonic; the remaining 88% were harmonic modes. This analysis was also applied to a 1 ns molecular dynamics simulation of human lysozyme in water, where it was found that only the first 5% of modes were anharmonic [61]. These results suggest that the harmonic component calculated from a single normal mode analysis may be separable from the anharmonic component. Under the assumption that all minima are identical, and using the "dual aspect model" of protein dynamics, contributions arising from transitions between minima can be separated from contributions from harmonic motion within minima in an expression for the variance–covariance matrix of atomic fluctuations [61]. This model, the jumping-among-minima, or JAM, model, can be used to determine many features of the multiple-minima energy surface, including the distribution of minima and barrier heights.

VI. CONCLUSIONS

One of the main attractions of normal mode analysis is that the results are easily visualized. One can sort the modes in terms of their contributions to the total MSF and concentrate on only those with the largest contributions. Each individual mode can be visualized as a collective motion that is certainly easier to interpret than the welter of information generated by a molecular dynamics trajectory. Figure 4 shows the first two normal modes of human lysozyme analyzed for their dynamic domains and hinge axes, showing how clean the results can sometimes be. However, recent analytical tools for molecular dynamics trajectories, such as the principal component analysis or essential dynamics method [25,62–64], promise also to provide equally clean, and perhaps more realistic, visualizations. That said, molecular dynamics is also limited in that many of the functional motions in biological molecules occur in time scales well beyond what is currently possible to simulate.

Various techniques exist that make possible a normal mode analysis of all but the largest molecules. These techniques include methods that are based on perturbation methods, reduced basis representations, and the application of group theory for symmetrical oligomeric molecular assemblies. Approximate methods that can reduce the computational load by an order of magnitude also hold the promise of producing reasonable approximations to the methods using conventional force fields.

Evidence exists that some of the softest normal modes can be associated with experimentally determined functional motions, and most studies apply normal mode analysis to this purpose. Owing to the veracity of the concept of the normal mode important subspace, normal mode analysis can be used in structural refinement methods to gain dynamic information that is beyond the capability of conventional refinement techniques.

Ironically, the normal mode analysis method can be used to determine properties of the multiple-minima energy surface of proteins.

Although not discussed in detail here, the normal mode analysis method has been used to calculate the electron transfer reorganization spectrum in *Ru*-modified cytochrome c [65,66]. In this application the normal mode analysis fits comfortably into the theory of electron transfer.

Despite its obvious limitations, normal mode analysis has found varied and perhaps unexpected applications in the study of the dynamics of biological molecules. In many

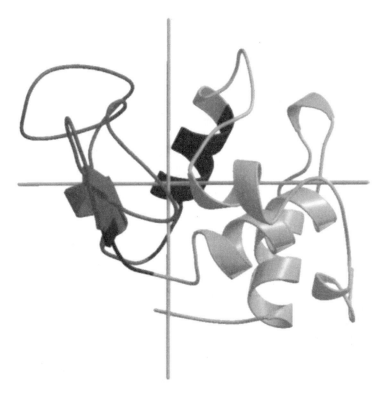

Figure 4 DynDom [67] analysis of the first two normal modes of human lysozyme. Dark grey and white indicate the two dynamic domains, separated by the black hinge bending region. The vertical line represents a hinge axis that produces a closure motion in the first normal mode. The horizontal line represents a hinge axis that produces a twisting motion in the second normal mode. (Adapted from Ref. 68.) The DynDom program is available from the Internet at http://md. chem.rug.nl/~steve/dyndom.html.

of these applications its weaknesses, in comparison to other simulation techniques, appear still to be outweighed by its strengths.

ACKNOWLEDGMENT

I wish to thank Dr. Akio Kitao for reading the manuscript.

REFERENCES

1. N Go, T Noguti, T Nishikawa. Proc Natl Acad Sci USA 80:3696–3700, 1983.
2. M Levitt, C Sander, PS Stern. Int J Quant Chem: Quant Biol Symp 10:181–199, 1983.
3. B Brooks, M Karplus. Proc Natl Acad Sci USA 80:6571–6575, 1983.
4. O Marques, Y-H Sanejouand. Proteins 23:557–560, 1995.
5. J Ma, M Karplus. J Mol Biol 274:114–131, 1997.
6. A Thomas, MJ Field, D Perahia. J Mol Biol 261:490–506, 1996.

7. L Mouawad, D Perahia. J Mol Biol 258:393–410, 1996.
8. RH Austin, KW Beeson, L Eisenstein, H Frauenfelder, IC Gunsalus. Biochemistry 14(24): 5355–5373, 1975.
9. R Elber, M Karplus. Science 235:318–321, 1987.
10. D Janezic, RM Venable, BR Brooks. J Comput Chem 16:1554–1566, 1995.
11. S Hayward, A Kitao, N Go. Proteins 23:177–186, 1995.
12. BN Parlett. The Symmetric Eigenvalue Problem. Englewood Cliffs, NJ: Prentice-Hall, 1980.
13. BR Brooks, D Janezic, M Karplus. J Comput Chem 16(12):1522–1542, 1995.
14. J-F Gibrat, J Garnier, N Go. J Comput Chem 15(8):820–834, 1994.
15. B Brooks, M Karplus. Proc Natl Acad Sci USA 82:4995–4999, 1985.
16. L Mouawad, D Perahia. Biopolymers 33:599–611, 1993.
17. M Hao, SC Harvey. Biopolymers 32:1393–1405, 1992.
18. M Hao, HA Scheraga. Biopolymers 34:321–335, 1994.
19. P Durand, G Trinquier, Y-H Sanejouand. Biopolymers 34:759–771, 1994.
20. E Bright Wilson Jr, JC Decius, PC Cross. Molecular Vibrations: The Theory of Infrared and Raman Vibrational Spectra. New York: McGraw-Hill, 1955.
21. B Roux, M Karplus. Biophys J 53:297–309, 1988.
22. T Simonson, D Perahia. Biophys J 61:410–427, 1992.
23. D Lin, A Matsumoto, N Go. J Chem Phys 107(9):3684–3690, 1997.
24. T Noguti, N Go. J Phys Soc(Jpn) 52(10):3685–3690, 1983.
25. A Kitao, N Go. J Comput Chem 12(3):359–368, 1991.
26. A Kitao, S Hayward, N Go. Biophys Chem 52:107–114, 1994.
27. S Sunada, N Go. J Comput Chem 16(3):328–336, 1995.
28. S Sunada, N Go, P Koehl. J Chem Phys 104(12):4768–4775, 1996.
29. T Nishikawa, N Go. Proteins 2:308–329, 1987.
30. M Levitt, C Sander, PS Stern. J Mol Biol 181:423–447, 1985.
31. D Janezic, BR Brooks. J Comput Chem 16(12):1543–1553, 1995.
32. N Go. Biophys Chem 35:105–112, 1990.
33. A Kidera, K Inaka, M Matsushima, N Go. J Mol Biol 225:477–486, 1992.
34. T Horiuchi, N Go. Proteins 10:106–116, 1991.
35. J-F Gibrat, N Go. Proteins 8:258–279, 1990.
36. MM Tirion, D ben-Avraham. J Mol Biol 230:186–195, 1993.
37. T Yamato, J Higo, N Go. Proteins 16:327–340, 1993.
38. Y Seno, N Go. J Mol Biol 216:95–109, 1990.
39. Y Seno, N Go. J Mol Biol 216:111–126, 1990.
40. T Ikura, N Go. Proteins 16:423–436, 1993.
41. H Wako, M Tachikawa, A Ogawa. Proteins 26:72–80, 1996.
42. H Ishida, Y Jochi, A Kidera. Proteins 32:324–333, 1998.
43. S Nakamura, J Doi. Nucleic Acids Res 22(3):514–521, 1994.
44. M Tomimoto, N Go, H Wako. J Comput Chem 17(7):910–917, 1996.
45. MM Tirion. Phys Rev Lett 77(9):1905–1908, 1996.
46. I Bahar, AR Atilgan, B Erman. Folding Des 2:173–181, 1997.
47. A Kidera, N Go. Proc Natl Acad Sci USA 87:3718–3722, 1990.
48. A Kidera, N Go. J Mol Biol 225:457–475, 1992.
49. R Diamond. Acta Cryst A46:425–435, 1990.
50. R Brueschweiler, DA Case. Phys Rev Lett 72(6):940–943, 1994.
51. A Kidera, K Inaka, M Matsushima, N Go. Protein Sci 3:92–102, 1994.
52. A Kitao, F Hirata, N Go. Chem Phys 158:447–472, 1991.
53. S Hayward, A Kitao, F Hirata, N Go. J Mol Biol 234:1207–1217, 1993.
54. A Kitao, F Hirata, N Go. J Phys Chem 97:10231–10235, 1993.
55. G Lamm, A Szabo. J Chem Phys 85(12):7334–7348, 1986.
56. J Kottalam, DA Case. Biopolymers 29:1409–1421, 1990.

57. J Smith, S Cusak, B Tidor, M Karplus. J Chem Phys 93(5):2974–2991, 1990.
58. JC Smith. Quart Rev Biophys 24(3):227–291, 1991.
59. JE Straub, D Thirumalai. Proc Natl Acad Sci USA 90:809–813, 1993.
60. JE Straub, J-K Choi. J Phys Chem 98(42):10978–10987, 1994.
61. A Kitao, S Hayward, N Go. Proteins, 33:496–517, 1998.
62. AE Garcia. Phys Rev Lett 68(17):2696–2699, 1992.
63. A Amadei, ABM Linssen, HJC Berendsen. Proteins 17:412–425, 1993.
64. S Hayward, N Go. Annu Rev Phys Chem 46:223–250, 1995.
65. G Basu, A Kitao, A Kuki, N Go. J Phys Chem B 102:2076–2084, 1998.
66. G Basu, A Kitao, A Kuki, N Go. J Phys Chem B 102:2085–2094, 1998.
67. S Hayward, HJC Berendsen. Proteins 30:144–154, 1998.
68. S Hayward, A Kitao, HJC Berendsen. Proteins 27:425–437, 1997.

9
Free Energy Calculations

Thomas Simonson
Centre National de la Recherche Scientifique, Strasbourg, France

I. INTRODUCTION

Sir Isaac Newton spent much of his life pursuing an elusive dream, the transmutation of base materials* into gold. Though he was not successful during his lifetime, he did manage to discover the equations of motion that, three centuries later, make alchemy possible on a computer. To perform this feat, Newton's equations need only be supplemented by the modern technology of free energy simulations.

The calculation of free energy differences is one of the most interesting applications of biomolecular simulations. Indeed, free energy calculations using molecular dynamics or Monte Carlo simulations provide a direct link between the microscopic structure and fluctuations of a system and its most important equilibrium thermodynamic property, the free energy. The earliest biological applications were calculations of changes in binding constants associated with chemical changes in inhibitors, substrates, and/or their protein targets [1–3]. Since then, many of the early difficulties have been resolved and the methodology has considerably matured. The basic theory has been described in several monographs and reviews [4–8], and applications to biological macromolecules have been reviewed [9–12].

The method relies on two fortunate circumstances. First, the free energy is a state function, which does not depend on the manner in which a particular equilibrium state is reached or prepared. Second, the energy function can be modified and manipulated with enormous flexibility in computer simulations, allowing a known system to be transformed into a wide range of other systems of interest with relative ease. For example, by gradually changing a few non-bonded and/or stereochemical force field parameters, a protein side chain can be alchemically ''mutated'' from one residue type into another in the course of a simulation. By proceeding slowly and reversibly, one can estimate, in principle, the resulting free energy change. Repeating the process, one can obtain the relative free energies of binding of a ligand to variants of a given protein or of a series of similar ligands to the same protein. This provides a potential route for rational protein engineering and for lead improvement in drug design. Applications in which charged groups are introduced

---

* It is not known whether he experimented with assistant professors.

or removed can provide detailed insights into the electrostatic properties of biological macromolecules, including mechanisms of transition state stabilization by enzymes [2,9]. Such calculations, in which a molecule is chemically changed in a way that is often impossible to accomplish experimentally, are referred to as "alchemical" free energy calculations.

The second major class of biological applications are calculations of free energy changes due to conformational rearrangements. Indeed, many biological macromolecules possess several distinct conformations that have functional relevance, such as the R and T forms of allosteric enzymes or the B and Z forms of DNA. The statistical weights, or relative probabilities, of these forms are determined by their relative free energies. By introducing suitable conformational restraints (or constraints) in the energy function, a biomolecule can be driven reversibly from one conformation into another and the associated free energy change obtained; recent applications include folding–unfolding studies of small proteins [13].

For reasons of space and because of their prime importance, we focus here on free energy calculations based on detailed molecular dynamics (MD) or Monte Carlo (MC) simulations. However, several other computational approaches exist to calculate free energies, including continuum dielectric models and integral equation methods [4,14].

To obtain reliable free energy estimates from simulations, sufficient conformational sampling must be achieved, not only of the starting and final states but also of many (usually less interesting) intermediate states. Because of the ruggedness of protein and nucleic acid energy surfaces, this represents a considerable challenge. Many techniques to enhance sampling have been proposed, and others are being developed. The most relevant are described below; others are covered elsewhere in this book. Accurate calculation of electrostatic free energies also requires that long-range forces be included, using specialized techniques such as continuum reaction field or Ewald summation methods [15]; these must be incorporated into the free energy formalism.

Section II introduces the use of thermodynamic cycles and covers the basic theoretical and technical aspects of free energy calculations. Section III is then devoted to standard binding free energies. Section IV describes the theoretical basis of conformational free energy calculations. "Electrostatic" free energy calculations, i.e., those associated with charge insertion and deletion are described in Section V. Section VI then describes techniques to enhance sampling in alchemical or conformational free energy calculations. Finally, Section VII briefly discusses directions for future development.

II. GENERAL BACKGROUND

A. Thermodynamic Cycles for Solvation and Binding

To describe solvation and binding of one or more ligands and their receptors it is convenient to introduce the thermodynamic cycles shown in Figure 1 [16]. Figure 1a describes the vapor $\rightarrow$ water transfer of two solutes S and S′ as well as their binding to a receptor R (e.g., a protein). Figure 1b describes the binding of a ligand S to a native receptor R and a mutant receptor R′. In each case, vertical legs correspond to processes that can usually be studied experimentally (solvation or binding), and horizontal legs correspond to a chemical transmutation of the ligand or protein that usually cannot be performed experimentally. The vertical legs can often be accomplished in a simulation, particularly if the solutes S, S′ are not too large; however, the horizontal legs usually involve smaller

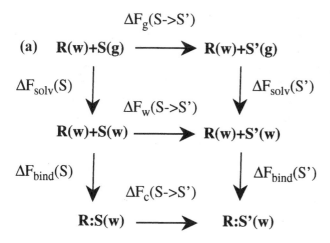

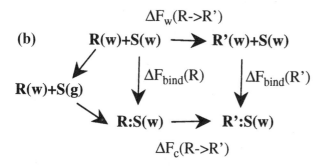

Figure 1 Thermodynamic cycles for solvation and binding. (a) Solutes S and S′ in the gas phase (g) and solution (w) and bound to the receptor R in solution. (b) Binding of *S* to the receptors *R* and R′. The oblique arrows on the left remove S to the gas phase, then transfer it to its binding site on R. This pathway allows the calculation of absolute binding free energies.

structural changes and are more straightforward (see below). Since the free energy *F* is a state function, the horizontal and vertical legs both provide routes to the solvation free energy difference between S and S′, for example:

$$\Delta \Delta F_{solv} = \Delta F_{solv}(S') - \Delta F_{solv}(S) = \Delta F_w(S \to S') - \Delta F_g(S \to S')$$

or to the difference in binding free energies of S and S′ to R, i.e.,

$$\Delta \Delta F_{bind} = \Delta F_{bind}(S') - \Delta F_{bind}(S) = \Delta F_c(S \to S') - \Delta F_w(S \to S')$$

(the notations are defined in Figure 1); similarly for the binding free energy difference associated with the R → R′ change. Experimental numbers will usually be obtained from the vertical legs, whereas simulation numbers will often come from the horizontal legs. Precise comparison between free energy simulations and experimental binding constants or partition coefficients requires that the simulations mimic conditions that have a simple relationship to the experimental standard states, as discussed below.

In addition to relative binding or solvation free energies, the left side of Figure 1b shows a pathway (oblique arrows) that can be used to compute absolute binding free energies [17]. The solute S is first removed from solution to the gas phase; this is accomplished by scaling the solute–water interactions gradually to zero in a simulation. Next, the solute is moved from the gas phase into the protein binding site by gradually scaling its interactions from 0 to 1 in a simulation of the protein–solute complex. The removal of S to and from the gas phase is computationally much less expensive than a simulation of the gradual association of R and S in solution. Simulating reversible association would require starting out with the receptor and ligand separated by tens of angstroms in a very large box of water and moving the ligand toward the receptor very slowly, so as to minimize departure of the system from equilibrium.

B. Thermodynamic Perturbation Theory

Free energy calculations rely on the following thermodynamic perturbation theory [6–8]. Consider a system A described by the energy function $E_A = U_A + T_A$. $U_A = U_A(r^N)$ is the potential energy, which depends on the coordinates $r^N = (r_1, r_2, \ldots, r_N)$, and T_A is the kinetic energy, which (in a Cartesian coordinate system) depends on the velocities v^N. For concreteness, the system could be made up of a biomolecule in solution. We limit ourselves (mostly) to a classical mechanical description for simplicity and reasons of space. In the canonical thermodynamic ensemble (constant N, volume V, temperature T), the classical partition function Z_A is proportional to the configurational integral Q_A, which in a Cartesian coordinate system is

$$Q_A(N, V, T) = \int \exp\left[-\frac{U_A(r^N)}{kT}\right] dr^N \tag{1}$$

where k is Boltzmann's constant, $dr^N = dr_1 dr_2 \cdots dr_N$, and the integral is over all possible conformations (of both the biomolecule and the solvent). The absolute Helmholtz free energy is

$$F_A = -kT \ln Z_A = -kT \ln Q_A + c(N, V, T) \tag{2}$$

where $c(N, V, T)$ is a constant arising from the velocity portion of Z_A. The canonical ensemble is not always the most relevant experimentally. In the isothermal-isobaric (N, p, T) ensemble, $Q_A(N, V, T)$ is replaced by

$$Q_A(N, p, T) = \int \exp\left[-\frac{U_A + pV}{kT}\right] dr^N dV \tag{3}$$

and the Helmholtz free energy is replaced by the Gibbs free energy $G_A = -kT \ln Q_A(N, p, T) + c(N, p, T)$.

The Helmholtz free energy can be rearranged to read

$$F_A = -kT \ln \frac{V^{N/3} \int \exp(-U_A/kT) dr}{\int \exp(U_A/kT) \exp(-U_A/kT) dr} + c(N, V, T) \tag{4}$$

$$= kT \ln\langle \exp(U_A/kT)\rangle_A + c(N, V, T) - kT \ln V^{N/3}$$

where the brackets $\langle\rangle_A$ indicate an average over the ensemble of system A and V is the volume. Although in theory the average could be obtained from an MD or MC simulation of A, in practice the quantity to be averaged, $\exp(U_A/kT)$, is largest where the Boltzmann weight $\exp(-U_A/kT)$ is smallest, and vice versa. Therefore, a reliable average cannot be obtained, and absolute free energies cannot normally be calculated from a simulation.

Practical calculations always consider differences between two or more similar systems. Suppose we effect a change in the system such that the potential energy function is changed into

$$U_B = U_A + V_{BA} \tag{5}$$

where V_{BA} denotes an additional, "perturbing," potential energy term. The free energy difference between A and B is

$$F_B - F_A = -kT \ln \frac{Z_B}{Z_A} = -kT \ln \frac{Q_B}{Q_A} \tag{6}$$

which can be rearranged to give [18,19]

$$F_B - F_A = -kT \ln \frac{\int \exp(-U_A/kT)\, \exp(-V_{BA}/kT)dr^N}{\int \exp(-U_A/kT)dr^N} \tag{7}$$

or

$$F_B - F_A = -kT \ln \left\langle \exp\left[-\frac{V_{BA}}{kT} \right] \right\rangle_A \tag{8}$$

The brackets on the right indicate an average over the ensemble of the starting system A, i.e., with Boltzmann weights $\exp(-U_A/kT)$. The last equality also holds for the Gibbs free energy in the (N, p, T) ensemble. Equation (8) leads to practical computation schemes, since the required ensemble average can be calculated (in favorable cases) from a simulation of the starting system, A. Equation (8), which is exact, is nevertheless referred to as a free energy perturbation formula, because it connects the perturbed system B to the reference system A. Changes in the temperature or pressure can be treated as perturbations in the same sense as above, and perturbation formulas analogous to Eq. (8) can easily be derived. Equation (8) also holds for a quantum system if the perturbing Hamiltonian V_{BA} commutes with the Hamiltonian of A (a common situation).

Equations (5)–(8) assume that the energy functions U_A and U_B operate on the same conformation space; i.e., A and B must have the same number N of degrees of freedom. In practice, this almost always implies that A and B have the same number of atoms or particles. Most biochemical changes of interest (e.g., point mutations of a protein) do not obey this requirement, but they can often be made to do so artificially through the use of dummy atoms (see below).

From the integral in the numerator on the right of Eq. (7), the conformations that contribute most to the free energy difference $F_B - F_A$ are those where V_{BA} is large and negative and at the same time the Boltzmann factor $\exp(-U_A/kT)$ is large. If systems A and B are similar, they will occupy similar regions of conformation space; regions with large Boltzmann weights will thus have small $|V_{BA}|$, while regions where V_{BA} is large and negative will tend to have low Boltzmann weights. The regions important for the averaging process will result from a compromise between these two effects and typically lie near the "edge" of the regions sampled by system A; illustrative examples are given in Figure

2. This suggests that only in cases where the perturbation energy V_{BA} is very small on average (on the order of $kT \sim 1$ kcal/mol) will Eq. (8) be directly applicable. More often, the transformation from A to B must be divided into several discrete steps, say n, to each of which corresponds a perturbation energy on the order of V_{BA}/n, which can be made arbitrarily small by increasing n.

The free energy perturbation formula (8) can be expanded in powers of V_{BA}, or equivalently in powers of $-1/kT$. Truncation at various orders gives free energy perturbation formulas in the true sense. Although high order terms are difficult to calculate because of sampling problems similar to those described above, expansions to low orders (1)–(4) are often more robust numerically than the original formula (8) and are especially useful for treating many small perturbations of a single reference system [20–22]. Because $\langle e^{-V_{BA}/kT} \rangle_A$ has the form of a moment-generating function [23], the coefficients of the expansion involve the cumulants C_n of V_{BA}:

Figure 2 (a) Mutation of argon into xenon in aqueous solution. Illustration of the averaging used to obtain $\langle V_{BA} \rangle_A$ [the first cumulant C_1, Eqs. (9) and (13); adapted from Ref. 73]. The only solute–solvent interactions are van der Waals interactions between argon/xenon and the water oxygens (the hydrogen van der Waals radius is zero for the force field used here). The mutation thus consists in changing the van der Waals parameters of the solute from those of argon to those of xenon, and $V_{BA} = U_{vdW}$ (xenon) $- U_{vdW}$ (argon), the difference between the solute–solvent interaction calculated with the argon and xenon van der Waals parameters. It is useful to write $V_{BA} = \sum_i v_{BA}(r_i)$, where the summation is over all water molecules, $v_{BA}(r_i)$ is the contribution of water molecule i, and r_i its distance from the solute. To obtain, e.g., C_1, averaging is performed over an argon–water simulation. For this simple mutation, $\langle V_{BA} \rangle_A$ depends only on the radial distribution of water oxygens around the solute. The density of water at a distance r from the solute is a function only of r, $\rho g(r)$, where ρ is the bulk water density and $g(r)$ is known as the radial distribution function. One can show that $\langle V_{BA} \rangle_A = \int_0^\infty 4\pi r^2 g(r) \rho v_{BA}(r)$. The functions $g(r)$, $v_{BA}(r)$ (in kT units), and $r^2 g(r) v_{BA}(r)$ are shown in the plot. To accurately integrate the latter function, the reference state simulation must provide adequate sampling of the region near the function's peak; i.e., the peak must overlap sufficiently with $g(r)$. The sampling required to properly average $e^{-V_{BA}/kT}$ cannot be analyzed quite as simply here; however, analogous considerations apply; see (b). (b) Charge insertion on an atom in a protein; illustration of the averaging used in the perturbation formula [Eq. (8), upper panel] and in the second derivative of the free energy [Eq. (12), lower panel]. The data shown correspond to charge insertion on an α-carbon of cytochrome c. (Adapted from Ref. 22.) The perturbation energy is $V_{BA} = q\Phi_q$, where Φ_q is the electrostatic potential at the site of charge insertion. Ensemble averages $\langle \rangle$ are over the reference state, i.e., before charge insertion. The probability distribution g of $\Delta V = V_{BA} - \langle V_{BA} \rangle$ is shown as dots, along with a Gaussian fit (short dashes). The averages sought are $\langle e^{-\Delta V/kT} \rangle = \int e^{-\Delta V/kT} g(\Delta V) d\Delta V$ (upper panel) and $\langle \Delta V^2 \rangle = \int \Delta V^2 g(\Delta V) d\Delta V$. For accurate integration, the simulation must sample the configurations that contribute most to the functions being integrated. With $q = e/10$, the peak in $e^{-\Delta V/kT} g(\Delta V)$ is indeed adequately sampled by the simulation. The tails of $g(\Delta V)$ are not sampled; their contributions to the ensemble averages $\langle e^{-\Delta V/kT} \rangle$ and $\langle \Delta V^2 \rangle$ can be estimated by assuming the tails are Gaussian (shaded in figure) and represent only $\approx 10\%$ and 3%, respectively. For larger perturbing charges ($q > e/10$), the contribution of the negative tail to the exponential average grows very rapidly, and the exponential formula [Eq. (8)] becomes unreliable. The first few free energy derivatives, in contrast, continue to be well sampled. However, as q increases, the contribution of higher derivatives (which are more difficult to sample) to the free energy increases. (From Ref. 22.)

(a)

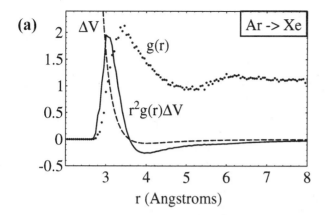

(b)

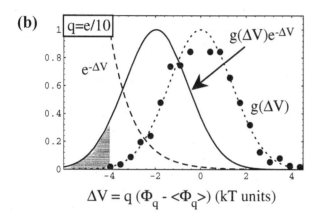

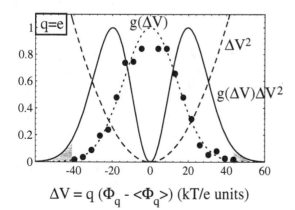

$$F_B - F_A = -kT \sum_{i=1}^{\infty} \frac{C_n}{n!} \left(\frac{-1}{kT} \right)^n \tag{9}$$

The cumulants [23] are simple functions of the moments of the probability distribution of V_{BA}, e.g., $C_1 = \langle V_{BA} \rangle_A$, $C_2 = \langle (V_{BA} - \langle V_{BA} \rangle_A)^2 \rangle_A$, $C_3 = \langle (V_{BA} - \langle V_{BA} \rangle_A)^3 \rangle_A$, $C_4 = \langle (V_{BA} - \langle V_{BA} \rangle_A)^4 \rangle_A - 3C_2^2$. The rate of convergence of the expansion is determined by the deviations of V_{BA} from its mean. Truncation of the expansion at order 2 corresponds to a linear response approximation [22] and is equivalent to assuming that V_{BA} is Gaussian (with zero moments and cumulants beyond order 2). To this order, the mean and width of the distribution determine the free energy; to higher orders, the detailed shape of the distribution contributes.

Two examples of "alchemical" perturbations are shown schematically in Figure 2. They involve, respectively, the transformation of an argon atom into xenon in solution and the insertion of a point charge onto a single atom in a protein. The expression of the perturbing energy V_{BA} is given in each case, assuming the potential energy function has the typical form used in many current biomolecular force fields [4,5]. These simple examples are chosen because the perturbation V_{BA} depends essentially on a single variable in each case: the radial distribution of water density around the argon atom and the electrostatic potential on the charge insertion site. This allows a simple graphical analysis of the averaging required to use the perturbation formula (8) and/or to obtain the first few cumulants [Eq. (9)].

Although the thermodynamic perturbation approach [Eq. (8)] is important conceptually, it is usually not the most efficient numerically. A more fruitful approach in practice is to introduce an explicit transformation of the potential energy function such that it is gradually changed from its starting form U_A into the final form U_B through the variation of one or a few convenient parameters or "coupling coordinates." The simplest such approach is linear interpolation between U_A and U_B, i.e., we introduce the "hybrid" energy function

$$U(r^N; \lambda) = (1 - \lambda)U_A(r^N) + \lambda U_B(r^N) \tag{10}$$

which represents a system that is a mixture of A and B. This hybrid system is usually such that it would be impossible to prepare experimentally yet is straightforward to prepare in a simulation model. By slowly varying the coupling coordinate λ from 0 to 1, the hybrid system is gradually changed from A into B. More than one coupling parameter and more complicated (e.g., nonlinear) functional forms $U(\lambda_1, \lambda_2, \ldots)$ can be used as long as the starting energy function is U_A and the final one is U_B. For example, λ_1 and λ_2 could be coupling parameters applied respectively to van der Waals and electrostatic terms in the energy function. To obtain a practical computation scheme, we notice that the derivative of the free energy with respect to (one of) the coupling coordinate(s) has the form

$$\frac{\partial F}{\partial \lambda}(\lambda) = \frac{\int \partial U(r^N; \lambda)/\partial \lambda \, \exp[-U(r^N; \lambda)/kT]dq}{\int \exp[-U(r^N; \lambda)/kT]dq} = \left\langle \frac{\partial U}{\partial \lambda} \right\rangle_\lambda \tag{11}$$

where the brackets $\langle \rangle_\lambda$ indicate an average over the ensemble corresponding to the hybrid energy function $U(r^N; \lambda)$. Second and higher derivatives can be obtained similarly. In the general case of several coupling parameters $\lambda_1, \lambda_2, \ldots$, the second derivatives are

$$\frac{\partial^2 F}{\partial\lambda_i\partial\lambda_j}(\lambda_1, \lambda_2, \ldots) = \left\langle \frac{\partial^2 U(r^N; \lambda_1, \lambda_2, \ldots)}{\partial\lambda_i\partial\lambda_j} \right\rangle_{\lambda_1,\lambda_2,\ldots}$$

$$- \frac{1}{kT}\left[\left\langle \frac{\partial U}{\partial\lambda_i}\frac{\partial U}{\partial\lambda_j} \right\rangle_{\lambda_1,\lambda_2,\ldots} - \left\langle \frac{\partial U}{\partial\lambda_i} \right\rangle_{\lambda_1,\lambda_2,\ldots}\left\langle \frac{\partial U}{\partial\lambda_j} \right\rangle_{\lambda_1,\lambda_2,\ldots} \right] \quad (12)$$

The same relations (11) and (12) hold for the Gibbs free energy in the (N, p, T) ensemble. Equation (11) is also valid for a quantum mechanical system. Note that for a linear coupling scheme such as Eq. (10), the first term on the right of Eq. (12) is zero; the matrix of second derivatives can then be shown to be definite negative, so that the free energy is a concave function of the λ_i.

The free energy derivatives are also related to the coefficients in a Taylor expansion of the free energy with respect to λ. In the case of linear coupling, we let $V_{BA} = \lambda(U_B - U_A)/kT$ in Eq. (9); we obtain

$$\frac{\partial^n F}{\partial\lambda^n} = kTc_n \quad (13)$$

where c_n is a cumulant of $(U_B - U_A)/kT$ (in the A ensemble).

To compute derivatives of F at the point $(\lambda_1, \lambda_2, \ldots)$ numerically, a simulation is performed with the hybrid energy function $U(r^N; \lambda_1, \lambda_2, \ldots)$ and the appropriate energy derivatives are averaged. This procedure is repeated for a few discrete values of the coupling parameter(s), spanning the interval from 0 to 1. The resulting free energy derivatives are then interpolated and integrated numerically to yield $F_B - F_A$. Many applications use a linear coupling and rely only on first derivatives calculated at evenly spaced points; however, efficiency can be improved by using somewhat more complicated schemes [20,24]. This general approach is known as *thermodynamic integration*.

C. Dummy Atoms and Endpoint Corrections

Figure 3 represents an illustrative biological application: an Asp $\rightarrow$ Asn mutation, carried out either in solution or in complex with a protein [25,26]. The calculation uses a hybrid amino acid with both an Asp and an Asn side chain. For convenience, we divide the system into subsystems or "blocks" [27]: Block 1 contains the ligand backbone as well as the solvent and protein (if present); block 2 is the Asp moiety of the hybrid ligand side chain; block 3 is the Asn moiety. We effect the "mutation" by making the Asn side chain gradually appear and the Asp side chain simultaneously disappear. We choose initially the hybrid potential energy function to have the form

$$U(\lambda) = U_{11} + U_{22} + U_{33} + (1 - \lambda)U_{12} + \lambda U_{13} \quad (14)$$

where U_{ii} represents the interactions within block i and U_{1i} represents the interactions between blocks 1 and i. Blocks 2 and 3 (the two side chains) do not interact. At the Asp endpoint ($\lambda = 0$), the Asn side chain has no interactions with its environment but retains its internal interactions; similarly for Asp at the Asn endpoint ($\lambda = 1$). At intermediate values of λ, the interactions between each side chain moiety and block 1 are weighted by λ or $1 - \lambda$.

This protocol has an important feature: Neither endpoint corresponds exactly to the biomolecule of interest. Each endpoint represents an artificial construct involving several

(a)

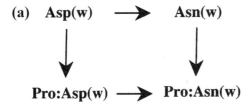

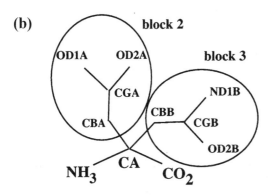

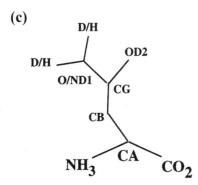

Figure 3   Mutation of a ligand Asp into Asn in solution and bound to a protein. (a) Thermodynamic cycle. (b) Dual topology description: a hybrid ligand with two side chains. "Blocks" are used to define the hybrid energy function [Eq. (14)]. Only the ligand is shown; the environment is either solvent or the solvated protein. (c) Single-topology description.

"dummy" atoms, i.e., atoms that have no interactions with their environment. An important question is, therefore, How can we relate the endpoint systems to the exact systems of interest? The use of dummy atoms is quite general in alchemical free energy calculations; despite the great flexibility with which the potential energy function can be manipulated, it is often difficult, inefficient, or impossible to connect the exact biochemical systems of interest through a thermodynamic perturbation or thermodynamic integration calculation.

In the present case, each endpoint involves—in addition to the fully interacting solute—an intact side chain fragment without any interactions with its environment. This fragment is equivalent to a molecule in the gas phase (acetamide or acetate) and contributes an additional term to the overall free energy that is easily calculated from ideal gas statistical mechanics [18]. This contribution is similar but not identical at the two endpoints. However, the corresponding contributions are the same for the transformation in solution and in complex with the protein; therefore, they cancel exactly when the upper and lower legs of the thermodynamic cycle are subtracted (Fig. 3a).

Although the above protocol leads to a simple endpoint correction, it poses sampling problems in the simulation steps immediately before the endpoints. Indeed, as $\lambda \to 0$, the CA—CBB bond force constant is reduced to a very small value [similarly for CA—CBA (Fig. 3b) as $\lambda \to 1$], and the corresponding side chain fragment begins to wander extensively. However, its excursions will never approach the free wandering of an ideal gas fragment, because of the finite (albeit small) force constant and the limited length of the simulation. Thus, the system will not approach the $\lambda = 0$ or $\lambda = 1$ endpoints closely or smoothly enough, and a discontinuity will occur in the calculated free energy. Mathematically, the free energy derivative can be shown in this example to go to infinity at the endpoints as $1/\lambda$ [or $1/(1 - \lambda)$]; i.e., the free energy varies very rapidly and has a singularity at each endpoint [28].

A better protocol scales the non-bonded interactions U_{12}^{nb}, U_{13}^{nb} but leaves the covalent terms intact [25]. This removes the sampling problems discussed (and the free energy singularity) but substantially complicates the form of the endpoint correction. Indeed, at each endpoint, the ''dummy'' side chain fragments remain attached to other interacting atoms through covalent energy terms. Therefore, they contribute to the partition function at the endpoints through both kinetic and potential energy terms. Although the corresponding free energy contribution usually cannot be calculated rigorously, approximate calculations can be made. In the present example, if one neglects (or turns off) dihedral terms that couple the interacting and noninteracting parts of the ligand, then the dummy atoms are coupled to the other atoms only through bond and angle terms, and these can be factored out of the configurational integral [28]. The resulting free energy contribution then cancels exactly when the upper and lower legs of the thermodynamic cycle of Figure 3a are subtracted. Other protocols make use of positional restraints for the noninteracting groups at the endpoint, tethering them to a specific point or region instead of to interacting atoms [29,30]. The free energy associated with this tethering can be calculated exactly and can be shown to cancel in a well-chosen thermodynamic cycle. For these reasons, there is usually no need to calculate such endpoint corrections explicitly in applications that compare two legs of a cycle. However, when a single leg of a cycle is being treated, considerable care must be used to correct for dummy atom contributions. For a detailed discussion of the finer points associated with dummy atoms and endpoint corrections, see Ref. 28.

The ''annihilation'' of a particle in a condensed environment to give a dummy atom can lead to another endpoint singularity, arising from the van der Waals energy term. Indeed, if the van der Waals interaction energy is scaled by a factor λ^n, the free energy derivative goes to infinity as $\lambda^{n/4-1}$ when $\lambda \to 0$ [31]. If $n = 1$ [linear coupling, as in Eq. (10)], there is a free energy singularity and the free energy derivative must be extrapolated to the endpoint $\lambda = 0$ with care (e.g., using the theoretical form $\lambda^{-3/4}$). If $n \geq 4$, the free energy derivative remains finite and there is no singularity.

Because the preceding protocols use a ligand with two side chains, they are often referred to as "dual topology" methods. A slightly different approach is shown in Figure 3c. Here, the ligand has a single side chain; the transformation changes the OD2 atom of Asp into the ND2 of Asn, and the Asn hydrogens HD21, HD22 are represented by dummy atoms in Asp. This is known as a "single-topology" approach; i.e., a single side chain is used and the number of dummy atoms is kept to a minimum. This approach imposes the same position on the two side chains, which can be a problem if the two ligands occupy different rotameric states in the complex, for example. The internal parameters of the side chain (bond, angle, and dihedral parameters; atomic van der Waals parameters; and charges) must all be altered during the mutation. A change in the length of a covalent bond, for example, contributes the following term to the free energy derivative [32]:

$$\frac{\partial F}{\partial \lambda} = -2k_{\mathrm{h}}(\langle b \rangle_\lambda - b_0(\lambda))\frac{db_0}{d\lambda} \tag{15}$$

where k_{h} is the bond force constant, b the instantaneous bond length, and b_0 the target bond length, which is a function of the coupling parameter λ. If the bond is stretched on average $[\langle b \rangle_\lambda > b_0(\lambda)]$, then the free energy to increase b_0 is negative. Changes in covalent bond lengths require careful sampling, as a bond's vibrational frequency is usually distant from the frequencies of most of the degrees of freedom around it, making equipartition of energy with the bond's surroundings inefficient. The use of Langevin dynamics can significantly improve the sampling. We return to free energy changes as a function of a conformational coordinate in Section IV; for a detailed discussion of single- vs. dual topology approaches, see Refs. 28 and 33.

D. Other Thermodynamic Functions

The free energy is the most important equilibrium thermodynamic function, but other quantities such as the enthalpy and entropy are also of great interest. Thermodynamic integration and perturbation formulas can be derived for them as well. For example, the derivative of the entropy can be written [24]

$$-T\frac{\partial S}{\partial \lambda} = \frac{1}{kT}\left(\left\langle \frac{\partial U}{\partial \lambda}U \right\rangle_\lambda - \left\langle \frac{\partial U}{\partial \lambda} \right\rangle_\lambda \langle U \rangle_\lambda \right) \tag{16}$$

In the (N, p, T) ensemble, U is replaced by $U + pV$ in the right-hand terms. For the energy $\overline{E} = \langle E \rangle = F + TS$, one obviously has

$$\frac{\partial \overline{E}}{\partial \lambda} = \frac{\partial F}{\partial \lambda} + T\frac{\partial S}{\partial \lambda} \tag{17}$$

More directly,

$$\overline{E}_\mathrm{B} - \overline{E}_\mathrm{A} = \langle U_\mathrm{B} \rangle_\mathrm{B} - \langle U_\mathrm{A} \rangle_\mathrm{A} \tag{18}$$

(A and B have the same number of degrees of freedom, so the mean kinetic energies cancel). In the (N, p, T) ensemble, the enthalpy change is

$$H_\mathrm{B} - H_\mathrm{A} = \langle U_\mathrm{B} \rangle_\mathrm{B} - \langle U_\mathrm{A} \rangle_\mathrm{A} + p(\langle V_B \rangle_\mathrm{B} - \langle V_\mathrm{A} \rangle_\mathrm{A}) \tag{19}$$

The second pV term is normally negligible [atmospheric pressure corresponds to an enthalpy of 1.5×10^{-5} kcal/(mol. Å^3)]. Thus, the enthalpy can safely be calculated in practice by integrating Eq. (17). Calculations of S, H, and $\bar{E}$ are intrinsically less precise than free energy calculations; for example, the direct enthalpy calculation [Eq. (19)] involves subtracting two large energies averaged over two separate simulations. However, with increasing computer power, such calculations are rapidly becoming routine [12].

E. Free Energy Component Analysis

In current force fields, the potential energy U is usually a sum of pairwise non-bonded terms (electrostatic and van der Waals) and bonded terms involving groups of two to four atoms. Thus, the energy can be directly decomposed into contributions from groups of atoms and from different energy terms (electrostatic, van der Waals, bonded). From Eq. (11), the free energy derivative $\partial F / \partial \lambda$ can be decomposed in the same way. This leads to a decomposition of the free energy based on groups of atoms and/or different energy terms [27]. For the Asp $\rightarrow$ Asn example of Figure 3, we have

$$U(\lambda) = \sum_i [U_{i2}^{nb}(\lambda) + U_{i3}^{nb}(\lambda)] + U_{12}^{b} + U_{13}^{b} + U_{11} + U_{22} + U_{33} \tag{20}$$

where the indices 1, 2, 3 represent the blocks of the system (defined above) and the sum is over atoms of block 1. Superscripts b and nb refer to bonded and non-bonded energy terms, respectively. Atom i of block 1 contributes a term

$$\frac{\partial F_i}{\partial \lambda} = \left\langle \frac{\partial (U_{i2}^{nb} + U_{i3}^{nb})}{\partial \lambda} \right\rangle_\lambda$$

to the free energy derivative and a term

$$F_i = \int_0^1 \frac{\partial F_i}{\partial \lambda} d\lambda$$

to the free energy difference. Summing over groups of atoms, one obtains the free energy contribution, or component, associated with each group. Unlike the total free energy, the free energy components are not state functions and thus are path-dependent [32,34]; i.e., they depend on the exact way U_A is transformed into U_B in a particular computation. Nevertheless, a number of studies have shown that when treated with care they can provide important insights into the microscopic interactions important for binding; see, e.g., Ref. 26 and references therein.

III. STANDARD BINDING FREE ENERGIES

The study of receptor–ligand binding is one of the most important applications of free energy simulations [35]. To study receptor–ligand binding theoretically, one must first partition the conformational space into "bound" and "unbound" states. There is no unique way to do this, but in practical situations there is often a natural choice. Thus, conformations where the ligand is within a well-defined binding pocket could be labeled "bound." Because there is likely to be an energy barrier at the boundary of such a pocket, ligand conformations near the boundary will often have high energies and low statistical weights. Therefore, they will not contribute greatly to thermodynamic properties such as the binding constant, which will consequently be robust with respect to the exact definition of the pocket. In addition, when the binding of two similar ligands to a receptor is being compared, there will be some cancellation of the boundary region contributions of each ligand.

The equilibrium binding constant is

$$K_b = \frac{\rho_{RL}}{\rho_R \rho_L} \tag{21}$$

where ρ_R, ρ_L, and ρ_{RL} are the concentrations (or number densities) of receptor, ligand, and complex and K_b has units of volume. The chemical potential of each species in solution is [18,35]

$$\mu_A = kT \ln\left[\frac{\rho_A}{\rho^0}\right] - kT \ln\left[\frac{Z_A}{Z_0 V \rho^0}\right] \tag{22}$$

where A = RL, R, or L; ρ^0 is the standard state concentration, V the volume of the system, Z_A the partition function of A in solution, and Z_0 the partition function of the solution without A. The condition for equilibrium is

$$-kT \ln K_b \rho^0 = \Delta F_b^0 \tag{23}$$

where $\Delta F_b^0 = \mu_{RL}^0 - \mu_R^0 - \mu_L^0$ is the standard binding free energy—the free energy to bring two single molecules R and L together to form a complex RL when the concentrations of all species are fixed at ρ^0.

To relate the standard binding free energy to free energies that can be obtained from simulations, we use

$$\Delta F_b^0 = -kT \ln\left[\frac{Z_{RL} Z_0 V \rho^0}{Z_R Z_L}\right] = -kT \ln\left[\frac{Q_{RL} Q_0 V \rho^0}{Q_R Q_L}\right] \tag{24}$$

$$= -kT \ln\left[\frac{Q_{RL} \rho^0}{Q_R Q_{L0}/V}\right] + kT \ln\left[\frac{Q_L}{Q_{L0} Q_0}\right]$$

where the second equality takes into account a cancellation of the velocity partition functions and Q_{L0} is the configuration integral of the ligand alone (i.e., in the gas phase). The second term in Eq. (24) is the free energy to "annihilate" L in solution, i.e., the free energy to reversibly turn off its interactions with the surrounding solution, effectively transferring it to the gas phase. This operation can be done with the ligand either free to move or positionally constrained (fixed center of mass); the free energy is the same, by translational invariance of the solution. The first term is the free energy to "annihilate" the ligand in the binding site, with its center of mass fixed [29]. The standard concentration ρ^0 appears explicitly here. Because the binding site does not have translational symmetry, this free energy takes the form of an average over all positions in the active site (see Ref. 29). To compute it, the ligand could in principle be fixed and annihilated in each of these positions, but this is impractical. A much better scheme does the annihilation in two steps [29,36]. First, the ligand's interactions with its surroundings are reversibly turned off, and at the same time a harmonic restraining potential is turned on, which confines the ligand to a region centered on the binding site and roughly equal to or slightly larger than it. This free energy does not depend on the standard concentration. Second, the free energy difference between the restrained and fixed ligands is calculated analytically; it has the form $-kT \ln \rho^0 (2\pi kT/k_h)^{3/2}$, where k_h is the force constant for the harmonic restraint, and the standard concentration appears explicitly. This scheme is illustrated in Figure 4. For a moderately large ligand (i.e., significantly larger than a water molecule), it becomes necessary to also introduce rotational restraints; their contribution to the free energy is obtained analytically in close analogy to the positional restraint [30]. The use of translational and rotational restraints makes it possible to estimate the "cratic" contribution to the binding free energy [30]. For large ligands, the endpoints of the above "annihilation" processes require very careful sampling and a proper extrapolation of the free energy derivatives.

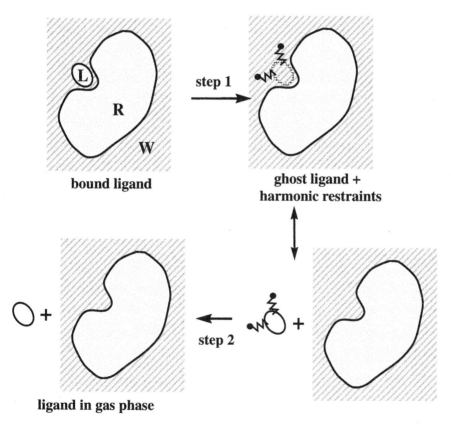

Figure 4 Thermodynamic pathway for the calculation of a standard binding free energy. In the first step, the interactions of the ligand (L) with its environment (receptor R + solvent W) are gradually turned off; at the same time, one or more harmonic restraints are turned on, restricting the translation and possibly the rotation of the ligand. The restrained, ghost ligand (at right in figure) is strictly equivalent to the restrained ligand in the gas phase. The second step removes the harmonic restraints; the corresponding free energy has a simple analytical form (see text).

Many applications are concerned only with binding free energy differences. Comparing the binding of two ligands, L and L′, to the receptors R and R′, we have

$$\Delta\Delta F_b^0 (L, L') = \Delta F_b^0 (RL') - \Delta F_b^0(RL) = -kT \ln\left[\frac{Z_{RL'}}{Z_{RL}}\right] + kT \ln\left[\frac{Z_{L'}}{Z_L}\right] \quad (25)$$

$$\Delta\Delta F_b^0 (R, R') = \Delta F_b^0(R'L) - \Delta F_b^0(RL) = -kT \ln\left[\frac{Z_{R'L}}{Z_{RL}}\right] + kT \ln\left[\frac{Z_{R'}}{Z_R}\right] \quad (26)$$

Thus, the standard state concentration cancels from these double free energy differences. The calculation can be done by mutating L to L′ (or R to R′) both in the complex and in solution (horizontal legs of Fig. 1a).

IV. CONFORMATIONAL FREE ENERGIES

Free energy changes associated with conformational changes are the second major application of free energy calculations. Simple examples are the free energy profile for rotating a protein side chain around one or more dihedral torsion angles or for modifying the length of an individual covalent bond. Recent applications have been as complex as the unfolding of a protein [13]. In all cases, a reaction coordinate q is defined, involving one or more conformational degrees of freedom. The Helmholtz free energy $W(q)$ along this coordinate is a configuration integral over all other degrees of freedom and takes the form

$$W(q) = -kT \ln P(q) \quad (27)$$

where $P(q)$ is the reaction coordinate probability density. $W(q)$ is known as the potential of mean force (pmf). When comparing two or a few conformations separated by very low energy barriers ($\sim kT \sim 1$ kcal/mol), the relative probabilities of each conformation can be estimated from an ordinary simulation, and Eq. (27) can be used directly to obtain the relative free energies. When the conformations are separated by larger barriers, barrier crossings in a simulation will be rare and $P(q)$ statistically unreliable. The system must then be driven along q with an appropriate set of constraints or restraints. The formalism is simpler in the case of restraints, so this case is treated first.

A. Conformational Restraints or Umbrella Sampling

To bias the sampling toward a region of interest that would not otherwise be significantly populated, a restraining potential $U_r(q)$ is added to the potential energy of the system. U_r is often referred to as an umbrella potential [37]. For concreteness, we assume the harmonic form

$$U_r(q; \lambda) = k_h[q - q_0(\lambda)]^2 \quad (28)$$

where k_h is a force constant, $q_0(\lambda)$ is a target value of q, and λ a coupling parameter. However, umbrella potentials are by no means limited to a harmonic form (see below and Section VI.B). The reaction coordinate q could be a dihedral angle, a distance between two selected atoms, or a more complicated, collective degree of freedom such as a normal mode amplitude. $q_0(\lambda)$ is constructed so that as the coupling coordinate λ varies, $q_0(\lambda)$ traverses the region or regions of interest.

The free energy difference between two stable conformations can be obtained by a thermodynamic integration approach [38,39]. Let q_A and q_B represent the centers of the two corresponding energy wells. The free energy derivative is seen to be

$$\frac{\partial F}{\partial \lambda}(\lambda) = -2k_h(\langle q \rangle_\lambda - q_0(\lambda))\frac{dq_0}{d\lambda}(\lambda) \tag{29}$$

which can be obtained from a simulation with the restraining potential $U_r(q; \lambda)$. Equation (29) is a generalization of Eq. (15) (where $q = b$). Integrating between q_A and q_B gives the free energy difference between the two wells, but with restraints present at each endpoint. Additional steps are needed in which the restraints are removed at the endpoints. The corresponding free energies can be obtained from the thermodynamic perturbation formula (8),

$$\Delta F(\text{restrained} \rightarrow \text{unrestrained}) = -kT \ln\langle \exp(U_r/kT) \rangle_{\lambda_A} \tag{30}$$

where the averaging is performed over the restrained endpoint simulation, i.e., $q_0(\lambda_A) = q_A$; a similar calculation is made at the B endpoint. This approach is easily generalized to nonharmonic restraint terms and to cases where several restraint terms are used. Thus, if restraints are applied to several dihedral angles or several interatomic distances, each will contribute a term of the form (29) to the free energy derivative.

Application of Eq. (30) corrects the free energies of the endpoints but not those of the intermediate conformations. Therefore, the above approach yields a free energy profile between q_A and q_B that is altered by the restraint(s). In particular, the barrier height is not that of the natural, unrestrained system. It is possible to correct the probability distributions P_r observed all along the pathway (with restraints) to obtain those of the unrestrained system [8,40]. From the relation $P(q)Z_{ur} = P_r(q)Z_r \exp(U_r/kT)$ and Eqs. (6)–(8), one obtains

$$P(q) = P_r(q)e^{U_r(q)/kT}/\langle e^{U_r/kT} \rangle_r \tag{31}$$

$$= P_r(q) e^{U_r(q)/kT}\langle e^{-U_r/kT} \rangle_{ur} \tag{32}$$

where the subscripts r and ur refer to the restrained and unrestrained systems, respectively. For reasons already discussed [see Fig. 2 and the discussion following Eq. (7)], the resulting $P(q)$ is expected to be accurate only if there is a large overlap between probable conformations of the unrestrained system and conformations where the restraint energy is small. Thus in practice, $q_0(\lambda)$ must be close to a stable energy minimum of the unrestrained system, and $P(q)$ will be accurate only close to $q_0(\lambda)$. To obtain $P(q)$ over a broader range, a series of umbrella potentials is required, covering a range of q_0 values. Let $U_{r'}$ be a second umbrella potential, corresponding to a $q_0(\lambda')$ slightly displaced relative to $q_0(\lambda)$. One can show that [40]

$$P(q) = P_{r'}(q) e^{U_{r'}(q)/kT}/(\langle e^{U_r/kT} \rangle_r \langle e^{(U_{r'}-U_r)/kT} \rangle_{r'}) \tag{33}$$

This formula is expected to be accurate close to $q_0(\lambda')$. Continuing in this manner, one can obtain accurate formulas for $P(q)$ over a broad range of q, provided the regions sampled with the successive umbrellas overlap.

For many problems, the ideal umbrella potential would be one that completely flattens the free energy profile along q, i.e., $U_r(q) = -W(q)$. Such a potential cannot be determined in advance. However, iterative approaches exist that are known as adaptive umbrella sampling [41]. Such approaches are especially important for large-scale sampling of many very different conformations, such as folded and unfolded conformations of a protein or peptide. Recent applications to protein folding have used the potential energy as a "reaction coordinate" q, building up an umbrella potential that leads to a flat probability distribution and smooth sampling over a broad range of potential energies [42–45].

B.   Weighted Histogram Analysis Method

It is often of interest to investigate not a one-dimensional but a two- or higher dimensional reaction coordinate. Free energy maps of polypeptides as a function of a pair of (ϕ, ψ) backbone torsion angles are an example. Equation (33) can be used to explore more than one coordinate q by using sets of umbrellas whose minima span a two-dimensional grid covering the range of interest. However, as the number of dimensions increases, propagation of error through Eq. (33) increases rapidly, and this approach becomes increasingly difficult. The weighted histogram analysis method (WHAM) is an alternative approach designed to minimize propagation of error by making optimal use of the information from multiple simulations with different umbrella potentials [42,46].

We consider the case of a two-dimensional reaction coordinate (q, s) first. R simulations are carried out, each having its own restraint energy term $U_j(q, s)$. The (q, s) values observed in each simulation j are binned and counted, giving a series of R two-dimensional histograms. Let the bins along q be indexed by k and those along s by l; let the number of counts in each bin in simulation j be $n_{j,kl}$, and let $N_j = \sum_{kl} n_{j,kl}$ be the total counts in simulation j. Let $c_{j,kl} = \exp[-U_j(q_k, s_l)/kT]$, where q_k and s_l are the centers of the bins k and l. The problem is to combine the histograms to obtain an estimate of the probability distribution $p_{kl}^0 = P(q_k, s_l)$ of the unrestrained system. Making use of Eq. (31), assuming the observed counts $n_{j,kl}$ follow a multinominal distribution, and maximizing a likelihood function, one obtains [42,46] the WHAM equations:

$$p_{kl}^0 = \frac{\sum_j n_{j,kl}}{\sum_j N_j f_j c_{j,kl}} \tag{34}$$

$$f_j = \frac{1}{\sum_{kl} c_{j,kl} p_{kl}^0} \tag{35}$$

Here, j runs over all simulations and k, l run over all bins. These equations can be solved iteratively, assuming an initial set of f_j (e.g., $f_j = 1$), then calculating p_{kl}^0 from Eq. (34) and updating the f_j by Eq. (35), and so on, until the p_{kl}^0 no longer vary, i.e., the two equations are self-consistent. From the $p_{kl}^0 = P(q_k, s_l)$ and Eq. (27), one then obtains the free energy of each bin center (q_k, s_l). Error estimates are also obtained [46]. The method can be applied to a one-dimensional reaction coordinate or generalized to more than two dimensions and to cases in which simulations are run at several different temperatures [46]. It also applies when the reaction coordinates are alchemical coupling coordinates (see below and Ref. 47).

C. Conformational Constraints

The foregoing approaches used an umbrella potential to restrain q. The pmf $W(q)$ can also be obtained from simulations where q is constrained to a series of values spanning the region of interest [48,49]. However, the introduction of rigid constraints complicates the theory considerably. Space limitations allow only a brief discussion here; for details, see Refs. 8 and 50–52.

To obtain thermodynamic perturbation or integration formulas for changing q, one must go back and forth between expressions of the configuration integral in Cartesian coordinates r^N and in suitably chosen generalized coordinates u^N [51]. This introduces Jacobian factors

$$J(q) = \det \left[\frac{\partial r_i}{\partial u_j}(q) \right]$$

into the formulas (where $i, j = 1, \ldots, N$ and "det" represents the matrix determinant). Furthermore, it becomes necessary to perform averages in an ensemble where q is fixed (at some value q_0) but the conjugate momentum p_q is unconstrained [50,52]. Indeed, we seek the probability distribution $P(q)$ of the natural system, whose momentum is not subjected to any particular constraints. This is not a problem in a Monte Carlo simulation, where the configurational degrees of freedom can be sampled without any assumptions about the velocities. But in a molecular dynamics simulation, fixing $q = q_0$ immediately constrains the conjugate momentum to be zero. Averages over such a simulation must therefore be corrected to remove the biasing effect of the momentum constraint. This introduces factors containing the mass-metric tensor; if q is one-dimensional, this tensor is a scalar function

$$Z(r^N; q) = \sum_i^N \frac{1}{m_i} \left(\frac{\partial q}{\partial r_i} \right)^2$$

where m_i is the mass of the particle corresponding to the coordinate r_i [50]. The free energy to change the reaction coordinate from $q - \delta q$ to $q + \delta q$ takes the rather formidable form [51]

$$W(q + \delta q) - W(q - \delta q)$$
$$= -kT \ln \left[\frac{\langle J(q)^{-1}(q + \delta q)Z(r^N; q)^{-1/2} \exp(-\Delta U^+/kT) \rangle_q}{\langle J(q)^{-1}(q - \delta q)Z(r^N; q)^{-1/2} \exp(-\Delta U^-/kT) \rangle_{q'}} \right] \quad (36)$$

where ΔU^+ and ΔU^- represent the potential energy difference required to change q into $q + \delta q$ or $q - \delta q$, respectively, with all other coordinates unchanged. The brackets indicate averages over a simulation where q is constrained (and the conjugate momentum is consequently zero).

A tractable example is the pmf between two particular particles in a macromolecule as a function of their separation q. The free energy to increase q by δq becomes

$$W(q + \delta q) - W(q) = -kT \ln \left[\left\langle \exp \left[-\frac{\Delta U^+}{kT} \right] \right\rangle_q \right] - 2kT \ln \left[\frac{q + \delta q}{q} \right] \quad (37)$$

where ΔU^+ is defined as above and the angled brackets indicate an ensemble average over the system with the separation fixed at q. The first term is the ordinary free energy perturbation expression, and the second arises from the unbiasing of the velocities. An equivalent formulation can be given that uses the constraint forces (the forces needed to maintain the constraints), which are readily available in many simulation programs [8,53].

If additional, auxiliary constraints are present that are not part of the reaction coordinate (e.g., constraints on covalent bond lengths), the formulas are much more complicated, and the algebra becomes rapidly prohibitive. The same is true when q is a multidimensional coordinate (e.g., a set of dihedrals). Umbrella sampling approaches (discussed in previous sections) are vastly simpler in such cases and appear to be the method of choice for all but the simplest reaction coordinates.

V. ELECTROSTATIC FREE ENERGIES

Many important biochemical processes involve charge separation or transfer. Examples include proton and electron transfer, ion binding, and point mutations that replace a neutral residue with a charged one. To study such processes, alchemical free energy calculations are frequently performed in which a net charge is created, deleted, or displaced (e.g., see Figs. 2b and 3). This poses specific problems, because long-range interactions usually make a significant contribution to the free energy change. Two main families of approximations have been used to treat long-range interactions in electrostatic free energy calculations. The first treats distant regions as a simple dielectric medium, i.e., either as a continuum or as a lattice of polarizable dipoles [9,14]. The second does not introduce a solvent continuum; rather, it assumes periodic boundary conditions and calculates electrostatic interactions over an infinite distance range through lattice summation methods such as the Ewald or particle mesh methods [15]. We discuss these two approaches in turn.

A. Dielectric Reaction Field Approaches

Consider an alchemical transformation of a particle in water, where the particle's charge is changed from 0 to q (e.g., neon $\rightarrow$ sodium; $q = 1$). Let the transformation be performed first with the particle in a spherical water droplet of radius R (formed of explicit water molecules), and let the droplet then be transferred into bulk continuum water. From dielectric continuum theory, the transfer free energy is just the Born free energy to transfer a spherical ion of charge q and radius R into a continuum with the dielectric constant ε_w of water:

$$\Delta G_{\text{Born}} = \frac{q^2}{2R}\left(\frac{1}{\varepsilon_w} - 1\right) \tag{38}$$

This estimate should be accurate if the droplet is sufficiently large (a few tens of angstroms).

The idea of a finite simulation model subsequently transferred into bulk solvent can be applied to a macromolecule, as shown in Figure 5a. The alchemical transformation is introduced with a molecular dynamics or Monte Carlo simulation for the macromolecule, which is solvated by a limited number of explicit water molecules and otherwise surrounded by vacuum. Then the finite model is transferred into a bulk solvent continuum and the transfer free energy is obtained from continuum electrostatics. This involves performing finite-difference Poisson–Boltzmann calculations that take into account the detailed shape of the macromolecule and an infinite bulk solvent continuum [25,26]. This protocol is computationally very efficient, because the macromolecule does not have to be fully solvated in the alchemical simulation step. It includes all electrostatic interactions (provided the alchemical simulation of the finite model is done without a cutoff). The approximation for the long-range interactions is well-defined (in contrast to cutoff treatments), and it can be systematically improved by increasing the size of the explicit solvent region.

A very simple version of this approach was used in early applications. An alchemical charging calculation was done using a distance-based cutoff R_c for electrostatic interactions, either with a finite or a periodic model. Then a cut-off correction equal to the Born free energy, Eq. (38), was added, with the spherical radius taken to be $R = R_c$. This is a convenient but ill-defined approximation, because the system with a cutoff is not equivalent to a spherical charge of radius R_c. A more rigorous cutoff correction was derived recently that is applicable to sufficiently homogeneous systems [54] but appears to be impractical for macromolecules in solution.

An approach widely used in liquid simulations is to include the bulk solvent medium in the alchemical simulation step, calculating the reaction field it produces on-the-fly [15], thus eliminating the need for a subsequent transfer step. This approach can be implemented for a macromolecule in several ways. If the biomolecule is fully solvated with periodic boundary conditions, then a standard liquid simulation approach can be used, in principle. The reaction field on each charge q_i (belonging either to the biomolecule or to a water molecule) due to charges beyond a certain cutoff distance r_{RF} is calculated with a continuum approximation, which assumes that the medium beyond r_{RF} is a homogeneous dielectric polarized by the inner medium [55]. The reaction field has a simple analytical form, but the homogeneity assumption will be accurate only if the system is predominantly made up of solvent and if r_{RF} is greater than the biomolecule's diameter (see Fig. 5b); this implies a large and costly model. Nevertheless, with rapidly increasing computer power, such protocols will become increasingly feasible. The homogeneous medium assumption can be dropped, but this implies a much more complicated and expensive reaction field calculation, which takes into account the exact distribution of solvent and solute in the simulation cell (see Fig. 5b).

A variant of this approach (Fig. 5c) uses a finite (usually spherical) simulation region surrounded by an infinite dielectric continuum [56,57]. The reaction field calculation, again, remains simple as long as the medium outside the simulation region is homogeneous. This implies, again, a large explicit solvent region completely surrounding the biomolecule. The surrounding dielectric continuum has sometimes been replaced by a large lattice of polarizable dipoles that follow a simplified Brownian dynamics [9].

Another variant that may turn out to be the method of choice performs the alchemical free energy simulation with a spherical model surrounded by continuum solvent, neglecting portions of the macromolecule that lie outside the spherical region. The reaction field due to the outer continuum is easily included, because the model is spherical. Additional steps are used to change the dielectric constant of that portion of the macromolecule that lies in the outer region from its usual low value to the bulk solvent value (before the alchemical simulation) and back to its usual low value (after the alchemical simulation); the free energy for these steps can be obtained from continuum electrostatics [58].

(a)

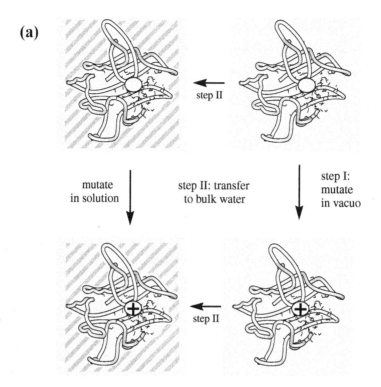

mutate
in solution

step II: transfer
to bulk water

step I:
mutate
in vacuo

(b)

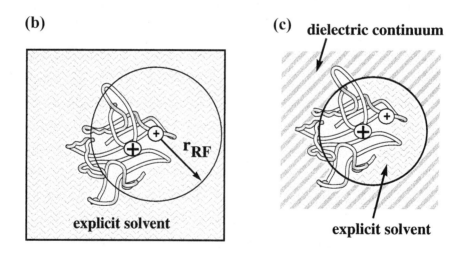

r_{RF}

explicit solvent

(c) **dielectric continuum**

explicit solvent

B. Lattice Summation Methods

Lattice summation methods, particularly the Ewald summation, are used increasingly in biomolecular simulations, as reviewed by Darden [15] and elsewhere [12]. They have been used recently in free energy simulations of charge creation in liquid water [57,59]. The Ewald summation method assumes periodic boundary conditions; for each partial charge q_i in the unit cell, a Gaussian charge distribution centered on q_i but with opposite sign is added to the system, screening the charge–charge interactions and making them short-ranged. The same Gaussian distribution is subtracted to recover the original system; interactions due to the subtracted Gaussians are handled using Fourier transforms.

For systems with a net charge, it is necessary to include a uniform background charge to neutralize the system. Such a uniform neutralizing charge does not affect the forces in the system and contributes a constant term to the potential energy. It is also trivial to implement, so it is usually not a technical concern. However, it must be correctly accounted for when the net charge of the system is modified, as in a charging free energy calculation. The electrostatic part of the potential energy has the form

$$U_{elec} = \frac{1}{2} \sum_{i \neq j} q_i q_j \phi_{Ew}(\mathbf{r}_{ij}) + \frac{1}{2} \sum_i q_i^2 \xi_{Ew} \tag{39}$$

where the sums are over all charges or pairs of charges q_i, q_j; $\mathbf{r}_{ij}$ is the vector connecting charges i and j; ϕ_{Ew} is the Ewald interaction potential, and ξ_{Ew} is the "self-potential." The Ewald potential ϕ_{Ew} includes short-range interactions between the screened partial charges and long-range interactions between the partial charges and the array of subtracted Gaussians. The self-potential represents the interaction of a unit charge with both the periodic images of its two associated Gaussians and the neutralizing background charge [59]. Therefore, to correctly account for the free energy associated with the neutralizing background, it is necessary and sufficient to take into account the self-term in either thermodynamic perturbation or thermodynamic integration formulas. For example, if q_i is

---

Figure 5 Continuum reaction field approaches for electrostatic free energies. (a) A two-step approach. The mutation introduces a positive charge near the center of a protein (shown in tube representation). The mutation in the fully solvated protein (left) is decomposed into two steps. Step I: The mutation is performed with a finite cap of explicit water molecules (shown in stick representation); the system is otherwise surrounded by vacuum. Step II: The two finite models (before and after mutation) are transfered into bulk solvent, treated as a dielectric continuum. The transfer free energy is obtained from continuum electrostatics. (From Ref. 25.) (b) Molecular dynamics with periodic boundary conditions: on-the-fly reaction field calculation. One simulation cell is shown. For each charge q_i, interactions with groups within r_{RF} are calculated in microscopic detail; everything beyond r_{RF} is viewed as a homogeneous dielectric medium, producing a reaction field on q_i [55]. The mutation is introduced using MD or MC simulations. As shown, for many of the charges the medium beyond r_{RF} is not truly homogeneous, being made up of both solvent and solute groups. (c) Spherical boundary conditions with continuum reaction field [56]. The region within the sphere (large circle) is simulated with MD or MC and explicit solvent; the region outside is treated as a dielectric continuum, which produces a reaction field on each charge within the sphere. If the sphere is smaller than the protein (as here), the outer region is heterogeneous and the reaction field calculation is rather difficult.

scaled by a coupling parameter λ, the second term in Eq. (39) contributes a constant term $\lambda q_i^2 \xi_{Ew}$ to the free energy derivative.

Although lattice summation methods avoid the introduction of an electrostatic cutoff, they impose a periodicity at all times that does not exist in any real system (even a crystal, let alone a liquid). This affects the polarization in a nonrandom way, e.g., the alignment of dipoles was shown to be overstabilized at long (~ 10 Å) distances with common Ewald protocols ("tinfoil" boundary conditions) [60]. Periodicity artifacts appear to be small for the free energy of charge creation in water [59] but have not yet been estimated for a macromolecule in solution. Methods to correct for them have been proposed for simple system geometries, which calculate the free energy difference between the periodic lattice and the nonperiodic system of interest from a dielectric continuum model; see, e.g., Refs. 12 and 61. With increasing computer power and simulation cell sizes, such artifacts will decrease.

VI. IMPROVING SAMPLING

The fundamental difficulty in free energy calculations lies in obtaining adequate sampling of conformations. Because of the ruggedness of the energy landscapes of proteins and nucleic acids, many energy barriers cannot be crossed in simulations spanning even a few nanoseconds. Therefore, specific strategies are needed to identify and sample all the energy basins, or substates, that contribute to a given free energy difference. The general problem of exploring and characterizing complex energy surfaces is much too vast to be discussed in detail here; see, e.g., Ref. 62. Even the techniques developed specifically for free energy calculations are so numerous that only a brief overview can be given.

A. Multisubstate Approaches

An alchemical free energy calculation compares two systems A and B, each of which usually possesses several slightly different, stable conformations. Thus Asp and Asn (Fig. 3) each possess three distinct stable rotamers around the χ_1 torsion angle as well as multiple shallow energy basins corresponding to different orientations of the backbone groups and the χ_2 torsion angle [26]. Whereas the latter basins are separated by small energy barriers ($\sim kT$), the χ_1 wells are separated by barriers of ≥ 3 kcal/mol, which are rarely crossed on the 100–1000 ps time scale. Therefore, it is best to view each system (Asp in solution, Asn in solution) as a superposition of three conformational substates, identified by the side chain χ_1 rotamer. The A $\rightarrow$ B free energy calculation can then be based on a thermodynamic cycle analogous to the one in Figure 6 [38,63]. The free energy of Asp (system A) can be written (to within a constant $c(N, V, T)$ [see Eq. (4)]),

$$F_A = -kT \ln\left(\int_1 e^{-U_A/kT} dr^N + \int_2 e^{-U_A/kT} dr^N + \int_3 e^{-U_A/kT} dr^N \right) \tag{40}$$

$$= -kT \ln(e^{-F_{A1}/kT} + e^{-F_{A2}/kT} + e^{-F_{A3}/kT})$$

Here, the configuration integral Q_A has been split into three integrals; the integration $\int_i$ is over all conformations where χ_1 is in the ith rotameric state. F_{Ai} is the "configurational

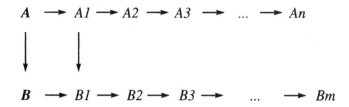

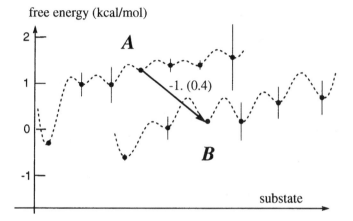

Figure 6 Thermodynamic cycle for multi-substate free energy calculation. System *A* has *n* substates; system *B* has *m*. The free energy difference between A and B is related to the substate free energy differences through Eq. (41). A numerical example is shown in the graph (from Ref. 39), where A and B are two isomers of a surface loop of staphylococcal nuclease, related by cis–trans isomerization of proline 117. The cis → trans free energy calculation took into account 20 substates for each isomer; only the six or seven most stable are included in the plot.

free energy'' of that state. More precisely, F_{Ai} is the free energy [to within the constant $c(N, V, T)$] of a hypothetical system where the potential energy inside the *i*th χ_1 energy well is the same as for A, but the potential energy outside the well is infinite. Whereas the absolute free energies F_{Ai} are difficult to compute [see discussion following Eq. (4)], the relative free energies of the three rotameric states are readily obtained by the methods of Section IV. The same calculation is performed for system B (Asn).

Finally, an alchemical free energy simulation is needed to obtain the free energy difference between any one substate of system A and any one substate of system B, e.g., $F_{B1} - F_{A1}$. In practice, one chooses two substates that resemble each other as much as possible. In the alchemical simulation, it is necessary to restrain appropriate parts of the system to remain in the chosen substate. Thus, for the present hybrid Asp/Asn molecule, the Asp side chain should be confined to the Asp substate 1 and the Asn side chain confined to its substate 1. Flat-bottomed dihedral restraints can achieve this very conveniently [38], in such a way that the most populated configurations (near the energy minimum) are hardly perturbed by the restraints. Note that if the substates A1 and B1 differ substantially, the transformation will be difficult to perform with a single-topology approach.

The A $\rightarrow$ B free energy change takes the final form

$$F_B - F_A = F_{B1} - F_{A1}$$

$$- kT \ln\left[\frac{1 + \exp[-\Delta F_B(1 \rightarrow 2)/kT] + \exp[-\Delta F_B(1 \rightarrow 3)/kT]}{1 + \exp[-\Delta F_A(1 \rightarrow 2)/kT] + \exp[-\Delta F_A(1 \rightarrow 3)/kT]}\right] \qquad (41)$$

where $F_A(1 \rightarrow 2) = F_{A2} - F_{A1}$, and the other notations are defined similarly. Illustrative applications of this technique are found in, e.g., Refs. 38, 39, and 63–65.

The multisubstate approach requires initially identifying all important substates, a difficult and expensive operation. In cases of moderate complexity (e.g., a nine-residue protein loop), systematic searching and clustering have been used [39,66]. For larger systems, methods are still being developed.

B. Umbrella Sampling

A powerful and general technique to enhance sampling is the use of umbrella potentials, discussed in Section IV. In the context of alchemical free energy simulations, for example, umbrella potentials have been used both to bias the system toward an experimentally determined conformation [26] and to promote conformational transitions by reducing dihedral and van der Waals energy terms involving atoms near a mutation site [67].

Similar to the approaches described in Section IV, free energies for the unbiased system can be recovered from the biased simulations in at least two ways. First, one can introduce steps where the umbrella potential is turned on (initially) and off (at the end) and compute the corresponding free energies in analogy to Eq. (30) [67]. Second, although the configurational probabilities are modified by the umbrella potential [Eqs. (31), (33)], it is possible in principle to recover ensemble averages for the system of interest, i.e., the system without the umbrella potential [37]. For an observable O, we obtain [e.g., by integrating Eq. (31) over q]

$$\langle O \rangle = \langle Oe^{U_r/kT} \rangle_r / \langle e^{U_r/kT} \rangle_r \qquad (42)$$

where the brackets $\langle \rangle_r$ and $\langle \rangle$ indicate averages over the system with and without the umbrella potential, respectively. In particular, the free energy derivatives $\partial F/\partial \lambda$, $\partial^2 F/\partial \lambda^2$, ... are ensemble averages [Eqs. (11), (12)] and can be obtained in this way from simulations with the umbrella potential [26]. Equation (42) can be generalized [37] to cases where simulations are run at a higher temperature than that of the system of interest (e.g., to promote conformational transitions further).

C. Moving Along λ

Another way to improve sampling for some problems is to treat the coupling coordinate or coordinates as dynamic variables. Thus, free energy simulations have been done where changes in a coupling coordinate λ were treated as Monte Carlo moves instead of being determined ahead of time [41,68]. More recently, coupling coordinates were included in the simulation as coordinates participating in the molecular dynamics, with artificial masses, akin to "pseudoparticles" [47,69] . An umbrella potential was used to drive the coupling coordinates from 0 to 1. The alchemical free energy calculation is thus treated as a pmf calculation along the coupling coordinate(s). Data from such a "λ-dynamics" simulation can be efficiently processed with the WHAM approach (above), where the coupling coordinates $\lambda_1, \lambda_2, \ldots$ play the role of the multidimensional reaction coordinates $q, s, \ldots$.

These approaches can be used to simulate several ligands simultaneously, either in solution or in a receptor binding site. Each ligand i is associated with its own coupling constant or weight, λ_i, and with a term $\lambda_i U_i$ in the energy function. The different weights obey $\Sigma_i \lambda_i = 1$. As the system evolves, the weights tend to adjust spontaneously in such a way that the most favorable ligand has the largest weight. Alternatively, the ligands can be made equiprobable by incorporating their free energies F_i into the energy function: Each term $\lambda_i U_i$ is replaced by $\lambda_i (U_i - F_i)$. F_i is not known ahead of time but can be determined iteratively [47]. This provides a new route to determining the relative solvation or binding free energies of two or more ligands, which was found to be more efficient than traditional thermodynamic perturbation or integration protocols in applications to simple systems. The variation of the λ_i with time implies that the system is never truly at equilibrium; to limit this effect, sufficient pseudomasses are needed for the λ_i; large masses in turn slow the exploration along each λ_i and limit efficiency. The performance for macromolecules has yet to be determined.

VII. PERSPECTIVES

With significant advances in recent years, free energy simulations can now be performed reliably for many biochemical problems if sufficient computing resources are available. Only a few representative applications could be mentioned above, including point mutations of buried residues [64,65], the creation of net charges in proteins [9,26], and conformational changes as large as the unfolding of small proteins [13]. References to many other applications can be found in papers just cited and in the review articles cited in Section I. Calculations of enthalpy and entropy changes are also becoming common. With increasing computer power, it will become straightforward to study charge creation with fully solvated simulation cells, lattice summation or reaction field methods, and force fields including atomic polarizability. All these calculations provide a direct connection between the macroscopic thermodynamics and the microscopic interactions of the investigated system.

Several important developing areas could only be touched on. One is the use of simplified free energy techniques to rapidly screen series of ligands or receptors [11,21,70]. These make use of a single simulation of a reference state and obtain the relative free energy of other, related molecules from a perturbation formula such as Eq. (9), e.g., truncated at second order (linear response). A recent twist has been to simulate a mixture of ligands with adjustable weights simultaneously, in analogy to a competitive binding experiment in solution [47,68]. Such calculations are parallelizable and will eventually be applicable to much larger, truly combinatorial libraries of ligands [71].

The protein folding problem is a central problem in computational biophysics that requires global optimization of the free energy. Although the prediction of structure from sequence alone is still out of reach, progress is being made in developing techniques to search broad ranges of conformations and estimate their free energies at different levels of approximation [62]. Some were mentioned above and are being used to study the folding and unfolding of proteins of known structures [13,43–45].

A final important area is the calculation of free energies with quantum mechanical models [72] or hybrid quantum mechanics/molecular mechanics models (QM/MM) [9]. Such models are being used to simulate enzymatic reactions and calculate activation free energies, providing unique insights into the catalytic efficiency of enzymes. They are reviewed elsewhere in this volume (see Chapter 11).

REFERENCES

1. C Wong, J McCammon. J Am Chem Soc 108:3830–3832, 1986.
2. A Warshel, F Sussman, G King. Biochemistry 25:8368–8372, 1986.
3. P Bash, U Singh, F Brown, R Langridge, P Kollman. Science 235:574–576, 1987.
4. C Brooks, M Karplus, M Pettitt. Adv Chem Phys 71:1–259, 1987.
5. J McCammon, S Harvey. Dynamics of Proteins and Nucleic Acids. Cambridge, UK: Cambridge Univ Press, 1987.
6. D Beveridge, F DiCapua. Annu Rev Biophys Biophys Chem 18:431–492, 1989.
7. W van Gunsteren, P Weiner, eds. Computation of Free Energy for Biomolecular Systems. Leiden: Escom Science, 1989.
8. W van Gunsteren, T Beutler, F Fraternali, P King, A Mark, P Smith. In: W van Gunsteren, P Weiner, A Wilkinson, eds. Computer Simulation of Biomolecular Systems. Leiden: ESCOM Science, 1993, pp 315–348.
9. A Warshel. Computer Modelling of Chemical Reactions in Enzymes and Solutions. New York: Wiley, 1991.
10. P Kollman. Chem Rev 93:2395, 1993.
11. ML Lamb, W Jorgensen. Curr Opin Chem Biol 1:449–457, 1997.
12. R Levy, E Gallicchio. Annu Rev Phys Chem 49:531, 1998.
13. FB Sheinerman, CL Brooks, Proc Natl Acad Sci USA 95:1562–1567, 1998.
14. B Roux, T Simonson, eds. Implicit Solvent Models for Biomolecular Simulations. Special Issue of Biophys Chem Amsterdam: Elsevier, 1999.
15. T Darden. This volume, Chapter 5.
16. B Tembe, J McCammon. Comput Chem 8:281–283, 1984.
17. W Jorgensen, K Buckner, S Boudon, J Tirado-Rives. J Chem Phys 89:3742–3746, 1988.
18. RH Fowler, EA Guggenheim. Statistical Thermodynamics. Cambridge, UK: Cambridge Univ Press, 1939.
19. F Zwanzig. J Chem Phys 22:1420, 1954.
20. G Hummer, A Szabo. J Chem Phys 105: 2004–2010, 1996.
21. H Liu, A Mark, W van Gunsteren. J Phys Chem 100:9485–9494, 1996.
22. T Simonson, D Perahia. J Am Chem Soc 117:7987–8000, 1995.
23. R von Mises. Mathematical Theory of Probability and Statistics. New York: Academic Press, 1964.
24. M Mezei. Mol Simul 10:225–239, 1993.
25. T Simonson, G Archontis, M Karplus. J Phys Chem B 101:8349–8362, 1997.
26. G Archontis, T Simonson, D Moras, M Karplus. J Mol Biol 275:823–846, 1998.
27. J Gao, K Kuczera, B Tidor, M Karplus. Science 244:1069–1072, 1989.
28. S Boresch, M Karplus. J Phys Chem A 103:103–118, 1999.
29. B Roux, M Nina, R Pomes, J Smith. Biophys J 71:670–681, 1996.
30. J Hermans, L Wang. J Am Chem Soc 119:2702–2714, 1997.
31. T Simonson. Mol Phys 80:441–447, 1993.
32. T Simonson, AT Brünger, Biochemistry 31:8661–8674, 1992.
33. S Boresch, M Karplus. J Phys Chem A 103:119–136, 1999.
34. S Boresch, G Archontis, M Karplus. Proteins 20:25–33, 1994.

35. M Gilson, J Given, B Bush, J McCammon. Biophys J 72:1047–1069, 1997.
36. J Hermans, S Shankar. Isr J Chem 27:225–227, 1986.
37. J Valleau, G Torrie. In: B Berne, ed. Modern Theoretical Chemistry, Vol. 5. New York: Plenum Press, 1977, pp 137–194.
38. J Hermans, RH Yun, AG Anderson. J Comput Chem 13:429–442, 1992.
39. A Hodel, L Rice, T Simonson, RO Fox, AT Brünger. Protein Sci 4:634–654, 1995.
40. C Haydock, J Sharp, F Prendergast. Biophys J 57:1269–1279, 1990.
41. M Mezei. Mol Simul 3:301–313, 1989.
42. C Bartels, M Karplus. J Comput Chem 18:1450, 1997.
43. C Bartels, M Karplus. J Phys Chem B 102:865–880, 1998.
44. N Nakajima, H Nakamura, A Kidera. J Phys Chem B 101:817–824, 1997.
45. M Hao, H Scheraga. J Chem Phys 102:1334–1348, 1995.
46. S Kumar, D Bouzida, R Swendsen, P Kollman, J Rosenberg. J Comput Chem 13:1011–1021, 1992.
47. X Kong, CL Brooks. J Chem Phys 105:2414–2423, 1996.
48. DJ Tobias, CL Brooks. Chem Phys Lett 142:472–476, 1987.
49. R Elber. J Chem Phys 93:4312–4320, 1990.
50. E Carter, G Ciccotti, J Hynes, R Kapral. Chem Phys Lett 156:472–477, 1989.
51. E Paci, G Ciccotti, M Ferrario, R Kapral. Chem Phys Lett 176:581–587, 1991.
52. WD Otter, W Briels. J Chem Phys 109:4139–4146, 1998.
53. T Straatsma, M Zacharias, J McCammon. Chem Phys Lett 196:297–301, 1992.
54. RH Wood. J Chem Phys 103:6177–6187, 1995.
55. H Schreiber, O Steinhauser. J Mol Biol 228:909–923, 1992.
56. D Beglov, B Roux. J Chem Phys 100:9050–9063, 1994.
57. T Darden, D Pearlman, L Pedersen. J Chem Phys 109:10921–10935, 1998.
58. T Simonson. J Phys Chem B 104:6509–6513, 2000.
59. G Hummer, L Pratt, A Garcia. J Phys Chem 100:1206–1215, 1996.
60. S Boresch, O Steinhauser. Ber Bunsenges Phys Chem 101:1019–1029, 1997.
61. G Hummer, L Pratt, A Garcia. J Phys Chem B 101:9275, 1997.
62. J Straub. In: R Elber, ed. New Developments in Theoretical Studies of Proteins. Singapore: World Scientific, 1999.
63. DJ Tobias, CL Brooks, SH Fleischman. Chem Phys Lett 156:256–260, 1989.
64. R Wade, J McCammon. J Mol Biol 225:679–712, 1992.
65. J Zeng, M Fridman, H Maruta, HR Treutlein, T Simonson. Prot Sci 8:50–64, 1999.
66. M Head, J Given, M Gilson. J Phys Chem A 101:1609–1618, 1997.
67. A Mark, W van Gunsteren, H Berendsen. J Chem Phys 94:3808–3816, 1991.
68. B Tidor. J Phys Chem 97:1069–1073, 1993.
69. R Pomes, E Eisenmesser, C Post, B Roux, J Chem Phys 111:3387–3395, 1999.
70. J Åqvist, C Median, J Samuelsson. Protein Eng 7:385–391, 1994.
71. E Duffy, W Jorgensen. J Am Chem Soc 122:2878–2888, 2000.
72. R Stanton, S Dixon, K Merz Jr. In: A Warshel, G Naray-Szabo, eds. Computational Approaches to Biochemical Reactivity. New York: Kluwer, 1996.
73. T Straatsma, H Berendsen, J Postma. J Chem Phys 85:6720, 1986.

10
Reaction Rates and Transition Pathways

John E. Straub
Boston University, Boston, Massachusetts

I. INTRODUCTION

For 25 years, molecular dynamics simulations of proteins have provided detailed insights into the role of dynamics in biological activity and function [1–3]. The earliest simulations of proteins probed fast vibrational dynamics on a picosecond time scale. Fifteen years later, it proved possible to simulate protein dynamics on a nanosecond time scale. At present it is possible to simulate the dynamics of a solvated protein on the microsecond time scale [4]. These gains have been made through a combination of improved computer processing (Moore's law) and clever computational algorithms [5].

In spite of these millionfold advances in the length of computer simulation, many dynamic processes remain outside the reach of direct dynamic simulation. Typically these processes involve the crossing of one or more energy barriers that stand between the reactant and product states. To overcome high energy barriers the system must concentrate an amount of energy far greater than the thermal energy in one or a few degrees of freedom. Such fluctuations are highly improbable and are known as "rare events"—they occur only infrequently. Processes of this kind include ligand rebinding in heme proteins, proton transfer in enzymatic catalysis, and the configurational reorganization of a folding protein.

A. Defining Reactant and Product "States"

The basic chemical description of rare events can be written in terms of a set of phenomenological equations of motion for the time dependence of the populations of the reactant and product species [6–9]. Suppose that we are interested in the dynamics of a conformational rearrangement in a small peptide. The concentration of reactant states at time t is $N_R(t)$, and the concentration of product states is $N_P(t)$. We assume that we can define the reactants and products as distinct "macrostates" that are separated by a "transition state" dividing surface. The transition state surface is typically the location of a significant energy barrier (see Fig. 1).

For example, when the energy barrier is high compared to the thermal energy, we can assume that when a reactant state is prepared there will be many oscillations in the reactant well before the system concentrates enough energy in the "reaction coordinate"

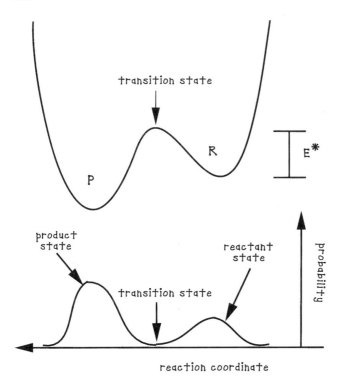

Figure 1   Double well potential for a generic conformational transition showing the regions of reactant and product states separated by the transition state surface.

to mount the barrier and cross from the reactant to the product state. Moreover, once the barrier is crossed and the system loses energy it will spend many oscillations in the product well before recrossing the barrier. When we examine the equilibrium probability distribution along such a reaction coordinate, we will note two macrostate maxima in the probability distribution separated by a probability minimum. The position of this minimum is often a good choice for the transition state dividing surface.

One possible definition of a transition state dividing surface would be to divide a single energy basin down the middle, recognizing that the reactants sit to the left of the minimum and the products to the right. This would be a bad idea. Such a choice of the "transition state" dividing surface would lead to fast oscillations between the reactant and product state populations as the system moves left to right and back again in the energy basin. Therefore, a condition is needed: There must be a separation in time scales between the fast, transient motion *within* a macrostate and the slow, activated dynamics of motion *between* macrostates.

B. Phenomenological Rate Equations

When it is possible to recognize distinct macrostates, we can write the phenomenological rate equations

$$\frac{dN_R}{dt} = -k_{PR}N_R(t) + k_{RP}N_P(t) \tag{1}$$

$$\frac{dN_{\rm P}}{dt} = k_{\rm PR}N_{\rm R}(t) - k_{\rm RP}N_{\rm P}(t) \tag{2}$$

where $k_{\rm PR}$ is the rate of transition from reactant to product while $k_{\rm RP}$ is the rate of reaction from product to reactant. It is these macroscopic population changes that are usually measured in chemical kinetics experiments.

These rate equations are easily solved. At long times, the chemical dynamics reaches a stationary "equilibrium" and the populations of reactants and products cease to change. The relative populations of reactants and products are given by the condition of "detailed balance," where the rate of transition from products to reactants equals the rate of transition from reactants to products, or

$$k_{\rm PR}N_{\rm R}^{\rm eq} = k_{\rm RP}N_{\rm P}^{\rm eq} \tag{3}$$

This relation defines the equilibrium constant between reactants and products, $K_{\rm eq} = N_{\rm P}^{\rm eq}/N_{\rm R}^{\rm eq} = k_{\rm PR}/k_{\rm RP}$.

At short times, there is a relaxation of the reactant and product state populations to their equilibrium values. For example, the deviation from the equilibrium concentration of products is given by

$$\delta N_{\rm P}(t) = N_{\rm P}(t) - N_{\rm P}^{\rm eq} = \delta N_{\rm P}(0) \exp\left[-(k_{\rm PR} + k_{\rm RP})t\right] \tag{4}$$

where $\delta N_{\rm P}(0)$ is a measure of the initial deviation from the equilibrium population of product. What is measured experimentally as the rate of change in the population of the product states is the kinetic rate constant $k = k_{\rm PR} + k_{\rm RP}$, which is the sum of the forward and backward rate constants.

The next step in understanding the chemical kinetics of this system is the calculation of the kinetic rate constant from a knowledge of the energetics of the reaction system.

II. TRANSITION STATE THEORY

The original microscopic rate theory is the transition state theory (TST) [10–12]. This theory is based on two fundamental assumptions about the system dynamics. (1) There is a transition state dividing surface that separates the short-time *intrastate* dynamics from the long-time *interstate* dynamics. (2) Once the reactant gains sufficient energy in its reaction coordinate and crosses the transition state the system will lose energy and become deactivated product. That is, the reaction dynamics is activated crossing of the barrier, and every activated state will successfully react to form product.

A. Building the TST Rate Constant

Given the foregoing assumptions, it is a simple matter to construct an expression for the transition state theory rate constant as the probability of (1) reaching the transition state dividing surface and (2) having a momentum along the reaction coordinate directed from reactant to product. Stated another way, $k_{\rm PR}^{\rm TST}$ is the equilibrium flux of reactant states across the transition state surface toward the product states. A typical trajectory that satisfies the assumptions of transition state theory is shown in Figure 2.

The transition state theory rate constant can be constructed as follows. The total flux of trajectories across the transition state dividing surface will be equal to the rate of transition k_{PR}^{TST} times the population of reactants at equilibrium N_R^{eq}, or

$$\text{flux}(R \to P) = k_{PR}^{TST} N_R^{eq}. \tag{5}$$

That means that the transition rate is equal to the relative probability of being an activated reactant state times the average forward flux

$$k_{PR}^{TST} = \frac{N^{\ddagger}}{N_R^{eq}} \left\langle \frac{p}{\mu} \right\rangle_+ = \left(\frac{k_B T}{2\pi\mu} \right)^{1/2} \left(\frac{N^{\ddagger}}{N_R^{eq}} \right) \tag{6}$$

where the average $\langle \cdots \rangle_+$ is taken over the positive momenta only. p is the momentum conjugate to the reaction coordinate, and μ is the reduced mass. Here we take the simple case of a linear reaction coordinate.

The average flux across the transition state dividing surface will be proportional to the relative probability of being found at the transition state,

$$N^{\ddagger} = \frac{1}{Z} \int dQ \, dq \, \delta \, (q - q^{\ddagger}) e^{-\beta \gamma(Q,q)} \tag{7}$$

where $q^{\ddagger}$ is the location of the transition state surface along the reaction coordinate q and $\delta(q - q^{\ddagger})$ counts only those phase points at the transition state. Z is a normalization constant proportional to the canonical partition function for the total system. The total

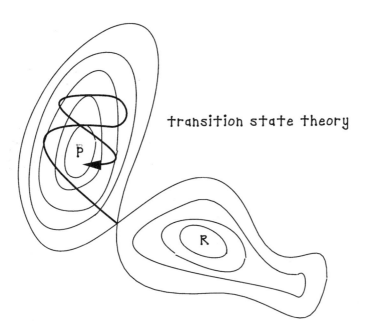

Figure 2 A typical trajectory satisfying the assumptions of transition state theory. The reactive trajectory crosses the transition state surface once and only once on its way from activated reactant to deactivated product.

energy of the system is taken to be a Hamiltonian function of the positions and momenta of both the reactive (q, p) and nonreactive (Q, P) degrees of freedom as $\mathcal{H}(\Gamma, \gamma) = \mathcal{T}(P, p) + \mathcal{V}(Q, q)$. $d\Gamma \, d\gamma$ is an incremental volume of the total phase space of nonreactive $d\Gamma = dQ \, dP$ and reactive $d\gamma = dq \, dp$ degrees of freedom. Similarly we find that

$$N_R^{eq} = \frac{1}{Z}\int dQ \, dq\theta(q^{\ddagger} - q)e^{-\beta\mathcal{V}(Q,q)} = \frac{1}{Z_\gamma}\int dq\theta(q^{\ddagger} - q)e^{-\beta W(q)} \tag{8}$$

where $\theta(q^{\ddagger} - q)$, the Heaviside step function, counts only those configurations in which the reaction coordinate is to the reactant side of the transition state dividing surface. Note that the nonreactive Q coordinates orthogonal to the reaction coordinate q have been averaged over to define the potential of mean force $W(q)$ as

$$\frac{1}{Z_\gamma}e^{-\beta W(q)} = \frac{1}{Z}\int dQ \, e^{-\beta\mathcal{V}(Q,q)} \tag{9}$$

where Z_γ is a normalization constant. The probability of finding the system along the reaction coordinate q will be proportional to $\exp[-\beta W(q)]$. The potential $W(q)$ can be used to define the probability distribution of the reduced system. It is possible to compute an exact transition state theory rate constant using $W(q)$. However, if the dynamics of the reaction coordinate is studied on the effective potential $W(q)$, the assumption is that the degrees of freedom Q are averaged over during the time scale for the essential dynamics of the reaction coordinate q.

It is a remarkable fact that the microscopic rate constant of transition state theory depends only on the equilibrium properties of the system. No knowledge of the system dynamics is required to compute the transition state theory estimate of the reaction rate constant k^{TST}.

B. Some Details

We counted the contribution of only those trajectories that have a positive momentum at the transition state. Trajectories with negative momentum at the transition state are moving from product to reactant. If any of those trajectories were deactivated as products, their contribution would need to be subtracted from the total. Why? Because those trajectories are ones that originated from the product state, crossed the transition state *twice*, and were deactivated in the product state. In the TST approximation, only those trajectories that originate in the reactant well are deactivated as product and contribute to the reactive flux. We return to this point later in discussing dynamic corrections to TST.

A few more comments are in order. The backward rate constant can be computed from the condition of detailed balance

$$k_{PR}^{TST}N_R^{eq} = k_{RP}^{TST}N_P^{eq} \tag{10}$$

Suppose that the reactant well can be approximated as harmonic and the activation energy is much larger than the thermal energy. In that case we can approximate the rate constant as

$$k_{PR}^{TST} = \frac{\omega_0}{2\pi}e^{-\beta\epsilon_{PR}^{\ddagger}} \tag{11}$$

where $\varepsilon^{\ddagger}_{PR}$ is the activation energy required to pass from the reactant state to the product state. The angular frequency of the reactant well is ω_0.

C. Computing the TST Rate Constant

What knowledge of the system is necessary to compute k^{TST}_{RP}?

1. We need to have a good estimate of the energy of the system as a function of the positions and momenta of all atoms in the system [13,14].
2. It is necessary to compute the average over the phase space of the system.
3. We must be able to define the reaction coordinate along which the transition state theory dividing surface is defined.

Each of these requirements can be difficult to meet for a complex biomolecular system. Each of these points is addressed in this chapter.

A variety of methods for finding reaction paths in simple chemical systems have been proposed. Good review articles summarizing those methods can be found [8,15,16]. An excellent historical overview of these methods is provided by Anderson [17]. Here we focus our discussion on those methods that have had the widest application to large-scale biomolecular systems and that hold the greatest promise for further development.

III. CORRECTIONS TO TRANSITION STATE THEORY

The assumptions of transition state theory allow for the derivation of a kinetic rate constant from equilibrium properties of the system. That seems almost too good to be true. In fact, it sometimes is [8,18–21]. Violations of the assumptions of TST do occur. In those cases, a more detailed description of the system dynamics is necessary for the accurate estimate of the kinetic rate constant. Keck [22] first demonstrated how molecular dynamics could be combined with transition state theory to evaluate the reaction rate constant (see also Ref. 17). In this section, an attempt is made to explain the essence of these dynamic corrections to TST.

Transition state theory assumes that once the system reaches the transition state the system dynamics will carry the activated reactant to product, where it will be deactivated. It assumes that the process of converting activated reactants to products is perfectly efficient. Thinking about the system dynamics, we understand that that will not always be the case. For a reaction system in an environment in which the collision rate (or friction) is low, the activated system may cross into the product state and recross the transition state surface back to the reactant state many times before undergoing collisions, losing energy, and becoming deactivated. Alternatively, when the collision rate (or friction) is very high, the activated system may be kicked back and forth across the transition state surface many times before being deactivated. Dynamics typical of both regimes are depicted in Figure 3. These dynamic processes in the low and high friction regimes can be effectively studied by using molecular dynamics simulations.

Either of the mechanisms of recrossing leads to inefficiency in converting reactant to product. How does this affect the reaction rate constant? Fewer activated reactants form

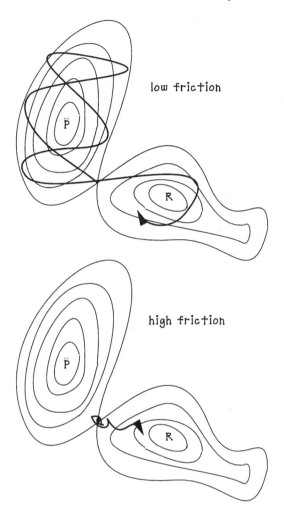

Figure 3 Dynamic recrossings in the low and high friction regimes. Recrossings back to the reactive state lead to a lowering of the rate constant below the transition state theory value.

products, so the rate constant will be lower than the TST estimate. This is summarized in the formula for the actual rate constant,

$$k^{\text{exact}} = \kappa k^{\text{TST}} \tag{12}$$

where κ is the transmission coefficient—a positive number less than or equal to 1.

A. Computing κ Using the Reactive Flux Method

In practice, we can compute κ as follows [19,23]. We start with a set of trajectories at the transition state $q = q^{\ddagger}$. The momenta have initial conditions distributed according to the normalized distribution functions

$$P^{(\pm)}(\Gamma, \gamma) \sim \delta(q - q^{\ddagger})\theta(\pm p)p \exp(-\beta\mathcal{H}) \tag{13}$$

where Γ and γ are the nonreactive and reactive phase space degrees of freedom, respectively. In one set, $P^{(+)}(\Gamma, \gamma)$, the trajectories initially have positive momenta (and at first move into the product well). In the complementary set, $P^{(-)}(\Gamma, \gamma)$, the trajectories initially have negative momenta (and at first move into the reactant well).

Using these distribution functions, we can write the reactive flux correlation function in the compact form

$$\hat{k}(t) = \int d\Gamma \, d\gamma [P^{(+)}(\Gamma, \gamma) - P^{(-)}(\Gamma, \gamma)]\theta[q(t) - q^{\ddagger}] \tag{14}$$

Note that $\hat{k}(t = 0) = 1$. What does this function measure? The function $\theta[q(t) - q^{\ddagger}]$ follows each trajectory and counts 1 if the trajectory is in the product well and 0 otherwise.

The calculation begins at $t = 0$ with a number of trajectories distributed according to $P^{(+)}(\Gamma, \gamma)$ and an equal number according to $P^{(-)}(\Gamma, \gamma)$. The trajectories are followed in time until they are deactivated in the reactant or product well. This is illustrated in Figure 4. Initially there may be rapid recrossings of the transition state, and this can lead to a rapid decay or ringing in $\hat{k}(t)$. After a time, all of the transient decay will have passed and only motion on the longest time scale—the time scale for activated barrier crossing—will be active. Eventually, $\hat{k}(t)$ will decay to zero on that long time scale. However, at the intermediate time scale, longer than the time scale for transient decay and shorter than the time scale for activated barrier crossing, the function $\hat{k}(t)$ will equal the transmission coefficient κ. Once κ is known, the total rate constant can be computed by multiplying κ by the TST rate constant.

If the assumptions underlying the TST are satisfied, the trajectories with initially positive momenta will be trapped in the product well and those with initially negative momentum will be trapped in the reactant well. That will result in a value of $\hat{k}(t) = 1$ and the rate constant $k = k^{\text{TST}}$.

If there are recrossings of the transition state, this will cause the positive contribution to $\hat{k}(t)$ to be somewhat less than 1 and the negative contribution to be somewhat greater

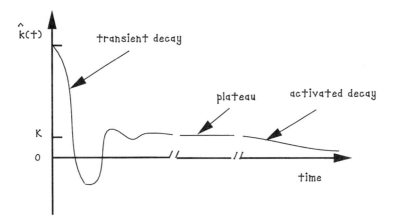

Figure 4 Reactive flux calculation for dynamics at low friction.

than 0, leading to a decay in $\hat{k}(t) < 1$ and a rate constant $k = \kappa k^{\mathrm{TST}}$ less than the transition state theory estimate.

B. How Dynamic Recrossings Lower the Rate Constant

Consider a symmetrical double well (but the argument is easily generalized to the asymmetrical case). The evolution of a set of trajectories is schematically shown in Figure 5. In TST all trajectories crossing the transition state surface from reactants to products contribute to the reactive flux. However, when there are recrossings this is not the case. Some trajectories that are part of the ensemble of forward-moving trajectories actually *originated* as products. The contribution of those trajectories must be *subtracted*. Other trajectories started as reactant but will be deactivated as reactant. Those trajectories should not count at all. Here is how we can perform the counting that is done when computing the reactive flux $\hat{k}(t)$.

We assume that when the activated reactants cross the transition state a fraction P are deactivated as product and the remaining fraction $1 - P$ *recross* the transition state surface [8,24]. If each fraction has roughly the same distribution of momenta as the original fraction, we can say that of the fraction $1 - P$ that recross, $P(1 - P)$ will be deactivated in the reactant well and the remaining $(1 - P)^2$ will recross the transition state into the

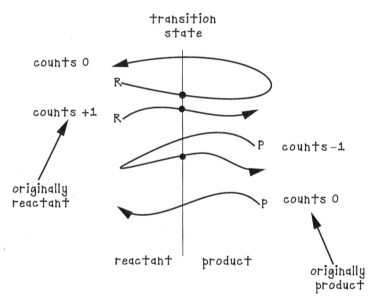

Figure 5 The transition state ensemble is the set of trajectories that are crossing the transition state from reactant to product at equilibrium (shown as black dots). There are four types of trajectories, shown top to bottom in the diagram. (1) Starting as reactant, the trajectory crosses and recrosses the transition state and is deactivated as reactant. It does not add to the reactive flux. (2) Starting as reactant, the trajectory is deactivated as product. It adds +1 to the reactive flux. (3) Starting as product, the trajectory crosses and recrosses the transition state and is deactivated as product. Such a trajectory must be subtracted from the ensemble, so it counts −1 to the reactive flux. (4) Starting as product, the trajectory is deactivated as reactant. It does not contribute to the reactive flux.

product well, where a fraction $P(1 - P)^2$ will be deactivated as product. And so on. Adding up all contributions we find that a total fraction

$$P + P(1 - P)^2 + P(1 - P)^4 + \cdots = \frac{1}{2 - P} \tag{15}$$

is deactivated as product.

But this is not the whole story! We not only need to know that a trajectory that crosses the transition state surface is eventually deactivated as product, we also need to know whether it *originated* from the reactant well! A trajectory that originates from the product well and ends up as product won't contribute to the forward rate of reaction. Some of the trajectories did originate as product. We need to find that fraction and subtract it.

What is that negative contribution? We can follow the trajectories backward in time to find the well from which they originated. Of the number of trajectories initially moving from product to reactant, a fraction P is deactivated as reactant and a fraction $1 - P$ recross the TST due to inertial motion or frequent collisions. A fraction $P(1 - P)$ will then be deactivated as product, and the remaining $(1 - P)^2$ will recross. And so on. The total fraction that is deactivated as product is

$$P(1 - P) + P(1 - P)^3 + \cdots = \frac{1 - P}{2 - P} \tag{16}$$

Now we can compute the transmission coefficient [17,24]. It will be the difference between the positive and negative contributions, or

$$\kappa = \frac{1}{2 - P} - \frac{1 - P}{2 - P} = \frac{P}{2 - P} \tag{17}$$

Note that when $P = 1$ we find that the assumptions of TST are met and $\kappa = 1$. As the number of recrossings of the transition state increases, both P and κ decrease.

C.   An Efficient Method for Computing Small Values of κ

In the very high and very low friction regimes, it might be that most trajectories do recross the transition state. It may also be that it takes a very long time to follow the system dynamics to the plateau region where κ can be measured. In such cases, it is also possible to compute the transmission coefficient in an approximate but accurate manner [24]. By placing an ''absorbing boundary'' at the transition state and simply following the trajectories until they recross or are deactivated, it is possible to estimate the value of κ with much less computational effort. Trajectories are started in the normal way as in the calculation of the reactive flux. However, when a trajectory recrosses the transition state, the run is stopped. By computing the fraction of trajectories (P_+) that recross the transition state from the $P^{(+)}(\Gamma, \gamma)$ distribution and the fraction (P_-) that recross from the $P^{(-)}(\Gamma, \gamma)$ distribution [24]

$$\kappa = \frac{P_+ P_-}{P_+ + P_- - P_+ P_-} \tag{18}$$

When there are many recrossings, P_+ and P_- may be much less than 1, few trajectories are integrated for long times, and the computational saving can be great. When the potential energy is a symmetrical double well, $P_+ = P_- = P$, and this result reduces to $\kappa = P/(2 - P)$ as we found earlier.

IV. FINDING GOOD REACTION COORDINATES

In simple chemical systems, it is often possible to make a good first guess at the dominant reaction pathway [25–28]. An example of such a reaction is the chair-to-boat isomerization in cyclohexane. In that pathway, a clever combination of two torsion angles provides an excellent reaction coordinate for the isomerization reaction [29,30].

In more complex systems, the dominant reaction pathway may be less than obvious. This may be the case even for a modest configurational rearrangement. A fine example is the transport of a Na^+ ion through the gramicidin ion channel. Even for such a well-defined process, with a highly constrained pathway for the ion to follow, a variety of reaction coordinates can be found with significantly different energetic features [31,32]. Moreover, even a minimal energy reaction pathway may include motion of atoms of the ion channel and water molecules in addition to the ion itself. Such a transition pathway may be difficult to express in terms of local atomic coordinates or simply defined collective modes.

A second example is the isomerization of a tyrosine residue about the ξ_2 torsion angle in the protein bovine pancreatic trypsin inhibitor [33]. Largely buried in the protein core, the tyrosine 35 ''ring flip'' is an infrequent event occurring on the time scale of seconds owing to an activation energy barrier estimated on the order of 13 kcal/mol. The first guess for a good reaction coordinate, the ξ_2 torsion angle, is inadequate. Exploring along this poor reaction coordinate, one grossly underestimates the energy barrier. What is missing? A section of backbone adjacent to the tyrosine ring is displaced in the transition. In fact, the barrier to rotation is largely determined by the steric contact between the rotating phenyl ring and the adjacent amide group, which must be included in the definition of the reaction coordinate [34].

If the motion of this group is ignored in the definition of the reaction coordinate, a search along the ξ_2 angle alone can provide a misleading estimate of the reaction energetics. This is indicated in Figure 6. The potential is essentially two-dimensional. If the reaction coordinate is described too simply (using a local Cartesian or internal coordinate rather than a globally optimized reaction pathway), the potential energy may indicate an artificially low energy barrier. The discrete representation of the reaction path in terms of a set of intermediate structures renders the ''motion'' along the reaction pathway essentially discontinuous. (A poor choice of the reaction coordinate q, one that might not be a single-valued function of the computed reaction pathway $l(\mathbf{r})$, can also lead to essential discontinuities in the reaction coordinate—points where $dq/dl = 0$.) In the region of the transition state the ''true'' reaction coordinate is practically orthogonal to the chosen reaction coordinate. A significant change in energy along the true reaction coordinate occurs (a large jump upward), with almost no change in the chosen reaction coordinate. Proper use of both coordinates in the definition of the reaction coordinate will lead to a continuous pathway that represents well the energetics of dynamic barrier crossing.

In other cases, it may be impossible to describe the kinetics properly using a single reaction path. A variety of pathways may contribute to the reaction kinetics. One or more paths may be dominant at low temperature, whereas other paths may be dominant at high temperatures. This results in a temperature-dependent reaction mechanism. In such situa-

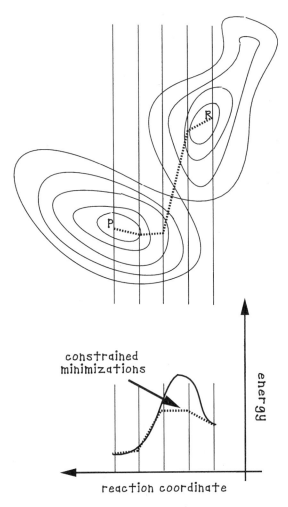

Figure 6   A poor choice of reaction coordinate can lead to a poor estimate of the activation energy and related rate constant. Because of the discrete nature of the reaction pathway, it is possible to step over the barrier. This leads to an underestimate of the activation energy.

tions, the goal is to be able to characterize the dominant reaction pathway or pathways at a given temperature [35–37].

V.   GLOBAL SEARCHES FOR IMPORTANT TRANSITION PATHWAYS

In a typical dynamic trajectory, the initial position is well controlled but the endpoint of the trajectory is unknown. For chemical reaction dynamics, we are interested in trajectories that link known initial (reactant) and final (product) states so that both the initial conditions and the final conditions of the trajectory are fixed.

A. Variational Methods for Computing Reaction Paths

A most successful paradigm for isolating reaction pathways in complex systems begins with a definition of the reaction pathway as a continuous line $l(\mathbf{r})$ that connects known reactant $\mathbf{r}_R$ and product $\mathbf{r}_P$ configurations. We then define an *integrated cost* functional

$$\text{cost}[l(\mathbf{r})] = \int_{\mathbf{r}_R}^{\mathbf{r}_P} \mathscr{C}(\mathbf{r})dl(\mathbf{r}) \tag{19}$$

as a function of the path $l(\mathbf{r})$ leading from the reactant configuration $\mathbf{r}_R$ to the product configuration $\mathbf{r}_P$. The goal is to minimize this functional in the space of all possible paths. That is, we want to find the extremum of this function by varying the path $l(\mathbf{r})$. The term $\mathscr{C}(\mathbf{r})dl(\mathbf{r})$ assigns a particular *differential cost* or penalty for motion about the position $\mathbf{r}$ over an increment $dl(\mathbf{r})$. We usually think of the cost function as a positive real number.

B. Choice of a Differential Cost Function

How can the continuous transition pathway $l(\mathbf{r})$ be represented by a computer for a complex molecular system? $l(\mathbf{r})$ can be approximately represented as a set of configurations of the system $\{\mathbf{r}_k\}$.

An example is shown in Figure 7 for the case of the coil-to-helix transition. The endpoints of the calculation are an unstructured coil $\mathbf{r}_R$ and helix $\mathbf{r}_P$. Intermediate peptide structures correspond to transition intermediates defining the pathway $l(\mathbf{r})$.

Elber and Karplus [38] presented an effective set of numerical methods for computing the reaction paths based on this approximation. First the path is discretized—it is expressed as a chain of intermediate configurations of the system $\{\mathbf{r}_k\}$. The line integrals of Eq. (19) are then written as

$$\text{cost}[l(\mathbf{r})] = \sum_{k=0}^{M-1} \mathscr{C}(\mathbf{r}_k)|\mathbf{r}_{k+1} - \mathbf{r}_k| \tag{20}$$

This discretized path allows us to represent the transition pathway numerically as a set of discrete configurations of the system.

1. Elber and Karplus Reaction Paths

What is the best choice of differential cost function? A variety of definitions of the cost function have been proposed. One stems form the highly original work of Elber and Karplus [38] and Czerminski and Elber [39], where

$$\mathscr{C}(\mathbf{r}) = \frac{1}{L} U(\mathbf{r}) \tag{21}$$

$U(\mathbf{r})$ is the potential energy of the molecular system, and L is the total length of the path $L = \int dl(\mathbf{r})$. In this case, the cost is the average potential energy along the pathway. The best path is defined as the path of minimum mean potential energy. This definition has been applied to compute the reaction paths of a large number of molecular systems including peptide conformational rearrangement [40] and ligand migration in heme proteins. Similar methods have been shown to have greater numerical efficiency in applications to large systems [41].

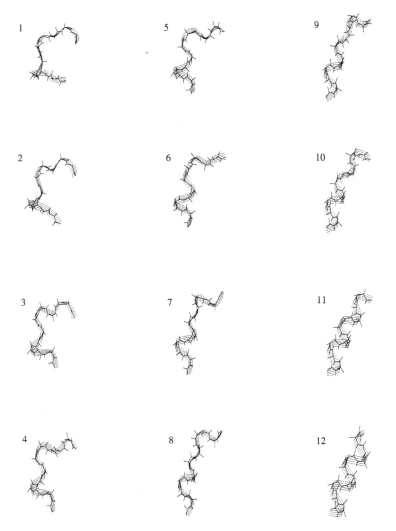

Figure 7 The discretized reaction path is represented by a number of intermediate configurations of the system connecting the fixed reactant (1, coil) and product (12, helix) states.

2. MaxFlux Reaction Paths

An alternative definition of the cost function was proposed by Huo and Straub [35] based on an earlier suggestion of Berkowitz et al. [42],

$$C(\mathbf{r}) = e^{\beta U(\mathbf{r})} \tag{22}$$

This result comes from the idea of a variational rate theory for a diffusive dynamics. If the dynamics of the reactive system is overdamped and the effective friction is spatially isotropic, the time required to pass from the reactant to the product state is expected to be proportional to the integral over the path of the inverse Boltzmann probability.

An important feature of this method is the connection to variational rate theory: The best reaction pathway is one that minimizes the reaction time or maximizes the reaction rate. Minimizing this cost functional as a function of the path is approximately equivalent to minimizing the average time required to pass from reactant to product. A second important feature is that the temperature is included in the cost function. The reaction pathway found to be the pathway of maximum flux may depend on the temperature [35]. The temperature dependence may manifest itself simply as saddle point avoidance or more globally through a complete change in the dominant reaction mechanism.

This method has been applied to derive a multitude of paths for the coil-to-helix transition in polyalanine using a continuum solvation model [36].

C. Diffusional Paths

An alternative method for computing reaction pathways is based on the idea of diffusion [43] or a ''noisy'' dynamics [44,45]. A prescient review article by Wolynes [8] anticipated many of the most impressive advances in this area.

Consider a diffusion process in which a molecule is initially in a conformation defined by the coordinates $\mathbf{r}_k$. A transition probability can be constructed using a product of joint probabilities $p(\mathbf{r}_{k+1}|\mathbf{r}_k)$ for moving between intermediate positions $\mathbf{r}_k$ and $\mathbf{r}_{k+1}$ integrated over all intermediate states,

$$p(\mathbf{r}_P|\mathbf{r}_R) \sim \int d\mathbf{r}_{M-1} \cdots d\mathbf{r}_2 \, d\mathbf{r}_1 \, p(\mathbf{r}_P|\mathbf{r}_{M-1}) \cdots p(\mathbf{r}_2|\mathbf{r}_1)p(\mathbf{r}_1|\mathbf{r}_R) \tag{23}$$

Pratt [43] made the innovative suggestion that transition pathways could be determined by maximizing the cumulative transition probability connecting the known reactant and product states. That is, the most probable transition pathways would be expected to be those with the largest conditional probability.

In the presence of a potential $U(\mathbf{r})$ the system will feel a force $F(\mathbf{r}_k) = -\nabla_r U(\mathbf{r})|\mathbf{r}_k$. There will also be a stochastic or random force acting on the system. The magnitude of that stochastic force is related to the temperature, the mass of the system, and the diffusion constant D. For a short time, it is possible to write the probability that the system has moved to a new position $\mathbf{r}_{k+1}$ as being proportional to the Gaussian probability [43]

$$p(\mathbf{r}_{k+1}|\mathbf{r}_k) \sim \exp\left[-\frac{1}{4D\delta t}[\mathbf{r}_{k+1} - \mathbf{r}_k - D\delta t\beta F(\mathbf{r}_k)]^2\right] \tag{24}$$

In the absence of an external force, the probability of moving to a new position is a spherically symmetrical Gaussian distribution (where we have assumed that the diffusion is spatially isotropic).

The diffusion constant should be small enough to damp out inertial motion. In the presence of a force the diffusion is biased in the direction of the force. When the friction constant is very high, the diffusion constant is very small and the force bias is attenuated—the motion of the system is strongly overdamped. The distance that a particle moves in a short time δt is proportional to $\sqrt{D\delta t}$.

Recently, Chandler and coworkers [46,47] revisited this idea and developed an elegant and promising methodology for the computation of reaction pathways and transition rates in molecular systems.

D. Onsager–Machlup Paths

Olender and Elber [45] made a novel suggestion that dynamic trajectories for long time processes having known initial and final states may be computed by using ''noisy'' dynamic trajectories. They define a probability that a trajectory would move from position $\mathbf{r}_k$ to $\mathbf{r}_{k+1}$ as

$$p(\mathbf{r}_{k+1}|\mathbf{r}_k) \sim \exp\left[-\frac{1}{2\sigma^2}[m(\mathbf{r}_{k+1} - 2\mathbf{r}_k + \mathbf{r}_{k-1}) - \delta t^2 F(\mathbf{r}_{k-1})]^2\right] \tag{25}$$

The highest probability paths will make the argument of the exponential small. That will be true for paths that follow Newtonian dynamics where $m\ddot{\mathbf{r}} = F(\mathbf{r})$. Olender and Elber [45] demonstrated how large values of the time step δt can be used in a way that projects out high frequency motions of the system and allows for the simulation of long-time molecular dynamics trajectories for macromolecular systems.

Elber et al. [48] applied this method to explore the dynamics of the C-peptide in water with impressive results. More than 30 trajectories of C-peptide were generated, and the process of helix formation in water was examined. Remarkably, a time step of 500 ps was used, which allowed for the study of peptide folding on extended time scales.

VI. HOW TO CONSTRUCT A REACTION PATH

In the computation of characteristic reaction pathways, it is essential to include a number of constraints and restraints on the reaction pathway. For example, for conformational transitions in a macromolecule, rigid-body translation and rotation should be constrained. In addition, to avoid clustering of intermediate configurations in potential energy minima, the distances between intermediate points along the path should be restrained to be roughly equal. Such a restraint forces the system to take steps of regular size along the reaction pathway.

In many cases, it is also helpful to have the path repel itself so that the transition pathway is *self-avoiding*. An actual dynamic trajectory may oscillate about a minimum energy configuration prior to an activated transition. In the computed restrained, self-avoiding path, there will be no clusters of intermediates isolated in potential energy minima and no loops or redundant segments. The self-avoidance restraint reduces the ''wasted effort'' in the search for a characteristic reaction pathway. The constraints and restraints are essential components of the computational protocol.

A. The Use of Constraints and Restraints

Following the computational protocol of Czerminski and Elber [39,40] a number of restraints and constraints are added to (1) encourage the mean-square distances between adjacent structures to be approximately constant,

$$C_A(\mathbf{R}) = \kappa \sum_{k=1}^{M} [(\mathbf{r}_k - \mathbf{r}_{k-1})^2 - d_{\text{ave}}^2]^2 \tag{26}$$

where $d_{\text{ave}}^2 = \sum_{k=1}^{M} (\mathbf{r}_k - \mathbf{r}_{k-1})^2/M$, (2) discourage intermediates from coming close to one another,

$$C_R(\mathbf{R}) = \frac{\rho}{\lambda} \sum_{j>k+1} \exp\left[-\frac{\lambda(\mathbf{r}_j - \mathbf{r}_k)^2}{\langle d \rangle^2}\right] \tag{27}$$

and (3) eliminate rigid-body translations and rotations [39]

$$\sum_{\mu=1}^{N} m_{\mu}(\mathbf{r}_{\mu} - \mathbf{r}^{\text{fix}}_{\mu}) = 0 \tag{28}$$

and

$$\sum_{\mu=1}^{N} m_{\mu}\mathbf{r}_{\mu} \times \mathbf{r}^{\text{fix}}_{\mu} = 0 \tag{29}$$

where N is the number of atoms in the system, m_{μ} is the atomic mass, and $\mathbf{r}_{\mu}$ gives the Cartesian coordinates for the μth atom. $\{\mathbf{r}^{\text{fix}}_{\mu}\}_{\mu=1,N}$ is the arithmetic average of the coordinates of the atoms in the reactant and product configurations. In the final refinement, these terms do not add significantly to the integrated cost function. These constraints have been applied in a variety of contexts. Values for the various parameters can be found in a number of works [15,35].

B. Variationally Optimizing the Cost Function

The goal is to find the global minimum value of the reaction time function in the space of all possible reaction paths subject to the restraints. This is a computationally demanding task in dealing with a large biomolecule. One method is to minimize the function of reaction time using a conjugate gradient minimization of the chain of intermediates with the reactant ($\mathbf{r}_R$) and product ($\mathbf{r}_P$) configurations remaining fixed. Such a "focal method" leads to a solution that is strongly dependent on the goodness of the initial guess.

Alternative algorithms employ global optimization methods such as simulated annealing that can explore the set of all possible reaction pathways [35]. In the MaxFlux method it is helpful to vary the value of β (temperature) that appears in the differential cost function from an initially low β (high temperature), where the effective surface is smooth, to a high β (the reaction temperature of interest), where the reaction surface is more rugged.

Using a differential cost function such as that of Elber and Karplus, the potential energy is averaged over the path by including a factor of $1/L$. In other definitions, such as the one employed in the MaxFlux method, there is no such normalization. Therefore, if the potential is set to zero, the MaxFlux method will find that the best path is the straight line path connecting reactants and products. However, methods where the differential cost is proportional to $1/L$ will find that all paths are equally good.

VII. FOCAL METHODS FOR REFINING TRANSITION STATES

Using the global search methods described above, it is possible to identify one or a number of important transition pathways. The coarseness of the search (the number of intermediate structures used to define the path) imposes a limit on smoothness of the pathway (see Fig. 8). In general, we can search along the reaction pathway and find the region of greatest energy curvature, $|\nabla_q^2 E|_{\max}$. There will also be an average distance between intermediates along the reaction coordinate δq^2. The product of these numbers provides an estimate of the error in the energy of the pathway:

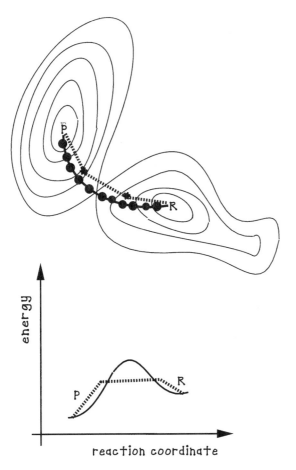

Figure 8   An accurate estimate of the barrier height can be found by adding a sufficient number of intermediate points in the discretized transition pathways. The solid line in the graph represents the energy profile for a reaction path described by 11 intermediate configurations of the system. The dashed line shows a coarse pathway described by only two intermediate configurations. The latter path underestimates the true energy barrier.

$$\text{Maximum error} \sim |\nabla_q^2 E|_{\max}\delta q^2 \tag{30}$$

For a simple bistable reaction potential, it is clear that maximum curvature along the reaction pathway will occur near the extrema—the minima and the barrier top. The path endpoints are typically chosen to sit in the reactant and product minima, and in such a case the maximum error will result from the path "straddling" the barrier top as in Figure 8. Of course, this is the error made in a single segment of the pathway. For a general potential the pathway will consist of multiple segments and may have many barriers.

The curvature along the reaction coordinate is fixed by the energy function. It is possible to reduce the error by adding intermediate structures and limiting the error by reducing δq^2, but there is an associated computational cost. It may be wiser to employ a focal method to refine the transition pathway—a local search for transition states and minima in the neighborhood of the globally defined pathway. A good number of very fine methods exist for refining the positions of transition states [26,27,49–52]. Once the position of the transition state is refined, a steepest descent pathway can be constructed [53,54].

VIII. HEURISTIC METHODS

An alternative approach to the construction of a smooth reaction pathway with a well-refined transition state is the "conjugate peak refinement" (CPR) method of Fischer and Karplus [55]. As in the global methods, the path is optimized as a whole and self-consistently. However, all points along the path are not treated equally. The computational effort is always directed at bringing the highest energy segment of the path closer to the valley of the energy surface. Starting from some initial guess at the path, a simple set of rules known as a heuristic is applied in each cycle of CPR, when one path point is either added, improved, or removed. This is repeated until the only remaining high energy path points are the actual saddle points of the transition pathway.

For example, the method may proceed as follows:

1. Start from a straight line path as the first guess at the transition pathway connecting known reactant and product structures.
2. Search along that pathway to isolate the highest energy point.
3. From that highest energy point, called the "peak," search *conjugate* to the direction of the reaction pathway and find the lowest energy point.
4. Make that lowest energy point a permanent intermediate point on the transition pathway.
5. Return to step 1 and refine the two intermediate segments in the same manner, and so on, until the desired level of detail is obtained.

The result of such a series of steps is depicted in Figure 9. This method has proved effective in isolating reaction pathways for conformational transitions involving localized torsional transitions including those involving a subtle isomerization mechanism [56].

On high dimensional energy surfaces, a newly added path point does not always reach the bottom of the valley in a single CPR cycle. Such a point is improved during a later CPR cycle, because eventually it will itself become the energy "peak" along the path, $\mathbf{r}_k$. In such a case, each intermediate structure $\mathbf{r}_k$ is connected to two nearest neighbor structures $\mathbf{r}_{k-1}$ and $\mathbf{r}_{k+1}$, which define the unit displacement vectors $\hat{u}_k = (\mathbf{r}_k - \mathbf{r}_{k-1})/|\mathbf{r}_k - \mathbf{r}_{k-1}|$ and $\hat{u}_{k+1} = (\mathbf{r}_{k+1} - \mathbf{r}_k)/|\mathbf{r}_{k+1} - \mathbf{r}_k|$. We can then define the tangent vector at point $\mathbf{r}_k$ as $\mathbf{t}_k = \hat{u}_{k+1} + \hat{u}_k$. The path point $\mathbf{r}_k$ is improved by performing an energy minimization conjugate to the tangent vector $\mathbf{t}_k$, which first involves maximizing the energy locally along $\mathbf{t}_k$. Sometimes, when this local maximization is not possible, the intermediate point $\mathbf{r}_k$ is simply removed and the path is rebuilt from $\mathbf{r}_{k-1}$ to $\mathbf{r}_{k+1}$. Following this process, the intermediate segments can be continually refined by following the protocol described above until the desired level of detail is achieved. This process of local refinement of a globally defined reaction pathway leads to the optimal use of each algorithm.

CPR can be used to find continuous paths for complex transitions that might have hundreds of saddle points and need to be described by thousands of path points. Examples of such transitions include the quaternary transition between the R and T states of hemoglobin [57] and the reorganization of the retinoic acid receptor upon substrate entry [58]. Because CPR yields the exact saddle points as part of the path, it can also be used in conjunction with normal mode analysis to estimate the vibrational entropy of activation

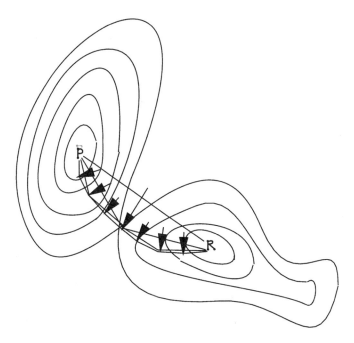

Figure 9 The refinement of an initial straight line path to a smooth transition pathway using the conjugate peak refinement algorithm. The initial guess is a straight line path. That path is refined by the addition of an intermediate point (the long-stemmed arrow). Two additional intermediates are added to create a path of three intermediates before four more intermediates are inserted. The process can be continued until the desired level of smoothness in the transition pathway is obtained.

[59] and to analyze enzymatic mechanisms [60]. The efficiency and robustness of the CPR method make it an efficient means of mapping the topology of complex energy surfaces [61].

IX. SUMMARY

In the study of long-time-scale processes in macromolecular systems, the greatest challenge remains the isolation of one or more characteristic reaction pathways. These pathways define the reaction mechanism. They are also the starting points for the computation of rates of reaction. Once the potential of mean force along the reaction coordinate is known, the reaction rate constant can be computed by using transition state theory. Subsequently, the system dynamics can be followed to compute the transmission coefficient. It is currently possible to carry out such a series of steps not only on a modest chemical system [62] but also on a complex biomolecular system [31]. The greatest uncertainty in the computed value of the rate constant is due to the uncertainty in the activation energy. Potentially, the greatest uncertainty in the activation energy lies in the choice of the reaction coordinate.

Significant advances are being made in the development of effective methods for determining reaction pathways in complex systems. These methods appear to be the most promising techniques for the study of long-time processes such as protein folding that continue to stand outside the range of direct molecular dynamics simulation.

ACKNOWLEDGMENT

I am grateful to Vio Buchete, Ron Elber, Stefan Fischer, and the editors for helpful comments.

REFERENCES

1. CL Brooks, M Karplus, M Pettitt. Proteins. Adv Chem Phys, Vol 71. New York: Wiley, 1988.
2. JA McCammon, SC Harvey. New York: Cambridge Univ Press, 1987.
3. R Elber, ed. Recent Developments in Theoretical Studies of Proteins. Singapore: World Scientific, 1997.
4. Y Duan, PA Kollman. Science 282:740–744, 1998.
5. BJ Berne, JE Straub. Curr Opin Struct Biol 7:181–189, 1997.
6. JT Hynes. Theory of Chemical Reaction Dynamics. Boca Raton, FL: CRC Press, 1985, pp 171–234.
7. BJ Berne, M Borkovec, JE Straub. J Phys Chem 92:3711–3725, 1988.
8. PG Wolynes. In: DL Stein, ed. Complex Systems: SFI Studies in the Science of Complexity, Reading, MA: Addison-Wesley, 1989, pp 355–387.
9. P Hänggi, P Talkner, M Borkovec. Rev Mod Phys 62:251–341, 1990.
10. S Glasstone, KJ Laidler, H Eyring. The Theory of Rate Processes. New York: McGraw-Hill, 1941.
11. P Pechukas. In: WH Miller, ed. Dynamics of Molecular Collisions, Part B. New York: Plenum Press, 1976, pp 269–322.
12. DG Truhlar, BC Garrett, SJ Klippenstein. J Phys Chem 100:12771–12800, 1996.
13. HA Scheraga. Carlsberg Res Commun 49:1–55, 1984.
14. AD Mackerell Jr, D Bashford, M Bellott, RL Dunbrack Jr, JD Evanseck, MJ Field, S Fischer, J Gao, H Guo, S Ha, D Joseph-McCarthy, L Kuchnir, K Kuczera, FTK Lau, C Mattos, S Michnick, T Ngo, DT Nguyen, B Prodhom, WE Reiher III, B Roux, M Schlenkrich, JC Smith, R Stote, JE Straub, M Watanabe, J Wiokiewicz-Kuczera, D Yin, M Karplus. J Phys Chem B 102:3586–3616, 1998.
15. R Elber. In: R Elber, ed. Recent Developments in Theoretical Studies of Proteins. Singapore: World Scientific; 1997, pp 65–136.
16. K Müller. Angew Chem Int Ed Engl 19:1–13, 1980.
17. JB Anderson. Adv Chem Phys 91:381–431, 1995.
18. HA Kramers. Physica 7:284–304, 1940.
19. D Chandler. J Chem Phys 68:2959–2970, 1978.
20. BJ Berne. In: P Hanggi, G Fleming, eds. Theoretical and Numerical Methods in Rate Theory. London: World Scientific, 1993.
21. M Karplus. J Phys Chem B 104: 11–27, 2000.
22. JC Keck. Disc Faraday Soc 33:173–182, 1962.
23. BJ Berne. Multiple Time Scales. New York: Academic Press, 1985, pp 419–436.
24. JE Straub, BJ Berne. J Chem Phys 83:1138–1139, 1985; JE Straub, DA Hsu, BJ Berne. J Phys Chem 89:5188–5191, 1985.
25. K Müller, LD Brown. Theor Chim Acta (Berl) 53:75–93, 1979.
26. CJ Cerjan, WH Miller. J Chem Phys 75:2800–2806, 1981.
27. DT Nguyen, DA Case. J Phys Chem 89:4020–4026, 1985.

28. T Lazaridis, DJ Tobias, CL Brooks III, ME Paulaitis. J Chem Phys 95:7612–7625, 1991.
29. HM Pickett, HL Strauss. J Am Chem Soc 105:7281–7290, 1970.
30. BD Ross, NS True. J Am Chem Soc 105:4871–4875, 1983.
31. B Roux, M Karplus. Annu Rev Biophys Biomol Struct 23:731–761, 1994.
32. R Elber, DP Chen, D Rojewska, R Eisenberg. Biophys J 68:906–924, 1995.
33. JA McCammon, M Karplus. Proc Natl Acad Sci USA 76:3585–3589, 1979.
34. SH Northrup, MR Pear, C-Y Lee, JA McCammon, M Karplus. Proc Natl Acad Sci USA 79: 4035–4039, 1982.
35. S Huo, JE Straub. J Chem Phys 105:5000–5006, 1997.
36. S Huo, JE Straub. Proteins 36:249–261, 1999.
37. R Elber, D Shalloway. J Chem Phys 112:5539–5545, 2000.
38. R Elber, M Karplus. Chem Phys Lett 139:375–380, 1987.
39. R Czerminski, R Elber. Int J Quant Chem 24:167–186, 1990.
40. R Czerminski, R Elber. J Chem Phys 92:5580–5601, 1990.
41. G Mills, H Jónsson, GK Schenter. Surf Sci 324:305–337, 1995.
42. M Berkowitz, JD Morgan, JA McCammon, SH Northrup. J Chem Phys 79:5563–5565, 1983.
43. LR Pratt. J Chem Phys 85:5045–5048, 1986.
44. AR Panchenko, J Wang, GU Nienhaus, PG Wolynes. J Phys Chem 99:9278–9282, 1995.
45. R Olender, R Elber. J Chem Phys 105:9299–9315, 1996.
46. C Dellago, PG Bolhuis, FS Csajika, D Chandler. J Chem Phys 108:1964–1977, 1998.
47. FS Csajika, D Chandler. J Chem Phys 109:1125–1133, 1998.
48. R Elber, J Meller, R Olender. J Phys Chem B 103:899–911, 1999.
49. TA Halgren, WN Lipscomb. Chem Phys Lett 49:225–232, 1977.
50. S Bell, JS Crighton, R Fletcher. Chem Phys Lett 82:122–126, 1981.
51. S Bell, JS Crighton. J Chem Phys 80:2464–2475, 1984.
52. JE Sinclair, R Fletcher. J Phys C 7:864–870, 1974.
53. A Ulitsky, R Elber. J Chem Phys 92:1510–1511, 1990.
54. R Olender, R Elber. THEOCHEM. J Mol Struct 398:63–71, 1997.
55. S Fischer, M Karplus. Chem Phys Lett 96:252–261, 1992.
56. S Fischer Jr, RL Dunbrack, M Karplus. J Am Chem Soc 116:11931–11937, 1994.
57. K Olsen, S Fischer, M Karplus. Biophys J 78:394A, 2000.
58. A Blondel, J-P Renaud, S Fischer, D Moras, M Karplus. J Mol Biol 291:101–115, 1999.
59. S Fischer, CS Verma, RE Hubbard. J Phys Chem 102:1797–1805, 1998.
60. S Fischer, S Michnick, M Karplus. Biochemistry, 32:13830–13837, 1993.
61. Y Levy, OM Becker. Phys Rev Lett 81:1126–1129, 1998.
62. JE Straub, M Borkovec, BJ Berne. J Chem Phys 89:4833–4847, 1988.

11

Computer Simulation of Biochemical Reactions with QM–MM Methods

Paul D. Lyne
Biogen, Inc., Cambridge, Massachusetts

Owen A. Walsh
Oxford University, Oxford, England

I. INTRODUCTION

Enzymes are phenomenally proficient molecular machines that are able to catalyze chemical reactions many orders of magnitude faster than the corresponding noncatalyzed reactions in solution [1]. How enzymes achieve these large rate enhancements is a matter of continuing research [2]. In recent years our understanding of the function of enzymes has been greatly increased by advances in structural biology. Current experimental techniques such as X-ray crystallography and nuclear magnetic resonance have produced atomic resolution structures of many enzymes. From the structures of these enzymes it is possible to infer mechanisms from the orientation of amino acids at the active site of the enzyme. In some cases, using the technique of Laue diffraction [3], it is possible to study an enzymatic reaction crystallographically, although this technique is far from being widely applicable. Despite having a detailed three-dimensional view of the active sites of many enzymes, it is often not possible to unambiguously favor one reaction mechanism over another on the basis of the structure. Other experimental methods such as kinetic studies or mutagenesis experiments can augment the information available, but currently it is not possible with experimental techniques to study the entire energy profile of a reaction in the active site of an enzyme. Nor is it possible to identify and quantify on a molecular level the interactions between the substrate and the enzyme as the reaction pathway is traversed.

Computer simulation techniques offer the ability to study the potential energy surfaces of chemical reactions to a high degree of quantitative accuracy [4]. Theoretical studies of chemical reactions in the gas phase are a major field and can provide detailed insights into a variety of processes of fundamental interest in atmospheric and combustion chemistry. In the past decade theoretical methods were extended to the study of reaction processes in mesoscopic systems such as enzymatic reactions in solution, albeit to a more approximate level than the most accurate gas-phase studies.

The best computational approach to the study of chemical reactions uses quantum mechanics; however, in practice the size of the enzyme system precludes the use of traditional quantum mechanical methods [5]. A typical simulation of a solvated enzyme could include approximately 10,000 atoms, which makes the simulation using quantum mechanics intractable even with current computing facilities. A similification of the problem can be achieved by considering only the substrate and selected active site residues while neglecting the remainder of the enzyme and solvent. However, the remainder of the enzyme and solvent, although not having a covalent effect on the reaction, can significantly alter the energetics of the system and thus need to be included in the simulation.

The emergence of hybrid quantum mechanical–molecular mechanical (QM–MM) methods in recent years addresses this problem. Pioneering studies of this type were made by Warshel and Levitt [6]. The method entails the division of the system of interest into a small region that is treated quantum mechanically, with the remainder of the system treated with computationally less expensive classical methods. The quantum region includes all the atoms that are directly involved in the chemical reaction being studied, and the remainder of the system, which is believed to change little during the reaction, is treated with a molecular mechanics force field [7]. The atoms in each system influence the other system through a coupled potential that involves electrostatic and van der Waals interactions [8–18]. Several molecular mechanics programs have been adapted to perform hybrid QM–MM simulations. In the majority of the implementations the quantum region has been treated either by empirical valence bond methods [19] or with a semiempirical method (usually AM1 [20]). These implementations have been applied, for example, to study solvation [12,21], condensed phase spectroscopy [22], conformational flexibility [23], and chemical reactivity in solution [24,25], in enzymes [18,26–36], and in DNA [37].

Although semiempirical methods have the advantage of being computationally inexpensive, they have a number of limitations [38–40]. The major limitations concern their accuracy and reliability. In general, they are less accurate than high level methods, and since they have been parametrized to reproduce the ground-state properties of molecules, they are often not well suited to the study of chemical reactions. A further disadvantage of the semiempirical methods is the limited range of elements for which parameters have been determined.

To overcome these limitations, the hybrid QM–MM potential can employ ad initio or density function methods in the quantum region. Both of these methods can ensure a higher quantitative accuracy, and the density function methods offer a computaitonally less expensive procedure for including electron correlation [5]. Several groups have reported the development of QM–MM programs that employ ab initio [8,10,13,16] or density functional methods [10,41–43].

This chapter presents the implementaiton and applicable of a QM–MM method for studying enzyme-catalyzed reactions. The application of QM–MM methods to study solution-phase reactions has been reviewed elsewhere [44]. Similarly, empirical valence bond methods, which have been successfully applied to studying enzymatic reactions by Warshel and coworkers [19,45], are not covered in this chapter.

II. BACKGROUND

A. QM–MM Methodology

In the combined QM–MM methodology the system being studies is partitioned into a quantum mechanical region and a molecular mechanical region (Fig. 1). The quantum

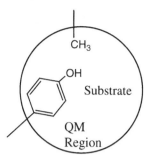

Figure 1 Schematic diagram depicting the partitioning of an enzymatic system into quantum and classical regions. The side chains of a tyrosine and valine are treated quantum mechanically, whereas the remainder of the enzyme and added solvent are treated with a classical force field.

region will normally include the substrate, side chains of residues believed to be involved in the reaction, and any cofactors. The remainder of the protein and solvent is included in the molecular mechanics region. For the QM region, the wave function of the system, Ψ, is a Slater determinant of one-electron molecular orbitals, ψ_i (or Kohn–Sham orbitals in the case of density functioned theory [46]),

$$|\Psi(r, R_q, R_c)\rangle = |\psi_1 \alpha(1) \psi_1 \beta(2) \cdots \psi_N \beta(2N)\rangle \tag{1}$$

where α and β refer to spin eigenfunctions, r the coordinates of the electrons, R_q the coordinates of the QM nuclei, and R_c the coordinates of the atoms in the MM region. The total energy of the system is evaluated by solving the Schrödinger equation with an effective Hamiltonian for the system:

$$\hat{H}_{\text{eff}} \Psi(r, R_q, R_c) = E(R_q, R_c)\Psi(r, R_q, R_c) \tag{2}$$

For QM–MM methods it is assumed that the effective Hamiltonian can be partitioned into quantum and classical components by writing [9]

$$\hat{H}_{\text{eff}} = \hat{H}_{\text{QM}} + \hat{H}_{\text{MM}} + \hat{H}_{\text{QM–MM}} \tag{3}$$

where $\hat{H}_{\text{QM}}$ is the pure quantum Hamiltonian, $\hat{H}_{\text{MM}}$ is the classical Hamiltonian, and $\hat{H}_{\text{QM–MM}}$ is the hybrid QM–MM Hamiltonian. From Eqs. (2) and (3) the energy of the system is given as

$$E(R_q, R_c) = \frac{\langle\Psi|\hat{H}_{\text{QM}}|\Psi\rangle + \langle\Psi|\hat{H}_{\text{QM–MM}}|\Psi\rangle}{\langle\Psi|\Psi\rangle} + E_{\text{MM}} \tag{4}$$

$$E(R_q, R_c) = E_{\text{QM}} + E_{\text{QM–MM}} + E_{\text{MM}} \tag{5}$$

To date the majority of QM–MM applications have employed density functional methods ab initio or semiempirical methods in the quantum region. The energy terms evaluated in these methods are generally similar, but there are specific differences. The relevant equations for the density functional based methods are described first, and this is followed by a description of the specific differences associated with the other methods.

For density functional based QM–MM methods the electronic energy terms depend explicitly on the electron density, $\rho(r)$, of the atoms in the quantum region [46]. The electron density is determined by solving the one-electron Kohn–Sham equations [46, 47]

$$\hat{H}_{DF}\psi_i(r) = e_i\psi_i(r), \qquad i = 1, \ldots, n \tag{6}$$

where e_i are the eigenfunctions associated with the Kohn–Sham orbitals. The quantum Hamiltonian $\hat{H}_{DF}$ is given by

$$\hat{H}_{DF} = -\sum_i \frac{\bar{h}^2}{2m_i} \nabla_i^2 - \sum_{i,q} \frac{Z_q}{r_{iq}} - \sum_{i,c} \frac{q_c}{r_{ic}}$$
$$+ \sum_{q<q'} \frac{Z_q Z_q'}{R_{qq'}} + \int \frac{\rho(r')}{|r - r'|} dr' + \frac{\partial E_{XC}}{\partial \rho(r)} \tag{7}$$

In this equation E_{XC} is the exchange correlation functional [46], q_c is the partial charge of an atom in the classical region, Z_q is the nuclear charge of an atom in the quantum region, r_{iq} is the distance between an electron and quantum atom q, r_{ic} is the distance between an electron and a classical atom c; $R_{qq'}$ is the distance between two quantum nuclei, and r' is the coordinate of a second electron. Once the Kohn–Sham equations have been solved, the various energy terms of the DF–MM method are evaluated as

$$E_{DF} = -\frac{\bar{h}^2}{2m} \int \sum_i \psi_i(r)\nabla^2\psi_i(r)\, dr - \int \sum_q \frac{Z_q}{|r_q - r|} \rho(r)\, dr$$
$$+ \frac{1}{2} \iint \frac{\rho(r)\rho(r')}{|r - r'|}\, dr\, dr' + E_{XC}[\rho(r)] + \sum_{q<q'} \frac{Z_q Z_q'}{R_{qq'}} \tag{8}$$

and

$$E_{DF-MM} = -\int \sum_c \frac{q_c}{|r_c - r|} \rho(r)dr + \sum_{q,c} \frac{Z_q c_q}{R_{qc}} + \sum_{q,c} V_{qc} \tag{9}$$

where V_{qc} is the van der Waals interaction energy between the quantum and classical regions. This Lennard-Jones term is given as

$$V_{qc} = 4\varepsilon_{qc} \left[\left(\frac{\sigma_{qc}}{R_{qc}}\right)^{12} - \left(\frac{\sigma_{qc}}{R_{qc}}\right)^6 \right] \tag{10}$$

This term is essential to obtain the correct geometry, because there is no Pauli repulsion between quantum and classical atoms. The molecular mechanics energy term, E_{MM}, is calculated with the standard potential energy term from CHARMM [48], AMBER [49], or GROMOS [50], for example.

The implementation of the method using ab initio methods for the quantum region is straightforward. The analogous equations for the electronic Hamiltonian and the corresponding energies in this case are [51]

$$E = \frac{1}{2} \sum_{\mu} \sum_{V} P_{\mu V} (H_{\mu V}^{\text{core}} + F_{\mu V}) + \sum_{\alpha} \sum_{\beta > \alpha} \frac{Z_\alpha Z_\beta}{R_{\alpha\beta}} + \sum_{\alpha,C} \frac{q_C Z_\alpha}{R_{\alpha C}} + \sum_{\alpha,C} V_{\alpha C} \qquad (11)$$

where $H_{\mu V}^{\text{core}}$ is defined as

$$H_{\mu V}^{\text{core}} \int \phi_\mu^* \left[-\frac{1}{2} \nabla^2 \right] \phi_V dr + \int \phi_\mu^* \left[-\left(\sum_\alpha \frac{Z_\alpha}{|R_\alpha - r|} + \sum_c \frac{q_C}{|R_C - r|} \right) \right] \phi_V dr \qquad (12)$$

The indices μ and v refer to the basis set orbitals, ϕ, and $P_{\mu v}$ and $F_{\mu v}$ are elements of the density and Fock matrices, respectively.

For QM–MM methods employing semiempirical methods in the quantum region, the implementaiton is not as straightforward. In semiempirical methods such s MNDO, AM1, or PM3 [38], a minimum basis set representation of the valence levels of each atom is employed. Thus, for example, for first row main group elements, the basis set comprises just four atomic orbitals, namely one s and three p orbitals. The remainder of the electrons in the atoms are grouped into core terms. Therefore, the added complication for QM–MM methods involving semiempirical Hamiltonians lies in how to treat the interactions between the core terms on the quantum atoms and the partial charges on atoms in the classical region. The treatment that is currently used follows that suggested by Field et al. [9], where the interaction between the semiempirical cores and the classical atoms are treated in the same manner as pure quantum interactions, with the atoms in the classical regions carrying an s orbital for this purpose. For Am1 or PM3 the interaction between the classical atoms and the cores of the quantum atoms is given as [20]

$$E_{\text{QM-MM}}^{\text{core}} = q_c Z_q (S_q S_q | S_c S_c)[1 + \exp(-\alpha_c R_{qc}) + \exp(-\alpha_q R_{qc})]$$

$$+ \frac{q_c Z_q}{R_{qc}} \sum_{i=1}^{4} \left\{ K_i^c \exp[-L_i^m (R_{qc} - M_i^m)^2] + K_i^q \exp[-L_i^c (R_{qc} - M_i^q)^2] \right\} \qquad (13)$$

From this equation it is seen that parameters have been introduced into the QM–MM method, with K_i^c, L_i^c, M_i^c, and α_c corresponding to the pseudo s orbital on the classical atom. These parameters can be optimized to reproduce experimental or high level theoretical data. Field et al. [9] performed extensive investigations of the values of these extra parameters and suggested that the parameters K_i^c, L_i^c, and M_i^c ($i = 1, \ldots, 4$) can be set to zero and that α_c should take a value of 5.0. These are generally the values used in most current QM–MM implementations that employ semiempirical methods in the quantum region.

Finally, the parametrization of the van der Waals part of the QM–MM interaction must be considered. This applies to all QM–MM implementations irrespective of the quantum method being employed. From Eq. (9) it can be seen that each quantum atom needs to have two Lennard-Jones parameters associated with it in order to have a van der Walls interaction with classical atoms. Generally, there are two approaches to this problem. The first is to derive a set of parameters, ε_q and σ_q, for each common atom type and then to use this standard set for any study that requires a QM–MM study. This is the most common aproach, and the ''derived'' Lennard-Jones parameters for the quantum atoms are simply the parameters found in the MM force field for the analogous atom types. For example, a study that employed a QM–MM method implemented in the program CHARMM [48] would use the appropriate Lennard-Jones parameters of the CHARMM force field [52] for the atoms in the quantum region.

The second approach is to derive Lennard-Jones parameters for the quantum atoms that are specific to the problem in hand. This is a less common approach but has been shown to improve the quantitative accuracy of the QM–MM approach in specific cases [53,54]. The disadvantage of this approach, however, is that it is necessary to derive Lennard-Jones parameters for the quantum region for every different study. Since the derivation of Lennard-Jones parameters is not a trivial exercise, this method of finding van der Walls parameters for the QM–MM interaction has not been widely used.

B.   The Quantum/Classical Boundary

In the preceding section the essentials of the implementation of QM–MM methods were presented. For studies on systems where there is a discrete natural boundary between the quantum and classical regions, the QM–MM methods can be applied as outlined earlier. An example of such a system might be the study of a chemical reaction in water, such as the S_N2 substitution of CH_3Cl by Cl^-. In this case the quantum region could comprise chloroform and the chloride ion, with the solvent being treated by classical methods [24]. For QM–MM studies on the active sites of enzymes there is no natural boundary between the quantum and classical regions. In these situations the quantum region might comprise the enzyme substrate and the side chains of several active-site residues. The active-site residues thus span both the quantum and classical regions, where the side chain is in the quantum region and the main chain atoms are in the classical region. The boundary between the quantum and classical regions falls across a covalent bond between the α- and β-carbons of the residue (see Fig. 2). This is an extra complication for QM–MM implementations, and it is necessary to devise a method that will handle this circumstance.

Because the electrons on atoms in the classical region are not treated explicitly, a method has to be devised that allows the electron density along the QM–MM bonds to be terminated satisfactorily. There are two common approaches that have been proposed to deal with this situation. The first of these is termed the ''link atom approach.'' In this approach a dummy atom called a link atom is introduced into the quantum system at the location of the boundary between the quantum and classical regions. The link atom serves the purpose of satisfying the truncation of the electron density in the quantum region. The link atom is not seen by the atoms in the classical region. The implementation of the link atom approach depends on the nature of the Hamiltonian used in the quantum region. In its original implementation, Field et al. [55] used the link atom in conjunction with a

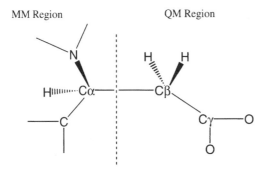

Figure 2   A glutamate side chain partitioned into quantum and classical regions. The terminal $CH_2CO_2^-$ group is treated quantum mechanically, and the backbone atoms are treated with the molecular mechanics force field.

semiempirical Hamiltonian. The protocol used there for evaluating the QM–MM interaction at the quantum/classical boundary is to perform a quantum calculation, with a link atom, using the equations shown in the previous section. Once a stable electron density has been determined, the contribution from the link atom to the QM–MM energy is removed. In the case of ab initio or density functional implementations the link atom is treated as a proper quantum atom and, as such, feels the charges from atoms in the classical region. [10]. The charges on the classical functional group or atom that is replaced by the link atom do not contribute to the one-electron integrals in the self-consistent calculation for the quantum region. For CH_3CH_2OH with OH treated by quantum mechanics and CH_3CH_2 treated by molecular mechanics as an example, the charges on the CH_2 group are not included in the QM Hamiltonian; all other partial charges on classical atoms (i.e., from the CH_3 group) are included. The link atom is initially positioned 1 Å along the original bond but is not constrained during the simulations.

The location of the quantum/classical boundary across a covalent bond also has implications for the energy terms evaluated in the E_{MM} term. Classical energy terms that involve only quantum atoms are not evaluated. These are accounted for by the quantum Hamiltonian. Classical energy terms that include at least one classical atom are evaluated. Referring to Figure 2, the $C_\alpha—C_\beta$ bond term; the $N—C_\alpha—C_\beta$, $C—C_\alpha—C_\beta$, $H_\alpha—C_\alpha—C_\beta$, $C_\alpha—C_\beta—H_{\beta 1}$, and $C_\alpha—C_\beta—H_{\beta 2}$ angle terms; and the proper dihedral terms involving a classical atom are all included.

An alternative approach to the link atom method is to use the frozen orbital approach developed by Rivail and coworkers [56] (the local self-consistent field, LSCF). The continuity of the electron density at the boundary region is maintained by a frozen orbital along the bond between the quantum and classical atoms. This frozen orbital is derived from calculaitons on model compounds, with the assumption that the orbitals from model compounds are transferable to the enzymatic system. A more generalized form of this implementation was presented by Gao et al. [57], in which a set of hybrid orbitals are used at the boundary region [this method is termed the generalized hybrid orbital (GHO) method by the authors]. The set is divided into auxiliary and active orbitals and acts as a basis set for the boundary atoms of the MM fragment. The active orbitals are optimized along with other orbitals on the QM atoms in the SCF cycle. In essence this method is an expansion of the approach of Rivail but has the advantage that the oribitals do not need to be parametrized for each specific problem.

There is concern in the field about current approaches for treating the boundary region in QM–MM calculations. The original implementation of the link atom aproach [9] results in arbitrary charge polarization in the QM region and the development of unrealistic partial charges on the link atom. The semiempirical implementation of the LSCF and GHO methods appears to improve the treatment of the boundary region compared to similar calculations using the link atom approach. However, the implementation of the link atom method for ab initio and density functional based QM–MM methods does not appear to introduce unphysical electrostatic perturbations into the QM region, and in a limited number of test cases it has been found to give very reasonable results for geometries, energetics, and charge distributions compared with pure quantum calculations [10].

III. APPLICATIONS

In this section the applications of QM–MM methods to three enzymes are presented with the intention of illustrating the techniques that are employed to simulate enzyme catalysis.

The specific enzymes chosen cover applications involving both semiempirical and ab initio methods in the quantum region.

A. Triosephosphate Isomerase

The chemical reaction catalyzed by triosephosphate isomerase (TIM) was the first application of the QM–MM method in CHARMM to the study of enzyme catalysis [26]. The study calculated an energy pathway for the reaction in the enzyme and decomposed the energetics into specific contributions from each of the residues of the enzyme. TIM catalyzes the interconversion of dihydroxyacetone phosphate (DHAP) and D-glyceraldehyde 3-phosphate (GAP) as part of the glycolytic pathway. Extensive experimental studies have been performed on TIM, and it has been proposed that Glu-165 acts as a base for deprotonation of DHAP and that His-95 acts as an acid to protonate the carbonyl oxygen of DHAP, forming an enediolate (see Fig. 3) [58].

Bash and coworkers [59] used QM–MM methods to follow the reaction path in TIM using a simulation protocol that has been employed in many subsequent simulations of enzyme catalysis with QM–MM methods. First, the initial coordinates used for the calculations came from a high resolution X-ray crystal structure [59]. This is an important point, because the validity of the results of a simulation of this nature on an enzyme is questionable in the absence of an accurate structure from crystallographic studies. The authors then chose to include only residues with a 16 Å radius sphere of the active site, with the remainder of the enzyme deleted. This saves on computation time while maintaining the bulk of the protein around the active site. In any given study of enzyme catalysis it is important to evaluate the effect on the results of different sized model systems (i.e., determine the radius of the sphere to include the study) in order to prevent any size dependence from being introduced to the study. Subsequently, Bash and coworkers surrounded the truncated protein with a sphere of preequilibrated water molecules and applied stochastic boundary conditions to the system to mimic the effects of bulk water. The reaction pathway was mapped adiabatically by fixing one degree of freedom corresponding to the reaction coordinate and fully minimizing all other degrees of freedom. The quantum region was chosen to include the substrate and the side chains of Glu-165 and His-95, with link atoms used at the quantum/classical boundary. The AM1 Hamiltonian was used for the quantum region, and the CHARMM force field for the classical region.

The results supported the proposal of Glu-165 as the general base and suggested the novel possibility of neutral histidine acting as an acid, contrary to the expectation that His-95 was protonated [26,58]. The conclusion that the catalytic His-95 is neutral has been confirmed by NMR spectroscopy [60]. The selection of neutral imidazole as the general acid catalyst has been discussed in terms of achieving a pK_a balance with the weakly acidic intermediate. This avoids the "thermodynamic trap" that would result from a too stable enediol intermediate, produced by reaction with the more acidic imidazolium [58].

Using the QM–MM method it is also possible to quantify the effects of individual residues on the reaction occurring at the active site. By performing a perturbative analysis involving sequential deletion of residues followed by a QM–MM interaction energy calculation, starting with the residue farthest from the active site and working in, an energy decomposition profile is generated [26]. This illustrates the effect that long-range electrostatics can have on the active site, with residues up to 14 Å away having an effect on the energy of the system in the active site. The most important contribution was found for

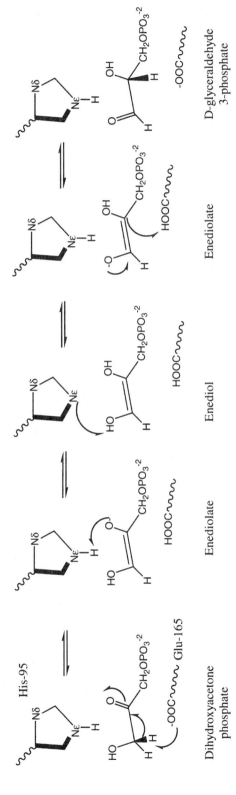

Figure 3 A possible mechanism for the isomerization of dihydroxyacetone phosphate (DHAP) to D-glyceraldehyde 3-phosphate (GAP) by the enzyme triosephosphate isomerase (TIM). The general acid (Glu-165) and general base (His-95) are shown.

Lys-12, which lies close to the active site, and the role of this residue is the stabilization of the enediolate. This explicitly explains the inactivity of a Lys-12 to Met mutation [61].

The QM–MM study of TIM was the first illustration of the potential of these methods for studying enzyme catalysis and has served as a reference for the protocol needed for subsequent studies of enzyme reactions.

B. Bovine Protein Tyrosine Phosphate

Recently, the study of the reaction catalyzed by bovine protein tyrosine phosphate (BPTP) using QM–MM methods was reported [30]. This study represents a progression from the techniques applied to the study of TIM and other enzymatic systems, because the reaction was followed by using molecular dynamics techniques and the QM–MM potential (QM–MD). QM–MD studies are a more powerful technique for studying chemical reactions in condensed phases because they allow for sampling of configuration space as the reaction pathway is followed and the generation of statistics that can be used to calculate reaction activation parameters such as the enthalpy and entropy of activation of various steps along the reaction pathway.

The use of QM–MD as opposed to QM–MM minimization techniques is computationally intensive and thus precluded the use of an ab initio or density functional method for the quantum region. This study was performed with an AM1 Hamiltonian, and the first step of the dephosphorylation reaction was studied (see Fig. 4). Because of the important role that phosphorus has in biological systems [62], phosphatase reactions have been studied extensively [63]. From experimental data it is believed that Cys-12 and Asp-129 residues are involved in the first step of the dephosphorylation reaction of BPTP [64,65]. Alahambra et al. [30] included the side chains of the phosphorylated tyrosine, Cys-12, and Asp-129 in the quantum region, with link atoms used at the quantum/classical boundaries. In this study the protein was not truncated and was surrounded with a 24 Å radius sphere of water molecules. Stochastic boundary methods were applied [66].

The reaction pathway in the enzyme was calculated by using the method of umbrella sampling, which has been widely used in the study of chemical reactions in solution [7,19]. The simulations were performed in 10–12 overlapping regions to cover the entire reaction coordinate, with 25 ps equilibration and 20 ps sampling in each window. A sampling time of 20 ps is rather short for fully sampling conformational space at each stage on the reaction coordinate. Nonetheless it represents a considerable improvement in simulation compared to simply minimizing the system at each point on the reaction pathway.

The study found that the transition state structure for the initial step of the dephosphorylation step is preferentially stabilized over the ground state through a Walden-inversion-enforced hydrogen-bonding mechanism at the active site [30]. It also suggested that a dianionic substrate is preferred in the reaction over a monoanionic mechanism, because the latter involves the breakage of a hydrogen bond between the nucleophile and the phosphoryl group, which causes the overall energy barrier to be raised.

This study is particularly noteworthy in the evolution of QM–MM studies of enzyme reactions in that a number of technical features have enhanced the accuracy of the technique. First, the authors explicitly optimized the semiempirical parameters for this specific reaction based on extensive studies of model reactions. This approach had also been used with considerable success in QM–MM simultation of the proton transfer between methanol and imidazole in solution.

Second, molecular dynamics techniques were employed that allowed the accurate

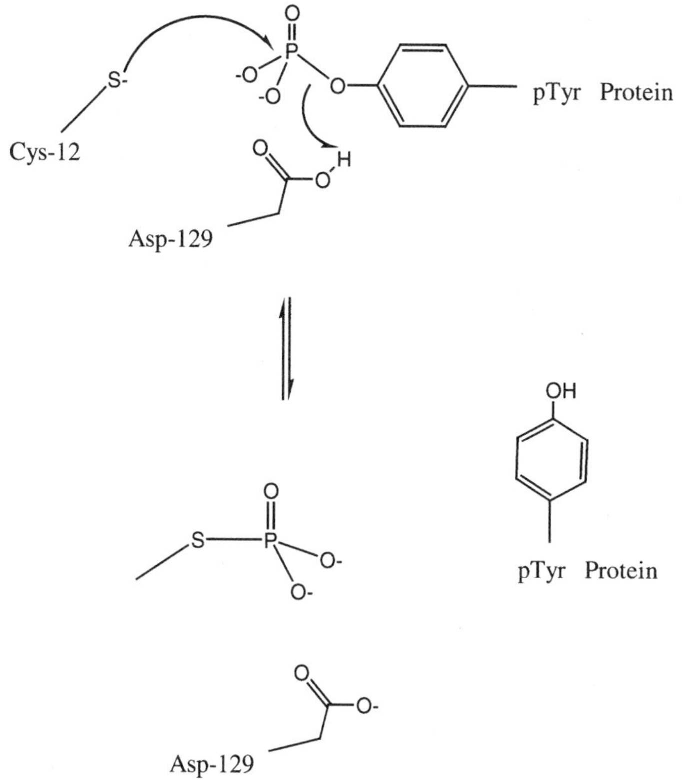

Figure 4 Schematic diagram of the first step of the reaction catalyzed by bovine protein tyrosine phosphatase (BPTP): formation of the cysteinyl phosophate intermediate.

determination of reaction activation parameters and the dynamics of active site residues to be followed along the trajectory of the reaction. This enabled the determination of a free energy of activation of 14 kcal/mol, which was in excellent agreement with the results from stopped-flow studies [30]. In addition, the dynamics of the hydrogen-bonding network of the active site could be followed throughout the reaction.

A free energy study of malate dehydrogenase [29] using semiempirical QM–MM methods has also been reported, and that study also attributes many of the benefits to simulation of enzyme reactions found in the BPTP study.

C. Citrate Synthase

The final application considered in this chapter is chosen to illustrate the application of a QM–MM study of an enzyme reaction that employs an ab initio Hamiltonian in the quantum region [67]. Because of the computational intensity of such calculations there are currently very few examples in the literature of QM–MM studies that use a quantum mechanical technique that is more sophisticated than a semiempirical method. Mulholland et al. [67] recently reported a study of part of the reaction catalyzed by citrate synthase (CS) in which the quantum region is treated by Hartree–Fock and MP2 methods [10,51], and this serves as a useful illustration of the current state of the art in the field of QM–MM applications to enzyme catalysis.

Citrate synthase catalyzes the metabolically important formation of citrate from acetyl-CoA and oxaloacetate [68]. Asp-375 (numbering for pig CS) has been shown to be the base for the rate-limiting deprotonation of acetyl-CoA (Fig. 5) [69]. An intermediate (which subsequently attacks the second substrate, oxaloacetate) is believed to be formed in this step; the intermediate is thought to be stabilized by a hydrogen bond with His-274. It is uncertain from the experimental data whether this intermediate is the enolate or enol of acetyl-CoA; related questions arise in several similar enzymatic reactions such as that catalyzed by triosephosphate isomerase. From the relative pK_a values of Asp-375

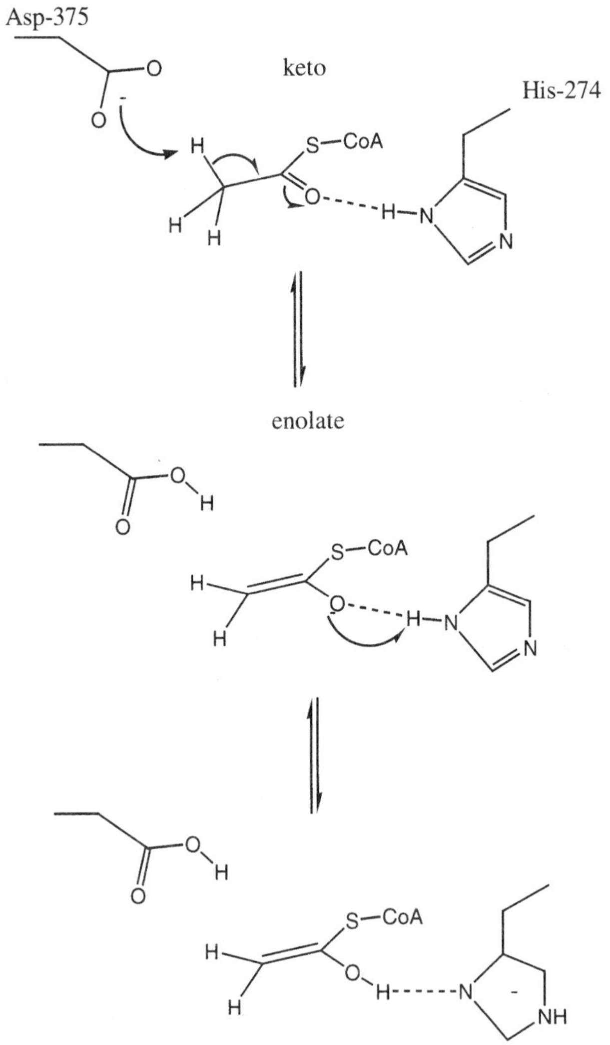

Figure 5   A suggested mechanism for the enolization of acetyl-CoA by the enzyme citrate synthase (CS). The keto, enolate, and enol forms of the substrate are shown.

and acetyl-CoA, it appears that the enolate (or enol) can be an intermediate in the enzymatic reaction only if it is stabilized by the enzyme. It has been proposed that the necessary stabilization is provided by a low barrier hydrogen bond (LBHB) in this and other enzymes [70,71]. A low barrier hydrogen bond is a covalent interaction between a hydrogen bond donor and the transition state of an enzymatic reaction, and it is believed to be the main source of catalysis. The hydrogen in an LBHB is almost equidistant from the heavy atoms, and the donor and acceptor are closer to each other than in normal hydrogen bonds. The potential energy surface for the transfer of the hydrogen between the donor and acceptor atoms is very small, leading to the appearance of a single broad energy well. For normal hydrogen bonds the potential energy for hydrogen transfer is characterized by two energy wells corresponding to the hydrogen being bonded to either of the heavy atoms. The suggested LBHB is betwen His-274, which is neutral, and an ''enolic'' acetyl-CoA intermediate [71,72]. To resolve the question of the nautre of the intermediate and its stabilization by the enzyme, the first reaction step in CS was investigated by ab initio QM–MM calculations with the CHARMM program.

The system contained all residues within 17 Å of the terminal carbon of acetyl-CoA, the *R*-malate substrate, and 23 crystallographic water molecules (the active site of CS is buried in the protein). All ab initio QM–MM calculations were performed using the CHARMM program interfaced to GAMESS [10]. Current computational resources preclude the use of molecular dynamics methods with an ab initio QM–MM potential, so the reaction pathway was followed by using the adiabatic mapping technique used previously for TIM and other enzymes. Even with this energy minimization approach, care is needed to avoid excessively long computational run times in calculating the reaction path. The approach used by Mulholland et al. [67] was to perform a series of calculations, beginning with a computationally inexpensive method and progressively increasing the level of calculation used in the quantum region. This is analogous to the approach often used in pure quantum mechanical studies that initially use a low level basis set to get a rough estimate of the potential energy surface and then refine this with a more accurate higher level basis set. The quantum region included the side chains of Asp-375, His-274, the thioester portion of acetyl-CoA, and the substrate.

For CS the reaction profile was initially extensively studied by QM–MM with an AM1 Hamiltonian [28]. The points along the reaction pathway were subsequently subjected to QM–MM minimization using RHF/3-21G* for the quantum region, and finally the minimization of the reaction points on the pathway was completed with RHF/6-31G* in the quantum region. This reaction pathway was further refined by performing single-point QM–MM calculations at the MP2/6-31G* level for the quantum region. This allowed the system to be fully minimized without incurring the large computational cost that would have resulted had the quantum region been treated at the RHF/6-31G* level from the start.

The study found that the enolate of acetyl-CoA is the intermediate in the rate-limiting step of the reaction, in agreement with previous experimental studies. The reaction catalyzed by CS has previously proposed to employ a mechanism that uses low barrier hydrogen bonds. Such bonds can be exceptionally strong in the gas phase and have been proposed to have energies of up to 20 kcal/mol in enzymes. However, the debate about their role in enzyme catalysis is controversial. A characteristic of low barrier hydrogen bonds is that the hydrogen is shared between atoms of approximately equal pK_a. In CS it has been proposed that the hydrogen bond between His-274 and an enolic (the proton is shared equally) intermediate is responsible for stabilizing the intermediate. The calculations on

CS indicated that the enolate of acetyl-CoA is significantly more stable than the enol or a proton-sharing enolic form and thus do not support the proposal that a low barrier hydrogen bond is involved in catalysis in CS. This study demonstrates the practial application of high level QM–MM studies to the elucidation of mechanistic details of an enzymatic reaction that are otherwise unclear.

IV. CONCLUSIONS

The field of QM–MM simulations of chemical reactions has grown considerably from the initial proposals of Warshel and Levitt [6] in the 1970s to a technique that can now deliver quantitatively accurate reaction pathways for reactions in the active sites of enzymes. Currently, the computational chemist has several options for treatment of the quantum region. Which method is employed in any given situation is dependent on a number of factors. The computational expense of the density functional and ab initio methods dictate studies using these methods need to be carried out on parallel computers. Naturally, these methods are more accurate for studying the chemistry of the process under consideration, and in the case of metalloenzymes with transition metals it is almost essential to use density functional methods. Nonetheless, it is possible to get quantitative accuracy with semiempirical QM–MM studies, particularly when the quantum atoms and the van der Waals parameters are parametrized for the specific reaction at hand. Additionally, semiempirical QM–MM methods allow a dynamic study of the chemical reaction, which is currently beyond the higher level methods, even with parallel computers. The field continues to expand, and it is to be expected that advances in the speed of the quantum calculations, the accuracy of the treatment of the quantum/classical boundary region, and computational speed will come over the next decade and enable further insight to be gained into the mechanisms of biochemical processes.

ACKNOWLEDGMENT

We are grateful to the Wellcome Trust for financial support.

REFERENCES

1. A Radzicka, R Wolfenden. Science 267:90–93, 1995.
2. AR Fersht. Structure and Mechanism in Protein Science: A Guide to Enzyme Catalysis and Protein Folding. New York: WH Freeman, 1999.
3. K Moffat, R Henderson. Curr Opin Struct Biol 5:656, 1995.
4. GC Schatz. J Phys Chem 100:12839, 1996.
5. M Head-Gordon. J Phys Chem 100:13213–13226, 1996.
6. A Warshel, M Levitt. J Mol Biol 103:227–249, 1976.
7. CL Brooks III, M Karplus, BM Pettitt. Proteins: A Theoretical Perspective of Dynamics, Structure and Thermodynamics. Adv Chem Phys Vol 71. New York: Wiley, 1988.
8. UC Singh, PA Kollman. J Comput Chem 7:718–730, 1986.
9. MJ Field, PA Bash, M Karplus. J Comput Chem 11:700–733, 1990.
10. PD Lyne, M Hodoscek, M Karplus. J Phys Chem A 103:3462, 1999.
11. BT Thole, PT van Duijnen. Biophys Chem 18:53–59, 1983.

12. J Gao, X Xia. Science 258:631, 1992.
13. RV Stanton, LR Little, KM Merz. J Phys Chem 99:17344–17348, 1995.
14. V Thery, D Rinaldi, J-L Rivail, B Maigret, GG Frenczy. J Comput Chem 15:269, 1994.
15. I Tunon, MTC Martins-Costa, C Millot, MF Ruiz-Lopez, JL Rivail. J Comput Chem 17:19–29, 1996.
16. F Maseras, K Morokuma. J Comput Chem 1995, In press.
17. MA Thompson. J Am Chem Soc 117:11341–11344, 1995.
18. MJ Harrison, NA Burton, IH Hillier. J Am Chem Soc 119:12285–12291, 1997.
19. A Warshel. Computer Modeling of Chemical Reactions in Enzymes and Solutions. New York: Wiley, 1991.
20. MJS Dewar, EG Zoebisch, EA Healy, JJP Stewart. J Am Chem Soc 107:3902–3909, 1985.
21. J Gao. J Phys Chem 96:537, 1992.
22. J Gao. J Am Chem Soc 116:9324, 1994.
23. H Liu, F Muller-Plathe, WF van Gunsteren. J Chem Phys 102:1722–1730, 1995.
24. PA Bash, MJ Field, MJ Karplus. J Am Chem Soc 109:8092–8094, 1987.
25. J Gao. J Am Chem Soc 116:1563, 1994.
26. PA Bash, MJ Field, RC Davenport, GA Petsko, D Ringe, M Karplus. Biochemistry 30:5826–5832, 1991.
27. PD Lyne, AJ Mulholland, WG Richards. J Am Chem Soc 117:11345–11350, 1995.
28. AJ Mulholland, WG Richards. Proteins: Struct Funct Genet 27:9–25, 1997.
29. MA Cunningham, LL Ho, DT Nguyen, RE Gillilan, PA Bash. Biochemistry 36:4800–4816, 1997.
30. C Alahambra, L Wu, Z-Y Zhang, J Gao. J Am Chem Soc 120:3858–3866, 1998.
31. S Antonczak, G Monard, MF Ruizlopez, JL Rivail. J Am Chem Soc 120:8825–8833, 1998.
32. P Varnai, WG Richards, PD Lyne. Proteins: Struc Funct Gen 37:218–227, 1999.
33. HY Liu, F Muller-Plathe, FF van Gunsteren. J Mol Biol 261:454–469, 1996.
34. S Ranganathan, JE Gready. J Phys Chem B 101:5614–5618, 1997.
35. DC Chatfield, KP Eurenius, BR Brooks. THEOCHEM 423:79–92, 1998.
36. J Bentzien, RP Muller, J Florian, A Warshel. J Phys Chem B 102:2293–2301, 1998.
37. AHE Elcock, PD Lyne, AJ Mulholland, A Nandra, WG Richards. J Am Chem Soc 117:4706, 1995.
38. JJP Stewart. J Comput Aided Mol Des 4:1–105, 1990.
39. HS Rzepa, M Yi. J Chem Soc Perkin Trans 2 1990:943–951, 1990.
40. MW Jurema, GC Shields. J Comput Chem 14:89–104, 1993.
41. I Tunon, MF Ruiz-Lopez, D Rinaldi, J Bertran. J Comput Chem 17:148–155, 1995.
42. RV Stanton, DS Hartsough, KM Merz Jr. J Comput Chem 16:113–128, 1995.
43. D Wei, DR Salahub. Chem Phys Lett 224:291, 1994.
44. J Gao, ed. Methods and Applications of Combined Quantum Mechanical and Molecular Mechanical Potentials, Vol. 7. New York: VCH, 1996.
45. A Warshel. Curr Opin Struct Biol 2:230–236, 1992.
46. RG Parr. Annu Rev Phys Chem 34:631–656, 1983.
47. W Kohn, LJ Sham. Phys Rev A 140:1133, 1965.
48. BR Brooks, RE Bruccoleri, BD Olafson, DJ States, S Swaminathan, M Karplus. J Comput Chem 4:187–217, 1983.
49. WD Cornell, P Ciepak, CI Bayly, IR Gould, KM Merz, DM Ferguson, DC Spellmeyer, T Fox, JW Caldwell, PA Kollman. J Am Chem Soc 118:2309, 1996.
50. WRP Scott, PH Hunenberger, IG Tironi, AE Mark, SR Billeter, J Fennen, AE Torda, T Huber, P Kruger, WF van Gunsteren. J Phys Chem A 103:3596–3607, 1999.
51. A Szabo, NS Ostlund. Modern Quantum Chemistry. New York: McGraw-Hill, 1989.
52. AD MacKerell Jr, D Bashford, M Bellott, RL Dunbrack Jr, JD Evanseck, MJ Field, S Fischer, J Gao, H Guo, S Ha, D Joseph-McCarthy, L Kuchnir, K Kuczera, FT Lau, C Mattos, S Michnick, T Ngo, DT Nguyen, B Prodhom, WE Reiher III, B Roux, M Schlenkrich, JC Smith,

R Stote, J Straub, M Watanabe, J Wiorkiewicz-Kuczera, D Yin, M Karplus. J Phys Chem B 102:3586–3616, 1998.

53. M Freindorf, J Gao. J Comput Chem 17:386–395, 1996.
54. PA Bash, LL Ho, AD Mackerell, D Levine, P Hallston. PNAS 93:3698–3703, 1996.
55. MJ Field, PA Bash, M Karplus. J Comput Chem 6:700, 1989.
56. G Monard, M Loos, V Thery, K Baka, J-L Rivail. Int J Quant Chem 58:153–159, 1996.
57. J Gao, P Amara, C Alahambra, MJ Field. J Phys Chem A 102:4714–4721, 1998.
58. JR Knowles. Phil Trans Roy Soc Lond B 332:115–121, 1991.
59. RC Davenport, PA Bash, BA Seaton, M Karplus, GA Petsko, D Ringe. Biochemistry 30: 5821–5826, 1991.
60. P Lodi, JR Knowles. Biochemistry 30:6948–6956, 1991.
61. PJ Lodi, LC Chang, JR Knowles, EA Komives. Biochemistry 33:2809–2814, 1994.
62. FH Westheimer. Science 235:1173–1177, 1987.
63. FH Westheimer. Acc Chem Res 1:70–78, 1968.
64. JA Stuckey, HL Schubert, EB Fauman, Z-Y Zhang, JE Dixon, MA Saper. Nature 370:571, 1994.
65. D Barford, AJ Flint, NK Tonks. Science 263:1397, 1994.
66. CL Brooks III, M Karplus. J Chem Phys 79:6312, 1983.
67. AJ Mulholland, PD Lyne, M Karplus. J Am Chem Soc 122:534–535, 2000.
68. G Pettersson, U Lill, H Eggerer. Eur J Biochem 182:119–124, 1989.
69. M Karpusas, B Branchaud, SJ Remington. Biochemistry 29:2213–2219, 1990.
70. JA Gerlt, PG Gassman. Biochemistry 32:11943–11952, 1993.
71. WW Cleland, MM Kreevoy. Science 264:1887, 1994.
72. JA Gerlt, PG Gassman. J Am Chem Soc 115:11552–11568, 1993.

12

X-Ray and Neutron Scattering as Probes of the Dynamics of Biological Molecules

Jeremy C. Smith
Interdisziplinäres Zentrum für Wissenschaftliches Rechnen der Universität Heidelberg, Heidelberg, Germany

I. INTRODUCTION

One of the major uses of molecular simulation is to provide useful theoretical interpretation of experimental data. Before the advent of simulation this had to be done by directly comparing experiment with analytical (mathematical) models. The analytical approach has the advantage of simplicity, in that the models are derived from first principles with only a few, if any, adjustable parameters. However, the chemical complexity of biological systems often precludes the direct application of meaningful analytical models or leads to the situation where more than one model can be invoked to explain the same experimental data.

Computer simulation gets around this problem by allowing more complicated, detailed, and realistic models of biological systems to be investigated. However, the price to pay for this is a degree of mathematical defeat, as the extra complexity prevents analytical solution of the relevant equations. We must therefore rely on numerical methods and hope that when allied with sufficient computer power they will lead to converged solutions of the equations we wish to solve. The simulation methods used involve empirical potential energy functions of the molecular mechanics type, and the equations of motion are solved using either stepwise integration of the full anharmonic function by molecular dynamics (MD) simulation or, for vibrating systems, by normal mode analysis involving a harmonic approximation to the potential function. In the case of molecular dynamics simulations of the dynamic properties of proteins, convergence is particularly difficult to achieve, because motions occur in proteins on time scales that are the same as or longer than the presently accessible time scale (nanoseconds). In the future things should improve, due to more efficient algorithms and the continuing rapid increase in computer power. Moreover, many dynamical phenomena of physical and biological interest occur on the subnanosecond time scale and thus can already be well sampled. The examples given in this chapter are mainly on this time scale.

Computer simulation can be used to provide a stepping stone between experiment and the simplified analytical descriptions of the physical behavior of biological systems. But before gaining the right to do this, we must first validate a simulation by direct comparison with experiment. To do this we must compare physical quantities that are measurable or derivable from measurements with the same quantities derived from simulation. If the quantities agree, we then have some justification for using the detailed information present in the simulation to interpret the experiments.

The spectroscopic techniques that have been most frequently used to investigate biomolecular dynamics are those that are commonly available in laboratories, such as nuclear magnetic resonance (NMR), fluorescence, and Mossbauer spectroscopy. In a later chapter the use of NMR, a powerful probe of local motions in macromolecules, is described. Here we examine scattering of X-ray and neutron radiation. Neutrons and X-rays share the property of being found in expensive sources not commonly available in the laboratory. Neutrons are produced by a nuclear reactor or "spallation" source. X-ray experiments are routinely performed using intense synchrotron radiation, although in favorable cases laboratory sources may also be used.

The X-ray and neutron scattering processes provide relatively direct spatial information on atomic motions via determination of the wave vector transferred between the photon/neutron and the sample; this is a Fourier transform relationship between wave vectors in reciprocal space and position vectors in real space. Neutrons, by virtue of the possibility of resolving their energy transfers, can also give information on the time dependence of the motions involved.

The comparison with experiment can be made at several levels. The first, and most common, is in the comparison of "derived" quantities that are not directly measurable, for example, a set of average crystal coordinates or a diffusion constant. A comparison at this level is convenient in that the quantities involved describe directly the structure and dynamics of the system. However, the obtainment of these quantities, from experiment and/or simulation, may require approximation and model-dependent data analysis. For example, to obtain experimentally a set of average crystallographic coordinates, a physical model to interpret an electron density map must be imposed. To avoid these problems the comparison can be made at the level of the measured quantities themselves, such as diffraction intensities or dynamic structure factors. A comparison at this level still involves some approximation. For example, background corrections have to made in the experimental data reduction. However, fewer approximations are necessary for the structure and dynamics of the sample itself, and comparison with experiment is normally more direct. This approach requires a little more work on the part of the computer simulation team, because methods for calculating experimental intensities from simulation configurations must be developed. The comparisons made here are of experimentally measurable quantities.

Having made the comparison with experiment one may then make an assessment as to whether the simulation agrees sufficiently well to be useful in interpreting the experiment in detail. In cases where the agreement is not good, the determination of the cause of the discrepancy is often instructive. The errors may arise from the simulation model or from the assumptions used in the experimental data reduction or both. In cases where the quantities examined agree, the simulation can be decomposed so as to isolate the principal components responsible for the observed intensities. Sometimes, then, the dynamics involved can be described by a simplified concept derived from the simulation.

In this chapter we present some basic equations linking scattering to dynamics that are of interest to computer simulators. Their derivations from first principles are quite long, and for these readers are referred to appropriate textbooks [1,2]. Examples of different properties that can be probed are given here from work involving our group. For more details on this work, readers are referred to the original references and to two reviews [3,4].

II. BASIC EQUATIONS RELATING ATOMIC POSITIONS TO X-RAY AND NEUTRON SCATTERING

Scattering experiments involve processes in which incident particles (X-rays or neutrons) with wave vector $\vec{k}_i$ (energy E_i) interact with the sample and emerge with wave vector $\vec{k}_f$ (energy E_f), obeying the conservation laws for momentum and energy,

$$\mathbf{k}_i - \mathbf{k}_f = \mathbf{Q} \tag{1}$$

$$E_i - E_f = \hbar\omega \tag{2}$$

The vectors $\mathbf{k}_i$, $\mathbf{k}_f$, and $\mathbf{Q}$ define the scattering geometry as illustrated in Figure 1.

We first examine the relationship between particle dynamics and the scattering of radiation in the case where both the energy and momentum transferred between the sample and the incident radiation are measured. Linear response theory allows dynamic structure factors to be written in terms of equilibrium fluctuations of the sample. For neutron scattering from a system of identical particles, this is [1,5,6]

$$S_{\text{coh}}(\vec{Q}, \omega) = \frac{1}{2\pi} \iint dt\, d^3r\, e^{i(\vec{Q}\cdot\vec{r}-\omega t)} G(\vec{r}, t) \tag{3}$$

$$S_{\text{inc}}(\vec{Q}, \omega) = \frac{1}{2\pi} \iint dt\, d^3r\, e^{i(\vec{Q}\cdot\vec{r}-\omega t)} G_s(\vec{r}, t) \tag{4}$$

where $\vec{Q}$ is the scattering wave vector, ω the energy transfer, and the subscripts coh and inc refer to coherent and incoherent scattering, discussed later. $G_s(\vec{r}, t)$ and $G(\vec{r}, t)$ are

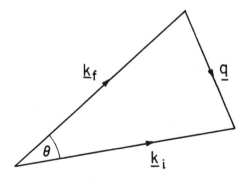

Figure 1 Scattering vector triangle.

van Hove correlation functions, which, for a system of N particles undergoing classical dynamics, are defined as follows:

$$G(\vec{r}, t) = \frac{1}{N} \sum_{i,j} \langle \delta(\vec{r} - \vec{R}_i(t) + \vec{R}_j(0)) \rangle \tag{5}$$

$$G_s(\vec{r}, t) = \frac{1}{N} \sum_i \langle \delta(\vec{r} - \vec{R}_i(t) + \vec{R}_i(0)) \rangle \tag{6}$$

where $\vec{R}_i(t)$ is the position vector of the ith scattering nucleus and $\langle \ldots \rangle$ indicates an ensemble average.

$G(\vec{r}, t)$ is the probability that, given a particle at the origin at time $t = 0$, any particle (including the original particle) is at $\vec{r}$ at time t. $G_s(\vec{r}, t)$ is the probability that, given a particle at the origin at time $t = 0$, the same particle is at $\vec{r}$ at time t.

Equation (3) has an analogous form in X-ray scattering, where the scattered intensity is given as [2]

$$|F(\vec{Q}, \omega)|^2 = \frac{1}{2\pi} \iint d^3r \, dt P(\vec{r}, t) e^{i(\vec{Q} \cdot \vec{r} - \omega t)} \tag{7}$$

where $P(\vec{r}, t)$ is the spatiotemporal Patterson function given by

$$P(\vec{r}, t) = \iint d\vec{R} \, dt \rho(\vec{R}, t) \rho(\vec{r} + \vec{R}, t + \tau) \tag{8}$$

and $\rho(\vec{r}, t)$ is the time-dependent electron density. Unfortunately, X-ray photons with wavelengths corresponding to atomic distances have energies much higher than those associated with thermal fluctuations. For example, an X-ray photon of 1.8 Å wavelength has an energy of 6.9 keV corresponding to a temperature of 8×10^7 K. Until very recently X-ray detectors were not sensitive enough to accurately measure the minute fractional energy changes associated with molecular fluctuations, so the practical exploitation of Eq. (7) was difficult. However, the use of new third-generation synchrotron sources has enabled useful inelastic X-ray scattering experiments to be performed on, for example, glass-forming liquids and liquid water (see, e.g., Ref. 7).

Although inelastic X-ray scattering holds promise for the investigation of biological molecular dynamics, it is still in its infancy, so in the subsequent discussion X-ray scattering is examined only in the case in which inelastic and elastic scattering are indistinguishable experimentally.

III. SCATTERING BY CRYSTALS

In an X-ray crystallography experiment the instantaneous scattered intensity is given by [2]

$$I_{hkl} = |F_{hkl}|^2 = \sum_{i=1}^N \sum_{j=1}^N f_i f_j^* \exp[i\vec{Q} \cdot (\vec{R}_i - \vec{R}_j)] \tag{9}$$

where F_{hkl} is the structure factor, $\vec{R}_i$ is the position vector of atom i in the crystal, and f_i is the X-ray atomic form factor. For neutron diffraction f_i is replaced by the coherent scattering length.

In an experimental analysis it is not feasible to insert into Eq. (9) the atomic positions for all the atoms in the crystal for every instant in the time of the experiment. Rather, the intensity must be evaluated in terms of statistical relationships between the positions. A convenient approach is to consider a real crystal as a superposition of an ideal periodic structure with slight perturbations. When exposed to X-rays the real crystal gives rise to two scattering components: the set of Bragg reflections arising from the periodic structure, and scattering outside the Bragg spots (diffuse scattering) that arises from the structural perturbations:

$$I_{hkl} = I_{hkl}^B + I_{hkl}^D \tag{10}$$

where I_{hkl}^B is the Bragg intensity found at integer values of h, k, and l, and I_{hkl}^D is the diffuse scattering, not confined to integer values of h, k, and l.

In terms of structure factors the various intensities are given by [2]

$$I_{hkl} = |F_{hkl}|^2 \tag{11}$$

$$I_{hkl}^B = |\langle F_{hkl} \rangle|^2 \tag{12}$$

and

$$I_{hkl}^D = |\Delta F_{hkl}|^2 \tag{13}$$

A. Bragg Diffraction

The Bragg peak intensity reduction due to atomic displacements is described by the well-known "temperature" factors. Assuming that the position $\vec{r}_i$ can be decomposed into an average position, $\langle \vec{r}_i \rangle$ and an infinitesimal displacement, $\vec{u}_i = \delta \vec{R}_i = \vec{R}_i - \langle \vec{R}_i \rangle$ then the X-ray structure factors can be expressed as follows:

$$F_{hkl} = \sum_{i=1}^{N} f_i(\vec{Q}) \exp[i\vec{Q} \cdot \langle \vec{R}_i \rangle] \exp[W_i(Q)] \tag{14}$$

where $W_i(Q) = -(1/3) \langle u_{i,Q}^2 \rangle Q^2$ and $\langle u_{i,Q}^2 \rangle$ is the mean-square fluctuation in the direction of Q. $W_i(Q)$ is the Debye–Waller factor.

Example: Low Temperature Vibrations in Acetanilide

Temperature factors are of interest to structural biologists mainly as a means of deriving qualitative information on the fluctuations of segments of a macromolecule. However, X-ray temperature factor analysis has drawbacks. One of the most serious is the possible presence of a static disorder contribution to the atomic fluctuations. This cannot be distinguished from the dynamic disorder due to the aforementioned absence of accurate energy analysis of the scattered X-ray photons. For quantitative work this problem can be avoided by choosing a system in which there is negligible static disorder and in which the harmonic approximation is valid. An example of such a system is acetanilide, $(C_6H_5\text{---}CONH\text{---}CH_3)$, at 15 K. The structure of acetanilide is shown in Figure 2. In recent work [8] a molecular mechanics force field was parametrized for this crystal, and normal mode analysis was performed in the full configurational space of the crystal i.e., including all intramolecular and intermolecular degrees of freedom. As a quantitative test of the accuracy of the force field, anisotropic quantum mechanical mean-square displacements of the hydrogen atoms were calculated in each Cartesian direction as a sum over the phonon normal modes of the crystal and compared with experimental neutron diffraction

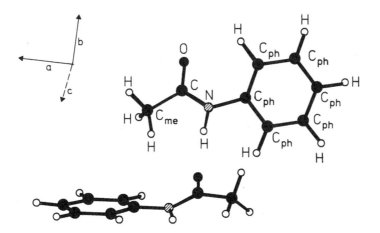

Figure 2 Crystalline acetanilide.

temperature factors [9]. The experimental and theoretical temperature factors are presented in Table 1. The values of the mean-square displacements are in excellent agreement.

B. X-Ray Diffuse Scattering

Any perturbation from ideal space-group symmetry in a crystal will give rise to diffuse scattering. The X-ray diffuse scattering intensity I_{hkl}^D at some point (hkl) in reciprocal space can be written as

$$I_{hkl}^D = N \sum_m \langle (F_n - \langle F \rangle)(F_{n+m} - \langle F \rangle)^* \rangle \exp(-\vec{Q} \cdot \vec{R}_m) \tag{15}$$

where F_n is the structure factor of the nth unit cell and the sum $\sum_m$ runs over the relative position vectors $\vec{R}_m$ between the unit cells. The correlation function $\langle (F_n - \langle F \rangle)(F_{n+m} - \langle F \rangle)^* \rangle$ is determined by correlations between atomic displacements.

   If the diffuse scattering of dynamic origin contributes significantly to the measured scattering, it may provide information on the nature of correlated motions in biological macromolecules that may themselves be of functional significance. To examine this possibility it is necessary to construct dynamic models of the crystal, calculate their diffuse scattering, and compare with experimental results. The advent of high intensity synchrotron sources and image plate detectors has allowed good quality X-ray diffuse scattering images to be obtained from macromolecular crystals.

   SERENA (scattering of X-rays elucidated by numerical analysis) is a program for calculating X-ray diffuse scattering intensities from configurations of atoms in molecular crystals [10]. The configurations are conveniently derived from molecular dynamics simulations, although in principle any collection of configurations can be used. SERENA calculates structure factors from the individual configurations and performs the averaging required in Eq. (11).

   Displacements correlated within unit cells but not between them lead to very diffuse scattering that is not associated with the Bragg peaks. This can be conveniently explored

Table 1   Mean-Square Displacements of Hydrogen Atoms in Crystalline Acetanilide at 15 K[a]

	I	II	III		I	II	III
Methyl H *a*		0.0359	0.0352	Phenyl H$_{para}$ *a*		0.0126	0.0118
b		0.0366	0.0438	*b*		0.0273	0.0278
c		0.0253	0.0258	*c*		0.0279	0.0258
isotropic		0.0326	0.0349	isotropic		0.0226	0.0218
Methyl H *a*		0.0150	0.0160	Phenyl H$_{meta}$ *a*		0.0215	0.0220
b		0.0482	0.0510	*b*		0.0189	0.0194
c		0.0328	0.0336	*c*		0.0265	0.0246
isotropic		0.0320	0.0335	isotropic		0.0223	0.0220
Methyl H *a*		0.0301	0.0286	Phenyl H$_{meta}$ *a*		0.0154	0.0162
b		0.0162	0.0172	*b*		0.0218	0.0210
c		0.0483	0.0606	*c*		0.0296	0.0256
isotropic		0.0315	0.0354	isotropic		0.0222	0.0209
Amide H *a*		0.0178	0.0192	Phenyl H$_{ortho}$ *a*		0.0159	0.0160
b		0.0110	0.0120	*b*		0.0189	0.0194
c		0.0258	0.0300	*c*		0.0286	0.0250
isotropic		0.0182	0.0204	isotropic		0.0211	0.0201
				Phenyl H$_{ortho}$ *a*		0.0203	0.0200
				b		0.0170	0.0164
				c		0.0264	0.0254
				isotropic		0.0212	0.0206

[a] Column I: Hydrogen atoms of acetanilide. Phenyl hydrogens are named according to their position relative to the N substitution site. *a*, *b*, *c* refer to the crystallographic directions. Column II: Anisotropic (*a*, *b*, *c* crystallographic directions) and isotropic mean-square displacements (Å^2) from neutron diffraction data [9]. Column III: Anisotropic (*a*, *b*, *c* crystallographic directions) and isotropic mean-square displacements (Å^2) from harmonic analysis.

using present-day simulations of biological macromolecules. However, motions correlated over distances larger than the size of the simulation model will clearly not be included.

Example: Correlated Motions in Lysozyme

Ligand binding and cooperativity often require conformational change involving correlated displacements of atoms [11]. A simple model for long distance transmission of information across a protein involves the activation and amplification of correlated motions that are present in the unperturbed protein. Although long-range correlated fluctuations are required for functional, dynamic information transfer, it is not clear to what extent they contribute to equilibrium thermal fluctuations in proteins. It is therefore important to know whether equilibrium motions in proteins can indeed be correlated over long distances or whether anharmonic and damping effects destroy such correlations.

To examine the dynamic origins of X-ray diffuse scattering by proteins, experimental scattering was measured from orthorhombic lysozyme crystals and compared to patterns calculated using molecular simulation [12]. The diffuse scattering was found to be approximately reproduced by both normal modes and molecular dynamics. More recently, a molecular dynamics analysis was performed of the dynamics of a unit cell of orthorhombic lysozyme, including four protein molecules [13]. Diffuse scattering calculated from the MD trajectory is compared in Figure 3 with that calculated from trajectories of rigid

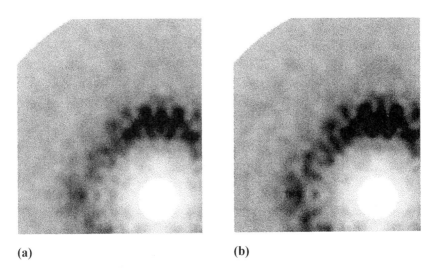

(a)   **(b)**

Figure 3   Calculated X-ray diffuse scattering patterns from (a) a full molecular dynamics trajectory of orthorhombic hen egg white lysozyme and (b) a trajectory obtained by fitting to the full trajectory rigid-body side chains and segments of the backbone. A full description is given in Ref. 13.

bodies obtained by fitting to the full simulation. The full simulation scattering is reproduced by the approximate representation to an agreement factor (*R*-factor) of 6%.

IV. NEUTRON SCATTERING

In contrast to X-rays, the mass of the neutron is such that the energy exchanged in exciting or deexciting picosecond time scale thermal motions is a large fraction of the incident energy and can be measured relatively precisely. A thermal neutron of 1.8 Å wavelength has an energy of 25 meV corresponding to $k_b T$ at 300 K. To further examine the neutron scattering case, we perform space Fourier transformation of the van Hove correlation functions [Eqs. (3) and (4)]:

$$S_{\text{coh}}(\vec{Q}, \omega) = \frac{1}{2\pi} \int_{-\infty}^{+\infty} dt\, e^{-i\omega t} I_{\text{coh}}(\vec{Q}, t) \tag{16}$$

$$I_{\text{coh}}(\vec{Q}, t) = \frac{1}{N} \sum_{i,j} b_{i,\text{coh}}^{*} b_{j,\text{coh}} e^{-i\vec{Q}\cdot\vec{R}_i(0)} \langle e^{i\vec{Q}\cdot\vec{R}_j(t)} \rangle \tag{17}$$

$$S_{\text{inc}}(\vec{Q}, \omega) = \frac{1}{2\pi} \int_{-\infty}^{+\infty} dt\, e^{-i\omega t} I_{\text{inc}}(\vec{Q}, t) \tag{18}$$

$$I_{\text{inc}}(\vec{Q}, t) = \frac{1}{N} \sum_{i} b_{i,\text{coh}}^{2} \langle e^{-i\vec{Q}\cdot\vec{R}_i(0)} e^{i\vec{Q}\cdot\vec{R}_i(t)} \rangle \tag{19}$$

Neutrons are scattered by the nuclei of the sample. Because of the random distribution of nuclear spins in the sample, the scattered intensity will contain a *coherent* part arising from the average neutron–nucleus potential and an *incoherent* part arising from fluctuations from the average. The coherent scattering arises from self- and cross-correlations of atomic motions, and the incoherent scattering, from single-atom motions. Each isotope has a coherent scattering length $b_{i,coh}$ and an incoherent scattering length $b_{i,inc}$ that define the strength of the interaction between the nucleus of the atom and the neutron. For more details on the origin of incoherent and coherent scattering see Ref. 1. We see from Eqs. (16) and (18) that the coherent and incoherent dynamic structure factors are time Fourier transforms of the coherent and incoherent intermediate scattering functions, $I_{coh}(\vec{Q}, t)$ and $I_{inc}(\vec{Q}, t)$; these are time-correlation functions [14]. $S_{inc}(\vec{Q}, \omega)$ and $S_{coh}(\vec{Q}, \omega)$ may contain elastic ($\omega = 0$) and inelastic ($\omega \neq 0$) parts. Elastic scattering probes correlations of atomic positions at long times, whereas the inelastic scattering process probes position correlations as a function of time.

A. Coherent Inelastic Neutron Scattering

The use of coherent neutron scattering with simultaneous energy and momentum resolution provides a probe of time-dependent pair correlations in atomic motions. Coherent inelastic neutron scattering is therefore particularly useful for examining lattice dynamics in molecular crystals and holds promise for the characterization of correlated motions in biological macro-molecules. A property of lattice modes is that for particular wave vectors there are well-defined frequencies; the relations between these two quantities are the phonon dispersion relations [1]. Neutron scattering is presently the most effective technique for determining phonon dispersion curves. The scattering geometry used is illustrated in Figure 4. The following momentum conservation law is obeyed:

$$\mathbf{k}_i - \mathbf{k}_f = \mathbf{Q} = \boldsymbol{\tau} + \mathbf{q} \tag{20}$$

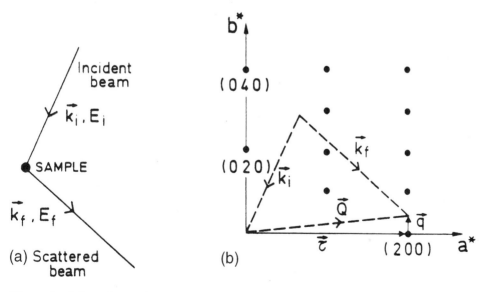

Figure 4 Schematic vector diagrams illustrating the use of coherent inelastic neutron scattering to determine phonon dispersion relationships. (a) Scattering in real space; (b) a scattering triangle illustrating the momentum transfer, $\mathbf{Q}$, of the neutrons in relation to the reciprocal lattice vector of the sample $\boldsymbol{\tau}$ and the phonon wave vector, $\mathbf{q}$. Heavy dots represent Bragg reflections.

The vibrational excitations have a wave vector q that is measured from a Brillouin zone center (Bragg peak) located at τ, a reciprocal lattice vector.

If the displacements of the atoms are given in terms of the harmonic normal modes of vibration for the crystal, the coherent one-phonon inelastic neutron scattering cross section can be analytically expressed in terms of the eigenvectors and eigenvalues of the harmonic analysis, as described in Ref. 1.

Example: Lattice Vibrations in L-Alanine

Zwitterionic L-alanine (^+H_3N—$C(CH_3)$—CO_2^-) is a dipolar molecule that forms large well-ordered crystals in which the molecules form hydrogen-bonded columns. The strong interactions lead to the presence of well-defined intra- and intermolecular vibrations that can usefully be described using harmonic theory.

Coherent inelastic neutron scattering experiments have been combined with normal mode analyses with a molecular mechanics potential function to examine the collective vibrations in deuterated L-alanine [15]. In Figure 5 are shown experimental phonon frequencies $\nu_i(\mathbf{q})(\nu = \omega/2\pi)$ for several modes propagating along the crystallographic direction $\mathbf{b}^*$. The solid lines represent the most probable paths for the dispersion curves $\nu_i(\mathbf{q})$. The theoretical dispersion curves are also given. The comparison between theory and experiment can be used to assess the accuracy with which the theory reproduces long-range interactions in the crystal.

B. Incoherent Neutron Scattering

Neutron scattering from nondeuterated organic molecules is dominated by incoherent scattering from the hydrogen atoms. This is largely because the incoherent scattering cross section ($4\pi b_{inc}^2$) of hydrogen is approximately 15 times greater than the total scattering cross section of carbon, nitrogen, or oxygen. The measured incoherent scattering thus essentially gives information on self-correlations of hydrogen atom motions. A program for calculating neutron scattering properties from molecular dynamics simulations has been published [16].

In practice, the measured incoherent scattering energy spectrum is divided into elastic, quasielastic, and inelastic scattering. Inelastic scattering arises from vibrations. Quasielastic scattering is typically Lorentzian or a sum of Lorentzians centered on $\omega = 0$ and arises from diffusive motions in the sample. Elastic scattering gives information on the self-probability distributions of the hydrogen atoms in the sample.

A procedure commonly used to extract dynamic data directly from experimental incoherent neutron scattering profiles is described in Ref. 17. It is assumed that the atomic position vectors can be decomposed into two contributions, one due to diffusive motion, $\vec{r}_{i,d}(t)$, and the other from vibrations, $\vec{u}_{i,v}(t)$, i.e.,

$$\vec{R}_i(t) = \vec{r}_{i,d}(t) + \vec{u}_{i,v}(t) \tag{21}$$

Combining Eq. (21) with Eq. (19) and assuming that $\vec{r}_{i,d}(t)$ and $\vec{u}_{i,v}(t)$ are uncorrelated, one obtains

$$I_{inc}(\vec{Q}, t) = I_d(\vec{Q}, t)I_v(\vec{Q}, t) \tag{22}$$

where $I_d(\vec{Q}, t)$ and $I_v(\vec{Q}, t)$ are obtained by substituting $\vec{R}_i(t)$ in Eq. (19) with $\vec{r}_{i,d}(t)$ and $\vec{u}_{i,v}(t)$, respectively.

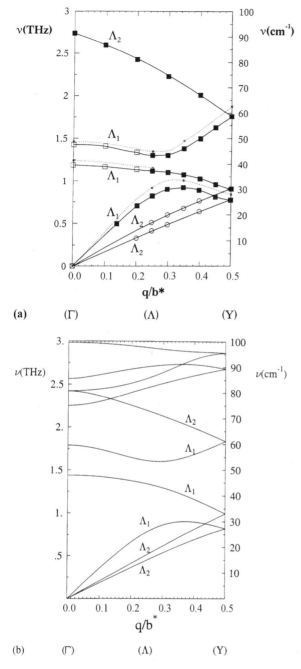

(a) (Γ)   (Λ)   (Y)

(b)   (Γ)   (Λ)   (Y)

Figure 5 (a) Dispersion curves for crystalline zwitterionic L-alanine at room temperature along the b* crystallographic direction determined by coherent inelastic neutron scattering. The ○ and ■ symbols are associated with phonon modes observed in predominantly transverse and purely longitudinal configurations, respectively, i.e., for vectors **Q** and **q** perpendicular and parallel to one another, respectively. They correspond to measurements performed around the strong Bragg reflections (200), (040), and (002). The □ symbols are neutron data points obtained around the (330), (103), and (202) reciprocal lattice points in a mixed configuration. Solid lines indicate the most probable connectivity of the dispersion curves, and dashed lines correspond to the measurements performed at low temperature $T = 100$ K. (b) Theoretical dispersion curves for L-alanine determined from normal mode analysis. (From Ref. 15.)

The Fourier transform of Eq. (22) gives

$$S(\vec{Q}, \omega) = S_d(\vec{Q}, \omega) \otimes S_v(\vec{Q}, \omega) \tag{23}$$

where $S_d(\vec{Q}, \omega)$ and $S_v(\vec{Q}, \omega)$ are obtained by Fourier transformation of $I_d(\vec{Q}, t)$ and $I_v(\vec{Q}, t)$ and the symbol $\otimes$ denotes the convolution product. Appropriate descriptions of $\vec{r}_{i,d}(t)$ and $\vec{u}_{i,v}(t)$ can be obtained from analytical theory or computer simulation.

$I_d(\vec{Q}, t)$ can be separated into time-dependent and time-independent parts as follows:

$$I_d(\vec{Q}, t) = A_0(\vec{Q}) + I'_d(\vec{Q}, t) \tag{24}$$

The elastic incoherent structure factor (EISF), $A_0(\vec{Q})$, is defined as [17]

$$A_0(\vec{Q}) = \lim_{t \to \infty} I_d(\vec{Q}, t) = \int d^3 r e^{i\vec{q} \cdot \vec{r}} \lim_{t \to \infty} G_d(\vec{r}, t) \tag{25}$$

where $G_d(\vec{r}, t)$ is the contribution to the van Hove self-correlation function due to diffusive motion. $A_0(\vec{Q})$ is thus determined by the diffusive contribution to the space probability distribution of the hydrogen nuclei.

Direct experiment–simulation quasielastic neutron scattering comparisons have been performed for a variety of small molecule and polymeric systems, as described in detail in Refs. 4 and 18–21. The combination of simulation and neutron scattering in the analysis of internal motions in globular proteins was reviewed in 1991 [3] and 1997 [4].

A dynamic transition in the internal motions of proteins is seen with increasing temperature [22]. The basic elements of this transition are reproduced by MD simulation [23]. As the temperature is increased, a transition from harmonic to anharmonic motion is seen, evidenced by a rapid increase in the atomic mean-square displacements. Comparison of simulation with quasielastic neutron scattering experiment has led to an interpretation of the dynamics involved in terms of rigid-body motions of the side chain atoms, in a way analogous to that shown above for the X-ray diffuse scattering [24].

Example: Change in Dynamics on Denaturing Phosphoglycerate Kinase

In this example we examine the change in the experimental dynamic neutron scattering signal on strong denaturation of a globular protein, phosphoglycerate kinase (PGK) [25]. Evidence for this comes from the EISF plotted in Figure 6. The main difference in the EISF is in the asymptote as $\mathbf{Q} \to \infty$, which is significantly lower in the case of the denatured protein. The asymptotic value can be shown to correspond to a nondiffusing fraction of the hydrogen atoms in the protein. Whereas the nondiffusing fraction is 40% in the native protein, it is reduced to 18% in the denatured protein.

Inelastic Incoherent Scattering Intensity. For a system executing harmonic dynamics, the transform in Eq. (4) can be performed analytically and the result expanded in a power series over the normal modes in the sample. The following expression is obtained [26]:

$$S_{\text{inc}}(\vec{Q}, \omega) = \sum_i b_{\text{inc}}^2 \exp[-2W_i(\vec{Q})] \tag{26}$$

$$\times \prod_\lambda \left[\sum_{n_\lambda} \exp\left(\frac{n_\lambda \hbar \omega_\lambda \beta}{2}\right) \mathbf{I}_{n_\lambda}\left(\frac{\hbar(\vec{Q} \cdot \vec{e}_{\lambda,i})^2}{2M\omega_\lambda \sinh(\hbar\omega_\lambda\beta/2)}\right) \right] \delta\left(\omega - \sum_\lambda n_\lambda \omega_\lambda\right)$$

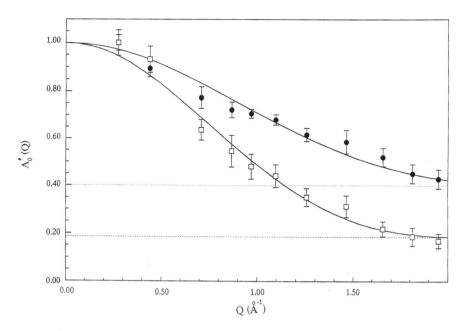

Figure 6 Apparent elastic incoherent structure factor $A'_0(Q)$ for ($\square$) denatured and ($\bullet$) native phosphoglycerate kinase. The solid line represents the fit of a theoretical model in which a fraction of the hydrogens of the protein execute only vibrational motion (this fraction is given by the dotted line) and the rest undergo diffusion in a sphere. For more details see Ref. 25.

In Eq. (26), M is the hydrogen mass, λ labels the mode, $\vec{e}_{\lambda,i}$ is the atomic eigenvector for hydrogen i in mode λ, and ω_λ is the mode angular frequency. n_λ is the number of quanta of energy $\hbar\omega_\lambda$ exchanged between the neutron and mode λ. I_{n_λ} is a modified Bessel function.

$W_i(\vec{Q})$ is the exponent of the Debye–Waller factor, $\exp[-2W_i(\vec{Q})]$, for hydrogen atom i and is given as follows:

$$2W_i(\vec{Q}) = \frac{1}{2NM} \sum_\lambda \frac{\hbar(\vec{Q}\cdot\vec{e}_{\lambda,i})^2}{\omega_\lambda[2n(\omega_\lambda) + 1]} = Q^2\langle u_{Q,i}^2\rangle \tag{27}$$

In Eq. (27), N is the number of modes, $n(\omega_\lambda)$ is the Bose occupancy, and $\langle u_{Q,i}^2\rangle$ is the mean-square displacement for atom i in the direction of $\vec{Q}$.

Equation (26) is an exact quantum mechanical expression for the scattered intensity. A detailed interpretation of this equation is given in Ref. 27. Inserting the calculated eigenvectors and eigenvalues into the equation allows the calculation of the incoherent scattering in the harmonic approximation for processes involving any desired number of quanta exchanged between the neutrons and the sample, e.g., one-phonon scattering involving the exchange of one quantum of energy $\hbar\omega_\lambda$, two-phonon scattering, and so on.

The label λ in Eq. (26) runs over all the modes of the sample. In the case of an isolated molecule, λ runs over the $3N - 6$ normal modes of the molecule, where N is the number of atoms.

Example: Vibrations in Staphylococcal Nuclease

Vibrations in proteins can be conveniently examined using normal mode analysis of isolated molecules. The results of such analyses indicate the presence of a variety of vibra-

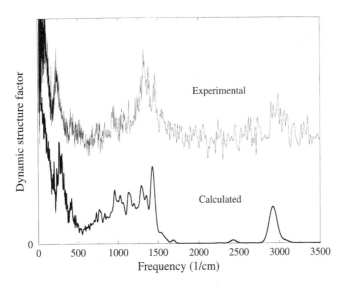

Figure 7 Experimental and theoretical inelastic neutron scattering spectrum from staphylococcal nuclease at 25 K. The experimental spectrum was obtained on the TFXA spectrometer at Oxford. The calculated spectrum was obtained from a normal mode analysis of the isolated molecule. (From Ref. 28.)

tions, with frequencies upward of a few inverse centimeters (cm^{-1}). Incoherent inelastic neutron scattering combined with normal mode analysis is well suited to examine low frequency vibrations in proteins. This is primarily due to the fact that large-amplitude displacements scatter neutrons strongly. Experiments on bovine pancreatic trypsin inhibitor (BPTI), combined with normal mode analysis of the isolated protein, demonstrated that low frequency underdamped vibrations do exist in the protein [3]. More recently, the TFXA spectrometer at the Rutherford-Appleton laboratory in Oxford was used to measure a spectrum of the high frequency local vibrations in the globular protein staphylococcal nuclease [28]. Figure 7 presents a comparison of the experimental dynamic structure factor at 25 K, with that calculated from a normal mode analysis of the protein. Comparison between the calculated and experimental profiles allows an assessment of the accuracy of the dynamical model and the assignment of the various vibrational features making up the experimental spectrum.

V. CONCLUSIONS

In this chapter, basic scattering properties have been described that can be measured for biological samples so as to obtain information on their internal motions. These properties were presented in such a way as to highlight their interface with computer simulation. As experimental intensities and resolutions improve and computer simulations become more and more powerful, it can be expected that the combination of simulation with X-ray and neutron scattering experiments will play an increasingly important role in elucidating the dynamic aspects of biological macromolecular folding and function.

REFERENCES

1. S Lovesey. Theory of Thermal Neutron Scattering from Condensed Matter. Int Seri Monogr Phys Vol 72. Oxford, UK: Oxford Science, 1984.
2. JM Cowley. Diffraction Physics. Amsterdam: North-Holland, 1975.
3. JC Smith. Quart Rev Biophys 24(3):227, 1991.
4. JC Smith. In: WF van Gunsteren, PK Weiner, AJ Wilkinson, eds. Computer Simulation of Biomolecular Systems: Theoretical and Experimental Applications, Vol 3. Dordrecht: Kluwer/ESCOM, 1997, pp 305–360.
5. L van Hove. Phys Rev 95:249, 1954.
6. L van Hove. Physica 24:404, 1958.
7. M Sampoli, G Ruocco, F Sete. Phys Rev Lett 79:1678, 1997.
8. RL Hayward, HD Middendorf, U Wanderlingh, JC Smith. J Chem Phys 102:5525, 1995.
9. M Barthes, H Kellouai, G Page, J Moret, SW Johnson, J Eckert. Physica D 68, 1993.
10. A Micu, JC Smith. Comput Phys Commun 91:331, 1995.
11. M Gerstein, AM Lesk, C Chothia. Biochemistry 33(22):6739, 1994.
12. P Faure, A Micu, AJ Doucet, JC Smith, J-P Benoît. Nature Struct Biol 2:124, 1994.
13. S Héry, D Genest, JC Smith. J Mol Biol 279:303–319, 1998.
14. DA McQuarrie. Statistical Mechanics. New York: Harper & Row, 1976.
15. A Micu, D Durand, M Quilichini, MJ Field, JC Smith. J Phys Chem 99:5645, 1995.
16. GR Kneller, V Keiner, VM Kneller, M Schiller. Comput Phys Commun 91:191, 1995.
17. M Bée. Quasielastic Neutron Scattering: Principles and Applications in Solid State Chemistry, Biology and Materials Science. Philadelphia: Adam Hilger, 1988.
18. M Souaille, F Guillaume, JC Smith. J Chem Phys 105(4):1516–1528, 1996.
19. M Souaille, F Guillaume, JC Smith. J Chem Phys 105(4):1529–1536, 1996.
20. M Souaille, JC Smith, F Guillaume. J Phys Chem 101:6753–6757, 1997.
21. N-D Morelon, GR Kneller, M Ferrand, A Grand, JC Smith, M Bée. J Chem Phys 109(7): 1–12.
22. W Doster, S Cusack, W Petry. Nature 337:754, 1989.
23. JC Smith, K Kuczera, M Karplus. Proc Natl Acad Sci USA. 87:1601, 1990.
24. GR Kneller, JC Smith. J Mol Biol 242:181, 1994.
25. V Receveur, P Calmettes, JC Smith, M Desmadril, G Coddens, D Durand. Proteins: Struct Funct Genet 28:380–387, 1997.
26. AC Zemach, RJ Glauber. Phys Rev 101:118–129, 1956.
27. JC Smith, S Cusack, B Brooks, U Pezzeca, M Karplus. J Chem Phys 85:3636, 1986.
28. AV Goupil-Lamy, JC Smith, J Yunoki, SF Parker, M Kataoka. J Am Chem Soc 119:9268–9273, 1997.

13

Applications of Molecular Modeling in NMR Structure Determination

Michael Nilges
European Molecular Biology Laboratory, Heidelberg, Germany

I. INTRODUCTION

High resolution liquid-state NMR emerged as a structure determination technique for biological macromolecules in 1985. From the beginning, molecular modeling has had a central place in the derivation of NMR solution structures [1–4]. There are several reasons for this. First, the energy parameters, typically derived from a molecular dynamics or molecular mechanics force field, play a central role in calculating and refining the structure. This is because experimental data are scarce, being available for only a fraction of the atoms (mostly the hydrogens). An additional difficulty is that most of the data describe relative positions of atoms and do not directly correspond to the global structure of the molecule. Second, models are not built manually but are automatically calculated by appropriate algorithms. In this way the conformational space consistent with the data is sampled randomly to test whether the data determine the structure uniquely. Consequently, a lot of effort has gone into the development of algorithms to fit the experimental data. The methods used for NMR structure calculations are usually adapted from algorithms originally developed for different purposes in molecular modeling. Third, the wealth of dynamic information obtained by NMR and the difficulties in interpreting it in structural terms have led to a close interaction with MD simulation [5–9].

II. EXPERIMENTAL DATA

A. Deriving Conformational Restraints from NMR Data

The principal sources of structural data [10] are the nuclear Overhauser effect (NOE), which gives information on the spatial proximity of protons (up to a distance of about 4 Å); coupling constants, which give information on dihedral angles; and residual dipolar couplings [11,12], which give information on the relative orientation of a bond vector to the molecule (e.g., to the chemical anisotropy tensor or an alignment tensor; see Fig. 1). With residual dipolar couplings one can, for example, define the relative orientation of domains. Because of the increasing number of experimental terms, we can get an increasingly complete description of the molecule in solution. The NOE is, however, still the

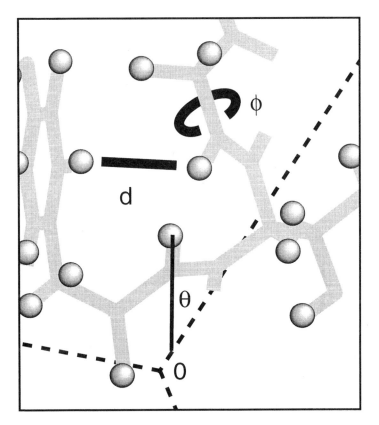

Figure 1 The principal sources of structural data are the NOEs, which give information on the spatial proximity *d* of protons; coupling constants, which give information on dihedral angles φ; and residual dipolar couplings, which give information on the relative orientation θ of a bond vector with respect to the molecule (to the magnetic anisotropy tensor or an alignment tensor). Protons are shown as spheres. The dashed line indicates a coordinate system rigidly attached to the molecule.

richest source of structural information and at the same time the most problematic to analyze. Therefore, in this chapter we mostly deal with the treatment of NOEs in determining NMR solution structures. The other energy terms are included in structure refinements in a very similar manner (see, e.g., the literature cited in Ref. 13).

The first step for any structure elucidation is the assignment of the frequencies (chemical shifts) of the protons and other NMR-active nuclei (^{13}C, ^{15}N). Although the frequencies of the nuclei in the magnetic field depend on the local electronic environment produced by the three-dimensional structure, a direct correlation to structure is very complicated. The application of chemical shift in structure calculation has been limited to final structure refinements, using empirical relations [14,15] for proton and ^{13}C chemical shifts and ab initio calculation for ^{13}C chemical shifts of certain residues [16].

In addition, hydrogen bonding can be deduced from NMR data by analyzing the exchange of labile protons [10]. Only the hydrogen bond donor can be determined in this way. The hydrogen bond acceptor is difficult to observe experimentally, and it has only recently been realized that scalar ("through-bond") couplings can be measured across hydrogen bonds [17,18]. Most often, the hydrogen bond acceptor is inferred from a preliminary structure (see, e.g., Ref. 19). Alternatively, it can be treated like an ambiguous NOE (see below) [20].

B. Distance Restraints

In an isolated two-spin system, the NOE (or, more accurately, the slope of its buildup) depends simply on d^{-6}, where d is the distance between two protons. The difficulties in the interpretation of the NOE originate in deviations from this simple distance dependence of the NOE buildup (due to spin diffusion caused by other nearby protons, and internal dynamics) and from possible ambiguities in its assignment to a specific proton pair. Molecular modeling methods to deal with these difficulties are discussed further below.

Usually, simplified representations of the data are used to obtain preliminary structures. Thus, lower and upper bounds on the interproton distances are estimated from the NOE intensity [10], using appropriate reference distances for calibration. The bounds should include the estimates of the cumulative error due to all sources such as peak integration errors, spin diffusion, and internal dynamics.

The dispersion of proton chemical shifts is usually incomplete in the one-dimensional spectrum of a macromolecule, resulting in many degenerate resonances. As a result, few NOEs can be assigned only on the basis of resonance assignments and without any knowledge of the structure of the molecule [21,22]. Unless ambiguities can be resolved by using additional information, such as the peak shape or data from heteronuclear experiments, the remaining NOEs are ambiguous and cannot be converted into restraints on distances between proton pairs.

Nevertheless, the information from ambiguous NOEs can be converted directly into structural restraints. The structure calculation or refinement with ambiguous data can proceed in a way directly analogous to refinement with standard distance restraints, restraining a "d^{-6}-summed distance" $\overline{D}$ by means of a distance target function (see Fig. 2). By analogy with standard unambiguous distance restraints between atom pairs, we call these "ambiguous distance restraints" (ADRs) [21]. Similar methods can be applied to ambiguities in other experimental data, such as hydrogen bonds [23,24], disulfide bridges [21], and paramagnetic shift broadening and chemical shift differences [25,26].

The distances in the structure are restrained to the upper and lower bounds derived from NOEs by "flat-bottom" potentials. The potential should be gradient-bounded and have an asymptotic region for large violations that is linear [27–29] (see Fig. 2). Then, for large restraint violations, the force approaches a maximum value or can even be decreased, depending on the parameters. This makes the optimization numerically more stable and seems to improve convergence by transiently allowing larger violations during the calculation, thus allowing the structure to gradually escape deep local minima.

The limitation of the gradient of the potential is particularly important for calculations with ADRs and for data sets that potentially contain noise peaks, since it facilitates the appearance of violations due to incorrect restraints. A standard harmonic potential would put a high penalty on large violations and would introduce larger distortions into the structure.

C. The Hybrid Energy Approach

Even if the set of data from NMR experiments is as complete as possible, it is insufficient to define the positions of all the atoms in the molecule, simply because most of the data

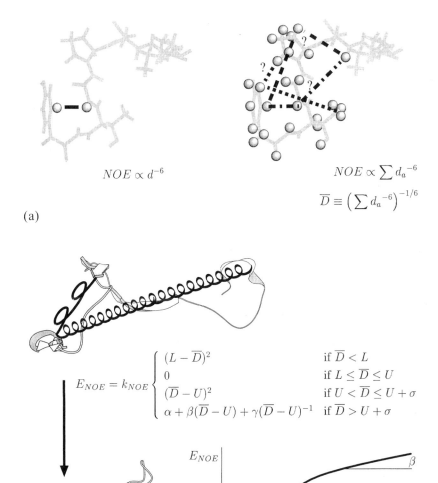

$NOE \propto d^{-6}$

$NOE \propto \sum d_a^{-6}$

$$\overline{D} \equiv \left(\sum d_a^{-6} \right)^{-1/6}$$

(a)

$$E_{NOE} = k_{NOE} \begin{cases} (L - \overline{D})^2 & \text{if } \overline{D} < L \\ 0 & \text{if } L \leq \overline{D} \leq U \\ (\overline{D} - U)^2 & \text{if } U < \overline{D} \leq U + \sigma \\ \alpha + \beta(\overline{D} - U) + \gamma(\overline{D} - U)^{-1} & \text{if } \overline{D} > U + \sigma \end{cases}$$

(b)

Figure 2 Use of unambiguous or ambiguous distance restraints in an optimization calculation. (a) The distance $\overline{D}$ that is restrained can be a distance measured between two protons in the molecule or a "$(\sum d^{-6})^{-1/6}$ summed distance" with contributions from many proton pairs, where the sum runs over all contributions to a cross-peak that are possible due to chemical shift degeneracy. The question marks indicate ambiguities in the assignment of the NOE. For clarity, a situation with only two assignment possibilities is shown. There can be many more possibilities with experimental data. (b) The restraining potential is gradient bounded to avoid large forces for large violations. k_{NOE} is the energy constant, and U and L are upper and lower bounds derived from the size of the NOE. The parameter σ determines the distance at which the potential switches from harmonic to asymptotic behavior, β is the asymptotic slope of the potential, and the coefficients α and γ are determined such that the potential is continuous and differentiable at $U + \sigma$. If D is between L and U, the energy and gradient are zero.

are measured for protons only. The positions of the other atoms have to be inferred, using values of bond lengths, bond angles, planarity, and van der Waals radii that are known a priori.

A molecular dynamics force field is a convenient compilation of these data (see Chapter 2). The data may be used in a much simplified form (e.g., in the case of metric matrix distance geometry, all data are converted into lower and upper bounds on interatomic distances, which all have the same weight). Similar to the use of energy parameters in X-ray crystallography, the parameters need not reflect the dynamic behavior of the molecule. The force constants are chosen to avoid distortions of the molecule when experimental restraints are applied. Thus, the force constants on bond angle and planarity are a factor of 10–100 higher than in standard molecular dynamics force fields. Likewise, a detailed description of electrostatic and van der Waals interactions is not necessary and may not even be beneficial in calculating NMR structures.

The problem of finding conformations of the molecule that satisfy the experimental data is then that of finding conformations that minimize a hybrid energy function E_{hybrid}, which contains different contributions from experimental data and the force field (see below). These contributions need to be properly weighted with respect to each other. However, if the chosen experimental upper and lower bounds are wide enough to avoid any geometrical inconsistencies between the force field and the data, this relative weight does not play a predominant role.

III. MINIMIZATION PROCEDURES

Finding the minimum of the hybrid energy function is very complex. Similar to the protein folding problem, the number of degrees of freedom is far too large to allow a complete systematic search in all variables. Systematic search methods need to reduce the problem to a few degrees of freedom (see, e.g., Ref. 30). Conformations of the molecule that satisfy the experimental bounds are therefore usually calculated with metric matrix distance geometry methods followed by optimization or by optimization methods alone.

Minimization is often not powerful enough for structure calculations of macromolecules unless it is used with an elaborate protocol (e.g., the "buildup method" [3]). More powerful approaches are based on global optimization of the hybrid energy function by molecular dynamics based simulated annealing [31–33]. Other optimization methods have been suggested for NMR structure calculation, notably Monte Carlo simulated annealing [34] and genetic algorithms [35]. Branch-and-bound algorithms have also been suggested for docking rigid monomers with ambiguous restraints [36] or with very sparse data sets [37]. An important feature of the latter is the addition of a hydrophobic potential [38] to the hybrid energy function, which serves to pack secondary structure elements.

Because the parameter-to-observable ratio is rather low, structures are calculated repeatedly with the same restraints. The aim is a random sampling of the conformational space consistent with the restraints. In metric matrix distance geometry, randomness is achieved by the random selection of distance estimates within the bounds. In optimization calculations, one achieves random searching by either selecting a starting conformation very far from the folded structure (e.g., an extended strand [4]) or by choosing starting conformations that are random (either in torsion angles or in Cartesian coordinates) [3,39].

A. Metric Matrix Distance Geometry

A distance geometry calculation consists of two major parts. In the first, the distances are checked for consistency, using a set of inequalities that distances have to satisfy (this part is called "bound smoothing"); in the second, distances are chosen randomly within these bounds, and the so-called metric matrix (M_{ij}) is calculated. "Embedding" then converts this matrix to three-dimensional coordinates, using methods akin to principal component analysis [40].

There are many extensive reviews on metric matrix distance geometry [41–44], some of which provide illustrative examples [45,46]. In total, we can distinguish five steps in a distance geometry calculation:

1. Bound smoothing
2. Distance selection and metrization
3. Construction of the metric matrix
4. Embedding
5. Refinement (optimization)

Bound smoothing serves two purposes: to check consistency of the distances and to transfer information between atoms. Distances have to satisfy the triangle inequalities in a metric space of any dimension (the sum of two sides of a triangle has to be larger than the third; see Fig. 3). To ensure consistency of the distances in three-dimensional space, more inequalities would be necessary (the triangle, tetrangle, pentangle, and hexangle inequalities) [41,47]. Only the tetrangle inequality is of practical use, and it is usually not employed because of high computational costs. This inequality transfers information from one diagonal of a tetrangle to the other; in two dimensions this is the parallelogram equation $|a + b| = |c + d|$.

The most important consequence of bound smoothing is the transfer of information from those atoms for which NMR data are available to those that cannot be observed directly in NMR experiments. Within the original experimental bounds, the minimal distance intervals are identified for which all N^3 triangle inequalities can be satisfied. A distance chosen outside these intervals would violate at least one triangle inequality. For example, an NOE between protons p_i and p_j and the covalent bond between p_j and carbon C_j imposes upper and lower bounds on the distance between p_i and C_j, although this distance is not observable experimentally nor is it part of E_{chem}.

The second step concerns *distance selection and metrization*. Bound smoothing only reduces the possible intervals for interatomic distances from the original bounds. However, the embedding algorithm demands a specific distance for every atom pair in the molecule. These distances are chosen randomly within the interval, from either a uniform or an estimated distribution [48,49], to generate a trial distance matrix. Uniform distance distributions seem to provide better sampling for very sparse data sets [48].

Note that although the bounds on the distances satisfy the triangle inequalities, particular choices of distances between these bounds will in general violate them. Therefore, if all distances are chosen within their bounds independently of each other (the method that is used in most applications of distance geometry for NMR structure determination), the final distance matrix will contain many violations of the triangle inequalities. The main consequence is a very limited sampling of the conformational space of the embedded structures for very sparse data sets [48,50,51] despite the intrinsic randomness of the tech-

bounds matrix

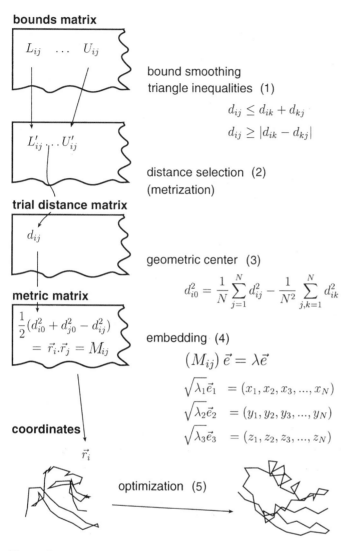

bound smoothing
triangle inequalities (1)

$$d_{ij} \leq d_{ik} + d_{kj}$$

$$d_{ij} \geq |d_{ik} - d_{kj}|$$

distance selection (2)
(metrization)

trial distance matrix

geometric center (3)

$$d_{i0}^2 = \frac{1}{N} \sum_{j=1}^{N} d_{ij}^2 - \frac{1}{N^2} \sum_{j,k=1}^{N} d_{ik}^2$$

metric matrix

embedding (4)

$$\left(M_{ij} \right) \vec{e} = \lambda \vec{e}$$

$$\sqrt{\lambda_1} \vec{e}_1 = (x_1, x_2, x_3, ..., x_N)$$

$$\sqrt{\lambda_2} \vec{e}_2 = (y_1, y_2, y_3, ..., y_N)$$

$$\sqrt{\lambda_3} \vec{e}_3 = (z_1, z_2, z_3, ..., z_N)$$

coordinates

optimization (5)

Figure 3 Flow of a distance geometry calculation. On the left is shown the development of the data; on the right, the operations. d_{ij} is the distance between atoms i and j; L_{ij} and U_{ij} are lower and upper bounds on the distance; L'_{ij} and U'_{ij} are the smoothed bounds after application of the triangle inequality; d_{i0} is the distance between atom i and the geometric center; N is the number of atoms; (M_{ij}) is the metric matrix; $\vec{r}_i$ is the positional vector of atom i; $\vec{e}_1$ is the first eigenvector of (M_{ij}) with eigenvalue λ_i; x_i, y_i, and z_i are the x-, y-, and z-coordinates of atom i. (1–5 correspond to the numbered list on pg. 258.)

nique. In spite of these limitations, the algorithm is remarkably stable in its simplest form.

Metrization guarantees that all distances satisfy the triangle inequalities by repeating a bound-smoothing step after each distance choice. The order of distance choice becomes important [48,49,51]; optimally, the distances are chosen in a completely random sequence

[49]. Metrization is a very computer-intensive operation. Computer time can be saved by using a partially random sequence [43] and terminating the process after $4N$ distances [51] (a three-dimensional object is completely specified by $4N - 10$ distances).

Metrization leads to a much better sampling of conformational space and dramatically improves the local quality of the structures when few long-range connectivities are present [48,51]. The better sampling of space comes at a certain price: The embedded structures may show errors in the topology that are not seen without metrization [31,43]. This may be due to the enforced propagation of an error in a distance choice to many other distances through the triangle inequality.

The *metric matrix* is the matrix of all scalar products of position vectors of the atoms when the geometric center is placed in the origin. By application of the law of cosines, this matrix can be obtained from distance information only. Because it is invariant against rotation but not translation, the distances to the geometric center have to be calculated from the interatomic distances (see Fig. 3). The matrix allows the calculation of coordinates from distances in a single step, provided that all $N_{atom}(N_{atom} - 1)/2$ interatomic distances are known.

Embedding is the calculation of coordinates from the metric matrix by methods akin to principal component analysis [40,52]. The eigenvectors of the metric matrix contain the principal coordinates of the atoms. If the distances correspond to a three-dimensional object, only three eigenvalues of the matrix are nonzero (see Refs. 41 and 53 for mathematical proofs), and the first eigenvector contains all x-coordinates, the second all y-coordinates, and the third all z-coordinates. If the distances are not consistent with a three-dimensional object (the usual situation with sparse NMR data, when the majority of distances come from the random number generator), there will be more than three positive eigenvalues. The eigenvector expansion is then truncated after the first three eigenvalues; this corresponds to a projection of a higher dimensional object into three-dimensional space.

Refinement of the embedded structures is always necessary to remove distortions in the structure. One shortcoming of the embedding algorithm is that data cannot be weighted according to their certainty in any way. During the projection, bond lengths are distorted in the same way as long-range distances guessed by the random number generator within possibly very wide bounds. Also, chirality information is completely absent during bound smoothing and embedding. The first step in the refinement is the selection of the correct enantiomer, which may be achieved on the basis of the chirality of Cα atoms [1], secondary structure elements, or partial refinement of both enantiomers and choice of the enantiomer with lower energy [51].

If the distances satisfy the triangle inequalities, they are embeddable in some dimension. One possible solution is therefore to try to start refinement in four dimensions and use the allowed deviation into the fourth dimension as an additional annealing parameter [43,54]. The advantages of refinement in higher dimensions are similar to those of soft atoms discussed below.

A time-saving variant of the distance geometry procedure described above is substructure embedding. Here, about a third of the atoms are chosen *after* the bound smoothing step and embedded. This procedure was originally used to improve the performance of the distance geometry algorithm by adding the distances from the embedded and partially refined structures back to the distance list [1]. The substructures can be refined directly with simulated annealing by filling in the missing atoms approximately in their correct positions [55].

B. Molecular Dynamics Simulated Annealing

In Cartesian coordinates, molecular dynamics-based simulated annealing (MDSA) refinement consists of the numerical solution of Newton's equations of motion (see Chapter 3). The specific advantage of molecular dynamics over energy minimization is the larger radius of convergence due to possible uphill motions over large energy barriers (Fig. 4). Together with variation of temperature or energy scales, very powerful minimization strategies can be implemented.

Scaling the temperature, the overall weight on E_{hybrid} or all masses m_i are formally equivalent [31]. The independent scaling of each contribution E_l by its weight factor w_l gives rise to a large number of possible simulated annealing schemes. We call annealing schemes that vary the w_l independently "generalized annealing schemes." The initial velocities are usually assigned from a Maxwell distribution at the desired starting temperature, and the temperature is controlled (e.g., by coupling to a heat bath [56]). For the use of MD as an optimization technique, it is convenient to use uniform masses $m_i \equiv m$ for all i [39]. This, in combination with uniform energy constants in the force field, allows the use of larger time steps in the molecular dynamics, because differences in vibrational frequencies are avoided (the time step is determined by the highest vibrational frequency).

Recently, MD constrained to torsion angle space [torsion angle dynamics (TAD)] was introduced to refinement calculations [33,57,58]. Earlier versions of the equations of

$$\frac{d^2 \vec{r}_i}{dt^2} = -\frac{c}{m_i} \frac{\partial}{\partial \vec{r}_i} E_{hybrid}$$

(a)

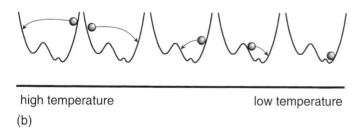

high temperature   low temperature

(b)

Figure 4   (a) Solving Newton's equations of motion at constant energy allows the molecule to overcome energy barriers in E_{hybrid}. The quantities $\vec{r}_i$ and m_i are the coordinate vectors and masses, respectively, of atom i, and E_{hybrid} is the target function of the minimization problem, containing different contributions from experimental data and from a priori knowledge (i.e., the force field). (b) With temperature variation, powerful minimization schemes can be implemented, allowing for large energy barriers to be crossed at high temperatures, ultimately leading to the identification of the "global" minimum.

motion for molecular dynamics in torsion angle space were very inefficient to solve owing to the need for a matrix inversion at every time step [59]. Newer algorithms break down the necessary operations into a series of multiplications of small matrices and are therefore much more efficient [60,61].

The application of TAD in standard MD calculations may require the development of dedicated force fields to emulate the missing flexibility by a reparametrization of the non-bonded potential. This is not necessary for its application in NMR structure calculation, because the energy parameters developed for this purpose already assume in most cases a rigid covalent geometry, either by employing high force constants or by using only torsion angles as degrees of freedom. The advantage of TAD is that the geometry of the molecule does not have to be maintained by high force constants, which lead to high vibrational frequencies. Therefore, longer time steps at higher temperatures can be used with TAD, and the refinement protocols are numerically more stable.

C. Folding Random Structures by Simulated Annealing

Various simulated annealing protocols have been suggested to fold random structures with experimental restraints. The choice of starting structure determines the optimal protocol. The most obvious choices are random distributions of dihedral angles (as indicated in Fig. 2). The minimization procedure has to try to avoid entanglement of the chain while properly relaxing large forces in the starting conformation, which could arise from overlapping atoms or distance restraints violated by a large amount. This is achieved by a combination of soft non-bonded interactions, a violation-tolerant form of the distance restraint potential, and high temperature dynamics.

To achieve convergence with an annealing protocol using Cartesian dynamics, multistage generalized annealing protocols were introduced (Fig. 5). The first stage is a high temperature search where the molecule adopts approximately the correct fold. In this stage, the non-bonded interactions are reduced to allow the chain to intersect itself, and the representation of the non-bonded interactions may be further simplified by computing them for only a fraction of the atoms. The protocol is also adaptable to ambiguous restraint lists by a specifically reduced weight w_{ambig} on the ADRs [20], which is varied independently of $w_{unambig}$ (see Fig. 5). A detailed description can be found elsewhere [20].

With mostly unambiguous data, this protocol has been successfully used for proteins with up to 160 residues [62]. Although virtually all structures converge to the correct fold for small proteins, we observe that approximately one-third of the structures are misfolded for larger proteins, or for low data density, or many ambiguities (see, e.g., Ref. 63). We have also used this protocol for most structure calculations with the automated NOE assignment method ARIA discussed in the next section.

Calculations starting from random Cartesian coordinates and using standard Newton dynamics illustrate the flexibility of the generalized annealing approach. The extremely bad geometry of the initial structures requires that the weights on the covalent geometry terms start with very low values, which are then slowly increased during the calculation. All torsion angle terms (dihedral angles, planarity, and chirality) are removed from E_{hybrid} because of the difficulty in calculating them for random Cartesian structures. Enantiomer selection and regularization are necessary with this protocol much as they are with MMDG embedded structures. The principal advantage of the use of random Cartesian coordinates over that of random dihedral coordinates is that the former give better sampling for highly ambiguous data. The initial structure does not bias toward intraresidue or sequential assignments of ambiguous NOEs.

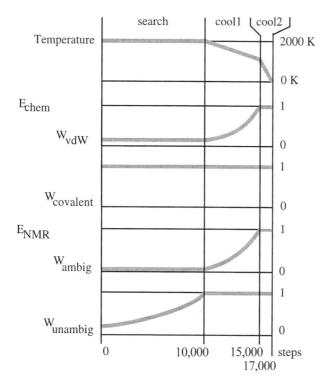

$$E_{hybrid} = \sum_l w_l E_l$$
$$= w_{bond} E_{bond} + w_{angle} E_{angle} + w_{improper} E_{improper}$$
$$+ w_{nonbonded} E_{nonbonded}$$
$$+ w_{unambig} E_{unambig} + w_{ambig} E_{ambig} + \ldots$$

Figure 5 Schematic representation of a Cartesian dynamics protocol starting from random torsion angles. The weights w for non-bonded (i.e., van der Waals) interactions, unambiguous distance restraints, and ambiguous distance restraints are varied independently. The covalent interactions are maintained with full weight, $w_{covalent}$, for the entire protocol. Weights for other experimental terms may be varied in an analogous way. Coupling constant restraints and anisotropy restraints are usually used only in a refinement stage.

A TAD protocol [58] may have a three-stage organization similar to that of the Cartesian MDSA protocol (Fig. 5), with two TAD stages (one high temperature, one cooling) and a final Cartesian cooling stage. The starting temperatures can be set to much higher values (up to 50,000 K). Weights on experimental and non-bonded terms differ in the different stages, with higher weights on the experimental terms in the high temperature stage, but the principal parameter that is varied during simulated annealing is the temperature. TAD protocols used with the program DYANA [64] are even simpler, with only temperature variation in the simulated annealing stage, which is followed by conjugate gradient minimization.

In general, TAD shows better convergence than Cartesian dynamics. For nucleic acid structures, for example, the convergence rate can be very low both for MMDG and for Cartesian dynamics owing to the low restraint density. The sampling of conformational space by TAD for very sparse data sets should be comparable to Cartesian dynamics protocols and better than for MMDG without metrization. Depending on the implementation, ambiguous distance restraints can be used throughout the protocol as with Cartesian dynamics. With its implementation in several NMR structure determination programs, including X-plor [65], CNS [66], and DYANA [33], the field seems to converge toward this calculation method.

IV. AUTOMATED INTERPRETATION OF NOE SPECTRA

The methods discussed in this section extend the original concept of deriving structures from experimental NMR data in two ways. First, during the structure calculation, part of the assignment problem is solved automatically. This allows specification of the NOE data in a form closer to the raw data, which makes the refinement similar to X-ray refinement. Second, the quality of the data is assessed. The methods have been recently reviewed in more detail [64,67].

A. Recognition of Incorrect Restraints: The Structural Consistency Hypothesis

Structure calculation algorithms in general assume that the experimental list of restraints is completely free of errors. This is usually true only in the final stages of a structure calculation, when all errors (e.g., in the assignment of chemical shifts or NOEs) have been identified, often in a laborious iterative process. Many effects can produce inconsistent or incorrect restraints, e.g., artifact peaks, imprecise peak positions, and insufficient error bounds to correct for spin diffusion.

Restraints due to artifacts may, by chance, be completely consistent with the correct structure of the molecule. However, the majority of incorrect restraints will be inconsistent with the correct structural data (i.e., the correct restraints and information from the force field). Inconsistencies in the data produce distortions in the structure and violations in some restraints. Structural consistency is often taken as the final criterion to identify problematic restraints. It is, for example, the central idea in the "bound-smoothing" part of distance geometry algorithms, and it is intimately related to the way distance data are usually specified: The error bounds are set wide enough that all data are geometrically consistent.

The problem in using violations to identify incorrect restraints is twofold. First, one has to distinguish between violations that appear because of insufficient convergence power of the structure calculation algorithm and violations due to incorrect restraints. Violations caused by incorrect restraints will be consistent (i.e., they will be present in the majority of structures), whereas insufficient convergence will produce violations that are randomly distributed. This reasoning has been formalized in the "self-correcting distance geometry" method [22,29], which calculates structures iteratively and modifies the list of restraints after each iteration. Consistent violations are identified by calculating the fraction of structures in which a particular restraint is violated by more than a threshold (e.g., 0.5 Å). If this fraction exceeds a certain value (e.g., 0.5), the restraint is removed from the list for the calculation in the next iteration.

Second, it is possible that an incorrect restraint produces a systematic violation of another restraint. Currently, this can be ruled out only by manually checking the results, where more information on the data can be used to evaluate the restraint (e.g., by inspecting the peak shape).

B. Automated Assignment of Ambiguities in the NOE Data

Assigning ambiguous NOEs is one of the major bottlenecks in NMR solution structure determination, comparable to map fitting in X-ray crystallography. In principle, ambiguous NOEs need not be explicitly assigned if they are used as ADRs, because the assignment is done implicitly in the structure calculation. This is because the summed distance $\overline{D}$ is strongly weighted toward the shortest of the contributing distances. If ADRs are used in the refinement of an already reasonably well determined structure, this weighting is expected to, in most cases, favor the really dominating contribution to the ambiguous NOE. The implicit assignments, achieved through weighting with the distances in the structure, will be mostly correct, and the path to the final structure satisfying all data will be relatively smooth.

The case is very different when ADRs are used for calculating structures ab initio, starting from random structures. Obviously, most of the initial implicit assignments from the random structures are incorrect, and the path toward the final structure is much more difficult owing to additional local minima in the energy. During the calculation, the interproton distances, and with them the weighting on different assignment possibilities, need to change. However, convergence can be achieved by treating the amiguous NOEs appropriately in generalized simulated annealing protocols (see Fig. 5).

C. Iterative Explicit NOE Assignment

The main difficulties with a fully automated method lie in defining rules for explicit assignment based on an ensemble of structures with possibly incorrect features and providing mechanisms for correcting incorrect assignments. Two fully automated iterative assignment methods have been proposed, one based on ADRs (ARIA: ambiguous restraints for iterative assignment) [20,23] and the other on self-correcting distance geometry (NOAH [22]).

Methods to assign ambiguous NOEs follow a sequence of steps (see Fig. 6). Structures are calculated and NOEs are assigned in an iterative way. In the first iteration, the assignments have to be based on the frequencies alone. In the following iterations, the assignments of ambiguous NOEs are derived from the structures by comparing interproton distances corresponding to each assignment possibility. In the "traditional" approach, one possibility would be chosen by hand, and a peak would not be used if an unambiguous assignment were impossible. In contrast, the automated methods use ambiguous peaks during the structure calculation. The key difference between these methods is in how ambiguous peaks are converted into distance restraints. The program ARIA generates one ADR for each ambiguous peak, whereas the program NOAH creates an unambiguous restraint for each assignment possibility.

In general, the advantages of using an automated method may be comparable to those of SA refinement in X-ray crystallography [68], where many of the operations necessary to refine a structure can be done automatically and the remaining manual interventions are easier because the SA refinement usually results in a more easily interpreted electron density map. Automated methods are usually used in combination with manual assignment. However, fully automated assignment of the NOEs is possible (see Fig. 7) [69].

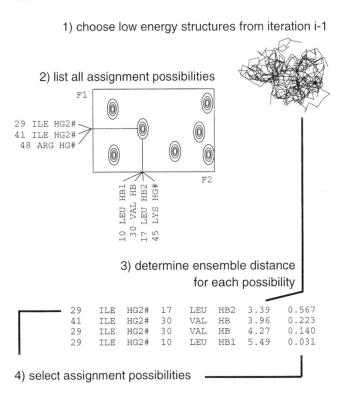

1) choose low energy structures from iteration i-1

2) list all assignment possibilities

3) determine ensemble distance
   for each possibility

```
29   ILE   HG2#   17   LEU   HB2   3.39   0.567
41   ILE   HG2#   30   VAL   HB    3.96   0.223
29   ILE   HG2#   30   VAL   HB    4.27   0.140
29   ILE   HG2#   10   LEU   HB1   5.49   0.031
```

4) select assignment possibilities

5) calibrate distances

6) calculate structures (iteration i)

Figure 6 Steps in automated assignment. (1) Select the S_{conv} lowest energy structures from iteration $i - 1$ that are used to interpret the spectra. (2) For each peak, list all possible assignments compatible with the resonances within a frequency range. (3) Extract a distance for each assignment possibility from the ensemble of structures. (4) Use the distances to assign ambiguous NOEs. (5) Calibrate the peak volumes to obtain distance restraints. (6) Calculate structures based on the new restraints.

D. Symmetrical Oligomers

Symmetrical oligomers present a special difficulty for NMR spectroscopy, because all symmetry-related hydrogens will have equivalent magnetic environments and therefore will be degenerate in chemical shift. Only one monomer is "seen" in the spectra. In principle, every single NOE peak in the spectrum is therefore ambiguous. This ambiguity that arises from the symmetry can be treated with the same concept as ambiguities due to limited spectral resolution with ambiguous distance restraints. Structure determinations of symmetrical oligomers were reviewed in detail recently [70].

The same principal ideas are incorporated in the calculation protocols for symmetrical oligomers as with asymmetrical systems, i.e., the weight is reduced specifically for ADRs or all distance restraints. In addition, the symmetry of the system restricts conforma-

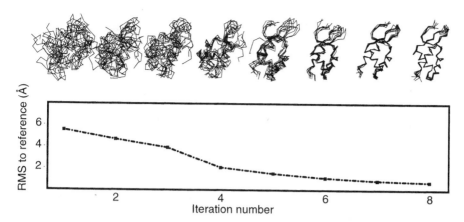

Figure 7 Example of a fully automated assignment. The structure ensemble of the seven lowest energy structures at each iteration is shown. These structures are used for the violation analysis and for a partial assignment of ambiguous NOEs. In the first iteration the structures are calculated with all restraints, where each restraint has all assignment possibilities. In each subsequent iteration, consistently violated restraints are removed, and assignment possibilities are selected with increasingly tight criteria such that at the end of the eight iterations most NOEs are unambiguously assigned.

tional space and is maintained during the calculation by additional restraints. An attractive potential between the monomers can be used in the beginning of the protocol to prevent them from drifting apart. The special difficulties with symmetrical oligomer calculations arise for two reasons: First, all NOEs are ambiguous a priori; second, the assignments of neighboring residues are strongly correlated. A minimization method such as simulated annealing that moves single atoms (or rigid parts of amino acids) may not be optimal for moving larger parts of the structure coherently if a whole set of NOEs needs to be implicitly reassigned. As a result, the structure calculation has a lower convergence rate than for asymmetrical systems. A combination of annealing calculations with other optimization approaches (e.g., a recent branch-and-bound algorithm [36]) may be a more efficient approach. However, the present approaches based on annealing alone have been successful in several cases for quite complex systems (up to a tetramer and a hexamer [70]).

V. TREATMENT OF SPIN DIFFUSION

Depending on experimental parameters, NOE intensities will be affected by spin diffusion (Fig. 8). Magnetization can be transferred between two protons via third protons such that the NOE between the two protons is increased and may be observed even when the distance between the two protons is above the usual experimental limit. This is a consequence of the d^{-6} distance dependence of the NOE. Depending on the conformation, it can be more efficient to move magnetization over intermediate protons than directly. The treatment of spin diffusion during structure refinement is reviewed in more detail in Refs. 31, and 71–73.

From a given structure, the NOE effect can be calculated more realistically by complete relaxation matrix analysis. Instead of considering only the distance between two protons, the complete network of interactions is considered (Fig. 8). Approximately, the

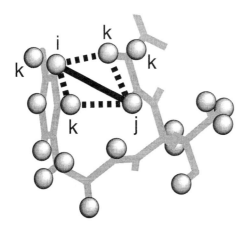

$$NOE_{ij} \propto \left[\exp\left(-(R_{kl})\tau_m\right)\right]_{ij}$$

$$R_{ij} = d_{ij}^{-6}\frac{\pi}{5}\gamma^4\hbar^2\left(\frac{6}{1+4\omega^2\tau^2}-1\right)$$

Figure 8 Effects of spin diffusion. The NOE between two protons (indicated by the solid line) may be altered by the presence of alternative pathways for the magnetization (dashed lines). The size of the NOE can be calculated for a structure from the experimental mixing time, τ_m, and the complete relaxation matrix, (R_{ij}), which is a function of all interproton distances d_{ij} and functions describing the motion of the protons. γ is the gyromagnetic ratio of the proton, $\hbar$ is the Planck constant, τ is the rotational correlation time, and ω is the Larmor frequency of the proton in the magnetic field. The expression for (R_{ij}) is an approximation assuming an internally rigid molecule.

relaxation matrix (R_{ij}) depends on all interproton distances and on parameters describing overall and local motion. The long mixing times often necessary in heteronuclear NOE experiments would make spin diffusion estimates especially valuable. To calculate hetero-nuclear NOEs realistically, the transfer efficiencies between protons and heteronuclei should be incorporated into the equations [74].

One approach to include spin diffusion corrections in a structure calculation is a direct refinement against NOE intensities, analogous to X-ray crystal structure refinement. In this approach, forces are calculated directly from the difference between the experimental NOE intensities and those calculated from the structure via the relaxation matrix. This necessitates, however, an expensive evaluation of derivatives of the simulated NOE spectra with respect to coordinates at every minimization step. Approximations and faster methods to evaluate the gradients [75] make this direct approach more feasible. Various pseudo-energy functions have been proposed. The simplest form (harmonic in the difference between experimental and calculated NOE) places a predominant weight on the largest intensities (shortest distances), which are most often due to intraresidue interactions and will therefore contribute little to determining the conformation of the molecule. By using pseudo-energy functions depending on the sixth root of the difference between calculated and experimental NOEs, the weight is distributed more equally.

Other approaches use complete relaxation matrix analysis to obtain spin diffusion corrected distances from the NOE intensities [71,72], which are then used in conventional distance-restrained optimization. This is more efficient in the use of CPU time, because the gradients do not have to be evaluated and a full relaxation matrix calculation is necessary only a few times in a refinement. These methods invert the calculation of the NOE intensities (Fig. 8) to calculate the relaxation matrix (R_{ij}) from a complete spectrum (NOE_{ij}), including the diagonal peaks. Since this is impossible to obtain experimentally for macromolecules, approximate iterative schemes are used. One method uses preliminary structures to calculate NOE intensities that are merged with the incomplete experimental data to obtain a complete spectrum [76]. A next generation of structures is then calculated with the distances, and an improved estimate of the relaxation matrix can be obtained. Another approach ''shortcuts'' the structure calculation and uses properties of the relaxation matrix itself (e.g., the relation of the diagonal elements to the off-diagonal elements) to iteratively correct the relaxation matrix [77]. Integration errors in the data have to be properly taken into account, otherwise they can lead to incorrect distance estimates [78].

VI. INFLUENCE OF INTERNAL DYNAMICS ON THE EXPERIMENTAL DATA

Internal dynamics of the macromolecule influences all experimental data that can be measured by NMR. The d^{-6} weighting of the NOE makes the averaging very nonlinear, and the measured distance may appear much shorter than the average distance (see Fig. 9).

The ''distance geometry approach'' to the problem is to use appropriately large error bounds for the distances and a rough estimate of dynamics from the diversity of the final ensemble of structures. Although this approach has given qualitatively satisfactory agreement with dynamics measurements [79] and theoretical calculations in some cases [63,80], it is somewhat unsatisfactory. The diversity reflects the distribution of experimental data. Internal dynamics does influence this distribution, but experimental artifacts and overlap are also important factors. In addition, the diversity will depend on exactly how the distance bounds are derived from the NOE data. Furthermore, local dynamics can result in locally conflicting data, while multiple conformations appear in the calculated structures predominantly in regions with little data.

The measured NOE is an average over time and a large ensemble of structures, whereas in a standard structure calculation the lower and upper bounds refer to instantaneous distances. Methods have been proposed to account for the averaging in the interproton distance by fitting either a dynamics trajectory to the measured distance, by means of time-averaged distance restraints [81], or an ensemble of structures [82]. Formally, an ensemble-averaged distance restraint is equivalent to an ambiguous distance restraint. The difference is just a scale factor. Therefore, we can understand an ensemble-averaged NOE as an NOE that is ambiguous between different conformers in the ensemble.

The most serious problem with ensemble average approaches is that they introduce many more parameters into the calculation, making the parameter-to-observable ratio worse. The effective number of parameters has to be restrained. This can be achieved by using only a few conformers in the ensemble and by determining the optimum number of conformers by cross-validation [83]. A more indirect way of restraining the effective number of parameters is to restrict the conformational space that the molecule can search

structure ensemble

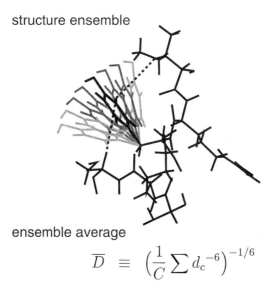

ensemble average

$$\overline{D} \;\equiv\; \left(\frac{1}{C}\sum d_c^{\,-6}\right)^{-1/6}$$

time average

$$\overline{D} \;\equiv\; \left(\frac{1}{T}\int_0^T d(t)^{-6}e^{-t/\tau}dt\right)^{-1/6}$$

Figure 9 Treating internal dynamics during the refinement process. Due to dynamics and the d^{-6} weighting of the NOE, the measured distance may appear much shorter than the average distance. This can be accounted for by using ensemble refinement techniques. In contrast to standard refinement, an average distance is calculated over an ensemble of C structures (ensemble refinement) or a trajectory (time-averaged refinement). The time-averaged distance is defined with an exponential window over the trajectory. T is the total length over the trajectory, t is the time, and τ is a ''relaxation time'' characterizing the width of the exponential window.

during refinement. For example, with a full molecular dynamics force field and low temperatures, only a small fraction of the conformational space is accessible. A more direct way to restrict the number of parameters would be to use motional models. Normal modes have been used, for example, to model NMR order parameters obtained from relaxation studies [84].

Another principal difficulty is that the precise effect of local dynamics on the NOE intensity cannot be determined from the data. The dynamic correction factor [85] describes the ratio of the effects of distance and angular fluctuations. Theoretical studies based on NOE intensities extracted from molecular dynamics trajectories [86,87] are helpful to understand the detailed relationship between NMR parameters and local dynamics and may lead to structure-dependent corrections. In an implicit way, an estimate of the dynamic correction factor has been used in an ensemble relaxation matrix refinement by including order parameters for proton–proton vectors derived from molecular dynamics calculations [72]. One remaining challenge is to incorporate data describing the local dynamics of the molecule directly into the refinement, in such a way that an order parameter calculated from the calculated ensemble is similar to the measured order parameter.

VII.   STRUCTURE QUALITY AND ENERGY PARAMETERS

The well-known difficulties in calculating three-dimensional structures of macromolecules from NMR data mentioned above (sparseness of the data, imprecision of the restraints due to spin diffusion and internal dynamics) also make the validation of the structures a challenging task. The quality of the data [88] and the energy parameters used in the refinement [89] can be expected to influence the quality of structures. Several principles can be used to validate NMR structures.

First, the structure should explain the data. Apart from the energy or target function value returned by the refinement program, this check can be performed with some independent programs (e.g., AQUA/PROCHECK-NMR [90], MOLMOL [91]). The analysis of the deviations from the restraints used in calculating the structures is very useful in the process of assigning the NOE peaks and refining the restraint list. As indicators of the quality of the final structure they are less powerful, because violations have been checked and probably removed. A recent statistical survey of the quality of NMR structures found weak correlations between deviations from NMR restraints and other indicators of structure quality [88].

A similar problem arises with present cross-validated measures of fit [92], because they also are applied to the final clean list of restraints. Residual dipolar couplings offer an entirely different and, owing to their long-range nature, very powerful way of validating structures against experimental data [93]. Similar to cross-validation, a set of residual dipolar couplings can be excluded from the refinement, and the deviations from this set are evaluated in the refined structures.

Second, the structures should satisfy the a priori information used in the refinement in the form of the energy parameters. Programs like PROCHECK-NMR check for deviation from expected geometries and close non-bonded contacts.

Finally, structural properties that depend directly neither on the data nor on the energy parameters can be checked by comparing the structures to statistics derived from a database of solved protein structures. PROCHECK-NMR and WHAT IF [94] use, e.g., statistics on backbone and side chain dihedral angles and on hydrogen bonds. PROSA [95] uses potentials of mean force derived from distributions of amino acid–amino acid distances.

VIII.   RECENT APPLICATIONS

Molecular modeling is an indispensable tool in the determination of macromolecular structures from NMR data and in the interpretation of the data. Thus, state-of-the-art molecular dynamics simulations can reproduce relaxation data well [9,96] and supply a model of the motion in atomic detail. Qualitative aspects of correlated backbone motions can be understood from NMR structure ensembles [63]. Additional data, in particular residual dipolar couplings, improve the precision and accuracy of NMR structures qualitatively [12].

Standard calculation methods developed for small proteins are sufficiently powerful to solve protein structures and complexes in the 30 kDa range and beyond [97,98] and protein–nucleic acid complexes [99]. Torsion angle dynamics offers increased convergence, in particular for nucleic acids, which are more difficult to calculate because of the sparseness of NMR data [100].

Examples of structures for which automated assignment methods were used from the start are still rare [69,101]. However, automated methods are being used increasingly as a powerful tool in structure determination in combination with manual assignment [102–105].

REFERENCES

1. T Havel, K Wüthrich. Bull Math Biol 46:673–698, 1984.
2. R Kaptein, ERP Zuiderweg, RM Scheek, WA van Gunsteren. J Mol Biol 182:179–182, 1985.
3. W Braun, N Go. J Mol Biol 186:611–626, 1985.
4. AT Brünger, GM Clore, AM Gronenborn, M Karplus. Proc Natl Acad Sci USA 83:3801–3805, 1986.
5. G Lipari, A Szabo, R Levy. Nature 300:197–198, 1982.
6. ET Olejniczak, CM Dobson, M Karplus, RM Levy. J Am Chem Soc 106:1923–1930, 1984.
7. R Brüschweiler, B Roux, M Blackledge, C Griesinger, M Karplus, R Ernst. J Am Chem Soc 114:2289–2302, 1992.
8. AG Palmer III. Curr Opin Struct Biol 7:732–737, 1997.
9. LM Horstink, R Abseher, M Nilges, CW Hilbers. J Mol Biol 287:569–577, 1999.
10. K Wüthrich. NMR of Proteins and Nucleic Acids. New York: Wiley, 1986.
11. JR Tolman, JM Flanagan, MA Kennedy, JH Prestegard. Proc Natl Acad Sci USA 92:9279–9283, 1995.
12. N Tjandra, A Bax. Science 278:1111–1114, 1997.
13. M Nilges. Curr Opin Struct Biol 6:617–621, 1996.
14. D Case. Curr Opin Struct Biol 8:624–630, 1998.
15. G Cornilescu, F Delaglio, A Bax. J Biomol NMR 13:289–302, 1999.
16. JG Pearson, JP Wang, JL Markley, B Le Hong, E Oldfield. J Am Chem Soc 117:8823–8829, 1995.
17. AJ Dingley, S Grzesiek. J Am Chem Soc 120:8293–8297, 1998.
18. F Cordier, S Grzesiek. J Am Chem Soc 121:1601–1602, 1999.
19. AM Gronenborn, GM Clore. Crit Rev Biochem Mol Biol 30:351–385, 1995.
20. M Nilges, MJ Macias, SI O'Donoghue, H Oschkinat. J Mol Biol 269:408–422, 1997.
21. M Nilges. J Mol Biol 245:645–660, 1995.
22. C Mumenthaler, W Braum. J Mol Biol 254:465–480, 1995.
23. A Kharrat, MJ Macias, T Gibson, M Nilges, A Pastore. EMBO J 14:3572–3584, 1995.
24. JG Omichinski, PV Pedone, G Felsenfeld, AM Gronenborn, GM Clore. Nature Struct Biol 4:122–132, 1997.
25. RHA Folmer, M Nilges, PJM Folkers, RNH Konings, CW Hilbers. J Mol Biol 240:341–357, 1994.
26. M Ubbink, M Ejdeback, BG Karlsson, DS Bendall. Structure 6:323–335, 1998.
27. M Nilges, AM Gronenborn, AT Brünger, GM Clore. Protein Eng 2:27–38, 1988.
28. J de Vlieg, RM Scheek, WF van Gunsteren, HJC Berendsen, R Kaptein, J Thomason. Proteins: Struct, Funct, Genet 3:209–218, 1988.
29. G Haenggi, W Braun. FEBS Lett 344:147–153, 1994.
30. RB Altman, O Jardetzky. Methods Enzymol 177:218–246, 1989.
31. AT Brünger, M Nilges. Quart Rev Biophys 26:49–125, 1993.
32. AT Brünger, PD Adams, LM Rice. Structure 5:325–336, 1997.
33. P Güntert, C Mumenthaler, K Wüthrich. J Mol Biol 17:283–298, 1997.
34. NB Ulyanov, U Schmitz, TL James. J Biomol NMR 3:547–568, 1993.

35. MJ Bayley, G Jones, P Willet, MP Williamson. Protein Sci 7:491–499, 1998.
36. CS Wang, T Lozano-Pérez, B Tidor. Proteins 32:26–42, 1998.
37. DM Standley, VA Eyrich, AK Felts, RA Friesner, AE McDermott. J Mol Biol 285:1691–1710, 1999.
38. G Casari, M Sippl. J Mol Biol 224:725–732, 1992.
39. M Nilges, GM Clore, AM Gronenborn. FEBS Lett 239:129–136, 1988.
40. GM Crippen, TF Havel. Acta Cryst A34:282–284, 1978.
41. TF Havel, ID Kuntz, GM Crippen. J Theor Biol 104:359–381, 1983.
42. ID Kuntz, JF Thomason, CM Oshiro. Methods Enzymol 177:159–204, 1989.
43. TF Havel. Prog Biophys Mol Biol 56:43–78, 1991.
44. GM Crippen, TF Havel. Distance Geometry and Molecular Conformation. Taunton, England: Research Studies Press, 1988.
45. AR Leach. Molecular Modeling: Principles and Applications. Harlow, UK: Longman, 1996, pp 426–434.
46. JM Blaney, JS Dixon. Distance geometry in molecular modeling. In: KB Lipkowitz, DB Boyd, eds. Reviews in Computational Chemistry, Vol 5. New York: VCH, 1994, pp 299–335.
47. GM Crippen. J Comput Phys 24:96–107, 1977.
48. TF Havel. Biopolymers 29(12–13):1565–1585, 1990.
49. ME Hodsdon, JW Ponder, DP Cistola. J Mol Biol 264:585–602, 1997.
50. W Metzler, D Hare, A Pardi. Biochemistry 28:7045–7052, 1989.
51. J Kuszewski, M Nilges, AT Brünger. J Biomol NMR 2:33–56, 1992.
52. JC Gower. Biometrika 53:325–338, 1966.
53. W Braun. Quart Rev Biophys 19:115–157, 1987.
54. RC van Schaik, HJ Berendsen, AE Torda, WF van Gunsteren. J Mol Biol 234:751–762, 1993.
55. M Nilges, GM Clore, AM Gronenborn. FEBS Lett 229:317–324, 1988.
56. HJC Berendsen, JPM Postma, WF van Gunsteren, A DiNola, J Haak. J Chem Phys 81:3684–3690, 1984.
57. LM Rice, AT Brünger. Proteins 19:277–290, 1994.
58. EG Stein, LM Rice, AT Brünger. J Magn Reson 124:154–164, 1997.
59. AK Mazur, RA Abagyan. J Biomol Struct Dyn 5:815–832, 1989.
60. DS Bae, EJ Haug. Mech Struct Mach 106:258–268, 1987.
61. A Jain, N Vaidehi, G Rodrigues. J Comput Phys 106:258–268, 1993.
62. RHA Folmer, RNH Konings, CW Hilbers, M Nilges. J Biomol NMR 9:245–258, 1997.
63. R Abseher, L Horstink, CW Hilbers, M Nilges. Proteins 31:370–382, 1998.
64. P Güntert. Quart Rev Biophys 31:145–237, 1998.
65. AT Brünger. X-PLOR. A System for X-Ray Crystallography and NMR. New Haven, CT: Yale Univ Press, 1992.
66. AT Brünger, PD Adams, GM Clore, WL DeLano, P Gros, RW Grosse-Kunstleve, J-S Jiang, J Kuszewski, M Nilges, NS Pannu, RJ Read, LM Rice, T Simonson, GL Warren. Acta Cryst D 54:905–921, 1998.
67. M Nilges, SI O'Donoghue. Prog NMR Spectrosc 32: 107–139, 1998.
68. AT Brünger, J Kuriyan, M Karplus. Science 235:458–460, 1987.
69. M Sunnerhagen, M Nilges, G Otting, J Carey. Nature Struct Biol 4:819–826, 1997.
70. SI O'Donoghue, M Nilges. Calculation of symmetric oligomer structures from NMR data. In: R Krishna, JL Berliner, eds. Modern Techniques in Protein NMR, vol. 17 of Biological Magnetic Resonance. New York: Kluwer Academic/Plenum, pp 131–161, 1999.
71. TL James. Curr Opin Struct Biol 1:1042–1053, 1991.
72. AMJJ Bonvin, R Boelens, R Kaptein. Determination of biomolecular structures by NMR: Use of relaxation matrix calculations. In: WF van Gunsteren, PK Weiner, AJ Wilkinson, eds. Computer Simulation of Biomolecular Systems: Theoretical and Experimental Applications, Vol 2. Leiden: ESCOM, 1993, pp 407–440.

73. D Case. New directions in NMR spectral simulation and structure refinement. In: WF van Gunsteren, PK Weiner, AJ Wilkinson, eds. Computer Simulation of Biomolecular Systems: Theoretical and Experimental Applications, Vol 2. Leiden: ESCOM, 1993, pp 382–406.

74. L Zhou, HJ Dyson, PE Wright. J Biomol NMR 11:17–29, 1998.

75. P Yip. J Biomol NMR 3:361–365, 1993.

76. R Boelens, TMG Koning, R Kaptein. J Mol Struct 173:299–311, 1989.

77. BA Borgias, TL James. J Magn Reson 87:475–487, 1990.

78. H Liu, HP Spielmann, NB Ulyanov, DE Wemmer, TL Jamew. J Biomol NMR 6:390–402, 1995.

79. C Redfield, J Boyd, LJ Smith, RAG Smith, CM Dobson. Biochemistry 31:10431–10437, 1992.

80. KD Berndt, P Güntert, K Wüthrich. Proteins 24:304–313, 1996.

81. WF van Gunsteren, RM Brunne, P Gros, RC van Schaik, CA Schiffer, AE Torda. Methods Enzymol 261:619–654, 1994.

82. J Kemmink, RM Scheek. J Biomol NMR 5:33–40, 1995.

83. AMJJ Bonvin, AT Brünger. J Mol Biol 250:80–93, 1995.

84. R Brüschweiler. J Am Chem Soc 114:5341–5344, 1992.

85. R Brüschweiler, D Case. Prog NMR Spectrosc 26:27–58, 1994.

86. R Brüschweiler, B Roux, M Blackledge, C Griesinger, M Karplus, RR Ernst. J Am Chem Soc 114:2289–2302, 1992.

87. T Schneider, AT Brünger, M Nilges. J Mol Biol 285:727–740, 1999.

88. JF Doreleijers, JA Rullmann, R Kaptein. J Mol Biol 281:149–164, 1998.

89. JP Linge, M Nilges. J Biomol NMR 13:51–59, 1999.

90. R Laskowski, J Rullman, M MacArthur, R Kaptein, J Thornton. J Biomol NMR 8:477–486, 1996.

91. R Koradi, M Billeter, K Wüthrich. J Mol Graph 14:29–32, 51–55, 1996.

92. AT Brünger, GM Clore, AM Gronenborn, R Saffrich, M Nilges. Science 261:328–331, 1993.

93. G Cornilescu, JL Marquardt, M Ottiger, A Bax. J Am Chem Soc 120:6836–6837, 1998.

94. G Vriend, C Sander. J Appl Cryst 26:47–60, 1993.

95. MJ Sippl. Proteins 17:355–362, 1993.

96. M Philippopoulos, AM Mandel, AG Palmer III, C Lim. Proteins 28:481–493, 1997.

97. DS Garrett, YJ Seok, A Peterkofsky, AM Gronenborn, GM Clore. Nature Struct Biol 6:166–173, 1999.

98. M Caffrey, M Cai, J Kaufman, SJ Stahl, PT Wingfield, DG Covell, AM Gronenborn, GM Clore. EMBO J 17:4572–4584, 1998.

99. FH Allain, PW Howe, D Neuhaus, G Varani. EMBO J 16:5764–5772, 1997.

100. MH Kolk, M van der Graaf, SS Wijmenga, CW Pleij, HA Heus, CW Hilbers. Science 280:434–438, 1998.

101. Y Xu, J Wu, D Gorenstein, W Braun. J Magn Reson 136:76–85, 1999.

102. FY Luh, SJ Archer, PJ Domaille, BO Smith, D Owen, DH Brotherton, AR Raine, X Xu, L Brizuela, SL Brenner, ED Laue. Nature 389:999–1003, 1997.

103. B Aghazadeh, K Zhu, TJ Kubiseski, GA Liu, T Pawson, Y Zheng, MK Rosen. Nature Struct Biol 5:1098–1107, 1998.

104. SC Li, C Zwahlen, SJ Vincent, CJ McGlade, LE Kay, T Pawson, JD Forman-Kay. Nature Struct Biol 5:1075–1083, 1998.

105. HR Mott, D Owen, D Nietlispach, PN Lowe, E Manser, L Lim, ED Laue. Nature 399:384–388, 1999.

14

Comparative Protein Structure Modeling

András Fiser, Roberto Sánchez, Francisco Melo, and Andrej Šali
The Rockefeller University, New York, New York

I. INTRODUCTION

The aim of comparative or homology protein structure modeling is to build a three-dimensional (3D) model for a protein of unknown structure (the target) based on one or more related proteins of known structure (the templates) (Fig. 1) [1–6]. The necessary conditions for getting a useful model are that the similarity between the target sequence and the template structures is detectable and that the correct alignment between them can be constructed. This approach to structure prediction is possible because a small change in the protein sequence usually results in a small change in its 3D structure [7]. Although considerable progress has been made in the ab initio protein structure prediction, comparative protein structure modeling remains the most accurate prediction method. The overall accuracy of comparative models spans a wide range. At the low end of the spectrum are the low resolution models whose only essentially correct feature is their fold. At the high end of the spectrum are the models with an accuracy comparable to medium resolution crystallographic structures [6]. Even low resolution models are often useful for addressing biological questions, because function can often be predicted from only coarse structural features of a model.

At this time, approximately one-half of all sequences are detectably related to at least one protein of known structure [8–11]. Because the number of known protein sequences is approximately 500,000 [12], comparative modeling could in principle be applied to over 200,000 proteins. This is an order of magnitude more proteins than the number of experimentally determined protein structures ($\sim$13,000) [13]. Furthermore, the usefulness of comparative modeling is steadily increasing, because the number of different structural folds that proteins adopt is limited [14,15] and because the number of experimentally determined structures is increasing exponentially [16]. It is predicted that in less than 10 years at least one example of most structural folds will be known, making comparative modeling applicable to most protein sequences [6].

All current comparative modeling methods consist of four sequential steps (Fig. 2) [5,6]. The first step is to identify the proteins with known 3D structures that are related to the target sequence. The second step is to align them with the target sequence and pick those known structures that will be used as templates. The third step is to build the model

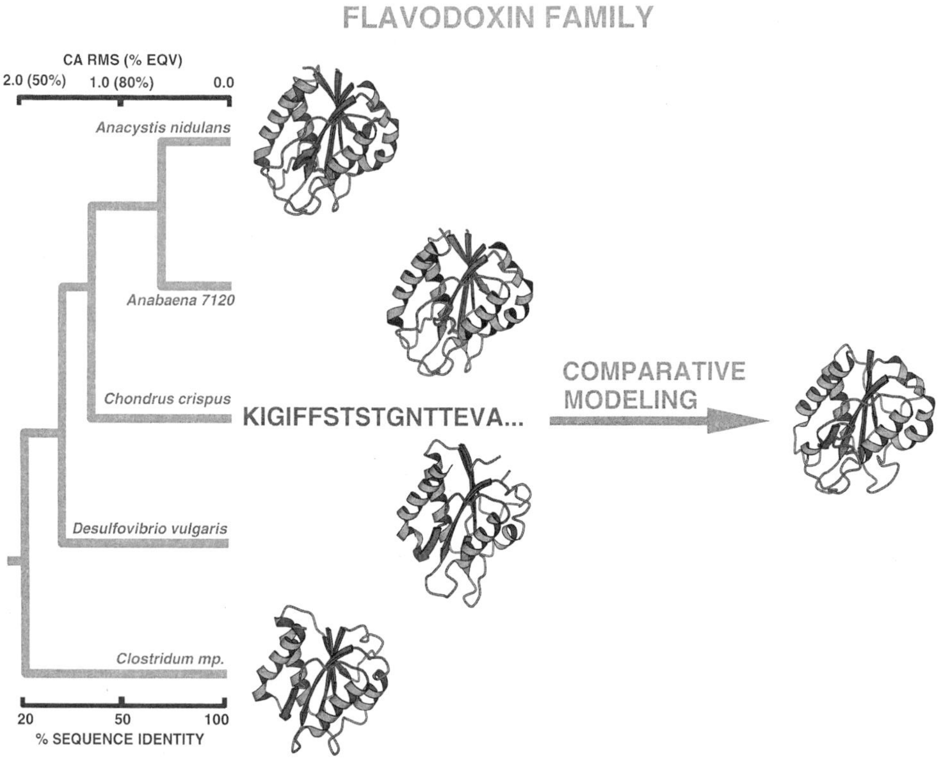

FLAVODOXIN FAMILY

Figure 1 The basis of comparative protein structure modeling. Comparative modeling is possible because evolution resulted in families of proteins, such as the flavodoxin family, modeled here, which share both similar sequences and 3D structures. In this illustration, the 3D structure of the flavodoxin sequence from *C. crispus* (target) can be modeled using other structures in the same family (templates). The tree shows the sequence similarity (percent sequence identity) and structural similarity (the percentage of the C_α atoms that superpose within 3.8 Å of each other and the RMS difference between them) among the members of the family.

for the target sequence given its alignment with the template structures. In the fourth step, the model is evaluated using a variety of criteria. If necessary, template selection, alignment, and model building are repeated until a satisfactory model is obtained. The main difference among the comparative modeling methods is in how the 3D model is calculated from a given alignment (the third step). For each of the steps in the modeling process, there are programs and servers available on the World Wide Web (Table 1).

We begin this chapter by describing the techniques for all the steps in comparative modeling (Section II). We continue by discussing the errors in model structures (Section IV) and methods for detecting these errors (Section V). We conclude by listing sample applications of comparative modeling to individual proteins (Section VI) and to whole genomes (Section VII). We emphasize our own work and experience, although we have profited greatly from the contributions of many others, cited in the list of references. The citations are not exhaustive, but exhaustive lists can be found in Refs. 5 and 6. The chapter highlights pragmatically the methods and tools for comparative modeling rather than the physical principles and rules on which the methods are based.

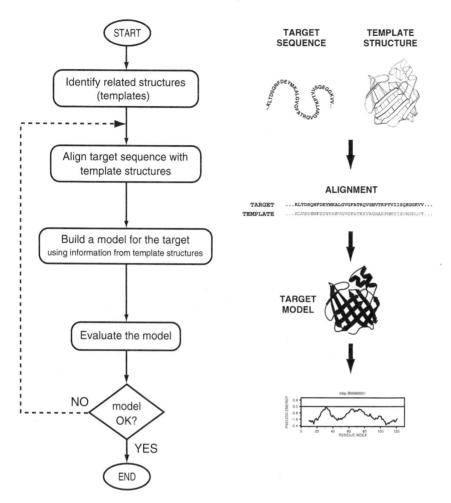

Figure 2 The flowchart for comparative protein structure modeling.

II. STEPS IN COMPARATIVE MODELING

A. Identifying Known Protein Structures Related to the Target Sequence

The first task in comparative modeling is to identify all protein structures related to the target sequence, some of which will be used as templates. This is greatly facilitated by databases of protein sequences and structures and by software for scanning those databases [16–19]. The target sequence can be searched against sequence databases such as PIR [20], GenBank [21], or TrEMBL/SWISS-PROT [12] and/or structure databases such as the Protein Data Bank [13,22], SCOP [23], DALI [24], and CATH [25] (Table 1). Searching against sequence databases can be useful even if it identifies only proteins of unknown structure, because such sequences can be used to increase the sensitivity of the search for the template structures (see below). At present, the probability of finding related proteins of known structure for a sequence picked randomly from a genome ranges from 20% to 70% [8–11].

Table 1   Web Sites Useful for Comparative Modeling

Databases	
NCBI	http://www.ncbi.nlm.nih.gov/
PDB	http://www.rcsb.org/pdb/
MSD	http://msd.ebi.ac.uk/
CATH	http://www.biochem.ucl.ac.uk/bsm/cath/
TrEMBL	http://www.expasy.ch/sprst/sprst-top.html
SCOP	http://scop.mrc-lmb.cam.ac.uk/scop/
PRESAGE	http://csb.stanford.edu/
ModBase	http://guitar.rockefeller.edu/modbase/
GeneCensus	http://bioinfo.mbb.yale.edu/genome
Template search, fold assignment	
BLAST	http://www.ncbi.nlm.nih.gov/BLAST/
FastA	http://fasta.bioch.virginia.edu/
DALI	http://www2.ebi.ac.uk/dali/
PRESAGE	http://presage.berkeley.edu
PhD, TOPITS	http://www.embl-heidelberg.de/predictprotein/predictprotein.html
THREADER	http://insulin.bmnel.ac.uk//threader/threader.html
123D	http://www-lmmb.ncifcrf.gov/~nicka/123D.html
UCLA-DOE	http://www.doe-mbi.ucla.edu/people/frsvr/frsvr.html
PROFIT	http://lore.came.sbg.ac.at/
Comparative modeling	
COMPOSER	http://www-cryst.bioc.cam.ac.uk/
CONGEN	http://www.cabm.rutgers.edu/~bruc
DRAGON	http://www.nimr.mrc.ac.uk/~mathbio/a-aszodi/dragon.html
MODELLER	http://guitar.rockefeller.edu/modeller/modeller.html
PrISM	http://honiglab.cpmc.columbia.edu/
SWISS-MODEL	http://www.expasy.ch/swissmod/SWISS-MODEL.html
WHAT IF	http://www.cmbi.kun.nl/whatif/
ICM	http://www.molsoft.com/
SCRWL	http://www.cmpharm.ucsf.edu/~dunbrack
InsightII	http://www.msi.com/
GENEMINE	http://www.bioinformatics.ucla.edu/genemine
SYBYL	http://www.tripos.com/
Model evaluation	
PROCHECK	http://www.biochem.ucl.ac.uk/~roman/procheck/procheck.html
WHATCHECK	http://www.sander.embl-heidelberg.de/whatcheck/
ProsaII	http://www.came.sbg.ac.at
ProCyon	http://www.horus.com/sippl/
BIOTECH	http://biotech.embl-ebi.ac.uk:8400/
VERIFY3D	http://www.doe-mbi.ucla.edu/Services/Verify3D.html
ERRAT	http://www.doe-mbi.ucla.edu/Services/Errat.html
ANOLEA	http://www.fundp.ac.be/pub/ANOLEA.html
AQUA	http://www-nmr.chem.ruu.nl/users/rull/aqua.html
SQUID	http://www.yorvic.york.ac.uk/~oldfield/squid
PROVE	http://www.ucmb.ulb.ac.be/UCMB/PROVE/

There are three main classes of protein comparison methods that are useful in fold identification. The first class compares the target sequence with each of the database sequences independently, using pairwise sequence–sequence comparison [26]. The performance of these methods in sequence searching [27] and fold assignments has been evaluated exhaustively [28]. The most popular programs in the class include Fasta [29] and BLAST [30]. Program MODELLER, which implements all the stages in comparative modeling [31], can also automatically search for proteins with known 3D structure that are related to a given sequence. It is based on the local dynamic programming method for pairwise sequence comparison [32].

The second class of methods rely on multiple sequence comparison to improve greatly the sensitivity of the search [10,33–36]. The best-known program in this class is PSI-BLAST [36]. Another similar approach that appears to perform even slightly better than PSI-BLAST has been described [10]. It begins by finding all sequences in a sequence database that are clearly related to the target and easily aligned with it. The multiple alignment of these sequences is the target sequence profile. Similar profiles are also constructed for all potential template structures. The templates are then found by comparing the target sequence profile with each of the template sequence profiles, using a local dynamic programming method that relies on the common BLOSUM62 residue substitution matrix [33]. These more sensitive fold identification techniques are especially useful for finding structural relationships when sequence identity between the target and the template drops below 25%. In fact, methods of this class, which rely on multiple sequence information, appear to be currently the most sensitive fully automated approach to detecting remote sequence–structure relationships [8,36–38].

The third class of methods rely on pairwise comparison of a protein sequence and a protein structure; that is, structural information is used for one of the two proteins that are being compared, and the target sequence is matched against a library of 3D profiles or threaded through a library of 3D folds. These methods are also called fold assignment, threading, or 3D template matching [39–43,238]. They are reviewed in Refs. 44–46 and evaluated in Ref. 47. These methods are especially useful when it is not possible to construct sequence profiles because there are not enough known sequences that are clearly related to the target or potential templates.

What similarity between the target and template sequences is needed to have a chance of obtaining a useful comparative model? This depends on the question that is asked of a model (Section VI). When only the lowest resolution model is required, it is tempting to use one of the statistical significance scores for a given match that is reported by virtually any sequence comparison program to select the best template. However, it is better to proceed with modeling even when there is only a remote chance that the best template is suitable for deriving a model with at least a correct fold. The usefulness of the template should be assessed by the evaluation of the calculated 3D model. This is the best approach, because the evaluation of a 3D model is generally more sensitive and robust than the evaluation of an alignment (Section V) [9].

B. Aligning the Target Sequence with the Template Structures

Once all the structures related to the target sequence are identified, the second task is to prepare a multiple alignment of the target sequence with all the potential template structures [16,48–50]. When the sequence identity between the target and the template is higher than approximately 40%, this is straightforward. The gaps and errors in the alignments are rare, whether they are prepared automatically or manually. However, at 30% sequence identity, the fraction of residues that are correctly aligned by pairwise sequence–sequence comparison methods is only 80% on average, and this number drops sharply with further decrease in sequence similarity [51]. Thus, an additional effort in obtaining a more accurate alignment is needed because comparative modeling cannot, at present, recover from an incorrect alignment; the quality of the alignment is the single most important factor determining the accuracy of the 3D model.

In the more difficult alignment problems, it is frequently beneficial to rely on the multiple structure and sequence information as follows [52]. First, the alignment of the potential templates is prepared by superposing their structures. Typically, all residues whose C_α atoms are within 3.5 Å of each other upon least-squares superposition are aligned. Next, the sequences that are clearly related to the templates and easy to align with them are added to the alignment. The same is done for the target sequence. And finally, the two profiles are aligned with each other, taking structural information into account as much as possible [53–55]. In principle, most sequence alignment and structure comparison methods can be used for these tasks [16,18,53,56]. In practice, it is frequently necessary to edit manually the positions of insertions and deletions to ensure that they occur in a reasonable structural context. For example, gaps are favored outside secondary structure segments, in exposed regions, and between residues that are far apart in space. Secondary structure prediction for the target sequence or its profile is also frequently useful in obtaining a more accurate alignment to the template structures [57]. Although 3D profile matching and threading techniques are relatively successful in identifying related folds, they appear to be somewhat less successful in generating correct alignments [47]. When there is an uncertainty about a region in the alignment, the best way to proceed is to generate 3D models for all alternative alignments, evaluate the corresponding models, and pick the best model according to the 3D model evaluation rather than the alignment score (Section V) [58,59].

Once a multiple alignment is constructed, matrices of pairwise sequence similarities are usually calculated and employed to construct a phylogenetic tree that expresses the relationships among the proteins in the family [60]. All significantly different structures in the cluster that contains the target sequence are usually used as templates in the subsequent model building [61], although other considerations should also enter into the template selection. For example, if the model is prepared to study the liganded state of a protein, then a template in the liganded state is preferred over a template without a ligand. Some methods allow short segments of known structure, such as loops [62], to be added to the alignment at this stage [31].

C. Model Building

1. Modeling by Assembly of Rigid Bodies

The first approach and one still widely used in comparative modeling is to assemble a model from a small number of rigid bodies obtained from the aligned protein structures [1,2,63]. This approach is based on the natural dissection of the protein structure into conserved core regions, variable loops that connect them, and side chains that decorate the backbone. For example, the following semiautomated procedure is implemented in the computer program COMPOSER [64] (Table 1). First, the template structures are selected and superposed. Second, the "framework" is calculated by averaging the coordinates of the C_α atoms of structurally conserved regions in the template structures. Third, the core main chain atoms of each core region in the target model are obtained by superposing on the framework the core segment from the template whose sequence is closest to that of the target. Fourth, the loops are generated by scanning a database of all known protein structures to identify the structurally variable regions that fit the anchor core regions and have a compatible sequence [65]. Fifth, the side chains are modeled based on their intrinsic conformational preferences and on the conformation of the equivalent side chains in the template structures [66]. And finally, the stereochemistry of the model is improved either by a restrained energy minimization or a molecular dynamics refinement. The accuracy of a model can be somewhat increased when more than one template structure is used to construct the framework and when the templates are averaged into the framework using weights corresponding to their sequence similarities to the target sequence [67]. For example, differences between the model and X-ray structures may be slightly smaller than the differences between the X-ray structures of the modeled protein and the homologs used to build the model. Possible future improvements of modeling by rigid-body assembly include incorporation of rigid-body shifts such as the relative shifts in the packing of α-helices [68].

2. Modeling by Segment Matching or Coordinate Reconstruction

The basis of modeling by coordinate reconstruction is the finding that most hexapeptide segments of protein structure can be clustered into only 100 structurally different classes [69]. Thus, comparative models can be constructed by using a subset of atomic positions from template structures as ''guiding'' positions, then identifying and assembling short all-atom segments that fit these guiding positions. The guiding positions usually correspond to the C_α atoms of the segments that are conserved in the alignment between the template structure and the target sequence. The all-atom segments that fit the guiding positions can be obtained either by scanning all the known protein structures, including those that are not related to the sequence being modeled [70,71], or by conducting a conformational search restrained by an energy function [72,73]. For example, a general method for modeling by segment matching is guided by the positions of some atoms (usually C_α atoms) to find the matching segments in the representative database of all known protein structures [74]. This method can construct both main chain and side chain atoms and can also model gaps. It is implemented in the program SEGMOD which is part of the Genemine package (Table 1). Even some side chain modeling methods [75] and the class of loop construction methods based on finding suitable fragments in the database of known structures [62] can be seen as segment-matching or coordinate reconstruction methods.

3. Modeling by Satisfaction of Spatial Restraints

The methods in this class begin by generating many constraints or restraints on the structure of the target sequence, using its alignment to related protein structures as a guide. The restraints are generally obtained by assuming that the corresponding distances between aligned residues in the template and the target structures are similar. These homology-derived restraints are usually supplemented by stereochemical restraints on bond lengths, bond angles, non-bonded atom–atom contacts, etc., which are obtained from a molecular mechanics force field. The model is then derived by minimizing the violations of all the restraints. This can be achieved by either distance geometry or real-space optimization. For example, an elegant distance geometry approach constructs all-atom models from lower and upper bounds on distances and dihedral angles [76,77]. Lower and upper bounds on C_α–C_α and main chain–side chain distances, hydrogen bonds, and conserved dihedral angles were derived for *E. coli* flavodoxin from four other flavodoxins; bounds were calculated for all distances and dihedral angles that had equivalent atoms in the template structures. The allowed range of values of a distance or a dihedral angle depended on the degree of structural variability at the corresponding position in the template structures. Distance geometry was used to obtain an ensemble of approximate 3D models, which were then exhaustively refined by restrained molecular dynamics with simulated annealing in water.

We now describe our own approach in more detail [31,58,78,79] (Fig. 3). The question addressed is, What is the most probable structure for a certain sequence, given its alignment with related structures? The approach was developed to use as many different types of data about the target sequence as possible. It is implemented in the computer program MODELLER (Table 1). The comparative modeling procedure begins with an alignment of the target sequence with related known 3D structures. The output, obtained with-

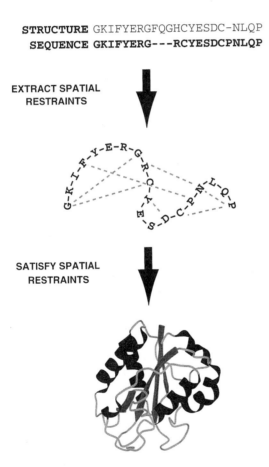

Figure 3 Model building by MODELLER [31]. First, spatial restraints in the form of atomic distances and dihedral angles are extracted from the template structure(s). The alignment is used to determine equivalent residues between the target and the template. The restraints are combined into an objective function. Finally, the model for the target is optimized until a model that best satisfies the spatial restraints is obtained. This procedure is technically similar to the one used in structure determination by NMR.

out any user intervention, is a 3D model for the target sequence containing all main chain and side chain non-hydrogen atoms.

In the first step of model building, distance and dihedral angle restraints on the target sequence are derived from its alignment with template 3D structures. The form of these restraints was obtained from a statistical analysis of the relationships between similar protein structures. The analysis relied on a database of 105 family alignments that included 416 proteins of known 3D structure [79]. By scanning the database of alignments, tables quantifying various correlations were obtained, such as the correlations between two equivalent $C_\alpha–C_\alpha$ distances or between equivalent main chain dihedral angles from two related proteins [31]. These relationships are expressed as conditional probability density functions (pdf's) and can be used directly as spatial restraints. For example, probabilities for different values of the main chain dihedral angles are calculated from the type of residue considered, from main chain conformation of an equivalent residue, and from sequence similarity between the two proteins. Another example is the pdf for a certain $C_\alpha–C_\alpha$ distance given equivalent distances in two related protein structures (Fig. 4). An important feature of the method is that the forms of spatial restraints were obtained empirically from a database of protein structure alignments.

In the second step, the spatial restraints and the CHARMM22 force field terms enforcing proper stereochemistry [80,81] are combined into an objective function. The general form of the objective function is similar to that in molecular dynamics programs such as CHARMM22 [80]. The objective function depends on the Cartesian coordinates of ~10,000 atoms (3D points) that form a *system* (one or more molecules):

$$F = F(\mathbf{R}) = F_{\text{symm}} + \sum_i c_i(\mathbf{f}_i, \mathbf{p}_i) \tag{1}$$

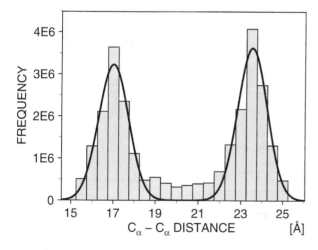

Figure 4 Sample spatial restraint in MODELLER. A restraint on a given $C_\alpha–C_\alpha$ distance, d, is expressed as a conditional probability density function that depends on two other equivalent distances ($d' = 17.0$ and $d'' = 23.5$): $p(d/d', d'')$. The restraint (continuous line) is obtained by least-squares fitting a sum of two Gaussian functions to the histogram, which in turn is derived from many triple alignments of protein structures. In practice, more complicated restraints are used that depend on additional information such as similarity between the proteins, solvent accessibility, and distance from a gap in the alignment.

where F_{symm} is an optional symmetry term that restrains several parts of the structure to the same conformation [53]. **R** are Cartesian coordinates of all atoms, c is a restraint term, **f** is a geometrical feature of a molecule, and $\mathbf{p}_i$ are parameters. For a 10,000 atom system there can be on the order of 200,000 restraints. The form of c is simple; it includes a quadratic function, harmonic lower and upper bounds, cosine, a weighted sum of a few Gaussian functions, Coulomb's law, Lennard-Jones potential, and cubic splines. The geometrical features presently include a distance; an angle; a dihedral angle; a pair of dihedral angles between two, three, four atoms and eight atoms, respectively; the shortest distance in the set of distances; solvent accessibility in square angstroms; and atomic density, which is expressed as the number of atoms around the central atom. A pair of dihedral angles can be used to restrain strongly correlated features such as the main chain dihedral angles Φ and Ψ. Each of the restraints also depends on a few parameters $\mathbf{p}_i$ that generally vary from restraint to restraint. Some restraints can be used to restrain pseudo-atoms such as the gravity center of several atoms.

Finally, the model is obtained by optimizing the objective function in Cartesian space. The optimization is carried out by the use of the variable target function method [82], employing methods of conjugate gradients and molecular dynamics with simulated annealing [83] (Fig. 5). Several slightly different models can be calculated by varying the initial structure, and the variability among these models can be used to estimate the lower bound on the errors in the corresponding regions of the fold.

Because modeling by satisfaction of spatial restraints can use many different types of information about the target sequence, it is perhaps the most promising of all comparative modeling techniques. One of the strengths of modeling by satisfaction of spatial restraints

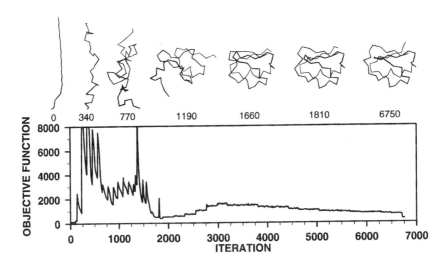

Figure 5 Optimization of the objective function in MODELLER. Optimization of the objective function (curve) starts with a random or distorted model structure. The iteration number is indicated below each sample structure. The first approximately 2000 iterations correspond to the variable target function method [82] relying on the conjugate gradients technique. This approach first satisfies sequentially local restraints, then slowly introduces longer range restraints until the complete objective function is optimized. In the remaining 4750 iterations, molecular dynamics with simulated annealing is used to refine the model [83]. CPU time needed to generate one model is about 2 min for a 250 residue protein on a medium-sized workstation.

is that constraints or restraints derived from a number of different sources can easily be added to the homology derived restraints. For example, restraints could be provided by rules for secondary structure packing [84], analyses of hydrophobicity [85] and correlated mutations [86], empirical potentials of mean force [87], nuclear magnetic resonance (NMR) experiments [88], cross-linking experiments, fluorescence spectroscopy, image reconstruction in electron microscopy, site-directed mutagenesis [89], intuition, etc. In this way, a comparative model, especially in the difficult cases, could be improved by making it consistent with available experimental data and/or with more general knowledge about protein structure.

D. Loop Modeling

In comparative modeling, target sequences often have inserted residues relative to the template structures or have regions that are structurally different from the corresponding regions in the templates. Thus, no structural information about these inserted or conformationally variable segments can be extracted from the template structures. These regions frequently correspond to surface loops. Loops often play an important role in defining the functional specificity of a given protein framework, forming the active and binding sites. The accuracy of loop modeling is a major factor determining the usefulness of comparative models in applications such as ligand docking. Loop modeling can be seen as a mini protein folding problem. The correct conformation of a given segment of a polypeptide chain has to be calculated mainly from the sequence of the segment itself. However, loops are generally too short to provide sufficient information about their local fold. Even identical decapeptides do not always have the same conformation in different proteins [90,91]. Some additional restraints are provided by the core anchor regions that span the loop and by the structure of the rest of a protein that cradles the loop. Although many loop modeling methods have been described, it is still not possible to model correctly and with high confidence loops longer than approximately eight residues [239].

There are two main classes of loop modeling methods: (1) the database search approaches, where a segment that fits on the anchor core regions is found in a database of all known protein structures [62,94], and (2) the conformational search approaches [95–97]. There are also methods that combine these two approaches [92,98,99].

The database search approach to loop modeling is accurate and efficient when a database of specific loops is created to address the modeling of the same class of loops, such as β-hairpins [100], or loops on a specific fold, such as the hypervariable regions in the immunoglobulin fold [94,101]. For example, an analysis of the hypervariable immunoglobulin regions resulted in a series of rules that allowed a very high accuracy of loop prediction in other members of the family. These rules were based on the small number of conformations for each loop and the dependence of the loop conformation on its length and certain key residues. There have been attempts to classify loop conformations into more general categories, thus extending the applicability of the database search approach to more cases [102–105]. However, the database methods are limited by the fact that the number of possible conformations increases exponentially with the length of a loop. As a result, only loops up to four to seven residues long have most of their conceivable conformations present in the database of known protein structures [106,107]. Even according to the more optimistic estimate, approximately 30% and 60% of all the possible eight- and nine-residue loop conformations, respectively, are missing from the database [106]. This is made even worse by the requirement for an overlap of at least one residue between the database fragment and the anchor core regions, which means that the modeling of a five-residue insertion requires at least a seven-residue fragment from the database [70]. Despite the rapid growth of the database of known structures, there is no possibility of covering most of the conformations of a nine-residue segment in the foreseeable future. On the other hand, most of the insertions in a family of homologous proteins are shorter than nine residues [108,239].

To overcome the limitations of the database search methods, conformational search methods were developed [95,96,109]. There are many such methods, exploiting different protein representations, objective function terms, and optimization or enumeration algorithms. The search algorithms include the minimum perturbation method [97], molecular dynamics simulations [92,110,111], genetic algorithms [112], Monte Carlo and simulated annealing [113,114], multiple copy simultaneous search [115–117], self-consistent field optimization [118], and an enumeration based on the graph theory [119].

We now describe a new loop modeling protocol in the conformational search class [239]. It is implemented in the program MODELLER (Table 1). The modeling procedure consists of optimizing the positions of all non-hydrogen atoms of a loop with respect to an objective function that is a sum of many spatial restraints. Many different combinations of various restraints were explored. The best set of restraints includes the bond length, bond angle, and improper dihedral angle terms from the CHARMM22 force field [80,81], statistical preferences for the main chain and side chain dihedral angles [31], and statistical preferences for non-bonded contacts that depend on the two atom types, their distance through space, and separation in sequence [120]. The objective function was optimized with the method of conjugate gradients combined with molecular dynamics and simulated annealing. Typically, the loop prediction corresponds to the lowest energy conformation out of the 500 independent optimizations. The algorithm allows straightforward incorporation of additional spatial restraints, including those provided by template fragments, disulfide bonds, and ligand binding sites. To simulate comparative modeling problems, the loop modeling procedure was evaluated by predicting loops of known structure in only approximately correct environments. Such environments were obtained by distorting the anchor regions corresponding to the three residues at either end of the loop and all the atoms within 10 Å of the native loop conformation for up to 2–3 Å by molecular dynamics simulations. In the case of five-residue loops in the correct environments, the average error was 0.6 Å, as measured by local superposition of the loop main chain atoms alone (C, N, C_α, O). In the case of eight-residue loops in the correct environments, 90% of the loops had less than 2 Å main chain RMS error, with an average of less than 1.2 Å (Fig. 6).

E. Side Chain Modeling

As for loops, side chain conformation is predicted from similar structures and from steric or energy considerations [5,121]. The geometry of disulfide bridges is modeled from disulfide bridges in protein structures in general [122,123] and from equivalent disulfide bridges in related structures [79]. Modeling the stability and conformation of point mutations by free energy perturbation simulations is not discussed here [124–127].

Vasquez [121] reviewed and commented on various approaches to side chain modeling. The importance of two effects on side chain conformation was emphasized. The first effect was the coupling between the main chain and side chains, and the second effect was the continuous nature of the distributions of side chain dihedral angles; for example,

Global superposition   Local superposition   Method accuracy

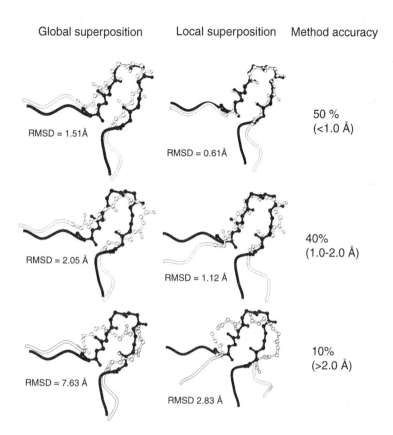

RMSD = 1.51Å

RMSD = 0.61Å

50 %
(<1.0 Å)

RMSD = 2.05 Å

RMSD = 1.12 Å

40%
(1.0-2.0 Å)

RMSD = 7.63 Å

RMSD 2.83 Å

10%
(>2.0 Å)

Figure 6   Oxidoreductase (2nac), loop residues 28–35. Anchor distortion = 1.2 Å. Sample models of varying accuracy for an eight-residue loop in an approximately correct protein environment. The calculated loops (shaded) are compared with the X-ray structure (black). Three levels of accuracy are illustrated: High accuracy corresponding to the backbone RMSD < 1 Å (top), medium accuracy corresponding to the backbone RMSD < 2 Å (middle), and low accuracy corresponding to the backbone RMSD > 2 Å (bottom). The panels on the left compare the loop backbone conformations after least-squares superposition of the complete protein structure. The panels on the right compare the loop backbone conformations after local superposition of the loops. The RMSD values are quoted for the main chain atoms only. The fraction of the loops modeled at each accuracy level is given in the rightmost column. The figure was prepared using MOLSCRIPT [236].

5–30% of side chains in crystal structures are significantly different from their rotamer conformations [128] and 6% of the χ_1 or χ_2 values are not within ±40° of any rotamer conformation [129]. Both effects appear to be important when correlating packing energies and stability [130]. The correct energetics may be obtained for the incorrect reasons; i.e., the side chains adopt distorted conformations to compensate for the rigidity of the backbone. Correspondingly, the backbone shifts may hinder the use of these methods when the template structures are related at less than 50% sequence identity [131]. This is consistent with the X-ray structure of a variant of λ repressor, which reveals that the protein accommodates the potentially disruptive residues with shifts in its α-helical arrangement and with only limited changes in side chain orientations [132]. Some attempts to include backbone flexibility in side chain modeling have been described [118,133,134], but the methods are not yet generally applicable.

Significant correlations were found between side chain dihedral angle probabilities and backbone Φ, Ψ values [129,135]. These correlations go beyond the dependence of side chain conformation on the secondary structure [136]. For example, the preferred rotamers can vary within the same secondary structure, with the changes in the Φ, Ψ dihedral angles as small as 20° [135]. Since these changes are smaller than the differences between closely related homologs, the prediction of the side chain conformation generally cannot be uncoupled from backbone prediction. This partly explains why the conformation of equivalent side chains in homologous structures is useful in side chain modeling [31]. A backbone-dependent rotamer library for amino acid side chains was developed and used to construct side chain conformations from main chain coordinates [135]. This automated method first places the side chains according to the rotamer library and then removes steric clashes by combinatorial energy minimization. It was also demonstrated that simple arguments based on conformational analysis could account for many features of the observed dependence of the side chain rotamers on the backbone [135]. Recently, the main chain–dependent side chain rotamer library was recalculated and extensively evaluated [129] (Table 1). The accuracy of the method was 82% for the χ_1 dihedral angle and 72% for both χ_1 and χ_2 dihedral angles when the backbones of templates in the range from 30% to 90% sequence identity were used; a prediction was deemed correct when it was within 40° of the target crystal structure value.

Chung and Subbiah [131,137] gave an elegant structural explanation for the rapid decrease in the conservation of side chain packing as the sequence identity decreases below 30%. Although the fold is maintained, the pattern of side chain interactions is generally lost in this range of sequence similarity [138]. Two sets of computations were done for two sample protein sequences: The side chain conformation was predicted by maximizing packing on the fixed native backbone and on a fixed backbone with approximately 2 Å RMSD from the native backbone; the 2 Å RMSD generally corresponds to the differences between the conserved cores of two proteins related at 25–30% sequence identity. The side chain predictions based on the two kinds of backbone turned out to be unrelated. Thus, inasmuch as packing reflects the true laws determining side chain conformation, a backbone with less than 30% sequence identity to the sequence being modeled is no longer sufficiently restraining to result in the correct packing of the buried side chains.

The solvation term is important for the modeling of exposed side chains [139–142]. It was also demonstrated that treating hydrogen bonds explicitly could significantly improve side chain prediction [135,143]. Calculations that do not take into account the solvent, either implicitly or explicitly, introduce errors into the hydrogen-bonding patterns even in the core regions of a protein [142]. Residues with zero solvent accessibility area can still have a significant interaction energy with the solvent atoms [144].

A recent survey analyzed the accuracy of three different side chain prediction methods [134]. These methods were tested by predicting side chain conformations on near-native protein backbones with <4 Å RMSD to the native structures. The three methods included the packing of backbone-dependent rotamers [129], the self-consistent mean-field approach to positioning rotamers based on their van der Waals interactions [145], and the segment-matching method of Levitt [74]. The accuracies of the methods were similar. They were able to predict correctly approximately 50% of χ_1 angles and 35% of both χ_1 and χ_2 angles. In typical comparative modeling applications where the backbone is closer to the native structures (<2 Å RMSD), these numbers increase by approximately 20% [146].

III. AB INITIO PROTEIN STRUCTURE MODELING METHODS

This section briefly reviews prediction of the native structure of a protein from its sequence of amino acid residues alone. These methods can be contrasted to the threading methods for fold assignment [Section II.A] [39–47,147], which detect remote relationships between sequences and folds of known structure, and to comparative modeling methods discussed in this review, which build a complete all-atom 3D model based on a related known structure. The methods for ab initio prediction include those that focus on the broad physical principles of the folding process [148–152] and the methods that focus on predicting the actual native structures of specific proteins [44,153,154,240]. The former frequently rely on extremely simplified generic models of proteins, generally do not aim to predict native structures of specific proteins, and are not reviewed here.

Although comparative modeling is the most accurate modeling approach, it is limited by its absolute need for a related template structure. For more than half of the proteins and two-thirds of domains, a suitable template structure cannot be detected or is not yet known [9,11]. In those cases where no useful template is available, the ab initio methods are the only alternative. These methods are currently limited to small proteins and at best result only in coarse models with an RMSD error for the C_α atoms that is greater than 4 Å. However, one of the most impressive recent improvements in the field of protein structure modeling has occurred in ab initio prediction [155–157].

Ab initio prediction relies on the thermodynamic hypothesis of protein folding [158]. The thermodynamic hypothesis suggests that the native structure of a protein sequence corresponds to its global free energy minimum state. Accordingly, ab initio prediction methods are generally formulated as optimizations. As such, they can be distinguished by the representation of a protein and its degrees of freedom, the function that defines the energy for each of the allowed conformations, and the optimization method that attempts to find the global minimum on a given energy surface.

Although the folding of short proteins has been simulated at the atomic level of detail [159,160], a simplified protein representation is often applied. Simplifications include using one or a few interaction centers per residue [161] as well as a lattice representation of a protein [162]. Some methods are hierarchical in that they begin with a simplified lattice representation and end up with an atomistic detailed molecular dynamics simulation [163].

The energy functions for folding simulations include atom-based potentials from molecular mechanics packages [164] such as CHARMM [81], AMBER [165], and ECEPP [166], the statistical potentials of mean force derived from many known protein structures [167], and simplified potentials based on chemical intuition [168–171]. Some methods also incorporate non-physical spatial restraints obtained from multiple sequence alignments and other considerations to reduce the size of the conformational space that needs to be explored [172–176].

Many different optimization methods [177,178]—even enumerations with some lattice models [171]—have been applied to the protein folding problem. These methods include molecular dynamics simulations [179,180], Monte Carlo sampling [173,181,182], the diffusion equation method [183], and genetic algorithm optimization [184–186]. A recent and particularly successful approach assembles the whole protein model from relatively short building blocks [187–189]. Many candidate blocks are obtained from known protein structures by relying on energetic, geometrical, and sequence similarity filters. The model of a whole protein is then assembled from such pieces by a Monte Carlo optimization of a statistical energy function [188].

There is scope for combining the comparative modeling and ab initio methods. The modeling of inserted loops in comparative prediction is based primarily on the sequence information alone. In addition, the alignment errors as well as large distortions of the target relative to the template require that such regions be modeled ab initio without relying on the template structure. It is likely that the ab initio approaches will help reduce some of the limitations of comparative modeling.

IV. ERRORS IN COMPARATIVE MODELS

The errors in comparative models can be divided into five categories [58] (Fig. 7):

1. Errors in side chain packing.
2. Distortions or shifts of a region that is aligned correctly with the template structures.
3. Distortions or shifts of a region that does not have an equivalent segment in any of the template structures.
4. Distortions or shifts of a region that is aligned incorrectly with the template structures.
5. A misfolded structure resulting from using an incorrect template.

Significant methodological improvements are needed to address all of these errors.

Errors 3–5 are relatively infrequent when sequences with more than 40% identity to the templates are modeled. For example, in such a case, approximately 90% of the main chain atoms are likely to be modeled with an RMS error of about 1 Å. In this range of sequence similarity, the alignment is mostly straightforward to construct, there are not many gaps, and structural differences between the proteins are usually limited to loops and side chains. When sequence identity is between 30% and 40%, the structural differences become larger, and the gaps in the alignment are more frequent and longer. As a result, the main chain RMS error increases to about 1.5 Å for about 80% of the residues. The rest of the residues are modeled with large errors because the methods generally fail to model structural distortions and rigid-body shifts and are unable to recover from misalignments. Below 40% sequence identity, misalignments and insertions in the target sequence become the major problems. Insertions longer than about eight residues cannot yet be modeled accurately, but shorter loops can frequently be modeled successfully [92,119,239]. When sequence identity drops below 30%, the main problem becomes the identification of related templates and their alignment with the sequence to be modeled (Fig. 8). In general, it can be expected that about 20% of residues will be misaligned and consequently incorrectly modeled with an error greater than 3 Å at this level of sequence similarity [51]. This is a serious impediment for comparative modeling because it appears

(a) **(b)** **(c)**

(d) **(e)**

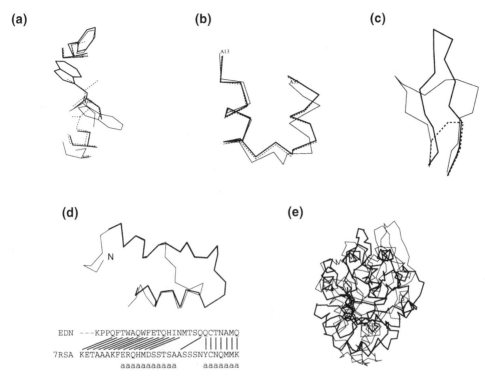

```
EDN  ---KPPQFTWAQWFETQHINMTSQQCTNAMQ
        ////////////      ||||||||
7RSA KETAAAKFERQHMDSSTSAASSSNYCNQMMK
        aaaaaaaaaaa      aaaaaaa
```

Figure 7 Typical errors in comparative modeling. (a) Errors in side chain packing. The Trp 109 residue in the crystal structure of mouse cellular retinoic acid binding protein I (thin line) is compared with its model (thick line) and with the template mouse adipocyte lipid-binding protein (broken line). (b) Distortions and shifts in correctly aligned regions. A region in the crystal structure of mouse cellular retinoic acid binding protein I (thin line) is compared with its model (thick line), and with the template fatty acid binding protein (broken line). (c) Errors in regions without a template. The C_α trace of the 112–117 loop is shown for the X-ray structure of human eosinophil neurotoxin (thin line), its model (thick line), and the template ribonuclease A structure (residues 111–117; broken line). (d) Errors due to misalignments. The N-terminal region in the crystal structure of human eosinophil neurotoxin (thin line) is compared with its model (thick line). The corresponding region of the alignment with the template ribonuclease A is shown. The black lines show correct equivalences, that is residues whose C_α atoms are within 5 Å of each other in the optimal least-squares superposition of the two X-ray structures. The ''a'' characters in the bottom line indicate helical residues. (e) Errors due to an incorrect template. The X-ray structure of α-trichosanthin (thin line) is compared with its model (thick line), which was calculated using indole-3-glycerophosphate synthase as the template. (From Ref. 146.)

that at least one-half of all related protein pairs are related at less than 30% sequence identity [9,190].

It has been pointed out that a comparative model is frequently more distant from the actual target structure than the closest template structure used to calculate the model [191]. However, at least for some modeling methods, this is the case only when there are errors in the template–target alignment used for modeling and when the correct structure-based template–target alignment is used for comparing the template with the actual target structure [58]. In contrast, the model is generally closer to the target structure than any of

(a)

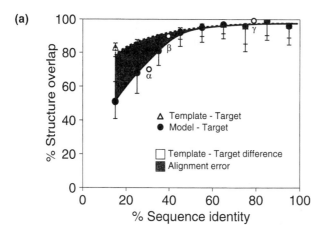

(b)

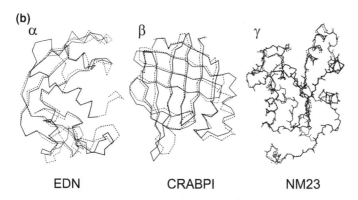

Figure 8   Average model accuracy as a function of the percentage identity between the target and template sequences. (a) The models were calculated entirely automatically, based on single template structures. As the sequence identity between the target sequence and the template structure decreases, the average structural similarity between the template and the target also decreases (dashed line, triangles). Structure overlap is defined as the fraction of equivalent C_α atoms. For comparison of the model with the actual structure (continuous line, circles), two C_α atoms were considered equivalent if they were within 3.5 Å of each other and belonged to the same residue. For comparison of the template structure with the actual structure (dashed line, triangles), two C_α atoms were considered equivalent if they were within 3.5 Å of each other after alignment and rigid-body superposition by the ALIGN3D command in MODELLER. (b) Three models (solid line) compared with their corresponding experimental structures (dotted line). The models were calculated with MODELLER in a completely automated fashion before the experimental structures were available [146]. When multiple sequence and structure information is used and the alignments are edited by hand, the models can be significantly more accurate than shown in this plot [58].

the templates if the modeling target–template alignment is used in evaluating the similarity between the actual target structure and the template [58]. As a result, using a model is generally better than using the template structure even when the alignment is incorrect, because the actual target structure, and therefore the correct template–target alignment, are not available in practical modeling applications.

To put the errors in comparative models into perspective, we list the differences among structures of the same protein that have been determined experimentally (Fig. 9). The 1 Å accuracy of main chain atom positions corresponds to X-ray structures defined at a low resolution of about 2.5 Å and with an *R*-factor of about 25% [192], as well as to medium resolution NMR structures determined from 10 interproton distance restraints per residue [193]. Similarly, differences between the highly refined X-ray and NMR structures of the same protein also tend to be about 1 Å [193]. Changes in the environment

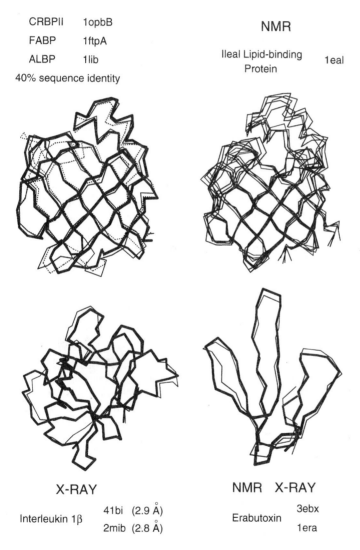

Figure 9 Relative accuracy of comparative models. Upper left panel, comparison of homologous structures that share ~40% sequence identity. Upper right panel, conformations of ileal lipid-binding protein that satisfy the NMR restraints set equally well. Lower left panel, comparison of two independently determined X-ray structures of interleukin 1β. Lower right panel, comparison of the X-ray and NMR structures of erabutoxin. The figure was prepared using the program MOLSCRIPT [236].

(e.g., oligomeric state, crystal packing, solvent, ligands) can also have a significant effect on the structure [194]. Overall, comparative modeling based on templates with more than 40% identity is almost as good as medium resolution experimental structures, simply because the proteins at this level of similarity are likely to be as similar to each other as are the structures for the same protein determined by different experimental techniques under different conditions. However, the caveat in comparative protein modeling is that some regions, mainly loops and side chains, may have larger errors.

A particularly informative way to test protein structure modeling methods, including comparative modeling, is provided by the biennial meetings on critical assessment of techniques for protein structure prediction (CASP) [191,195,196]. The most recent meeting was held in December 1998 [241]. Protein modelers are challenged to model sequences with unknown 3D structure and to submit their models to the organizers before the meeting. At the same time, the 3D structures of the prediction targets are being determined by X-ray crystallography or NMR methods. They become available only after the models are calculated and submitted. Thus, a bona fide evaluation of protein structure modeling methods is possible.

V. MODEL EVALUATION

Essential for interpreting 3D protein models is the estimation of their accuracy, both the overall accuracy and the accuracy in the individual regions of a model. The errors in models arise from two main sources, the failure of the conformational search to find the optimal conformation and the failure of the scoring function to identify the optimal conformation. The 3D models are generally evaluated by relying on geometrical preferences of the amino acid residues or atoms that are derived from known protein structures. Empirical relationships between model errors and target–template sequence differences can also be used. It is convenient to approach an evaluation of a given model in a hierarchical manner [9]. It first needs to be assessed if the model at least has the correct fold. The model will have a correct fold if the correct template is picked and if that template is aligned at least approximately correctly with the target sequence. Once the fold of a model is confirmed, a more detailed evaluation of the overall model accuracy can be performed based on the overall sequence similarity on which the model is based (Fig. 8). Finally, a variety of error profiles can be constructed to quantify the likely errors in the different regions of a model. A good strategy is to evaluate the models by using several different methods and identify the consensus between them. In addition, energy functions are in general designed to work at a certain level of detail and are not appropriate to judge the models at a finer or coarser level [197]. There are many model evaluation programs and servers [198,199] (Table 1).

A basic requirement for a model is that it have good stereochemistry. The most useful programs for evaluating stereochemistry are PROCHECK [200], PROCHECK-NMR [201], AQUA [201], SQUID [202], and WHATCHECK [203]. The features of a model that are checked by these programs include bond lengths, bond angles, peptide bond and side chain ring planarities, chirality, main chain and side chain torsion angles, and clashes between non-bonded pairs of atoms. In addition to good stereochemistry, a model also has to have low energy according to a molecular mechanics force field, such as that of CHARMM22 [80]. However, low molecular mechanical energy does not ensure that the model is correct [204,205]. Thus, distributions of many spatial features have been compiled from high resolution protein structures, and any large deviations from the most likely values have been interpreted as strong indicators of errors in the model. Such features include packing [206], formation of a hydrophobic core [207], residue and atomic solvent accessibilities [208–212], spatial distribution of charged groups [213], distribution of atom–atom distances [214], atomic volumes [215], and main chain hydrogen bonding [200].

Another group of methods for testing 3D models that implicitly take into account many of the criteria listed above involve 3D profiles and statistical potentials [87,216]. These methods evaluate the environment of each residue in a model with respect to the expected environment as found in the high resolution X-ray structures. Programs implementing this approach include VERIFY3D [216], PROSA [217], HARMONY [218], and ANOLEA [120].

An additional role of the model evaluation methods is to help in the actual modeling procedure. In principle, an improvement in the accuracy of a model is possible by incorporating the quality criteria into a scoring function being optimized to derive the model in the first place.

VI. APPLICATIONS OF COMPARATIVE MODELING

Comparative modeling is often an efficient way to obtain useful information about the proteins of interest. For example, comparative models can be helpful in designing mutants to test hypotheses about the protein's function [89,219]; identifying active and binding sites [220]; searching for, designing, and improving ligands for a given binding site [221]; modeling substrate specificity [222]; predicting antigenic epitopes [223]; simulating protein–protein docking [224]; inferring function from calculated electrostatic potential around the protein [225]; facilitating molecular replacement in X-ray structure determination [226]; refining models based on NMR constraints [227]; testing and improving a sequence–structure alignment [228]; confirming a remote structural relationship [59]; and rationalizing known experimental observations. For an exhaustive review of comparative modeling applications, see Ref. 3.

Fortunately, a 3D model does not have to be absolutely perfect to be helpful in biology, as demonstrated by the applications listed above. However, the type of question that can be addressed with a particular model does depend on the model's accuracy. At the low end of the accuracy spectrum, there are models that are based on less than 25% sequence identity and have sometimes less than 50% of their C_α atoms within 3.5 Å of their correct positions. However, such models still have the correct fold, and even knowing only the fold of a protein is frequently sufficient to predict its approximate biochemical function. More specifically, only nine out of 80 fold families known in 1994 contained proteins (domains) that were not in the same functional class, although 32% of all protein structures belonged to one of the nine superfolds [229]. Models in this low range of accuracy combined with model evaluation can be used for confirming or rejecting a match between remotely related proteins [9,58].

In the middle of the accuracy spectrum are the models based on approximately 35% sequence identity, corresponding to 85% of the C_α atoms modeled within 3.5 Å of their correct positions. Fortunately, the active and binding sites are frequently more conserved than the rest of the fold and are thus modeled more accurately [9]. In general, medium resolution models frequently allow a refinement of the functional prediction based on sequence alone, because ligand binding is most directly determined by the structure of the binding site rather than its sequence. It is frequently possible to predict correctly important features of the target protein that do not occur in the template structure. For example, the location of a binding site can be predicted from clusters of charged residues [225], and the size of a ligand may be predicted from the volume of the binding site cleft [222]. Medium resolution models can also be used to construct site-directed mutants with altered or destroyed binding capacity, which in turn could test hypotheses about the sequence–structure–function relationships. Other problems that can be addressed with medium resolution comparative models include designing proteins that have compact structures without long tails, loops, and exposed hydrophobic residues for better crystallization and designing proteins with added disulfide bonds for extra stability.

The high end of the accuracy spectrum corresponds to models based on 50% sequence identity or more. The average accuracy of these models approaches that of low resolution X-ray structures (3 Å resolution) or medium resolution NMR structures (10 distance restraints per residue) [58]. The alignments on which these models are based generally contain almost no errors. In addition to the already listed applications, high quality models can be used for docking of small ligands [221] or whole proteins onto a given protein [224,230].

We now describe two applications of comparative modeling in more detail: (1) Modeling of substrate specificity aided by a high accuracy model and (2) confirming a remote structural relationship based on a low accuracy model.

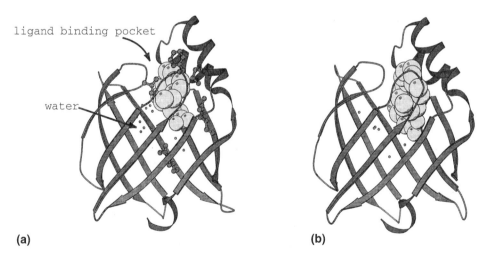

(a)   **(b)**

Figure 10 Models of complexes between BLBP and two different fatty acids. The fatty acid ligand is shown in the CPK representation. The small spheres in the ligand-binding cavity are water molecules. (a) Model of the BLBP–oleic acid complex, in which the cavity is not filled. (b) Model of the BLBP–docosahexaenoic acid complex, in which the cavity is filled. The figure was prepared using the program MOLSCRIPT [236].

A. Ligand Specificity of Brain Lipid-Binding Protein

Brain lipid-binding protein (BLBP) is a member of the family of fatty acid binding proteins that was isolated from brain [222]. The problem was to find out which one of the many fatty acids known to bind to fatty acid binding proteins in general is the likely physiological ligand of BLBP. To address this problem, comparative models of BLBP complexed with many fatty acids were calculated by relying on the structures of the adipocyte lipid-binding protein and muscle fatty acid binding protein, in complex with their ligands. The models were evaluated by binding and site-directed mutagenesis experiments [222]. The model of BLBP indicated that its binding cavity was just large enough to accommodate docosahexaenoic acid (DHA) (Fig. 10). Because DHA filled the BLBP binding cavity completely, it was unlikely that BLBP would bind a larger ligand. Thus, DHA was the ligand predicted to have the highest affinity for BLBP. The prediction was confirmed by the measurement of binding affinities for many fatty acids. It turned out that the BLBP–DHA interaction was the strongest fatty acid–protein interaction known to date. The binding affinities of

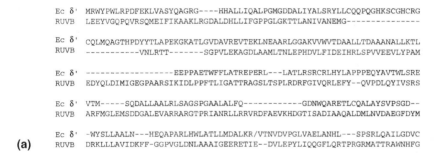

(a)

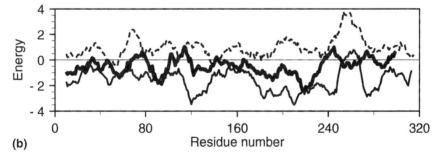

(b)

Figure 11 Confirming structural similarity between the *E. coli* δ′ subunit of DNA polymerase III and RuvB. (a) A sequence alignment between the δ′ subunit and RuvB. (b) ProsaII profiles for the X-ray structure of the δ′ subunit (thin continuous line), $Z = -11.0$; a model of RuvB based on its alignment to the δ′ subunit (thick line), $Z = -7.3$; and a test model based on an incorrect alignment (dashed line), $Z = -0.9$. The RuvB model based on the correct alignment has a significant Z-score and only a few positive peaks in the profile. This indicates that the model is plausible and that RuvB is indeed related structurally to the *E. coli* δ′ subunit. (From Ref. 217.)

the ligands correlated with the surface areas buried by the protein–ligand interactions, as calculated from the corresponding models, and explained why DHA had the highest affinity.

This case illustrates how a comparative model provides new information that cannot be deduced directly from the template structures despite their high (60%) sequence identity to BLBP. The two templates have smaller binding sites and consequently different patterns of binding affinities for the same set of ligands. The study also illustrated how new information is obtained relative to the target–template alignment even when the similarity between the target and the template sequences is high. The volumes and contact surfaces can be calculated only from a 3D model.

B. Finding Proteins Remotely Related to the *E. coli* δ′ Subunit

The structure of the δ′ subunit of the clamp–loader complex of *E. coli* DNA polymerase III was determined by X-ray crystallography [59]. Several biological considerations and extremely weak sequence patterns indicated that δ′ may be structurally related to the RuvB family of DNA helicases. However, the relationship was not possible to prove on the basis of the alignment of the corresponding sequences alone; the sequence identities ranged from only 9% to 21%. To substantiate the putative match, comparative models for several RuvB helicases were constructed using the crystal structure of the δ′ subunit as the template. The models were evaluated by calculating their PROSAII Z-scores and energy profiles [217] (Fig. 11). This evaluation indicated strongly that the model is plausible and that RuvB is indeed related structurally to the *E. coli* δ′ subunit.

VII. COMPARATIVE MODELING IN STRUCTURAL GENOMICS

In a few years, the genome projects will have provided us with the amino acid sequences of more than a million proteins—the catalysts, inhibitors, messengers, receptors, transporters, and building blocks of the living organisms. The full potential of the genome projects will be realized only when we assign and understand the function of these new proteins. This will be facilitated by structural information for all or almost all proteins. This aim will be achieved by structural genomics, a focused, large-scale determination of protein structures by X-ray crystallography and nuclear magnetic resonance spectroscopy, combined efficiently with accurate, automated, and large-scale comparative protein structure modeling techniques [231]. Given current modeling techniques, it seems reasonable to require models based on at least 30% sequence identity, corresponding to one experimentally determined structure per *sequence family* rather than fold family. Since there are 1000–5000 fold families and perhaps about five times as many sequence families [16], the experimental effort in structural genomics has to deliver at least 10,000 protein domain structures.

To enable the large-scale comparative modeling needed for structural genomics, the steps of comparative modeling are being assembled into a completely automated pipeline. Because many computer programs for performing each of the operations in comparative modeling already exist, it may seem trivial to construct a pipeline that completely automates the whole process. In fact, it is not easy to do so in a robust manner. For a good

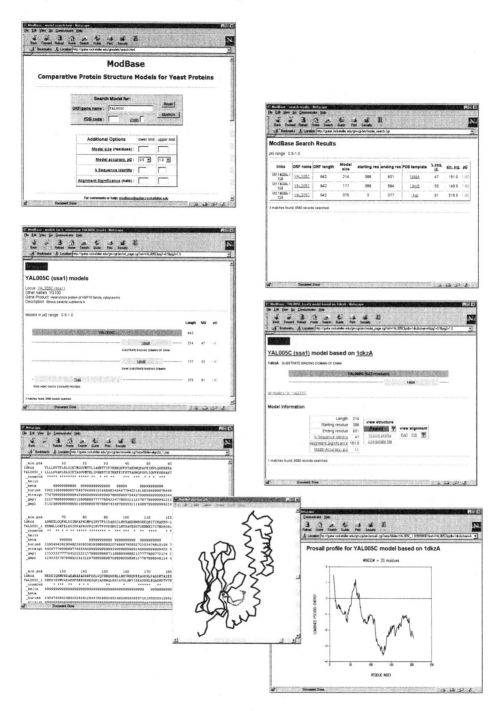

Figure 12 ModBase, a database of comparative protein structure models. Screenshots of the following ModBase panels are shown: A form for searching for the models of a given protein, summary of the search results, summary of the models of a given protein, details about a single model, alignment on which a given model was based, 3D model displayed by RASMOL [237], and a model evaluation by the ProsaII profile [217].

reason, most of the tasks in modeling of individual proteins, including template selection, alignment, and model evaluation, are typically performed with significant human intervention. This allows the use of the best tool for a particular problem at hand and consideration of many different sources of information that are difficult to take into account entirely automatically. Because large-scale modeling can be performed only in a completely automated manner, the main challenge is to build an automated and robust pipeline that approaches the performance of a human expert as much as possible.

Two applications of comparative modeling to complete genomes have been described. For the sequences encoded in the *E. coli* genome, models were built for 10–15% of the proteins using the SWISS-MODEL web server [232,233]. Peitsch et al. have recently also modeled many proteins in SWISS-PROT and made the models available on their SWISS-MODEL web site (see Table 1). Another large-scale modeling study was our own modeling of five prokaryotic and eukaryotic genomes [9]. The calculation resulted in the models for substantial segments of 17.2%, 18.1%, 19.2%, 20.4%, and 15.7% of all proteins in the genomes of *Saccharomyces cerevisiae* (6218 proteins in the genome); *Escherichia coli* (4290 proteins), *Mycoplasma genitalium* (468 proteins), *Caenorhabditis elegans* (7299 proteins, imcomplete), and *Methanococcus janaschii* (1735 proteins), respectively. An important feature of this study was an evaluation of all the models. This evaluation is important because most of the related protein pairs share less than 30% sequence identity, resulting in significant errors in the models. The models were assigned into the reliable or unreliable class by a procedure [9] that relies on the statistical potential function from PROSAII [217]. This allowed identification of those models that were likely to be based on correct templates and at least approximately correct alignments. As a result, 236 yeast proteins without any prior structural information were assigned to a particular fold family; 40 of these proteins did not have any prior functional annotation. The models were also evaluated more precisely by using a calibrated relationship between the model accuracy and the percentage sequence identity on which the model is based [9]. Almost half of the 1071 reliably modeled proteins in the yeast genome share more than approximately 35% sequence identity with their templates. All the alignments, models, and model evaluations are available in the ModBase database of comparative protein structure models (Fig. 12) [234]. Most recently, the combined use of PSI-BLAST [36] with the model building and a new model evaluation [9] allowed us to calculate reliable models for 50% of the proteins in the TrEMBL database (R. Sánchez, F. Mels, A. Šali, in preparation) [234].

Large-scale comparative modeling opens new opportunities for tackling existing problems by virtue of providing many protein models from many genomes. One example is the selection of a target protein for which a drug needs to be developed. A good choice is a protein that is likely to have high ligand specificity; specificity is important because specific drugs are less likely to be toxic. Large-scale modeling facilitates imposing the specificity filter in target selection by enabling a structural comparison of the ligand binding sites of many proteins, either human or from other organisms. Such comparisons may make it possible to select rationally a target whose binding site is structurally most different from the binding sites of all the other proteins that may potentially react with the same drug. For example, when a human pathogenic organism needs to be inhibited, a good target may be a protein whose binding site shape is different from related binding sites in all of the human proteins. Alternatively, when a human metabolic pathway needs to be regulated, the target identification could focus on that particular protein in the pathway that has the binding site most dissimilar from its human homologs.

VIII. CONCLUSION

Whereas an experimental structure or a comparative model is generally insufficient on its own to infer the biological function of a protein, it is often complementary to sequence analysis and direct experiment. Comparative modeling efficiently increases the value of sequence information from the genome projects, although it is not yet possible to model all proteins with useful accuracy. The main bottlenecks are the absence of structurally defined members in many protein families and the difficulties in detecting weak similarities, both for fold recognition and for sequence–structure alignment. The fraction of protein sequences that can be modeled with useful accuracy by comparative modeling is increasing rapidly. The main reasons for this improvement are the increases in the numbers of known folds and the structures per fold family [16] as well as the improvement in the fold recognition and comparative modeling techniques [196]. It has been estimated that globular protein domains cluster in only a few thousand fold families, approximately 800 of which have already been structurally defined [16,23]. Assuming the current growth rate in the number of known protein structures, the structure of at least one member of most globular folds will be determined in less than 10 years [16]. According to this argument, comparative modeling would be applicable to most of the globular protein domains before the expected completion of the human genome project. However, there are some classes of proteins, including membrane proteins, that will not be amenable to modeling without improvements in structure determination and modeling techniques. For example, it has been predicted that 839 (13.9%) of the yeast ORFs have at least two transmembrane helices [235]. To maximize the number of proteins that can be modeled reliably, a concerted effort toward structural determination of the new folds by X-ray crystallography and nuclear magnetic resonance spectroscopy is in order [242]. A combination of a more complete database of known protein structures with accurate modeling techniques will efficiently increase the value of sequence information from the genome projects.

ACKNOWLEDGMENTS

We are grateful to Dr. Azat Badretdinov and Mr. Eric Feyfant for many discussions about comparative protein structure modeling. AF is a Burroughs Wellcome Fellow. RS is a Howard Hughes Medical Institute predoctoral fellow. FM is a Norman and Rosita Winston Biomedical Research Foundation Fellow. AŠ is a Sinsheimer Scholar and an Alfred P. Sloan Research Fellow. The investigations have also been aided by grants from NIH (GM 54762) and NSF (BIR-9601845).

REFERENCES

1. TL Blundell, BL Sibanda, MJE Sternberg, JM Thornton. Knowledge-based prediction of protein structures and the design of novel molecules. Nature 326:347–352, 1987.
2. J Greer. Comparative modelling methods: Application to the family of the mammalian serine proteases. Proteins 7:317–334, 1990.
3. MS Johnson, N Srinivasan, R Sowdhamini, TL Blundell. Knowledge-based protein modelling. CRC Crit Rev Biochem Mol Biol 29:1–68, 1994.

4. J Bajorath, R Stenkamp, A Aruffo. Knowledge-based model building of proteins: Concepts and examples. Protein Sci 2:1798–1810, 1994.
5. A Šali. Modelling mutations and homologous proteins. Curr Opin Biotech 6:437–451, 1995.
6. R Sánchez, A Šali. Advances in comparative protein-structure modeling. Curr Opin Struct Biol 7:206–214, 1997.
7. C Chothia, AM Lesk. The relation between the divergence of sequence and structure in proteins. EMBO J 5:823–826, 1986.
8. M Huynen, T Doerks, F Eisenhaber, C Orengo, S Sunyaev, Y Yuan, P Bork. Homology-based fold predictions for Mycoplasma genitalium proteins. J Mol Biol 280:323–326, 1998.
9. R Sánchez, A Šali. Large-scale protein structure modeling of the Saccharomyces cerevisiae genome. Proc Natl Acad Sci USA 95:13597–13602, 1998.
10. L Rychlewski, B Zhang, A Godzik. Fold and function predictions for Mycoplasma genitalium proteins. Fold Des 3:229–238, 1998.
11. DT Jones. Genthreader: An efficient and reliable protein fold recognition method for genomic sequences. J Mol Biol 287:797–815, 1999.
12. A Bairoch, R Apweiler. The SWISS-PROT protein sequence data bank and its supplement TrEMBL in 1999. Nucleic Acids Res 27:49–54, 1999.
13. EE Abola, FC Bernstein, SH Bryant, TF Koetzle, J Weng. Protein data bank. In: FH Allen, G Bergerhoff, R Sievers, eds. Crystallographic Databases: Information, Content, Software Systems, Scientific Applications, Bonn: Data Commission of the International Union of Crystallography, 1987, pp 107–132.
14. C Chothia. One thousand families for the molecular biologist. Nature 360:543–544, 1992.
15. ZT Zhang. Relations of the numbers of protein sequences, families and folds. Protein Eng 10:757–761, 1997.
16. L Holm, C Sander. Mapping the protein universe. Science 273:595–602, 1996.
17. SF Altschul, MS Boguski, W Gish, JC Wootton. Issues in searching molecular sequence databases. Nature Genet 6:119–129, 1994.
18. GJ Barton. Protein sequence alignment and database scanning. In: MJE Sternberg, ed. Protein Structure Prediction: A Practical Approach. Oxford, UK: IRL Press at Oxford Univ Press, 1998.
19. GD Schuler. Sequence alignment and database searching. Methods Biochem Anal 39:145–171, 1998.
20. WC Barker, JS Garavelli, DH Haft, LT Hunt, CR Marzec, BC Orcutt, GY Srinivasarao, LSL Yeh, RS Ledley, HW Mewes, F Pfeiffer, A Tsugita. The PIR-International protein sequence database. Nucleic Acids Res 26:27–32, 1998.
21. DA Benson, MS Boguski, DJ Lipman, J Ostell, BFF Ouellette. GenBank. Nucleic Acids Res 26:1–7, 1997.
22. HM Berman. The past and future of structure databases. Curr Opin Biotech 10:76–80, 1999.
23. TJP Hubbard, B Ailey, SE Brenner, AG Murzin, C Chothia. SCOP: A structural classification of proteins database. Nucleic Acids Res 27:254–256, 1999.
24. L Holm, C Sander. Protein folds and families: Sequence and structure alignments. Nucleic Acids Res 27:244–247, 1999.
25. CA Orengo, FMG Pearl, JE Bray, AE Todd, AC Martin, L Lo Conte, JM Thornton. The CATH database provides insights into protein structure/function relationship. Nucleic Acids Res 27:275–279, 1999.
26. A Apostolico, R Giancarlo. Sequence alignment in molecular biology. J Comput Biol 5:173–196, 1998.
27. WR Pearson. Comparison of methods for searching protein sequence databases. Protein Sci 4:1145–1160, 1995.
28. SE Brenner, C Chothia, TJ Hubbard. Assessing sequence comparison methods with reliable

structurally identified distant evolutionary relationships. Proc Natl Acad Sci USA 95:6073–6078, 1998.

29. WR Pearson. Empirical statistical estimates for sequence similarity searches. J Mol Biol 276: 71–84, 1998.

30. SF Altschul, W Gish, W Miller, EW Myers, DJ Lipman. Basic local alignment search tool. J Mol Biol 215:403–410, 1990.

31. A Šali, TL Blundell. Comparative protein modelling by satisfaction of spatial restraints. J Mol Biol 234:779–815, 1993.

32. TF Smith, MS Waterman. Identification of common molecular subsequences. J Mol Biol 147:195–197, 1981.

33. S Henikoff, JG Henikoff. Protein family classification based on searching a database of blocks. Genomics 19:97–107, 1994.

34. A Krogh, M Brown, IS Mian, K Sjolander, D Haussler. Hidden Markov models in computational biology: Applications to protein modeling. J Mol Biol 235:1501–1531, 1994.

35. M Gribskov. Profile analysis. Methods Mol Biol 25:247–266, 1994.

36. SF Altschul, TL Madden, AA Schaffer, J Zhang, Z Zhang, W Miller, DJ Lipman. Gapped BLAST and PSI-BLAST: A new generation of protein database search programs. Nucleic Acids Res 25:3389–3402, 1997.

37. B Zhang, L Jaroszewski, L Rychlewski, A Godzik. Similarities and differences between nonhomologous proteins with similar folds: Evaluation of threading strategies. Fold Des 2: 307–317, 1998.

38. L Jaroszewski, L Rychlewski, B Zhang, A Godzik. Fold prediction by a hierarchy of sequence, threading, and modeling methods. Protein Sci 6:1431–1440, 1998.

39. TL Blundell, MS Johnson. Catching a common fold. Protein Sci 2:877–883, 1993.

40. JU Bowie, R Lüthy, D Eisenberg. A method to identify protein sequences that fold into a known three-dimensional structure. Science 253:164–170, 1991.

41. DT Jones, WR Taylor, JM Thornton. A new approach to protein fold recognition. Nature 358:86–89, 1992.

42. A Godzik, A Kolinski, J Skolnick. Topology fingerprint approach to the inverse protein folding problem. J Mol Biol 227:227–238, 1992.

43. MJ Sippl, S Weitckus. Detection of native-like models for amino acid sequences of unknown three-dimensional structure in a data base of known protein conformations. Proteins 13:258–271, 1992.

44. DT Jones. Progress in protein structure prediction. Curr Opin Struct Biol 7:377–387, 1997.

45. TF Smith, L Lo Conte, J Bienkowska, C Gaitatzes, RGJ Rogers, R Lathrop. Current limitations to protein threading approaches. J Comput Biol 4:217–225, 1997.

46. AE Torda. Perspectives in protein-fold recognition. Curr Opin Struct Biol 7:200–205, 1997.

47. M Levitt. Competitive assessment of protein fold recognition and alignment accuracy. Proteins Suppl 1:92–104, 1997.

48. WR Taylor. Multiple protein sequence alignment: Algorithms and gap insertion. Methods Enzymol 266:343–367, 1996.

49. P Briffeuil, G Baudoux, C Lambert, X De Bolle, C Vinals, E Feytmans, E Depiereux. Comparative analysis of seven multiple protein sequence alignment servers: Clues to enhance reliability of predictions. Bioinformatics 14:357–366, 1998.

50. AD Baxevanis. Practical aspects of multiple sequence alignment. Methods Biochem Anal 39:172–188, 1998.

51. MS Johnson, JP Overington. A structural basis for sequence comparisons: An evaluation of scoring methodologies. J Mol Biol 233:716–738, 1993.

52. WR Taylor, TP Flores, CA Orengo. Multiple protein structure alignment. Protein Sci 3:1858–1870, 1994.

53. A Šali, R Sánchez, AY Badretdinov, A Fiser, F Melo, JP Overington, E Feyfant, MA Martí-

Renom. MODELLER, A Protein Structure Modeling Program, Release 5. 1999. http://guitar.rockefeller.edu/

54. KK Koretke, Z Luthey-Schulten, PG Wolynes. Self-consistently optimized statistical mechanical energy functions for sequence structure alignment. Protein Sci 5:1043–1059, 1996.

55. F Jeanmougin, JD Thompson, M Gouy, DG Higgins, TJ Gibson. Multiple sequence alignment with clustal X. Trends Biochem Sci 23:403–405, 1998.

56. M Levitt, M Gerstein. A unified statistical framework for sequence comparison and structure comparison. Proc Natl Acad Sci USA 95:5913–5920, 1998.

57. B Rost, C Sander. Prediction of protein structure at better than 70% accuracy. J Mol Biol 232:584–599, 1993.

58. R Sánchez, A Šali. Evaluation of comparative protein structure modeling by MODELLER-3. Proteins Suppl 1:50–58, 1997.

59. B Guenther, R Onrust, A Šali, M O'Donnell, J Kuriyan. Crystal structure of the δ' subunit of the clamp-loader complex of *E. coli* DNA polymerase III. Cell 91:335–345, 1997.

60. J Felsenstein. Confidence limits on phylogenies: An approach using the bootstrap. Evolution 39:783–791, 1985.

61. MS Johnson, JP Overington, A Šali. Knowledge-based protein modelling: Human plasma kallikrein and human neutrophil defensin. In: JJ Villafranca, ed. Current Research in Protein Chemistry: Techniques, Structure and Function. San Diego: Academic Press, 1990, pp 567–574.

62. TH Jones, S Thirup. Using known substructures in protein model building and crystallography. EMBO J 5:819–822, 1986.

63. WJ Browne, ACT North, DC Phillips, K Brew, TC Vanaman, RC Hill. A possible three-dimensional structure of bovine α-lactalbumin based on that of hen's egg-white lysozyme. J Mol Biol 42:65–86, 1969.

64. MJ Sutcliffe, I Haneef, D Carney, TL Blundell. Knowledge based modelling of homologous proteins, Part I: Three dimensional frameworks derived from the simultaneous superposition of multiple structures. Protein Eng 1:377–384, 1987.

65. CM Topham, A McLeod, F Eisenmenger, JP Overington, MS Johnson, TL Blundell. Fragment ranking in modelling of protein structure. Conformationally constrained environmental amino acid substitution tables. J Mol Biol 229:194–220, 1993.

66. MJ Sutcliffe, FRF Hayes, TL Blundell. Knowledge based modeling of homologous proteins, Part II: Rules for the conformation of substituted side-chains. Protein Eng 1:385–392, 1987.

67. N Srinivasan, TL Blundell. An evaluation of the performance of an automated procedure for comparative modelling of protein tertiary structure. Protein Eng 6:501–512, 1993.

68. BVB Reddy TL Blundell. Packing of secondary structural elements in proteins. Analysis and prediction of inter-helix distances. J Mol Biol 233:464–479, 1993.

69. R Unger, D Harel, S Wherland, JL Sussman. A 3-D building blocks approach to analyzing and predicting structure of proteins. Proteins 5:355–373, 1989.

70. M Claessens, EV Cutsem, I Lasters, S Wodak. Modelling the polypeptide backbone with "spare parts" from known protein structures. Protein Eng 4:335–345, 1989.

71. L Holm, C Sander. Database algorithm for generating protein backbone and side-chain co-ordinates from C_α trace: Application to model building and detection of co-ordinate errors. J Mol Biol 218:183–194, 1991.

72. D Bassolino-Klimas, RE Bruccoleri. Application of a directed conformational search for generating 3-D coordinates for protein structures from α-carbon coordinates. Proteins 14:465–474, 1992.

73. CWG van Gelder, FJJ Leusen, JAM Leunissen, JH Noordik. A molecular dynamics approach for the generation of complete protein structures from limited coordinate data. Proteins 18:174–185, 1994.

74. M Levitt. Accurate modeling of protein conformation by automatic segment matching. J Mol Biol 226:507–533, 1992.

75. G Vriend, C Sander, PFW Stouten. A novel search method for protein sequence–structure relations using property profiles. Protein Eng 7:23–29, 1994.

76. TF Havel, ME Snow. A new method for building protein conformations from sequence alignments with homologues of known structure. J Mol Biol 217:1–7, 1991.

77. TF Havel. Predicting the structure of the flavodoxin from Eschericia coli by homology modeling, distance geometry and molecular dynamics. Mol Simul 10:175–210, 1993.

78. A Šali, TL Blundell. Definition of general topological equivalence in protein structures: A procedure involving comparison of properties and relationships through simulated annealing and dynamic programming. J Mol Biol 212:403–428, 1990.

79. A Šali, JP Overington. Derivation of rules for comparative protein modeling from a database of protein structure alignments. Protein Sci 3:1582–1596, 1994.

80. BR Brooks, RE Bruccoleri, BD Olafson, DJ States, S Swaminathan, M Karplus. CHARMM: A program for macromolecular energy minimization and dynamics calculations. J Comput Chem 4:187–217, 1983.

81. AD MacKerell Jr, D Bashford, M Bellott, RL Dunbrack Jr, JD Evanseck, MJ Field, S Fischer, J Gao, H Guo, S Ha, D Joseph-McCarthy, L Kuchnir, K Kuczera, FTK Lau, C Mattos, S Michnick, T Ngo, DT Nguyen, B Prodhom, WE Reiher III, B Roux, M Schlenkrich, JC Smith, R Stote, J Straub, M Watanabe, J Wiorkiewicz-Kuczera, D Yin, M Karplus. All-atom empirical potential for molecular modeling and dynamics studies of proteins. J Phys Chem B 102:3586–3616, 1998.

82. W Braun, N. Go. Calculation of protein conformations by proton—proton distance constraints: A new efficient algorithm. J Mol Biol 186:611–626, 1985.

83. GM Clore, AT Brünger, M Karplus, AM Gronenborn. Application of molecular dynamics with interproton distance restraints to 3D protein structure determination. J Mol Biol 191: 523–551, 1986.

84. FE Cohen, ID Kuntz. Tertiary structure prediction. In: GD Fasman, ed. Prediction of Protein Structure and the Principles of Protein Conformation. New York: Plenum Press, 1989, pp 647–705.

85. A Aszódi, WR Taylor. Secondary structure formation in model polypeptide chains. Protein Eng 7:633–644, 1994.

86. WR Taylor, K Hatrick. Compensating changes in protein multiple sequence alignments. Protein Eng 7:341–348, 1994.

87. MJ Sippl. Calculation of conformational ensembles from potentials of mean force. An approach to the knowledge-based prediction of local structures in globular proteins. J Mol Biol 213:859–883, 1990.

88. MJ Sutcliffe, CM Dobson, RE Oswald. Solution structure of neuronal bungarotoxin determined by two-dimensional NMR spectroscopy: Calculation of tertiary structure using systematic homologous model building, dynamical simulated annealing, and restrained molecular dynamics. Biochemistry 31:2962–2970, 1992.

89. JP Boissel, WR Lee, SR Presnell, FE Cohen, HF Bunn. Erythropoietin structure–function relationships. Mutant proteins that test a model of tertiary structure. J Biol Chem 268:15983–15993, 1993.

90. C Sander, R Schneider. Database of homology-derived protein structures and the structural meaning of sequence alignment. Proteins 9:56–68, 1991.

91. M Mezei. Chameleon sequences in the PDB. Protein Eng 11:411–414, 1998.

92. HWT van Vlijmen, M Karplus. PDB-based protein loop prediction: Parameters for selection and methods for optimization. J Mol Biol 267:975–1001, 1997.

93. S Mosimann, R Meleshko, MNG James. A critical assessment of comparative molecular modeling of tertiary structures of proteins. Proteins 23:301–317, 1995.

94. C Chothia, AM Lesk. Canonical structures for the hypervariable regions of immunoglobulins. J Mol Biol 196:901–917, 1987.

95. J Moult, MNG James. An algorithm for determining the conformation of polypeptide segments in proteins by systematic search. Proteins 1:146–163, 1986.

96. RE Bruccoleri, M Karplus. Prediction of the folding of short polypeptide segments by uniform conformational sampling. Biopolymers 26:137–168, 1987.

97. RM Fine, H Wang, PS Shenkin, DL Yarmush, C Levinthal. Predicting antibody hypervariable loop conformations. II: Minimization and molecular dynamics studies of MCP603 from many randomly generated loop conformations. Proteins 1:342–362, 1986.

98. ACR Martin, JC Cheetham, AR Rees. Modeling antibody hypervariable loops: A combined algorithm. Proc Natl Acad Sci USA 86:9268–9272, 1989.

99. C Chothia, AM Lesk, M Levitt, AG Amit, RA Mariuzza, SEV Phillips, RJ Poljak. The predicted structure of immunoglobulin d1.3 and its comparison with the crystal structure. Science 233:755–758, 1986.

100. BL Sibanda, TL Blundell, JM Thornton. Conformation of β-hairpins in protein structures: A systematic classification with applications to modelling by homology, electron density fitting and protein engineering. J Mol Biol 206:759–777, 1989.

101. C Chothia, AM Lesk, A Tramontano, M Levitt, SJ Smith-Gill, G Air, S Sheriff, EA Padlan, D Davies, WR Tulip, PM Colman, S Spinelli, PM Alzari, RJ Poljak. Conformation of immunoglobulin hypervariable regions. Nature 342:877–883, 1989.

102. CS Ring, DG Kneller, R Langridge, FE Cohen. Taxonomy and conformational analysis of loops in proteins. J Mol Biol 224:685–699, 1992.

103. SD Rufino, LE Donate, LHJ Canard, TL Blundell. Predicting the conformational class of short and medium size loops connecting regular secondary structures: Application to comparative modeling. J Mol Biol 267:352–367, 1997.

104. B Oliva, PA Bates, E Querol, FX Aviles, MJE Sternberg. An automated classification of the structure of protein loops. J Mol Biol 266:814–830, 1997.

105. J Wojcik, J-P Mornon, J Chomilier. New efficient statistical sequence-dependent structure prediction of short to medium-sized protein loops based on an exhaustive loop classification. J Mol Biol 289:1469–1490, 1999.

106. K Fidelis, PS Stern, D Bacon, J Moult. Comparison of systematic search and database methods for constructing segments of protein structure. Protein Eng 7:953–960, 1994.

107. U Lessel, D Schomburg. Similarities between protein 3D structures. Protein Eng 7:1175–1187, 1994.

108. SA Benner, GH Gonnet, MA Cohen. Empirical and structural models for insertions and deletions in the divergent evolution of proteins. J Mol Biol 229:1065–1082, 1993.

109. MJ Dudek, HA Scheraga. Protein structure prediction using a combination of sequence homology and global energy minimization. I. Global energy minimization of surface loops. J Comput Chem 11:121–151, 1990.

110. BR Bruccoleri, M Karplus. Conformational sampling using high temperature molecular dynamics. Biopolymers 29:1847–1862, 1990.

111. R Abagyan, M Totrov. Biased probability Monte Carlo conformational searches and electrostatic calculations for peptides and proteins. J Mol Biol 235:983–1002, 1994.

112. CS Ring, FE Cohen. Conformational sampling of loop structures using genetic algorithm. Isr J Chem 34:245–252, 1994.

113. V Collura, J Higo, J Garnier. Modeling of protein loops by simulated annealing. Protein Sci 2:1502–1510, 1993.

114. J Higo, V Collura, J Garnier. Development of an extended simulated annealing method: Application to the modeling of complementary determining regions of immunoglobulins. Biopolymers 32:33–43, 1992.

115. Q Zheng, R Rosenfeld, S Vajda, C DeLisi. Determining protein loop conformation using scaling-relaxation techniques. Protein Sci 2:1242–1248, 1993.

116. Q Zheng, R Rosenfeld, C DeLisi, DJ Kyle. Multiple copy sampling in protein loop modeling: Computational efficiency and sensitivity to dihedral angle perturbations. Protein Sci 3:493–506, 1994.

117. D Rosenbach, R Rosenfeld. Simultaneous modeling of multiple loops in proteins. Protein Sci 4:496–505, 1995.

118. P Koehl, M Delarue. A self consistent mean field approach to simultaneous gap closure and side-chain positioning in protein homology modelling. Nature Struct Biol 2:163–170, 1995.

119. R Samudrala, J Moult. A graph-theoretic algorithm for comparative modeling of protein structure. J Mol Biol 279:287–302, 1998.

120. F Melo, E Feytmans. Assessing protein structures with a non-local atomic interaction energy. J Mol Biol 277:1141–1152, 1998.

121. M Vásquez. Modeling side-chain conformation. Curr Opin Struct Biol 6:217–221, 1996.

122. JM Thornton. Disulphide bridges in globular proteins. J Mol Biol 151:261–287, 1981.

123. S-H Jung, I Pastan, B Lee. Design of interchain disulfide bonds in the framework region of the Fv fragment of the monoclonal antibody B3. Proteins 19:35–47, 1994.

124. S Boresch, G Archontis, M Karplus. Free energy simulations: The meaning of the individual contributions from a component analysis. Proteins 20:25–33, 1994.

125. TP Straatsma, JA McCammon. Computational alchemy. Annu Rev Phys Chem 43:407–430, 1992.

126. PA Kollman. Free energy calculations: Applications to chemical and biochemical phenomena. Chem Rev 93:2395–2417, 1992.

127. Y-Y Shi, AE Mark, C Wang, F Huang, HJC Berendsen, WF Van Gunsteren. Can the stability of protein mutants be predicted by free energy calculations? Protein Eng 6:289–295, 1993.

128. H Schrauber, F Eisenhaber, P Argos. Rotamers: To be or not to be? An analysis of amino acid side-chain conformations in globular proteins. J Mol Biol 230:592–612, 1993.

129. MJ Bower, FE Cohen, RL Dunbrack Jr. Prediction of protein side-chain rotamers from a backbone-dependent rotamer library: A new homology modeling tool. J Mol Biol 267:1268–1282, 1997.

130. C Lee. Testing homology modeling on mutant proteins: Predicting structural and thermodynamic effects in the Ala98 → Val mutants of T4 lysozyme. Folding Des 1:1–12, 1995.

131. SY Chung, S Subbiah. A structural explanation for the twilight zone of protein sequence homology. Structure 4:1123–1127, 1996.

132. WA Lim, A Hodel, RT Sauer, FM Richards. The crystal structure of a mutant protein with altered but improved hydrophobic core packing. Proc Natl Acad Sci USA 91:423–427, 1994.

133. PB Harbury, B Tidor, PS Kim. Repacking proteins cores with backbone freedom: Structure prediction for coiled coils. Proc Natl Acad Sci USA 92:8408–8412, 1995.

134. ES Huang, P Koehl, M Levitt, RV Pappu, JW Ponder. Accuracy of side-chain prediction upon near-native protein backbones generated by ab initio folding methods. Proteins 33:204–217, 1998.

135. RL Dunbrack, M Karplus. Prediction of protein side-chain conformations from a backbone conformation dependent rotamer library. J Mol Biol 230:543–571, 1993.

136. MJ McGregor, SA Islam, MJE Sternberg. Analysis of the relationship between side-chain conformation and secondary structure in globular proteins. J Mol Biol 198:295–310, 1987.

137. SY Chung, S Subbiah. The use of side-chain packing methods in modeling bacteriophage repressor and cro proteins. Protein Sci 4:2300–2309, 1995.

138. RB Russell, GJ Barton. Structural features can be unconserved in proteins with similar folds. An analysis of side-chain to side-chain contacts secondary structure and accessibility. J Mol Biol 244:332–350, 1994.

139. CA Schiffer, JW Caldwell, PA Kollman, RM Stroud. Prediction of homologous protein structures based on conformational searches and energetics. Proteins 8:30–43, 1990.

140. D Cregut, J-P Liautard, L Chiche. Homology modeling of annexin I: Implicit solvation improves side-chain prediction and combination of evaluation criteria allows recognition of different types of conformational error. Protein Eng 7:1333–1344, 1994.

141. C Wilson, LM Gregoret, DA Agard. Modeling side-chain conformation for homologous proteins using an energy-based rotamer search. J Mol Biol 229:996–1006, 1993.

142. RJ Petrella, T Lazaridis, M Karplus. Protein sidechain conformer prediction: A test of the energy function. Folding Des 3:353–377, 1998.

143. V De Filippis, C Sander, G Vriend. Predicting local structural changes that result from point mutations. Protein Eng 7:1203–1208, 1994.

144. T Lazaridis, G Archontis, M Karplus. Enthalpic contribution to protein stability: Atom-based calculations and statistical mechanics. Adv Protein Chem 47:231–306, 1995.

145. P Koehl, M Delarue. Application of a self-consistent mean field theory to predict protein side-chains conformation and estimate their conformational entropy. J Mol Biol 239:249–275, 1994.

146. A Šali, L Potterton, F Yuan, H van Vlijmen, M Karplus. Evaluation of comparative protein structure modeling by MODELLER. Proteins 23:318–326, 1995.

147. B Rost. Twilight zone of protein sequence alignments. Protein Eng 12:85–94, 1999.

148. DJ Wales, HA Scheraga. Global optimization of clusters, crystals, and biomolecules. Science 285:1368–1372, 1999.

149. CM Dobson, A Šali, M Karplus. Protein folding: A perspective from theory and experiment. Angew Chem Int Ed 37:868–893, 1998.

150. KA Dill, HS Chan. From Levinthal to pathways to funnels. Nature Struct Biol 4:10–19, 1997.

151. ND Socci, JN Onuchic, PG Wolynes. Protein folding mechanisms and the multi-dimensional folding funnel. Protein Eng 32:136–158, 1998.

152. EI Shakhnovich. Folding nucleus: Specific or multiple? Insights from lattice models and experiments. Folding Des 3:R108–R111, 1998.

153. RA Friesner, JR Gunn. Computational studies of protein folding. Annu Rev Biophys Biomol Struct 25:315–342, 1996.

154. M Levitt, M Gerstein, E Huang, S Subbiah, J Tsai. Protein folding: The endgame. Annu Rev Biochem 66:1368–1372, 1999.

155. M Sippl. Who solved the protein folding problem? Structure 7:R81–R83, 1999.

156. MJE Sternberg, PA Bates, LA Kelley, RM MacCallum. Progress in protein structure prediction: Assessment of CASP3. Curr Opin Struct Biol 9:368–373, 1999.

157. P Koehl, M Levitt. A brighter future for protein structure prediction. Nature Struct Biol 6:108–111, 1999.

158. CB Anfinsen. Principles that govern the folding of protein chains. Science 181:223–238, 1973.

159. EM Boczko, CL Brooks III. First-principles calculation of the folding free energy of a three-helix bundle protein. Science 269:393–396, 1996.

160. Y Duan, PA Kollman. Pathways to a protein folding intermediate observed in a 1-microsecond simulation in aqueous solution. Science 282:740–744, 1998.

161. M Levitt. A simplified representation of protein conformations for rapid simulation of protein folding. J Mol Biol 104:59–107, 1976.

162. J Skolnick, A Kolinski. Simulations of the folding of a globular protein. Science 250:1121–1125, 1990.

163. JD Hirst, M Vieth, J Skolnick, CL Brooks III. Predicting leucine zipper structures from sequence. Protein Eng 9:657–662, 1996.

164. IK Roterman, MH Lambert, KD Gibson, HA Scheraga. A comparison of the CHARMM, AMBER and ECEPP potentials for peptides. II. PHI-PSI maps for N-acetyl alanine N′-methyl amide: Comparisons, contrasts and simple experimental tests. J Biomol Struct Dyn 7:421–453, 198.

165. WD Cornell, P Cieplak, CI Bayly, IR Gould, KM Merz Jr, DM Fergusson, DC Spellmeyer, DC Fox, JW Caldwell, PA Kollman. A second generation force field for the simulation of proteins and nucleic acids. J Am Chem Soc 117:5179–5197, 1995.

166. G Némethy, KD Gibson, KA Palmer, CN Yoon, G Paterlini, A Zagari, S Rumsey, HA Scheraga. Energy parameters in peptides. Improved geometrical parameters and non-bonded interactions for use in the ECEPP/3 algorithm, with application to proline-containing peptides. J Phys Chem 96:6472–6484, 1992.

167. S Vajda, M Sippl, J Novotny. Empirical potentials and functions for protein folding and binding. Curr Opin Struct Biol 7:228–228, 1997.

168. DJE Callaway. Proteins, 1994.

169. DA Hinds, M Levitt. Exploring conformational space with a simple lattice model for protein structure. J Mol Biol 243:668–682, 1994.

170. K Yue, KA Dill. Folding proteins with a simple energy function and extensive conformational searching. Protein Sci 5:254–261, 1996.

171. ES Huang, S Subbiah, M Levitt. Recognizing native folds by the arrangements of hydrophobic and polar residues. J Mol Biol 252:709–720, 1995.

172. DM Standley, JR Gunn, RA Friesner, AE McDermott. Tertiary structure prediction of mixed alpha/beta proteins via energy minimization. Proteins 33:240–252, 1998.

173. AR Ortiz, A Kolinski, J Skolnick. Fold assembly of small proteins using Monte Carlo simulations driven by restraints derived from multiple sequence alignments. J Mol Biol 277:419–448, 1998.

174. A Aszódi, MJ Gradwell, WR Taylor. Global fold determination from a small number of distance restraints. J Mol Biol 251:308–326, 1995.

175. C Mumenthaler, W Braun. Predicting the helix packing of globular proteins by self-correcting distance geometry. Protein Sci 4:863–871, 1995.

176. S Sun, PD Thomas, KA Dill. A simple protein folding algorithm using a binary code and secondary structure constraints. Protein Eng 8:769–778, 1995.

177. M Vasquez, G Nemethy, HA Scheraga. Conformational energy calculations on polypeptides and proteins. Chem Rev 94:2183–2239, 1994.

178. BJ Berne, JE Straub. Novel methods of sampling phase space in the simulation of biological systems. Curr Opin Struct Biol 7:181–189, 1997.

179. M Levitt. Protein folding by constrained energy minimization and molecular dynamics. J Mol Biol 170:723–764, 1983.

180. C Wilson, S Doniach. A computer model to dynamically simulate protein folding: Studies with crambin. Proteins 6:193–209, 1989.

181. DG Covell. Folding protein α-carbon chains into compact forms by Monte Carlo methods. Proteins 14:409–420, 1992.

182. A Monge, R Friesner, B Honig. An algorithm to generate low-resolution protein tertiary structures from knowledge of secondary structure. Proc Natl Acad Sci USA 91:5027–5029, 1994.

183. J Kostrowicki, HA Scheraga. Application of the diffusion equiation method for global optimization to oligopeptides. J Phys Chem 18:7442–7449, 1992.

184. T Dandekar, P Argos. Folding the main chain of small proteins with the genetic algorithm. J Mol Biol 236:844–861, 1994.

185. S Sun. Reduced representation model of protein structure prediction: Statistical potential and genetic algorithms. Protein Sci 2:762–785, 1994.

186. Y Cui, RS Chen, WH Wong. Protein folding simulation with genetic algorithm and supersecondary structure constraints. Proteins 31:247–257, 1998.

187. M Vasquez, HA Scheraga. Calculation of protein conformation by the build-up procedure. Application to bovine pancreatic trypsin inhibitor using limited simulated nuclear magnetic resonance data. J Biomol Struct Dyn 5:705–755, 1988.

188. KT Simons, C Kooperberg, E Huang, D Baker. Assembly of protein tertiary structures from

fragments with similar local sequences using simulated annealing and Bayesian scoring functions. J Mol Biol 268:209–225, 1997.

189. C Bystroff, D Baker. Prediction of local structure in proteins using a library of sequence-structure motifs. J Mol Biol 281:565–577, 1998.

190. B Rost. Protein structures sustain evolutionary drift. Folding Des 2:S19–S24, 1997.

191. ACR Martin, MW MacArthur, JM Thornton. Assessment of comparative modeling in CASP2. Proteins Suppl 1:14–28, 1997.

192. DH Ohlendorf. Accuracy of refined protein structures. II. Comparison of four independently refined models of human interleukin 1β. Acta Cryst D50:808–812, 1994.

193. GM Clore, MA Robien, AM Gronenborn. Exploring the limits of precision and accuracy of protein structures determined by nuclear magnetic resonance spectroscopy. J Mol Biol 231: 82–102, 1993.

194. HR Faber, BW Matthews. A mutant T4 lysozyme displays five different crystal conformations. Nature 348:263–266, 1990.

195. J Moult, T Hubbard, SH Bryant, K Fidelis, JT Pedersen. Critical assessment of methods of protein structure prediction (CASP): Round II. Proteins Suppl 1:2–6, 1997.

196. RL Dunbrack Jr, DL Gerloff, M Bower, X Chen, O Lichtarge, FE Cohen. Meeting review: The second meeting on the critical assessment of techniques for protein structure prediction (CASP2), Asilomar, CA, Dec 13–16, 1996. Folding Des 2:R27–R42, 1997.

197. BH Park, ES Huang, M Levitt. Factors affecting the ability of energy functions to discriminate correct from incorrect folds. J Mol Biol 266:831–846, 1997.

198. RA Laskowski, MW MacArthur, JM Thornton. Validation of protein models derived from experiment. Curr Opin Struct Biol 5:631–639, 1998.

199. KS Wilson, Z Dauter, VS Lamsin, M Walsh, S Wodak, J Richelle, J Pontius, A Vaguine, RWW Hooft, C Sander, G Vriend, JM Thornton, RA Laskowski, MW MacArthur, EJ Dodson, G Murshudov, TJ Oldfield, R Kaptein, JAC Rullman. Who checks the checkers? Four validation tools applied to eight atomic resolution structures. J Mol Biol 276:417–436, 1998.

200. RA Laskowski, MW McArthur, DS Moss, JM Thornton. PROCHECK: A program to check the stereochemical quality of protein structures. J Appl Cryst 26:283–291, 1993.

201. RA Laskowski, JAC Rullmann, MW MacArthur, R Kaptein, JM Thornton. AQUA and PROCHECK-NMR: Programs for checking the quality of protein structures solved by NMR. J Biomol NMR 8:477–486, 1996.

202. TJ Oldfield. SQUID: A program for the analysis and display of data from crystallography and molecular dynamics. J Mol Graphics 10:247–252, 1992.

203. RWW Hooft, C Sander, G Vriend. Verification of protein structures: Side-chain planarity. J Appl Crystallogr 29:714–716, 1996.

204. J Novotny, R Bruccoleri, M Karplus. An analysis of incorrectly folded protein models: Implications for structural predictions. J Mol Biol 177:787–818, 1984.

205. J Novotny, AA Rashin, RE Bruccoleri. Criteria that discriminate between native proteins and incorrectly folded models. Proteins 4:19–30, 1988.

206. LM Gregoret, FE Cohen. Effect of packing density on chain conformation. J Mol Biol 219: 109–122, 1991.

207. SH Bryant, LM Amzel. Correctly folded proteins make twice as many hydrophobic contacts. Int J Peptide Protein Res 29:46–52, 1987.

208. L Chiche, LM Gregoret, FE Cohen, PA Kollman. Protein model structure evaluation using the solvation free energy of folding. Proc Natl Acad Sci USA 87:3240–3244, 1990.

209. L Holm, C Sander. Evaluation of protein models by atomic solvation preference. J Mol Biol 225:93–105, 1992.

210. G Baumann, C Frömmel, C Sander. Polarity as a criterion in protein design. Protein Eng 2: 239–334, 1989.

211. J Vila, RL Williams, M Vásquez, HA Scheraga. Empirical solvation models can be used to

differentiate from near-native conformations of bovine pancreatic trypsin inhibitor. Proteins 10:199–218, 1991.

212. P Koehl, M Delarue. Polar and nonpolar atomic environments in the protein core: Implication for folding and binding. Proteins 20:264–278, 1994.

213. SH Bryant, CE Lawrence. The frequency of ion-pair substructures in proteins is quantitatively related to electrostatic potential: A statistical model for nonbonded interactions. Proteins 9: 108–119, 1991.

214. C Colovos, TO Yeates. Verification of protein structures: Patterns of non-bonded atomic interactions. Protein Sci 2:1511–1519, 1993.

215. J Pontius, J Richelle, SJ Wodak. Deviations from standard atomic volumes as a quality measure for protein crystal structures. J Mol Biol 264:121–136, 1996.

216. R Lüthy, JU Bowie, D Eisenberg. Assessment of protein models with three-dimensional profiles. Nature 356:83–85, 1992.

217. MJ Sippl. Recognition of errors in three-dimensional structures of proteins. Proteins 17:355–362, 1993.

218. CM Topham, N Srinivasan, CJ Thorpe, JP Overington, NA Kalsheker. Comparative modelling of major house dust mite allergen der p I: Structure validation using an extended environmental amino acid propensity table. Protein Eng 7:869–894, 1994.

219. G Wu, A Fiser, B ter Kuile, A Šali, M Müller. Convergent evolution of Trichomonas vaginalis lactate dehydrogenase from malate dehydrogenase. Proc Natl Acad Sci USA 96:6285–6290, 1999.

220. Y Sheng, A Šali, H Herzog, J Lahnstein, S Krilis. Modelling, expression and site-directed mutagenesis of human β_2-glycoprotein I: Identification of the major phospholipid binding site. J Immunol 157:3744–3751, 1996.

221. CS Ring, E Sun, JH McKerrow, GK Lee, PJ Rosenthal, ID Kuntz, FE Cohen. Structure-based inhibitor design by using protein models for the development of antiparasitic agents. Proc Natl Acad Sci USA 90:3583–3587, 1993.

222. LZ Xu, R Sánchez, A Šali, N Heintz. Ligand specificity of brain lipid binding protein. J Biol Chem 271:24711–24719, 1996.

223. A Šali, R Matsumoto, HP McNeil, M Karplus, RL Stevens. Three-dimensional models of four mouse mast cell chymases. Identification of proteoglycan-binding regions and protease-specific antigenic epitopes. J Biol Chem 268:9023–9034, 1933.

224. IA Vakser. Evaluation of GRAMM low-resolution docking methodology on the hemagglutinin-antibody complex. Proteins, Suppl 1:226–230, 1997.

225. R Matsumoto, A Šali, N Ghildyal, M Karplus, RL Stevens. Packaging of proteases and proteoglycans in the granules of mast cells and other hematopoietic cells. A cluster of histidines in mouse mast cell protease-7 regulates its binding to heparin serglycin proteoglycan. J Biol Chem 270:19524–19531, 1995.

226. PL Howell, SC Almo, MR Parsons, J Hajdu, GA Petsko. Structure determination of turkey egg-white lysozyme using Laue diffraction data. Acta Crystallogr B, 48:200–207, 1992.

227. S Modi, MJ Paine, MJ Sutcliffe, L-Y Lian, WU Primrose, CR Wolfe, GCK Roberts. A model for human cytochrome p_{450} 2d6 based on homology modeling and NMR studies of substrate binding. Biochemistry 35:4540–4550, 1996.

228. E Wolf, A Vassilev, Y Makino, A Šali, Y Nakatani, SK Burley. Crystal structure of a GCN5-related N-acetyltransferase: Serratia marcescens aminoglycoside 3-N-acetyltransferase. Cell 94:51–61, 1998.

229. CA Orengo, DT Jones, JM Thornton. Protein superfamilies and domain superfolds. Nature 372:631–634, 1994.

230. M Totrov, R Abagyan. Detailed ab initio prediction of lysozyme-antibody complex with 1.6 *AA* accuracy. Nature Struct Biol 1:259–263, 1994.

231. A Šali. 100,000 protein structures for the biologist. Nature Struct Biol 5:1029–1032, 1998.

232. MC Peitsch, MR Wilkins, L Tonella, JC Sánchez, RD Appel, DF Hochstrasser. Large-scale

protein modelling and integration with the SWISS-PROT and SWISS-2DPAGE databases: The example of Escherichia coli. Electrophoresis 18:498–501, 1997.

233. MC Peitsch. PROMOD and SWISS-MODEL: Internet-based tools for automated comparative protein modeling. Biochem Soc Trans 24:274–279, 1996.

234. R Sánchez, A Šali. ModBase: A database of comparative protein structure models. Bioinformatics 15:1060–1061, 1999.

235. HW Mewes, K Albermann, M Bähr, D Frishman, A Gleissner, J Hani, K Heumann, K Kleine, A Maierl, SG Oliver, F Pfeiffer, A Zollner. Overview of the yeast genome. Nature 387(6634 Suppl):7–65, 1997.

236. P Kraulis. MOLSCRIPT: A program to produce both detailed and schematic plots of protein structure. J Appl Crystallogr 24:946–950, 1991.

237. R Sayle, EJ Milner-White. RasMol: Biomolecular graphics for all. Trends Biochem Sci 20: 374, 1995.

238. DT Jones. GenTHREADER: An efficient and reliable protein fold recognition method for genomic sequences. J Mol Biol 287:797–815, 1999.

239. A Fiser, RKG Do, A Šali. Modeling of loops in protein structures. Prot Sci 9:1753–1773, 2000.

240. D Baker. A surprising simplicity to protein folding. Nature 405:39–42, 2000.

241. J Moult, T Hubbard, K Fidelis, JT Pedersen. Critical assessment of methods of protein structure prediction (CASP): round III. Proteins, Suppl 3:2–6, 1999.

242. SK Burley, SC Almo, JB Bonano, M Capel, MR Chance, T Gaasterland, D Lin, A Šali, FW Studier, S Swaminathan. Structural genomics: Beyond the human genome project. Nat Genet 23:151–157, 1999.

15

Bayesian Statistics in Molecular and Structural Biology

Roland L. Dunbrack, Jr.
Institute for Cancer Research, Fox Chase Cancer Center, Philadelphia, Pennsylvania

I. INTRODUCTION

Much of computational biophysics and biochemistry is aimed at making predictions of protein structure, dynamics, and function. Most prediction methods are at least in part knowledge-based rather than being derived entirely from the principles of physics. For instance, in comparative modeling of protein structure, each step in the process—from homolog identification and sequence–structure alignment to loop and side-chain modeling—is dominated by information derived from the protein sequence and structure databases (see Chapter 14). In molecular dynamics simulations, the potential energy function is based partly on conformational analysis of known peptide and protein structures and thermodynamic data (see Chapter 2).

The biophysical and biochemical data we have available are complex and of variable quality and density. We have sequences from many different kinds of organisms and sequences for proteins that are expressed in very different environments in a single organism or even a single cell. Some sequence families are very large, and some have only one known member. We have structures from many protein families, from NMR spectroscopy and from X-ray crystallography, some of high resolution and some not. These structures can be analyzed on the level of bond lengths and angles, or dihedral angles, and interatomic distances, or in terms of secondary, tertiary, and quaternary structure. Some structural features are very common, such as α-helices, and some are relatively rare, such as valine residues with backbone dihedral $\phi > 0°$.

The amount of data is also increasing. The nonredundant protein sequence database available from GenBank now contains over 500,000 amino acid sequences, and there are at least 30 completed genomes from all three kingdoms of life. The number of unique sequences in the Protein Databank of experimentally determined structures is now over 3000 [1]. The number of known protein folds is at least 400 [2–4]. In the next few years, the databanks will continue to grow exponentially as the *Drosophila, Arabidopsis,* corn, mouse, and human genomes are completed. Several institutions are planning projects to determine as many protein structures as possible in target genomes, such as yeast, *Mycoplasma genitalium,* and *E. coli.*

To gain the most predictive utility as well as conceptual understanding from the sequence and structure data available, careful *statistical* analysis will be required. The statistical methods needed must be robust to the variation in amounts and quality of data in different protein families and for structural features. They must be updatable as new data become available. And they should help us generate as much understanding of the determinants of protein sequence, structure, dynamics, and functional relationships as possible.

In recent years, Bayesian statistics has come to the forefront of research among professional statisticians because of its analytical power for complex models and its conceptual simplicity. In the natural and social sciences, Bayesian methods have also attracted significant attention, including the fields of genetics [5], epidemiology [6,7], medicine [8], high energy physics [9], astrophysics [10,11], hydrology [12], archaeology [13], and economics [14]. Bayesian statistics have been used in molecular and structural biology in sequence alignment [15], remote homolog detection [16,17], threading [18,19], NMR spectroscopy [20–24], X-ray structure determination [25–27], and side-chain conformational analysis [28]. Its counterpart, *frequentist statistics*, has in turn lost ground. To see why, we need to examine their basic conceptual frameworks. In the next section, I compare the Bayesian and frequentist viewpoints and discuss the reasons Bayesian methods are superior in both their conceptual components and their practical aspects. After that, I describe some important aspects of Bayesian statistics required for its application to protein sequence and structural data analysis. In the last section, I review several applications of Bayesian inference in molecular and structural biology to demonstrate its utility and conceptual simplicity. A useful introduction to Bayesian methods and their applications in machine learning and molecular biology can be found in the book by Baldi and Brunak [29].

II. BAYESIAN STATISTICS

A. Bayesian Probability Theory

The goal of any statistical analysis is inference concerning whether on the basis of available data, some hypothesis about the natural world is true. The hypothesis may consist of the value of some parameter or parameters, such as a physical constant or the exact proportion of an allelic variant in a human population, or the hypothesis may be a qualitative statement, such as "This protein adopts an α/β barrel fold" or "I am currently in Philadelphia." The parameters or hypothesis can be unobservable or as yet unobserved. How the data arise from the parameters is called the *model* for the system under study and may include estimates of experimental error as well as our best understanding of the physical process of the system.

Probability in Bayesian inference is interpreted as the degree of belief in the truth of a statement. The belief must be predicated on whatever knowledge of the system we possess. That is, probability is always conditional, $p(X|I)$, where X is a hypothesis, a statement, the result of an experiment, etc., and I is any information we have on the system. Bayesian probability statements are constructed to be consistent with common sense. This can often be expressed in terms of a fair bet. As an example, I might say that "the probability that it will rain tomorrow is 75%." This can be expressed as a bet: "I will bet $3 that it will rain tomorrow, if you give me $4 if it does and nothing if it does not." (If I bet $3 on 4 such days, I have spent $12; I expect to win back $4 on 3 of those days, or $12).

At the same time, I would not bet \$3 on no rain in return for \$4 if it does not rain. This behavior would be inconsistent, since if I did both simultaneously I would bet \$6 for a certain return of only \$4. Consistent betting would lead me to bet \$1 on no rain in return for \$4. It can be shown that for consistent betting behavior, only certain rules of probability are allowed, as follows.

There are two central rules of probability theory on which Bayesian inference is based [30]:

1. The sum rule: $p(A|I) + p(\overline{A}|I) = 1$
2. The product rule: $p(A, B|I) = p(A|B, I)p(B|I) = p(B|A, I)p(A|I)$

The first rule states that the probability of A plus the probability of not-A ($\overline{A}$) is equal to 1. The second rule states that the probability for the occurrence of two events is related to the probability of one of the events occurring multiplied by the conditional probability of the other event given the occurrence of the first event. We can drop the notation of conditioning on I as long as it is understood implicitly that all probabilities are conditional on the information we possess about the system. Dropping the I, we have the usual expression of Bayes' rule,

$$p(A, B) = p(A|B)p(B) = p(B|A)p(A) \tag{1}$$

For Bayesian inference, we are seeking the probability of a hypothesis H given the data D. This probability is denoted $p(H|D)$. It is very likely that we will want to compare different hypotheses, so we may want to compare $p(H_1|D)$ with $p(H_2|D)$. Because it is difficult to write down an expression for $p(H|D)$, we use Bayes' rule to invert the probability of $p(D|H)$ to obtain an expression for $p(H|D)$:

$$p(H|D) = \frac{p(D|H)p(H)}{p(D)} \tag{2}$$

In this expression, $p(H)$ is referred to as the *prior probability* of the hypothesis H. It is used to express any information we may have about the probability that the hypothesis H is true before we consider the new data D. $p(D|H)$ is the *likelihood* of the data given that the hypothesis H is true. It describes our view of how the data arise from whatever H says about the state of nature, including uncertainties in measurement and any physical theory we might have that relates the data to the hypothesis. $p(D)$ is the *marginal distribution* of the data D, and because it is a constant with respect to the parameters it is frequently considered only as a normalization factor in Eq. (2), so that $p(H|D) \propto p(D|H)p(H)$ up to a proportionality constant. If we have a set of hypotheses that are exclusive and exhaustive, i.e., one and only one must be true, then

$$p(D) = \sum_i p(D|H_i)p(H_i) \tag{2a}$$

$p(H|D)$ is the *posterior distribution*, which is, after all, what we are after. It gives the probability of the hypothesis after we consider the available data and our prior knowledge. With the normalization provided by the expression for $p(D)$, for an exhaustive set of hypotheses we have $\sum_i p(H_i|D) = 1$, which is what we would expect from the sum rule axiom described above.

As an example of likelihoods and prior and posterior probabilities, we give the following example borrowed from Gardner [31].* The chairman of a statistics department has decided to grant tenure to one of three junior faculty members, Dr. A, Dr. B, or Dr. C. Assistant professor A decides to ask the department's administrative assistant, Mr. Smith, if he knows who is being given tenure. Mr. Smith decides to have fun with Dr. A and says that he won't tell her who is being given tenure. Instead, he will tell her which of Dr. B and Dr. C is going to be *denied* tenure. Mr. Smith does not yet know who is and who is not getting tenure and tells Dr. A to come back the next day. In the meantime, he decides that if A is getting tenure he will flip a coin and will tell A that B is not getting tenure if the coin shows heads, and that C is not getting tenure if it shows tails. If B or C is getting tenure, he will tell A that either C or B, respectively, is not getting tenure

Dr. A comes back the next day, and Mr. Smith tells A that C is not getting tenure. A then figures that her chances of tenure have now risen to 50%. Mr. Smith believes he has not in fact changed A's knowledge concerning her tenure prospects. Who is correct?

For prior probabilities, if H_A is the statement "A gets tenure" and likewise for H_B and H_C, we have prior probabilities $p(H_A) = p(H_B) = p(H_C) = 1/3$. We can evaluate the likelihood of S, that Mr. Smith will say "C is not getting tenure," if H_A, H_B, or H_C is true:

$$p(S|H_A) = 0.5; \quad p(S|H_B) = 1; \quad p(S|H_C) = 0$$

So the posterior probability that A will get tenure based on Mr. Smith's statement is

$$p(H_A|S) = \frac{p(S|H_A)p(H_A)}{\sum_{r=A,B,C} p(S|H_r)p(H_r)} \tag{3}$$

$$= \frac{(1/2) \times (1/3)}{[(1/2 \times 1/3)] + [1 \times (1/3)] + [0 \times (1/3)]} = \frac{1}{3}$$

Mr. Smith has not in fact changed A's knowledge, because her prior and posterior probabilities of getting tenure are both 1/3. Mr. Smith has, however, changed A's knowledge of B's prospects of tenure, which are now 2/3. Another way to think about this problem is that before Mr. Smith has told A anything, the probability of B or C getting tenure was 2/3. After his statement, the same 2/3 total probability applies to B and C, but now C's probability of tenure is 0 and B's has therefore risen to 2/3. A's posterior probability is unchanged.

B. Bayesian Parameter Estimation

Most often the hypothesis H concerns the value of a continuous parameter, which is denoted θ. The data D are also usually observed values of some physical quantity (temperature, mass, dihedral angle, etc.) denoted y, usually a vector. y may be a continuous variable, but quite often it may be a discrete integer variable representing the counts of some event occurring, such as the number of heads in a sequence of coin flips. The expression for the posterior distribution for the parameter θ given the data y is now given as

---

* The original story concerned three prisoners to be executed, one of whom is pardoned.

$$p(\theta|y) = \frac{p(y|\theta)p(\theta)}{\int_{\Theta} p(y|\theta)p(\theta)d\theta} \tag{4}$$

where the sum over hypotheses in Eq. (2a) has been replaced with an integral over Θ, the allowed values of the parameter θ. For example, if θ is a proportion, then the integral is from 0 to 1. The prior probability $p(\theta)$ is a *prior probability distribution*, a continuous function of θ that expresses any knowledge we have about likely values for θ before we consider the new data at hand. This probability distribution may be flat over the entire range of θ. Such a prior is referred to as *uninformative* (as with A, B, C's equal prior probability of tenure above). Or we may have some prior data so that we know that θ is more likely to be in a particular range of Θ than outside that range, so the probability should be higher in that range than outside it. This is an *informative* prior. Under ordinary circumstances, the prior should be *proper*, that is, normalized to integrate to 1 with respect to θ: $\int_{\Theta} p(\theta)\,d\theta = 1$. The likelihood, by contrast, should be normalized with respect to integration over the data for given θ: $\int_{y} p(y|\theta)dy = 1$. If y is discrete (count data), then this is a sum over all possible y. The posterior distribution $p(\theta|y)$ is a continuous function of the parameter θ for known data y. It can be used for any inference on the value of θ desired. For instance, the probability that θ is greater than some value θ_0 is $P(\theta > \theta_0|y)$ = $\int_{\theta_0}^{\theta_{max}} p(\theta|y)$. Ninety-five percent confidence intervals can be calculated easily by choosing θ_1 and θ_2 such that $\int_{\theta_1}^{\theta_2} p(\theta|y)\,d\theta = 0.95$. This alternative notation for parameters and data is common in the statistical literature, and we will use it throughout the rest of this chapter.

As an example of a continuous variable, we can calculate a posterior distribution for the proportion of red and green balls in a barrel sitting before us. You are asked to bet whether it is the red balls or the green balls that are more plentiful in the barrel. You have \$10 in your possession, and you decide to bet in proportion to your certainty in your opinion on the red or green majority. You get a very quick look at the open barrel and estimate that the number of red and green balls look approximately equal. As a prior probability you would use a probability distribution that is a bell-shaped curve centered around θ_{red} of 0.5. (The appropriate mathematical form is a beta distribution, described below.) But your look was very brief, and you do not have a lot of confidence in this prior view. You are given a sample of balls from the barrel, which consists of 7 red balls and 3 green balls. It is clear you should bet on red, but how much? The likelihood function in this case is a binomial distribution. It gives the probability of n_{red} and n_{green} for a sample of $N = n_{red} + n_{green}$ balls, given θ_{red} and $\theta_{green} = 1 - \theta_{red}$. The results are shown in Figure 1. The dashed line gives the posterior distribution of θ_{red}, the solid line gives the prior distribution, and the dotted line gives the likelihood of the data. The figure shows that the posterior is a compromise between the prior and the likelihood. We can integrate the posterior distribution to decide the amount of the bet from

$$\int_{0.5}^{1} p(\theta_{red}|n_{red} = 7; n_{green} = 3)d\theta_{red} = 0.72$$

So we bet \$7.20 on red.

C. Frequentist Probability Theory

It should be noted that the Bayesian conception of probability of a hypothesis and the Bayesian procedure for assessing this probability is the original paradigm for probabilistic

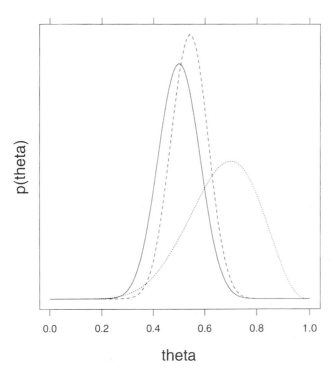

Figure 1   (——) Prior, (····) data (likelihood), and (- - -) posterior distributions for estimating a proportion of red and green balls in a barrel. The prior is based on a sample of 40 balls with 20 of them red, Beta (20,20). The likelihood is shown for a draw of 10 balls from the barrel, seven of them red. The posterior distribution is Beta (27,23).

inference. Both Bayes [32] and Laplace [33] used Bayes' rule to make probability statements for hypothetical statements given observed data, $p(H|D)$, by inverting the likelihood function, $p(D|H)$ [or, more accurately, determining $p(\theta|y)$ by inverting $p(y|\theta)$]. But in the nineteenth and early twentieth centuries the idea that a hypothesis could have a probability associated with it seemed too subjective to practitioners of the new science of statistics. They were also uncomfortable with the notion of prior probabilities, especially when applied to continuous parameters, $H = \theta$. To deal with these problems, they simply removed these elements of statistical reasoning.

In frequentist statistics, probability is instead a long-run relative occurrence of some event out of an infinite number of repeated trials, where the event is a possible outcome of the trial.* A hypothesis or parameter that expresses a state of nature cannot have a probability in frequentist statistics, because after an infinite number of experiments there can be no uncertainty in the parameter left. A hypothesis or parameter value is either

---

* The field of statistics as a separate discipline began in the early to mid nineteenth century among German, British, French, and Belgian social reformers, referred to as *statists*: i.e., those that were concerned with numbers related to the state, including crime rates, income distributions, etc. [34] The appeal of frequency-based interpretation of probability would have been natural in the study of large human populations.

always true or always false. Because a hypothesis cannot have a probability, frequentist probability is restricted to inference about data given a single hypothesis or a single value for the parameter. Data can be assigned probabilities because they can have varying values due to experimental error, even under a constant hypothesis.

The usual frequentist procedure comprises a number of steps [11]:

1. Define a procedure Π for selecting a hypothesis based on some characteristic(s) of the data, $S(y)$. S is called a *statistic*. Often this will be a *null hypothesis* that is deliberately at odds with $S(y)$. So, for instance, whereas the characteristic of the data might be the sample average $\bar{y}$ or the variance σ^2, the hypothesis H might be that the parameter θ (what y is measuring) has a value of 0.* Or if we are trying to show that two samples taken under different conditions are really different, we might define H to be the statement that the sample averages are in fact equal.
2. Because the data y are random, the statistics based on y, $S(y)$, are also random. For all possible data y (usually simulated) that can be predicted from H, calculate $p(S(y_{\text{sim}})|H)$, the probability distribution of the statistic S on simulated data y_{sim} given the truth of the hypothesis H. If H is the statement that $\theta = 0$, then y_{sim} might be generated by averaging samples of size N (a characteristic of the actual data) with variance $\sigma^2 = \sigma^2 (y_{\text{actual}})$ (yet another characteristic of the data).
3. Compare the statistic S calculated on the actual data y_{actual} to the distribution $p(S(y_{\text{sim}})|H)$. If $\int^{\infty} S(y_{\text{actual}}) \, p(S(y_{\text{sim}})|H) \, dS(y_{\text{sim}})$ is very small (<0.05, for instance), then reject the hypothesis H. If $S\,(\bar{y}_{\text{actual}})$ falls just to the right of 95% of the simulated $S(\bar{y})$, then we can conclude that $\theta > 0$ by rejecting the null hypothesis.†

If we do this over and over again, we will have done the right thing 95% of the time. Of course, we do not yet know the probability that, say, $\theta > 5$. For this purpose, confidence intervals for θ can be calculated that will contain the true value of θ 95% of the time, given many repetitions of the experiment. But frequentist confidence intervals are actually defined as the range of values for the data average that would arise 95% of the time from a single value of the parameter. That is, for normally distributed data,

$$\Pr\!\left(\bar{y} - \frac{\sigma}{\sqrt{N}} < \theta < \bar{y} + \frac{\sigma}{\sqrt{N}}\right) = 0.95$$

Bayesian confidence intervals, by contrast, are defined as the interval in which.

$$\Pr\!\left(\theta - \frac{\sigma}{\sqrt{N}} < \bar{y} < \theta + \frac{\sigma}{\sqrt{N}}\right) = 0.95$$

---

* Note that the bar above y in $\bar{y}$ in this section denotes the *average* of y. A bar over a statement or hypothesis A in the previous section was used to denote *not-A*. Both of these are standard notations in statistics and probability theory, respectively.

†The futility of frequentist testing of a point null hypothesis has been examined at length by Berger and Delampady [35].

The frequentist interval is often *interpreted* as if it were the Bayesian interval, but it is fundamentally defined by the probability of the data values given the parameter and not the probability of the parameter given the data.

D. Bayesian Methods Are Superior to Frequentist Methods

The Bayesian and frequentist theories can be considered two competing paradigms (in the sense of Kuhn [36]) of what the word "probability" means and how we should make inferential statements about hypotheses given data [37]. It is an unusual situation in the history of science that there should be two competing views on such basic notions, both of which have sizable entrenched camps of adherents, and that the controversy has lasted so long. Bayesian views fell out of favor until the book of Jeffreys in 1939 [38] and the work of Jaynes [30] and Savage [39] in the 1950s. Since then the Bayesian camp has increased tremendously in size. Because of some of the computational difficulties in evaluating posterior distributions, the advent of Markov chain Monte Carlo methods and fast computers has vastly increased the power and flexibility of Bayesian methods [40]. It is impossible to review the controversy in great depth here (see Refs. 9–11 and 37), but I will make some arguments in favor of the Bayesian view for molecular and structural biology.

1. The Bayesian View of Probability Corresponds to Most Scientists' Thinking

Bayesian inference is a process of taking current views of the probability that competing hypotheses about nature might be true and updating these beliefs in light of new data. It corresponds to the daily experience of scientists and nonscientists alike of judgments made on past experience and present facts. As Good [41] has argued, all animals must be at least informal Bayesians, even non-Bayesian statisticians, because they evaluate possibilities and outcomes of their actions in light of their probabilities for success. One might argue that sane dogs are *better* Bayesians than humans, given our propensity for foolish and destructive behavior, regardless of prior experience.

Bayesian methods are quite similar to most scientists' basic intuition about the nature of data and fundamental physical processes. The use of the prior distribution formalizes what we do naturally when we restrict our view of any parameter to a certain range of values or a certain number of outcomes and when we treat outliers with suspicion. If we get a value for an experiment that yields a parameter value that seems absurd compared to previous experiments, we are likely to repeat the experiment rather than publish the absurd result. Implicitly, we set the prior distribution of the parameters that are outside our range of interest or unphysical or unlikely on some ground to 0. The remaining probability density of the parameters must lie within the range we believe to be at least remotely possible and integrates to 1 within this range. Although basic statistics textbooks spend much time discussing null hypotheses, t-tests, F tests, etc., scientists rarely use this language in assessing data and inferred values of parameters based on the data. We do not usually ask whether the value of some parameter is exactly θ_0 or simply whether it is greater than some value θ_0. Rather we want to know what is the most likely range of values for the parameter, which is inherently a probability distribution over the parameter.

By contrast, the frequentist view is often contrary to common sense and common scientific practice. The classic example of this is the *stopping rule* problem [42]. If I am trying to decide whether a coin is fair, I can set up an experiment in two different ways. I can throw the coin some number of times N and then record the number of heads. Or I can keep throwing the coin until I observe some preset number of heads, n_{heads}. Suppose one person throws the coin $N = 100$ times and observes $n_{heads} = 55$ and another person throws the coin until 55 heads are observed and in fact it takes 100 throws. To a frequentist, these are quite different experiments and result in different inferences on the fairness of the coin. This is because the experiment that would be repeated an infinite number of times to obtain p_{heads} is different in each case (in the first case, a binomial distribution is used to make the inference; in the second case, a negative binomial is used). This seems absurd to most people, and to a Bayesian the probability of a fair coin is the same, because the data are in fact the same in each case.

The previous example also highlights the controversy of the subjectivity or objectivity of the two competing views of probability. Frequentist probability theory arose because of the apparent subjectivity of Bayesian prior distributions. But it replaced this subjectivity with a set of procedures based on test statistics. Inference is based on the probability of the test statistic, calculated on *hypothetical* data consistent with the null hypothesis, being as large as or larger than the test statistic calculated on the real data. But the stopping rule example exemplifies that the nature of the repeated experiment, the *sampling space*, is in itself arbitrary, as is the functional form of the test statistic. It is the emphasis on data *not* seen that makes frequentist statistics unintuitive and gives statistics much of its reputation of being difficult for beginning students.

2. Bayesian Methods Perform Better

In biology we are often faced with a set of situations that demand inference based on varying amounts of data. For instance, we may try to assign a protein fold to a new protein sequence based on very remote homology to some sequence in the Protein Data Bank (PDB). In some cases, we may have many sequences related to our target sequence from the genome projects, and the multiple alignment of sequences can help us establish a homology to the PDB sequence that may in fact be quite remote. In other cases, we may have only a small number of homologs in the sequence database, and establishing a remote homology may be quite difficult. Another example arises in side-chain conformational analysis. For some ranges of ϕ and ψ many examples exist of residues in the PDB, and reasonable estimates for the three χ_1 rotamer population can be calculated from the data. But the number of residues in some parts of the Ramachandran map is very small, but we would still like a good estimate of the three rotamer probabilities for protein folding simulations and comparative modeling [43].

Far from being a disadvantage, the need for a prior distribution can be a distinct advantage. First, it generally forces us to consider the full range of the possibilities for the hypothesis. Second, although in the absence of any prior information we can use uninformative priors, in some cases we may choose to use informative priors. Usually the prior corresponds in functional form to some number of data points, and we can choose to weight the prior in accordance with the strength of our belief in the prior information. Often a prior can be obtained by decoupling two or more parameters such that $p(\theta_1, \theta_2) \approx p(\theta_1)p(\theta_2)$. The two factors on the right-hand side might be obtained by pooling the data of y_1 regardless of y_2 and vice versa. In any case, the posterior is always a compromise between the prior and the likelihood. If the prior represents a larger sample than the data (our data set is quite small), then the prior will dominate the data. If the data sample is large, then the prior will have little effect on the posterior, which will resemble the likelihood. In cases where there are only a few data points, frequentist methods perform badly by sacrificing good short-term behavior in favor of good long-term behavior.

Another aspect in which Bayesian methods perform better than frequentist methods is in the treatment of nuisance parameters. Quite often there will be more than one parameter in the model but only one of the parameters is of interest. The other parameter is a *nuisance* parameter. If the parameter of interest is θ and the nuisance parameter is ϕ, then Bayesian inference on θ alone can be achieved by integrating the posterior distribution over ϕ. The *marginal probability* of θ is therefore

$$p(\theta|y) = \int_{\Phi} p(\theta, \phi|y)d\phi = \frac{\int_{\Phi} p(y|\theta, \phi)p(\theta, \phi)d\phi}{\int_{\Theta,\Phi} p(y|\theta, \phi)p(\theta, \phi)d\phi d\theta} \tag{5}$$

In frequentist statistics, by contrast, nuisance parameters are usually treated with point estimates, and inference on the parameter of interest is based on calculations with the nuisance parameter as a constant. This can result in large errors, because there may be considerable uncertainty in the value of the nuisance parameter.

In the next subsection, I describe how the basic elements of Bayesian analysis are formulated mathematically. I also describe the methods for deriving posterior distributions from the model, either in terms of conjugate prior likelihood forms or in terms of simulation using Markov chain Monte Carlo (MCMC) methods. The utility of Bayesian methods has expanded greatly in recent years because of the development of MCMC methods and fast computers. I also describe the basics of hierarchical and mixture models.

E. Setting Up Bayesian Models

Bayesian inference has three major components [44]:

1. Setting up the probability model for the data and parameters of the system under study. This entails defining prior distributions for all relevant parameters and a likelihood function for the data given the parameters.
2. Calculating the posterior distribution for the parameters given existing data. This calculation can sometimes be performed analytically, but in the general case it is performed by simulation.
3. Evaluating the model in terms of how well the model fits the data, including the use of posterior predictive simulations to determine whether data predicted from the posterior distribution resemble the data that generated them and look physically reasonable. Overfitting the data will produce unrealistic posterior predictive distributions.

The complexity of information that can be incorporated into the model gives Bayesian statistics much of its power.

We need a mathematical representation of our prior knowledge and a likelihood function to establish a model for any system to be analyzed. The calculation of the posterior distribution can be performed analytically in some cases or by simulation, which I describe later. For an analytical solution it is usually the case that we need prior distribution forms that are *conjugate* to the likelihood function. If the prior distribution and the likelihood function are conjugate, then by definition the posterior distribution will have the same mathematical form as the prior distribution. A description of the important types follows.

1. Binomial and Multinomial Models

Any data set that consists of discrete classification into outcomes or descriptors is treated with a binomial (two outcomes) or multinomial (three or more outcomes) likelihood function. For example, if we have y successes from n experiments, e.g., y heads from n tosses of a coin or y green balls from a barrel filled with red and green balls in unknown proportions, the likelihood function is a binomial distribution:

$$p(y|\theta) = \text{Bin}(y|n,\theta) = \frac{n!}{y!(n-y)!}\theta^y(1-\theta)^{n-y} \tag{6}$$

An informative conjugate prior distribution can be formulated in terms of a beta distribution:

$$p(\theta) = \text{Beta}(\theta|\alpha, \beta) = \frac{\Gamma(\alpha+\beta)}{\Gamma(\alpha)\Gamma(\beta)}\theta^{\alpha-1}(1-\theta)^{\beta-1} \tag{7}$$

We can think of the beta distribution as the likelihood of α prior successes and β failures out of $\alpha + \beta$ experiments. The Γ functions in front serve as a normalization constant, so that $\int_0^1 p(\theta)\,d\theta = 1$. Note that for an integer, $\Gamma(x+1) = x!$ The posterior distribution that results from multiplying together the right-hand sides of Eqs. (2) and (3) is also a beta distribution:

$$p(\theta|y) = p(y|\theta)p(\theta)/p(y) = \text{Beta}(\theta|\alpha+y, \beta+n-y)$$
$$= \frac{\Gamma(n+\alpha+\beta)}{\Gamma(\alpha+y)\Gamma(\beta+n-y)}\theta^{\alpha+y-1}\theta^{\beta+n-y-1} \tag{8}$$

We can see that the prior and posterior distributions have the same mathematical forms, as is required of conjugate functions. Also, we have an analytical form for the posterior, which is exact under the assumptions made in the model.

It is tempting to use Eq. (7) to derive Eq. (6), because they have similar forms given the relationship of the Γ function to the factorial. But the binomial and the beta distribution are not normalized in the same way. The beta is normalized over the values of θ, whereas the binomial is normalized over the counts, y given n. That is,

$$\int_0^1 \text{Beta}(\theta: \alpha, \beta)d\theta = 1 \tag{9a}$$

$$\sum_{y=0}^{n} \text{Bin}(y: n, \theta) = 1 \tag{9b}$$

It should be noted that the expected value of θ in a beta distribution is $\alpha/(\alpha+\beta)$, and the θ with maximum probability (the mode) is $(\alpha-1)/(\alpha+\beta-2)$. In terms of the expectation values, α and β behave as a total of $\alpha + \beta$ counts even though the exponents are $\alpha - 1$ and $\beta - 1$. The beta distribution is defined such that it is the expectation values and not the modes that correspond to counts of α and β.

If there are more than two outcomes, we can use the multinomial distribution for the likelihood:

$$p(y|\theta) = \left(\frac{n!}{\prod_i y_i!}\right) \prod_i \theta_i^{y_i} \tag{10}$$

where $\sum_{i=1}^m y_i = n$ for m possible outcomes of n experiments. For example, if a barrel contains red, blue, and green balls, then $m = 3$. We might make $n = 100$ draws from the barrel (returning the balls to the barrel) and get 30 green, 50 red, and 20 blue balls. The *conjugate* prior distribution is the Dirichlet distribution,

$$p(\theta) = \text{Dirichlet}(\{x_i\}) = \left(\frac{\Gamma(x_0)}{\prod_i \Gamma(x_i)}\right) \prod_{i=1}^m \theta_i^{x_i-1} \tag{11}$$

where $x_0 = \sum_{i=1} x_i$. The Dirichlet distribution can be considered a generalization of the multinomial distribution, where the counts are no longer restricted to be integers. The values of the *hyperparameters* x_i that define the prior distribution can be thought of as estimated counts for each of the m outcomes in some sample of size $x_0 = \sum_{i=1}^m x_i$. The total number of prior counts, x_0, can be scaled to any value to alter the dependence of the posterior distribution on the prior distribution. The larger x_0 is, the more precise the prior distribution is and the closer the posterior density is to values near $\theta_i = x_i/x_0$. The posterior distribution that results from Eqs. (10) and (11) is also Dirichlet with parameters $x_i + y_i$, i.e.,

$$p(\theta|y) = \left(\frac{\Gamma(x_0 + y_0)}{\prod_i \Gamma(x_i + y_i)}\right) \prod_i \theta_i^{x_i+y_i-1} \tag{12}$$

Again, the use of the conjugate prior distribution results in the analytical form of the posterior distribution and therefore also simple expressions for the expectation values for the θ_i, their variances, covariances, and modes:

$$E(\theta_i) = \frac{x_i + y_i}{x_0 + y_0} \tag{13a}$$

$$\text{mode}(\theta_i) = \frac{x_i + y_i - 1}{x_0 + y_0 - m} \tag{13b}$$

$$\text{var}(\theta_i) = \frac{E(\theta_i)[1 - E(\theta_i)]}{x_0 + y_0 + 1} \tag{13c}$$

$$\text{cov}(\theta_i, \theta_j) = -\frac{E(\theta_i) E(\theta_j)}{x_0 + y_0 + 1} \tag{13d}$$

A noninformative prior distribution could be formed by setting each x_i to 1.

2. Normal Models

The normal model can take a variety of forms depending on the choice of noninformative or informative prior distributions and on whether the variance is assumed to be a constant or is given its own prior distribution. And of course, the data could represent a single variable or could be multidimensional. Rather than describing each of the possible combinations, I give only the univariate normal case with informative priors on both the mean and variance. In this case, the likelihood for data y given the values of the parameters that comprise θ, μ (the mean), and σ^2 (the variance) is given by the familiar exponential

$$p(y|\mu, \sigma) = \frac{1}{\sigma\sqrt{2\pi}} \exp\left(\frac{-(y - \mu)^2}{2\sigma^2}\right) \tag{14}$$

This expression is valid for a single observation y. For multiple observations, we derive $p(y|\theta)$ from the fact that $p(y|\mu, \sigma^2) = \Pi_i p(y_i|\mu, \sigma^2)$. The result is that the likelihood is also normal with the average value of y, $\bar{y}$, substituted for y and σ^2/n substituted for σ^2 in Eq. (14). The conjugate prior distribution for Eq. (14) is

$$p(\mu, \sigma^2) = p(\mu|\sigma^2)p(\sigma^2)$$

$$= \frac{1}{\sigma\sqrt{2\pi}} \exp\left(\frac{-\kappa_0(\mu - \mu_0)^2}{2\sigma^2}\right) \times \left(\frac{2^{v_0/2}}{\Gamma(v_0/2)}\right) \sigma_0^{v_0/2} \sigma^{2-v_0} \exp\left(\frac{-v_0\sigma_0^2}{2\sigma^2}\right) \tag{15}$$

where the prior for μ is a normal distribution dependent on the value of σ^2 as well as two hyperparameters, the mean μ_0 and the scale κ_0, while the prior for σ^2 is a scaled inverse χ^2 distribution [to the right of the $\times$ sign in Eq. (15)] with two hyperparameters, the scale σ_0 and degrees of freedom v_0. The posterior distribution that results from Eqs. (14) and (15) has the same form as the prior (because the prior is conjugate to the likelihood), so that [44]

$$p(\mu, \sigma^2|y) = \frac{1}{\sigma\sqrt{2\pi}} \exp\left(\frac{-\kappa_n(\mu - \mu_n)^2}{2\sigma^2}\right) \times \frac{2^{v_n/2}}{\Gamma(v_n/2)} \sigma_n^{v_n/2} \sigma^{2-v_n} \exp\left(\frac{-v_n\sigma_n^2}{2\sigma^2}\right) \tag{16}$$

where

$$\mu_n = \frac{1}{\kappa_n}(\kappa_0\mu_0 + n\bar{y}) \tag{17a}$$

$$\kappa_n = \kappa_0 + n, \qquad v_n = v_0 + n \tag{17b}$$

$$v_n\sigma_n^2 = v_0\sigma_0^2 + (n - 1)s^2 + \frac{n\kappa_0}{\kappa_n}(\bar{y} - \mu_0)^2 \tag{17c}$$

$$s^2 = \frac{1}{n - 1} \sum_{i=1}^{n} (y_i - \bar{y})^2 \tag{17d}$$

Although this is a complicated expression, the results can be given a simple interpretation. The data sample size is n, whereas the prior sample size is κ_0, and therefore μ_n is the weighted average of the prior ''data'' and the actual data. σ_n^2 is the weighted average of the prior variance (σ_0^2), the data variance (s^2), and a term from the difference in the prior and data sample means. The weight of the prior variance is represented by the degree of freedom, ν_0, while the weight of the data variance is $n - 1$.

F. Simulation via Markov Chain Monte Carlo Methods

In practice, it may not be possible to use conjugate prior and likelihood functions that result in analytical posterior distributions, or the distributions may be so complicated that the posterior cannot be calculated as a function of the entire parameter space. In either case, statistical inference can proceed only if random values of the parameters can be drawn from the full posterior distribution:

$$p(\theta|y) = \frac{p(y|\theta)p(\theta)}{\int_{\Theta} p(y|\theta)p(\theta)d\theta} \tag{18}$$

We can also calculate expected values for any function of the parameters:

$$E[f(\theta|y)] = \frac{\int_{\Theta} f(\theta)p(y|\theta)p(\theta)d\theta}{\int_{\Theta} p(y|\theta)p(\theta)d\theta} \tag{19}$$

If we could draw directly from the posterior distribution, then we could plot $p(\theta|y)$ from a histogram of the draws on θ. Similarly, we could calculate the expectation value of any function of the parameters by making random draws of θ from the posterior distribution and calculating

$$E[f(\theta)] \approx \frac{1}{n} \sum_{t=1}^{n} f(\theta_t) \tag{20}$$

In some cases, we may not be able to draw directly from the posterior distribution. The difficulty lies in calculating the denominator of Eq. (18), the marginal data distribution $p(y)$. But usually we can evaluate the ratio of the probabilities of two values for the parameters, $p(\theta_t|y)/p(\theta_u|y)$, because the denominator in Eq. (18) cancels out in the ratio. The Markov chain Monte Carlo method [40] proceeds by generating draws from some distribution of the parameters, referred to as the *proposal* distribution, such that the new draw depends only on the value of the old draw, i.e., some function $q(\theta_t|\theta_{t-1})$. We accept the new draw with probability

$$\pi(\theta_t|\theta_{t-1}) = \min\left(1, \frac{p(\theta_t|y)q(\theta_{t-1}|\theta_t)}{p(\theta_{t-1}|y)q(\theta_t|\theta_{t-1})}\right) \tag{21}$$

and otherwise we set $\theta_t = \theta_{t-1}$. This is the Metropolis–Hastings method, first proposed by Metropolis and Ulam [45] in the context of equation of state calculations [46] and further developed by Hastings [47]. This scheme can be shown to result in a stationary distribution that asymptotically approaches the posterior distribution.

Several variations of this method go under different names. The Metropolis algorithm uses only symmetrical proposal distributions such that $q(\theta_t|\theta_{t-1}) = q(\theta_{t-1}|\theta_t)$. The expression for $\pi(\theta_t|\theta_{t-1})$ reduces to

$$\pi(\theta_t|\theta_{t-1}) = \min\left(1, \frac{p(\theta_t|y)}{p(\theta_{t-1}|y)}\right) \tag{22}$$

This is the form that chemists and physicists are most accustomed to. The probabilities are calculated from the Boltzmann equation and the energy difference between state t and state $t - 1$. Because we are using a ratio of probabilities, the normalization factor, i.e., the partition function, drops out of the equation. Another variant when θ is multidimensional (which it usually is) is to update one component at a time. We define $\theta_{t,-i} = \{\theta_{t,1}, \theta_{t,2}, \ldots, \theta_{t,i-1}, \theta_{t-1,i+1}, \ldots, \theta_{t-1,m}\}$, where m is the number of components in θ. So $\theta_{t,-i}$ contains all of the components except $\theta_{.,i}$ and all the components that precede the ith component have been updated in step t, while the components that follow have not yet been updated. The m components are updated one at a time with this probability:

$$\pi(\theta_{t,i}|\theta_{t,-i}) = \min\left(1, \frac{p(\theta_{t,i}|y, \theta_{t,-i})q(\theta_{t-1,i}|\theta_{t,i}, \theta_{t,-i})}{p(\theta_{t-1,i}|y, \theta_{t,-i})q(\theta_{t,i}|\theta_{t-1,i}, \theta_{t,-i})}\right) \tag{23}$$

If draws can be made from the posterior distribution for each component conditional on values for the others, i.e., from $p(\theta_{t,i}|y, \theta_{t,-i})$, then this conditional posterior distribution can be used as the proposal distribution. In this case, the probability in Eq. (23) is always 1, and all draws are accepted. This is referred to as Gibbs sampling and is the most common form of MCMC used in statistical analysis.

G. Mixture Models

Mixture models have come up frequently in Bayesian statistical analysis in molecular and structural biology [16,28] as described below, so a description is useful here. Mixture models can be used when simple forms such as the exponential or Dirichlet function alone do not describe the data well. This is usually the case for a multimodal data distribution (as might be evident from a histogram of the data), when clearly a single Gaussian function will not suffice. A mixture is a sum of simple forms for the likelihood:

$$p(y|\theta) = \sum_{i=1}^{n} q_i p(y|\theta_i) \tag{24}$$

where $\sum_{i=1}^{n} q_i = 1$ for the n components of the mixture. For instance, if the terms in Eq. (24) are normal, then each term is of the form (for a single data point y_j)

$$p(y_j|\theta_i) = \frac{1}{\sqrt{2\pi}\sigma_i} \exp\left(-\frac{(y_j - \mu_i)^2}{2\sigma_i^2}\right) \tag{25}$$

so each $\theta_i = \{\mu_i, \sigma_i^2\}$.

Maximum likelihood methods used in classical statistics are not valid to estimate the θ's or the q's. Bayesian methods have only become possible with the development of Gibbs sampling methods described above, because to form the likelihood for a full data set entails the product of many sums of the form of Eq. (24):

$$p(\{y_1, y_2, \ldots, y_N\}|\theta) = \prod_{j=1}^{N}\left(\sum_{i=1}^{n} q_i p(y_j|\theta_i)\right) \tag{26}$$

Because we are dealing with count data and proportions for the values q_i, the appropriate conjugate prior distribution for the q's is the Dirichlet distribution,

$$p(q_1, q_2, \ldots, q_k) = \text{Dirichlet}(\alpha_1, \alpha_2, \ldots, \alpha_k)$$

where the α's are prior counts for the components of the mixture. A simplification is to associate each data point with a single component, usually the component with the nearest location (i.e., μ_i). In this case, it is necessary to associate with each data point y_j a variable c_j that denotes the component that y_j belongs to. These variables c_j are unknown and are therefore called "missing data." Equation (26) now simplifies to

$$p(\{y_1, y_2, \ldots, y_N\} | \theta) = \sum_{j=1}^{N} p(y_j | \theta_{c_j}) \tag{27}$$

A straightforward Gibbs sampling strategy when the number of components is known (or fixed) is as follows [48].

Step 1. From a histogram of the data, partition the data into N components, each roughly corresponding to a mode of the data distribution. This defines the c_j. Set the parameters for prior distributions on the θ parameters that are conjugate to the likelihoods. For the normal distribution the priors are defined in Eq. (15), so the full prior for the n components is

$$p(\theta_1, \theta_2, \ldots, \theta_k) = \prod_{i=1}^{n} N(\mu_{0i}, \sigma_{0i}^2/\kappa_0) \text{ Inv } \chi^2(\nu_{0i}, \sigma_{0i}^2) \tag{28}$$

The prior hyperparameters, μ_{0i}, etc., can be estimated from the data assigned to each component. First define $N_i = \sum_{j=1}^{N} I(c_j = i)$, where $I(c_j = i) = 1$ if $c_j = i$ and is 0 otherwise. Then, for instance, the prior hyperparameters for the mean values are defined by

$$\mu_{0i} = \frac{1}{N_i} \sum_{j=1}^{N} I(c_j = i) y_j \tag{29}$$

The parameters of the Dirichlet prior for the q's should be proportional to the counts for each component in this preliminary data analysis. So we now have a collection of prior parameters $\{\theta_{0i} = (\mu_{0i}, \kappa_{0i}, \sigma_{0i}^2, \nu_{0i})\}$ and a preliminary assignment of each data point to a component, $\{c_j\}$, and therefore the preliminary number of data points for each component, $\{N_i\}$.

Step 2. Draw a value for each $\theta_i = \{\mu_i, \sigma_i^2\}$ from the normal posterior distribution for N_i data points with average $\bar{y}_i$,

$$p(\theta_i | \{y_i\}) = N(\mu_{N_i}, \sigma_{N_i}^2) \text{ Inv } \chi^2(\nu_{N_i}, \sigma_{N_i}^2) \tag{30}$$

where [as in Eq. (17)]

$$\mu_{N_i} = \frac{1}{\kappa_{N_i}} (\kappa_{0i} \mu_{0i} + N_i \bar{y}_i) \tag{31a}$$

$$\kappa_{N_i} = \kappa_{0i} + N_i, \qquad \nu_{N_i} = \nu_{0i} + N_i \tag{31b}$$

$$v_{N_i}\sigma_{N_i}^2 = v_{0i}\sigma_{0i}^2 + (N_i - 1)s_i^2 + \frac{N_i\kappa_{0i}}{\kappa_{N_i}}(\bar{y}_i - \mu_{0i})^2 \tag{31c}$$

$$s_i^2 = \frac{1}{N_i - 1}\sum_{k=1}^{N}(y_k - \bar{y}_i)^2 \tag{31d}$$

Draw $(q_1, q_2, \ldots, q_n)$ from Dirichlet $(\alpha_1 + N_1, \alpha_2 + N_2, \ldots, \alpha_n + N_n)$, which is the posterior distribution with prior counts α_i and data counts N_i.

Step 3.   Reset the c_j by drawing a random number u_j between 0 and 1 for each c_j and set c_j to i if

$$\frac{1}{Z}\sum_{i=1}^{i'-1}q_i p(y_j|\theta_i) < u_j \leq \frac{1}{Z}\sum_{i=i'+1}^{n}q_i p(y_j|\theta_i) \tag{32}$$

where $Z = \sum_{i=1}^{n}q_i p(y_j|\theta_i)$ is the normalization factor.

Step 4.   Sum up the N_i and calculate the averages $\bar{y}_i$ from the data and the values of c_j.

Step 5.   Repeat steps 2–4 until convergence.

The number of components necessary can usually be judged from the data, but the appropriateness of a particular value of n can be judged by comparing different values of n and calculating the entropy distance, or Kullback–Leibler divergence,

$$ED(g, h) = \int g(x) \ln \frac{g(x)}{h(x)} dx \tag{33}$$

where, for instance, g might be a three-component model and h might be a two-component model. If $ED(g, h) > 0$, then the model g is better than the model h.

H. Explanatory Variables

There is some confusion in using Bayes' rule on what are sometimes called *explanatory variables*. As an example, we can try to use Bayesian statistics to derive the probabilities of each secondary structure type for each amino acid type, that is $p(\mu|r)$, where μ is α, β, or γ (for coil) secondary structures and r is one of the 20 amino acids. It is tempting to write $p(\mu|r) = p(r|\mu)p(\mu)/p(r)$ using Bayes' rule. This expression is, of course, correct and can be used on PDB data to relate these probabilities. But this is not Bayesian statistics, which relate parameters that represent underlying properties with (limited) data that are manifestations of those parameters in some way. In this case, the parameters we are after are $\theta_\mu(r) \equiv p(\mu|r)$. The data from the PDB are in the form of counts for $y_\mu(r)$, the number of amino acids of type r in the PDB that have secondary structure μ. There are 60 such numbers (20 amino acid types $\times$ 3 secondary structure types). We then have for *each* amino acid type a Bayesian expression for the posterior distribution for the values of $\theta_\mu(r)$:

$$p(\theta(r)|y(r)) \propto p(y(r)|\theta(r))p(\theta(r)) \tag{34}$$

where θ and y are vectors of three components α, β, and γ. The prior is a Dirichlet distribution with some number of counts for the three secondary structure types for amino acid type r, i.e., Dirichlet $(n_\alpha(r), n_\beta(r), n_\gamma(r))$. We could choose the three $n_\mu(r)$ to be equal to some small number, say 10. Or we could set them equal to $100 \times p_\mu$, where p_μ is the proportion of each secondary structure type in the PDB. 100 would be the sample size for the prior. Because 100 is small compared to the data sample size from the PDB, the prior will have only a small influence on the posterior distribution. The likelihood is Multinom $(y_\alpha(r), y_\beta(r), y_\gamma(r): \theta_\alpha(r), \theta_\beta(r), \theta_\gamma(r))$. The posterior is Dirichlet($\theta_\alpha(r), \theta_\beta(r)$, $\theta_\gamma(r): y_\alpha(r) + n_\alpha(r), y_\beta(r) + n_\beta(r), y_\gamma(r) + n_\gamma(r))$.

III. APPLICATIONS IN MOLECULAR BIOLOGY

A. Dirichlet Mixture Priors for Sequence Profiles

One of the most important goals in bioinformatics is the identification and sequence alignment of proteins that are very distantly related by descent from a common ancestor. Such remote homolog detection methods can proceed in the absence of any structural information for the proteins involved, relying instead on multiple sequence alignments of protein families. Profile methods have a long history [49,50] and are based on the idea that even very distantly related proteins have particular patterns of hydrophobic and polar amino acids and some well-conserved amino acids at certain positions in the sequence. A profile consists of the probabilities of the 20 amino acids at each position in a multiple sequence alignment of a protein family. In some cases, a ''gap'' amino acid is also defined to indicate the probability of a gap at each position in the alignment. Physical information derived from the protein structure, such as hydrophobicity, can also be included [50].

One limiting factor in some cases is that there may be only a few sequences that are obviously related, and a profile built from their multiple alignment is not likely to be very accurate on statistical grounds of a small sample. For instance, for a buried residue, we might have only Val at a particular position in all of the known sequences. The question arises whether this position in an alignment of many more family members would be likely to have Ile and Leu as well as Val. If the conserved residue is Pro, what is the likelihood that in other related sequences this residue might be different? Bayesian statistics is ideally suited to treat the different situations we might encounter in sequence analysis within a single consistent framework. The key is to identify prior distributions for profiles that when combined with the data (the known relatives in the family) will produce the best estimate possible of the profile that might be derived from the full sequence family from a set of known sequences of any size.

A prior distribution for sequence profiles can be derived from mixtures of Dirichlet distributions [16,51–54]. The idea is simple: Each position in a multiple alignment represents one of a limited number of possible distributions that reflect the important physical forces that determine protein structure and function. In certain core positions, we expect to get a distribution restricted to Val, Ile, Met, and Leu. Other core positions may include these amino acids plus the large hydrophobic aromatic amino acids Phe and Trp. There will also be positions that are completely conserved, including catalytic residues (often Lys, Glu, Asp, Arg, Ser, and other polar amino acids) and Gly and Pro residues that are important in achieving certain backbone conformations in coil regions. Cys residues that form disulfide bonds or coordinate metal ions are also usually well conserved.

A prior distribution of the probabilities of the 20 amino acids at a particular position in a multiple alignment can be represented by a Dirichlet distribution, described in Section II.E. That is, it is an expression of the values of θ_r, the probabilities of each residue type r, where r ranges from 1 to 20, and $\sum_{r=1}^{20} \theta_r = 1$:

$$p(\theta_1, \theta_2, \ldots, \theta_{20} | \alpha_1, \alpha_2, \ldots, \alpha_{20}) = \text{Dirichlet } (\theta_1, \theta_2, \ldots, \theta_{20} | \alpha_1, \alpha_2, \ldots, \alpha_{20})$$

$$= \left(\frac{\Gamma(\alpha_0)}{\prod\limits_{r=1}^{20} \Gamma(\alpha_r)} \right) \prod_{r=1}^{20} \theta_r^{\alpha_r - 1} \tag{35}$$

$\alpha_0 = \sum_{r=1}^{20} \alpha_r$ represents the total number of counts that the prior distribution represents, and the α_r, the counts for each type of amino acid (not necessarily integers). Because different distributions will occur in multiple sequence alignments, the prior distribution for any position should be represented as a *mixture* of N Dirichlet distributions:

$$p(\theta_1, \theta_2, \ldots, \theta_{20} | \alpha_{1,1}, \alpha_{1,2}, \ldots, \alpha_{N,19}, \alpha_{N,20}; q_1, q_2, \ldots, q_N)$$

$$= \sum_{j=1}^{N} q_j \text{ Dirichlet } (\alpha_{j,1}, \alpha_{j,2}, \ldots, \alpha_{j,20}) \tag{36}$$

where the q_i represent the mixture coefficients and α_{ij} the count of the ith amino acid type in the jth component of the mixture. One can imagine components that represent predominantly a single amino acid type commonly conserved (Gly, Pro, Trp, Cys) and others that represent physical properties such as hydrophobicity and charge. We can establish such distributions from an understanding of what we expect to see in multiple sequence alignments, or the mixture can be derived from a set of sequence alignments and some optimization procedure.

Sjölander et al. [16] describe the process assumed in their model of sequence alignments, which is how the counts for a particular position in a multiple sequence alignment would arise from the mixture Dirichlet prior:

1. A component j is chosen from among the N Dirichlet components in Eq. (36) according to their respective probabilities, q_j.
2. A probability distribution $(\theta_1, \theta_2, \ldots, \theta_{20})$ for the 20 amino acids is chosen from component j according to Dirichlet $(\alpha_{j,1}, \alpha_{j,2}, \ldots, \alpha_{j,20})$.
3. A count vector $(n_1, n_2, \ldots, n_{20})$ for the position is drawn from the multinomial distribution given $(\theta_1, \theta_2, \ldots, \theta_{20})$.

For a single Dirichlet component, Dirichlet $(\alpha_{j,1}, \alpha_{j,2}, \ldots, \alpha_{j,20})$, the expected posterior probability for amino acid type i, given the counts for the 20 amino acids in a single position of the multiple alignment is

$$\hat{p}_i = \frac{n_i + \alpha_{j,i}}{n + \alpha_{j,0}} \tag{37}$$

where n is the total of the n_i and $\alpha_{j,0}$ is the total of the $\alpha_{j,i}$.

With a mixture we have to factor in the posterior probability that component j is the correct component:

$$\hat{p}_i = \sum_{j=1}^{N} p(\vec{\alpha}_j | \vec{n}) \frac{n_i + \alpha_{j,i}}{n + \alpha_{j,0}} \tag{38}$$

where

$$p(\vec{\alpha}_j | \vec{n}) = \frac{p(\vec{n} | \vec{\alpha}_j) p(\vec{\alpha}_j)}{p(\vec{n})} = \frac{p(\vec{n} | \vec{\alpha}_j) q_j}{p(\vec{n})} \tag{39}$$

$$= q_j \frac{\Gamma(n)\Gamma(\alpha_{j,0})}{\Gamma(n + \alpha_{j,0})} \prod_{i=1}^{20} \frac{\Gamma(n_i + \alpha_{j,i})}{\Gamma(n_i)\Gamma(\alpha_{j,i})}$$

The parameters $\{\alpha_{j,i}\}$ and $\{q_j\}$ are determined from multiple sequence alignments, e.g., those available from the BLOCKS [55], FSSP [3], or HSSP [56] databases. For each position in each alignment, there is a vector of counts, $\vec{n}_l$, and the objective is to find the set of α and q that maximizes the probability of the counts arising from the Dirichlet mixture, $\Pi_l p(\vec{n}_l | \alpha, n_l)$. As an example, in Table 1 we list the vectors of α for a mixture of Dirichlet distributions calculated from BLOCKS alignments by Kevin Karplus using an expectation maximization algorithm [16]. A notation at the bottom of each column describes the type of position most likely to be represented by this mixture component (hydrophobic, charged residues, etc.). The last three components represent conserved cysteine, proline, and glycine residues. The other components represent various physical properties.

To demonstrate how the use of a Dirichlet mixture regularizes a profile derived from a small number of sequences, we used the mixture in Table 1 on a set of 10 nuclear hormone receptor (nhr) sequences from *C.* elegans and compared this with the actual data distribution from the 10 sequences and the data and regularized distributions derived from 100 nhr sequences from C. elegans [57]. The results are shown in Table 2 for a short segment of nhr proteins beginning with the second conserved cysteine in the first zinc finger. Cys 21 (numbering is based on the *nhr*-25 sequence) is very conserved, which is evident in the data profiles (n_i/n) and the posterior modes of the regularized distributions $(\hat{p}_i)$ derived from Eq. (38). For 100 sequences, the data profile and regularized profiles are quite similar, which is what we would expect from a Bayesian analysis with a large amount of data. However, when there are only 10 sequences, the regularized profiles are different from the data profiles and are in fact more similar to the profiles derived from 100 sequences. This is the intent of the Bayesian analysis—to estimate parameter values from limited data that will be as close as possible to parameter values we would calculate if we had a large amount of data.

B. Bayesian Sequence Alignment

The primary tool of bioinformatics is sequence alignment, and numerous algorithms have been developed for this purpose, including variants of pairwise and multiple sequence alignments, profile alignments, local and global sequence alignments, and exact and heuristic methods. Sequence alignment programs usually depend on a number of fixed parameters, including the choice of sequence similarity matrix and gap penalty parameters. The correct choice of these parameters is a much debated topic, and the choice may depend on the purpose of the alignments: to determine whether two sequences are in fact related to each other or to determine the ''correct'' alignment between them, whatever that may be.

The Bayesian alternative to fixed parameters is to define a probability distribution for the parameters and simulate the joint posterior distribution of the sequence alignment and the parameters with a suitable prior distribution. How can varying the similarity matrix

Table 1 Dirichlet Mixture Components[a]

Comp.#	1	2	3	4	5	6	7	8	9	10	11	12
q	0.137	0.066	0.134	0.070	0.053	0.091	0.128	0.082	0.161	0.008	0.022	0.050
α_0	1.603	1.361	8.916	2.998	3.931	2.208	2.569	5.024	0.138	220.926	22.809	21.949
C	0.040	0.014	0.077	0.013	0.023	0.043	0.012	0.088	0.001	219.778	0.032	0.042
P	0.063	0.012	0.291	0.070	0.014	0.013	0.068	0.098	0.010	0.038	18.602	0.105
G	0.139	0.014	0.428	0.074	0.000	0.012	0.159	0.139	0.000	0.024	0.233	20.261
W	0.005	0.074	0.065	0.010	0.020	0.007	0.003	0.082	0.005	0.000	0.016	0.014
Y	0.017	0.359	0.198	0.041	0.038	0.015	0.022	0.247	0.005	0.000	0.044	0.032
F	0.014	0.406	0.144	0.029	0.203	0.057	0.012	0.293	0.005	0.000	0.071	0.010
M	0.016	0.034	0.123	0.033	0.249	0.083	0.013	0.235	0.002	0.000	0.029	0.012
I	0.035	0.053	0.245	0.051	0.378	0.637	0.018	0.599	0.003	0.045	0.188	0.009
V	0.091	0.043	0.355	0.073	0.220	0.784	0.038	0.737	0.005	0.000	0.314	0.047
L	0.044	0.116	0.348	0.092	2.530	0.280	0.034	0.644	0.008	0.000	0.301	0.059
A	0.381	0.029	0.769	0.129	0.064	0.116	0.164	0.385	0.008	0.148	0.706	0.262
T	0.211	0.018	0.730	0.132	0.051	0.091	0.136	0.343	0.010	0.145	0.285	0.049
S	0.329	0.033	0.902	0.140	0.028	0.025	0.245	0.255	0.006	0.358	0.508	0.241
H	0.012	0.059	0.253	0.091	0.016	0.006	0.057	0.100	0.008	0.000	0.095	0.046
Q	0.027	0.014	0.574	0.269	0.045	0.000	0.183	0.152	0.008	0.209	0.225	0.056
N	0.066	0.027	0.630	0.117	0.008	0.009	0.276	0.107	0.011	0.000	0.142	0.213
E	0.025	0.009	0.803	0.146	0.013	0.009	0.395	0.133	0.011	0.000	0.323	0.084
D	0.025	0.013	0.615	0.047	0.008	0.008	0.503	0.074	0.013	0.062	0.165	0.162
R	0.032	0.023	0.549	0.689	0.014	0.006	0.069	0.142	0.011	0.000	0.265	0.140
K	0.031	0.011	0.819	0.755	0.008	0.009	0.161	0.173	0.008	0.119	0.262	0.105
Description	Small, polar	Aromatic	Polar	Positive charge	Leu	Ile,Val	Negative charge	Hydro-Phobic	Polar	Cys	Pro	Gly

[a] This 12-component mixture was derived by Kevin Karplus (UC Santa Cruz), http://www.cse.ucsc.edu/research/compbio/dirichlets/index.html, from the BLOCKS database. The first line lists the q_j value for each component. The second line lists the total prior counts, α_0. The last line provides a rough description of each component.

Table 2   Raw Data and Posterior Modes from Dirichlet Mixtures for a Six Amino Acid Segment of Nuclear Hormone Receptors[a]

	Res. no.	Res. type	No. seq.	C	P	G	W	Y	F	M	I	V	L	A	T	S	H	Q	N	E	D	R	K
n_i/n	21	C	10	100	0	0	0	0	0	0	0	0	0	0	0	0	0	0	0	0	0	0	0
$\hat{p}_i$	21	C	10	99	0	0	0	0	0	0	0	0	0	0	0	0	0	0	0	0	0	0	0
n_i/n	21	C	100	100	0	0	0	0	0	0	0	0	0	0	0	0	0	0	0	0	0	0	0
$\hat{p}_i$	21	C	100	100	0	0	0	0	0	0	0	0	0	0	0	0	0	0	0	0	0	0	0
n_i/n	22	G	10	20	0	40	0	10	10	0	0	0	0	10	0	0	0	0	0	10	0	0	0
$\hat{p}_i$	22	G	10	13	1	27	0	8	8	1	3	3	3	9	3	3	1	2	2	9	2	2	2
n_i/n	22	G	100	4	0	45	0	3	2	0	0	0	1	4	2	12	3	2	5	6	6	3	2
$\hat{p}_i$	22	G	100	4	0	42	0	3	2	0	0	0	1	4	3	12	3	2	5	6	6	3	3
n_i/n	23	D	10	0	0	20	0	0	0	0	0	0	0	0	0	0	0	10	10	30	20	0	10
$\hat{p}_i$	23	D	10	0	1	16	0	0	0	0	0	1	1	2	2	3	1	9	10	25	18	1	9
n_i/n	23	D	100	0	0	8	0	1	0	1	1	0	7	5	3	4	1	11	8	7	33	6	3
$\hat{p}_i$	23	D	100	0	0	8	0	1	0	1	1	0	7	5	3	5	1	11	8	7	31	6	4
n_i/n	24	R	10	0	30	0	0	0	0	0	0	10	0	0	10	0	0	0	0	0	0	10	40
$\hat{p}_i$	24	R	10	0	22	1	0	1	0	0	1	8	1	2	9	2	1	2	2	2	1	11	33
n_i/n	24	R	100	1	27	2	0	0	1	0	2	7	0	4	4	8	3	2	2	7	2	13	15
$\hat{p}_i$	24	R	100	1	25	2	0	0	1	0	2	7	0	4	4	8	3	2	2	7	2	12	15
n_i/n	25	V	10	0	0	10	0	0	0	0	0	10	0	60	0	20	0	0	0	0	0	0	0
$\hat{p}_i$	25	V	10	0	1	10	0	0	0	0	0	9	0	54	2	20	0	0	1	0	0	0	0
n_i/n	25	V	100	1	6	7	0	1	0	0	1	5	0	48	7	15	1	0	2	2	1	2	1
$\hat{p}_i$	25	V	100	1	6	7	0	1	0	0	1	5	0	45	7	15	1	1	2	3	1	2	2
n_i/n	26	S	10	0	0	0	0	0	10	0	0	0	0	0	10	10	40	0	20	0	10	0	0
$\hat{p}_i$	26	S	10	0	1	2	0	1	7	1	1	2	2	3	9	10	25	3	15	4	9	2	4
n_i/n	26	S	100	0	0	2	0	0	4	0	4	3	0	9	11	16	24	1	13	1	7	3	2
$\hat{p}_i$	26	S	100	0	0	2	0	0	4	0	4	3	0	9	11	16	22	1	13	2	7	3	3

[a] Sequences are all from the *C. elegans* genome. Sequence and sequence numbering in the second and third columns refer to the *nhr*-25 sequence.

and gap penalties be justified, case by case, when these parameters in fact determine the alignment? Similarity matrices, such as the PAM series [58–60], for instance, are parametrized as a function of evolutionary distance between the sequences to be compared. When comparing two sequences, we do not yet know the distance between them, and it is hard to say that a PAM250 or PAM100 matrix should be used. But certainly after we look at a proposed alignment, we have a better idea of what the distance is and what the matrix should be, even if this alignment would change with a different choice of matrix. Similarly, gap penalties may be different for different sequence pairs, because some proteins may be tolerant to insertions whereas others may not, depending on the pressure of natural selection. For instance, if two proteins are orthologs, i.e., they perform the same function in different species, we might expect fewer and shorter insertions and deletions. If two proteins are paralogs, i.e., homologous proteins that perform different functions in one species (or potentially in different species), then there may be more insertions and deletions to account for the change in function and/or substrate. Thus the use of fixed matrix and gap penalties is not entirely justifiable, other than for convenience.

Zhu et al. [15] and Liu and Lawrence [61] formalized this argument with a Bayesian analysis. They are seeking a joint posterior probability for an alignment $\mathbf{A}$, a choice of distance matrix Θ, and a vector of gap parameters, Λ, given the data, i.e., the sequences to be aligned: $p(\mathbf{A}, \Theta, \Lambda | R_1, R_2)$. The Bayesian likelihood and prior for this posterior distribution is

$$
\begin{aligned}
p(\mathbf{A}, \Theta, \Lambda | R_1, R_2) &= \frac{p(R_1, R_2 | \mathbf{A}, \Theta, \Lambda) p(\mathbf{A}, \Theta, \Lambda)}{p(R_1, R_2)} \\
&= \frac{p(R_1, R_2 | \mathbf{A}, \Theta) p(\mathbf{A} | \Lambda) p(\Lambda) p(\Theta)}{p(R_1, R_2)}
\end{aligned}
\tag{40}
$$

The second line results from the first, because Θ and Λ are independent a priori, whereas the score of the alignment $\mathbf{A}$ depends on Λ, and the likelihood of the sequences, given the alignment, depends only on the scoring matrix Θ.

The usual alignment algorithm fixes Θ and Λ as Θ_0 and Λ_0, so that the prior is 1 when $\Theta = \Theta_0$ and $\Lambda = \Lambda_0$ and is 0 otherwise. Clearly, if experience justifies this choice or some other nonuniform choice, we can choose an informative prior that biases the calculation to certain values for the parameters or limits them to some likely range. The likelihood is well defined by the alignment model defined by using a similarity matrix and affine gap penalties, so that

$$
\ln p(R_1, R_2 | \mathbf{A}, \Theta) = \sum_{i,j} A_{ij} \Psi_{R_1(i), R_2(j)}
\tag{41}
$$

The effect of the gap penalties occurs through the factor $p(\mathbf{A} | \Lambda)$, which is given by

$$
p(\mathbf{A} | \Lambda) = p(\mathbf{A} | \lambda_0, \lambda_e) = \frac{\lambda_0^{k_g(A)} \lambda_e^{l_g(A) - k_g(A)}}{\sum_{A'} \lambda_0^{k_g(A')} \lambda_e^{l_g(A') - k_g(A')}}
\tag{42}
$$

where λ_0 and λ_e are the gap opening and extension parameters, respectively, and $k_g(A)$ and $l_g(A)$ are the total number and length of gaps in the alignment $\mathbf{A}$. To obtain marginal posterior distributions for the gap penalties and scoring matrix, Eq. (40) must be summed over all alignments. For the gap penalties alone, we need to sum over both the alignments and the scoring matrices:

$$p(\Lambda|R_1, R_2) = \sum_A \sum_\Theta \frac{p(R_1, R_2|\mathbf{A}, \Theta)p(\mathbf{A}|\Lambda)p(\Lambda)p(\Theta)}{p(R_1, R_2)} \tag{43}$$

The sum in Eq. (43) can be obtained by a recursion algorithm used commonly in dynamic programming [62].

C. Sequence–Structure Alignment

Threading methods have been developed by many groups over a number of years [18,49,63–74] and consist of three steps: (1) selecting a library of candidate folds usually derived from the Protein Data Bank (PDB) of experimentally determined protein structures; (2) selecting the most likely fold from the fold library for a protein sequence of unknown structure (the target sequence); and (3) simultaneously aligning the sequence of the target sequence to the structure (and sequence) from the fold library. Many factors can be considered in choosing the appropriate fold and aligning the sequence to the structure, including sequence similarity (with some amino acid substitution matrix such as the BLOSUM [75–77] or PAM [60] matrices); burial of hydrophobic amino acids; exposure of hydrophilic, especially charged, amino acids; *contacts* between hydrophobic amino acids (as distinct from simple burial from solvent); secondary structure predictions compared to the parent structure secondary structure; and others. Although the methods are designed to be applicable in cases where the target sequence is not, in fact, related to the parent by descent from a common ancestor (i.e., homologous), in practice most successful threadings are due to homologous relationships.

Lathrop and Smith [18] and Lathrop et al. [19], in a series of papers, have given a Bayesian theory of sequence–structure alignment. Their goal is to derive a model for protein threading that is Bayes-optimal, that is, selecting both the fold and sequence alignment that have the highest joint probability over all possible folds and sequence alignments. The advantage of the formalism they develop is that most other threading scoring functions can be expressed within their framework. The probability they wish to calculate is $p(C, \mathbf{t}|\mathbf{a}, n)$, where C is the protein fold or "core" selected from the fold library, $\mathbf{t}$ is the alignment of the target sequence to the fold sequence and structure, $\mathbf{a}$ is the sequence of the target (assumed known), and n is its length.

Before setting up priors and likelihoods, we can factor the joint probability of the core structure choice and the alignment $\mathbf{t}$ by using Bayes' rule:

$$p(C, \mathbf{t}|\mathbf{a}, n) = p(\mathbf{t}|\mathbf{a}, n, C)p(C|\mathbf{a}, n) \tag{44}$$

[This can be thought of as follows: If $p(C, \mathbf{t}) = p(\mathbf{t}|C)p(C)$, then we simply add the conditionality on $\mathbf{a}$, n to each term to achieve Eq. (44).] Lathrop and coworkers provide a likelihood and prior for each term in Eq. (44). For the first term, considering $y = \mathbf{a}$ as the data and $\theta = \mathbf{t}$ as the parameter we are looking for, and leaving n, C where they are on the right-hand side of the vertical bar [i.e., read Eq. (45) without the n, C, and it reads like the standard Bayesian equation $p(\theta|y) = p(y|\theta)p(\theta)/p(y)$], we have

$$p(\mathbf{t}|\mathbf{a}, n, C) = \frac{p(\mathbf{a}|\mathbf{t}, n, C)p(\mathbf{t}|n, C)}{p(\mathbf{a}|n, C)} \tag{45}$$

For the second term, we use the same trick: Leave n on the right-hand side of the vertical and use the Bayesian statistics equation to invert C and $\mathbf{a}$:

$$p(C|\mathbf{a}, n) = \frac{p(\mathbf{a}|C, n)p(C|n)}{p(\mathbf{a}|n)} \tag{46}$$

Multiplying out Eqs. (45) and (46) and canceling the $p(\mathbf{a}|n, C) = p(\mathbf{a}|C, n)$ terms, we have

$$p(C, \mathbf{t}|\mathbf{a}, n) = p(\mathbf{a}|\mathbf{t}, n, C) \frac{p(\mathbf{t}|n, C)p(C|n)}{p(\mathbf{a}|n)} \tag{47}$$

where the last term forms the prior distribution normalized by the probability of the data (i.e., the sequence and its length), and the first term on the right is the likelihood function of the sequence given the alignment, its length, and the choice of core structure. What remains is to decide functional forms for each of these terms.

For $p(C|n)$, Lathrop et al. provide a noninformative prior, $p(C|n) = 1/L$, where L is the size of the library. A slightly more complicated prior could take into account that some folds are more common than others and that some folds are more common in some phyla than others [78]. Given certain rules that constitute a proper threading, $\mathbf{t}$, we can establish how many threadings there can be for a given core structure C and sequence length n of the target. Their model for protein threading consists of core structures that contain only regular secondary structures (just helices and sheet strands, no loops or irregular secondary structure) and a constraint that no gap can occur within a secondary structure unit of the core structure. Under these constraints a finite number of threadings are possible, $T(n, C)$. The noninformative prior for $p(\mathbf{t}|n, C)$ is then just $1/T(n, C)$. A more elaborate prior for $p(\mathbf{t}|n, C)$ is described in [18,19] that takes into account likely loop lengths in real proteins and coverage of the core elements by residues from the sequence. Even in the absence of knowledge of the target sequence, some alignments are more likely than others. Finally, for the prior, $p(\mathbf{a}|n)$ is a constant and can be ignored because we are looking for relative probabilities of the choice of fold and alignment.

The likelihood function is an expression for $p(\mathbf{a}|\mathbf{t}, n, C)$, which is the probability of the sequence $\mathbf{a}$ (of length n) given a particular alignment $\mathbf{t}$ to a fold C. The expression for the likelihood is where most threading algorithms differ from one another. Since this probability can be expressed in terms of a pseudo free energy, $p(\mathbf{a}|\mathbf{t}, n, C) \propto \exp[-f(\mathbf{a}, \mathbf{t}, C)]$, any energy function that satisfies this equation can be used in the Bayesian analysis described above. The normalization constant required is akin to a partition function, such that

$$p(\mathbf{a}|\mathbf{t}, n, C) = \frac{\exp[-f(\mathbf{a}, \mathbf{t}, C)]}{\sum_i \exp[-f(\mathbf{a}, \mathbf{t}_i, C)]} \tag{48}$$

where the sum is over all possible alignments. The final result for the posterior distribution is

$$p(C, \mathbf{t}|\mathbf{a}, n) = \frac{\exp[-f(\mathbf{a}, \mathbf{t}, C)]}{\sum_i \exp[-f(\mathbf{a}, \mathbf{t}_i, C)]} \left(\frac{1}{LT(n, C)p(\mathbf{a}|n)} \right) \tag{49}$$

The success of any method using this formalism will be entirely dependent on the function f used and the search strategy. Lathrop et al. [19] use a branch-and-bound algorithm to search the space of legal threadings and a function f that does not depend on residue interactions. The details of these aspects are not inherently Bayesian in character and are outside the scope of this review.

IV. APPLICATIONS IN STRUCTURAL BIOLOGY

A. Secondary Structure and Surface Accessibility

A common use of statistics in structural biology is as a tool for deriving predictive distributions of structural parameters based on sequence. The simplest of these are predictions of secondary structure and side-chain surface accessibility. Various algorithms that can learn from data and then make predictions have been used to predict secondary structure and surface accessibility, including ordinary statistics [79], information theory [80], neural networks [81–86], and "Bayesian" methods [87–89]. A disadvantage of some neural network methods is that the parameters of the network sometimes have no physical meaning and are difficult to interpret.

Unfortunately, some authors describing their work as "Bayesian inference" or "Bayesian statistics" have not, in fact, used Bayesian statistics; rather, they used Bayes' rule to calculate various probabilities of one observed variable conditional upon another. Their work turns out to comprise derivations of informative prior distributions, usually of the form $p(\theta_1, \theta_2, \ldots, \theta_N) = \Pi_{i=1}^{N} p(\theta_i)$, which is interpreted as the posterior distribution for the θ_i without consideration of the joint distribution of the data variables in the likelihood.

For example, Stolorz et al. [88] derived a Bayesian formalism for secondary structure prediction, although their method does not use Bayesian statistics. They attempt to find an expression for $p(\mu|\text{seq}) = p(\text{seq}|\mu)p(\mu)/p(\text{seq})$, where μ is the secondary structure at the middle position of seq, a sequence window of prescribed length. As described earlier in Section II, this is a use of Bayes' rule but is not Bayesian statistics, which depends on the equation $p(\theta|y) = p(y|\theta)p(\theta)/p(y)$, where y is data that connect the parameters in some way to observables. The data are not sequences alone but the combination of sequence and secondary structure that can be culled from the PDB. The parameters we are after are the probabilities of each secondary structure type as a function of the sequence in the sequence window, based on PDB data. The sequence can be thought of as an explanatory variable. That is, we are looking for

$$p(\theta_\alpha(\mathbf{x}), \theta_\beta(\mathbf{x}), \theta_\gamma(\mathbf{x})|y_\alpha(\mathbf{x}), y_\beta(\mathbf{x}), y_\gamma(\mathbf{x}))$$
$$\propto p(y_\alpha(\mathbf{x}), y_\beta(\mathbf{x}), y_\gamma(\mathbf{x})|\theta_\alpha(\mathbf{x}), \theta_\beta(\mathbf{x}), \theta_\gamma(\mathbf{x})), p(\theta_\alpha(\mathbf{x}), \theta_\beta(\mathbf{x}), \theta_\gamma(\mathbf{x})) \quad (50)$$

where, for example, $\theta_\alpha(\mathbf{x})$ is the probability of an α-helix at the middle position of a string of amino acid positions in a protein with sequence $\mathbf{x}$. Because we are dealing with count data, the appropriate functional forms for Eq. (50) are a multinomial likelihood and Dirichlet functions for the prior and posterior distributions. For any window of length N there are 20^N sequences possible, so there are insufficient data in the PDB for these calculations for any $N > 5$. There are three possible solutions:

1. Reduce the size of the alphabet from 20 amino acids to a smaller alphabet of, say, six (aliphatic, aromatic, charged, polar, glycine, proline).

2. Use *very* informative priors, where perhaps the prior could be based on the product of individual probabilities for each secondary structure type and amino acid type.
3. Do both of the above.

In this case, we are looking for "counts" for each secondary structure type μ for each sequence $\mathbf{x}$, which might be derived from PDB data by

$$n_\mu \propto \prod_{i=1}^{N} p(\mu | r_i) \tag{51}$$

The constant of proportionality is based on normalizing the probability and establishing the size of the prior, that is, the number of data points that the prior represents. The advantage of the Dirichlet formalism is that it gives values for not only the modes of the probabilities but also the variances, covariances, etc. See Eq. (13).

Thompson and Goldstein [89] improve on the calculations of Stolorz et al. by including the secondary structure of the entire window rather than just a central position and then sum over all secondary structure segment types with a particular secondary structure at the central position to achieve a prediction for this position. They also use information from multiple sequence alignments of proteins to improve secondary structure prediction. They use Bayes' rule to formulate expressions for the probability of secondary structures, given a multiple alignment. Their work describes what is essentially a sophisticated prior distribution for $\theta_\mu(\mathbf{X})$, where $\mathbf{X}$ is a matrix of residue counts in a multiple alignment in a window about a central position. The PDB data are used to form this prior, which is used as the predictive distribution. No posterior is calculated with posterior = prior $\times$ likelihood.

A similar formalism is used by Thompson and Goldstein [90] to predict residue accessibilities. What they derive would be a very useful prior distribution based on multiplying out independent probabilities to which data could be added to form a Bayesian posterior distribution. The work of Arnold et al. [87] is also not Bayesian statistics but rather the calculation of conditional distributions based on the simple counting argument that $p(\sigma | r) = p(\sigma, r)/p(r)$, where σ is some property of interest (secondary structure, accessibility) and r is the amino acid type or some property of the amino acid type (hydrophobicity) or of an amino acid segment (helical moment, etc).

B. Side-Chain Conformational Analysis

Analysis and prediction of side-chain conformation have long been predicated on statistical analysis of data from protein structures. Early rotamer libraries [91–93] ignored backbone conformation and instead gave the proportions of side-chain rotamers for each of the 18 amino acids with side-chain dihedral degrees of freedom. In recent years, it has become possible to take account of the effect of the backbone conformation on the distribution of side-chain rotamers [28,94–96]. McGregor et al. [94] and Schrauber et al. [97] produced rotamer libraries based on secondary structure. Dunbrack and Karplus [95] instead examined the variation in χ_1 rotamer distributions as a function of the backbone dihedrals ϕ and ψ, later providing conformational analysis to justify this choice [96]. Dunbrack and Cohen [28] extended the analysis of protein side-chain conformation by using Bayesian statistics to derive the full backbone-dependent rotamer libraries at all values of ϕ and ψ. This library has proven useful in side-chain conformation prediction [43], NMR structure determination [98], and protein design [99]. The PDB has grown tremendously in the last 10 years, so whereas Ponder and Richard's library [93] was based on 19 chains in 1987, we based our most recent backbone-dependent rotamer library [1] on 814 chains from the PDB with resolution of 2.0 Å or better, *R*-factor less than 20%, and mutual sequence identity less than 50%.

Even though there are substantially more data for side chains than in the 1980s, the data are very uneven. Some residue types (Asp, Lys, Ser) are common, and some (Cys, Trp, Met) are not. The long side chains (Met, Glu, Gln, Lys, Arg) have either 27 or 81 possible rotamers, some of which have been seen very rarely or not at all because of the large steric energies involved. The uneven distribution of residues across the Ramachandran map is also problematic, because for prediction purposes we would like to have probabilities for the different rotamers even in unusual backbone conformations where there may be only a few cases in the PDB or again none at all. Bayesian statistics was the most logical choice for overcoming this unevenness in data, providing a unified framework for adjusting between data-rich situations and data-poor situations where an informative prior distribution could make up for the lack of data.

The parameters we are searching for are the probabilities as well as the average values of the χ angles for the χ_1, χ_2, χ_3, χ_4 rotamers given values for the explanatory variables ϕ and ψ. To distinguish between side-chain dihedrals and rotamers we denoted the χ_1 rotamer as r_1, the χ_2 rotamer as r_2, etc. For sp^3–sp^3 dihedrals, there are three well-defined rotameric positions: $\chi \sim +60°$ (g^+), $\chi \sim 180°$ (t), and $\chi \sim -60°$ (g^-). For sp^3–sp^2 dihedrals (e.g., in aromatic amino acids, Asn, Asp, Glu, Gln), the situation is more complicated. This work proposed what is essentially a mixture model for side-chain conformations, although not explicitly described as such in Ref. [28], using normal distribution functions where the μ's and σ's depend on the rotamer type and backbone conformation, as do the mixture coefficients, the amount of each component in the mixture. We denote the mixture coefficients for each region of the ϕ, ψ map as follows:

$$\theta_{ijkl|ab} \equiv p(r_1 = i, r_2 = j, r_3 = k, r_4 = l | \phi_a - 10° < \phi$$
$$\leq \phi_a + 10°, \psi_b - 10° < \psi \leq \psi_b + 10°) \tag{51a}$$

We denote the mean χ angles and their variances σ^2 similarly as $\mu_{ijkl|ab}^{(m)}$, where (m) represents the χ angle (1, 2, 3, or 4) for the $ijkl$th rotamer, and $\sigma_{ijkl|ab}^{2(m)}$ the variance. To simplify the model a bit, we set the covariances to 0. The full model is therefore

$$p(\chi_1, \chi_2, \chi_3, \chi_4 | \phi_a, \psi_b) = \sum_{i=1}^{3} \sum_{j=1}^{3} \sum_{k=1}^{3} \sum_{l=1}^{3} \theta_{ijkl|ab} \prod_{m=1}^{4} N(\mu_{ijkl|ab}^{(m)}, \sigma_{ijkl|ab}^{2(m)}) \tag{52}$$

where N signifies a normal distribution (Eq. 14). This is a likelihood function for the χ angles given their explanatory variables ϕ, ψ, i.e., $p(y|\theta)$, where the y are the χ angles in each ϕ, ψ region and the θ are the mixture coefficients and the mean and variance parameters in the normal distributions. Ordinarily, which component a particular data point (a single side chain conformation and its backbone conformation) belongs to is left as missing data. However, in our case we know that our χ angles are trimodal with well-spaced modes (near $+60°$, $180°$, and $-60°$), and we can assign each side chain to a particular rotamer unambiguously. We use Bayesian statistics to determine the parameters in Eq. (52).

Because each side chain can be identifiably assigned to a particular component, the mixture coefficients and the normal distribution parameters can be determined separately.

For the $\theta_{ijkl|ab}$ we use Dirichlet priors combined with a multinominal likelihood to determine a Dirichlet posterior distribution. The data in this case are the set of counts $\{n_{ijkl|ab}\}$. We determined these counts from PDB data (lists of values for $\{\phi, \psi, \chi_1, \chi_2, \chi_3, \chi_4\}$) by counting side chains in overlapping $20° \times 20°$ square blocks centered on (ϕ_a, ψ_b) spaced $10°$ apart. The likelihood is therefore of the form

$$p(\{n_{ijkl|ab}\}|\theta_{ijkl|ab}) = \frac{n_{ab}!}{\prod\limits_{i,j,k,l} n_{ijkl|ab}} \prod\limits_{i,j,k,l} \theta_{ijkl|ab}^{n_{ijkl|ab}} \tag{53}$$

The major benefit of using Bayesian statistics for side-chain conformational data is to account for the uneven distribution of data across the Ramachandran map and the ability to determine large numbers of parameters from limited data using carefully chosen informative prior distributions. Prior distributions can come from any source, either previous data or some pooling of data that in the true posterior would be considered distinct. For instance, if we considered it likely that the distributions of two different amino acid types are likely to be very similar, we might pool their conformations to determine the prior distribution and then determine separate posterior distributions for each type by adding in the data for that type only in the likelihood. We can use a process of building prior distributions by combining posterior distributions for subsets of the parameters. The prior distribution for the $\{\theta_{ijkl|ab}\}$ is a Dirichlet distribution where the counts, $\alpha_{ijkl|ab}$, are determined as follows:

$$\alpha_{ijkl|ab} = K\hat{\theta}_{ij|a}\hat{\theta}_{ij|b}\hat{\theta}_{kl|ij} \tag{54}$$

where K represents the total weight to the prior and the θ's come from the expectation value of the posterior distributions for these parameters. These variables have posterior distributions

$$p(\theta_{ij|a}) = \text{Dirichlet}(\{n_{ij|a} + \alpha_{ij|a}\}) \tag{55}$$

where the prior parameters $\{\alpha_{ij|a}\}$ are determined from the data as $\alpha_{ij|a} = K(n_{i|a}/n_a)(n_{j|i}/n_j)$, with K as some constant. In principle, the fractions in this expression could also be obtained from posterior distributions based on informative prior distributions, but at some point there are sufficient data to work directly from the data to determine the priors. The other factors in Eq. (54) are determined similarly.

The χ angle distributions are also determined by deriving posterior distributions from data and prior distributions that are themselves determined from posterior distributions and *their* prior distributions and data. The appropriate expressions are given in Eq. (17). In our previous work we did not put a prior distribution on the variances. This resulted in wildly varying values for the variances, especially for rotamers that are rare in the database. If there are only a few side chains in the count, their variance will be outweighed by the prior estimate if there is enough weight on the prior values in the Inv χ^2 distribution (the degrees of freedom, ν_0). Also, when there is only one count or no counts, a reasonable value for the variance is determined from the prior.

As an example of analysis of side-chain dihedral angles, the Bayesian analysis of methionine side-chain dihedrals is given in Table 3 for the $r_1 = g^-$ rotamers. In cases where there are a large number of data—for example, the (3, 3, 3) rotamer—the data and posterior distributions are essentially identical. These are normal distributions with the averages and standard variations given in the table. But in cases where there are few data,

Table 3 Bayesian Analysis of Methionine Side Chain Dihedral Angles[a]

Dist.	r_1	r_2	r_3	N	χ_1	σ_1	χ_2	σ_2	χ_3	σ_3
data	*3*	*1*	1	24	*-80.5*	*18.7*	*77.1*	*16.7*	74.1	17.3
prior					-78.1	22.0	71.9	11.5	72.6	15.4
post					-79.8	19.5	75.6	15.3	73.7	16.5
data	*3*	*1*	2	16	-70.4	27.8	77.1	29.4	173.5	25.5
prior					-78.1	22.0	73.2	14.0	-175.3	28.5
post					-73.4	25.4	75.6	24.0	177.8	26.8
data	*3*	*1*	*3*	7	*-87.7*	*15.0*	*66.3*	*19.7*	*-93.5*	*29.7*
prior					-78.1	22.0	73.8	15.9	-90.8	25.0
post					-82.1	19.7	70.7	17.3	-91.9	26.1
data	3	2	1	611	-69.2	11.1	178.2	12.4	70.8	17.6
prior					-68.3	11.0	179.3	9.2	71.4	18.0
post					-69.2	11.1	178.2	12.4	70.8	17.6
data	3	2	2	332	-67.3	11.2	179.8	12.1	-175.5	23.3
prior					-68.3	11.0	179.7	9.2	-178.2	24.0
post					-67.4	11.2	179.8	12.0	-175.5	23.3
data	3	2	3	443	-67.9	10.8	-176.8	13.4	-74.5	18.7
prior					-68.3	11.0	-178.9	9.3	-73.6	19.0
post					-67.9	10.8	-176.8	13.3	-74.5	18.7
data	3	*3*	*1*	145	-65.9	10.7	*-67.7*	*14.0*	*98.2*	*13.1*
prior					-65.7	11.4	-65.6	9.7	98.0	13.2
post					-65.8	10.7	-67.6	13.8	98.2	13.1
data	3	*3*	2	157	-64.5	11.7	-66.0	13.6	168.4	25.7
prior					-65.7	11.4	-65.8	9.9	170.0	26.3
post					-64.6	11.7	-66.0	13.3	168.5	25.7
data	3	*3*	3	601	-66.0	11.5	-62.2	12.7	-70.0	15.9
prior					-65.7	11.4	-64.1	9.8	-70.6	16.0
post					-66.0	11.5	-62.2	12.7	-70.0	15.9

[a] Average data dihedrals along with the prior and posterior modes and data, prior, and posterior standard deviations are given for the g-rotamer. Rotamers with *syn*-pentane steric interactions are italicized.

e.g., the (3, 1, n) rotamers, the prior exhibits a strong pull on the data to yield a posterior distribution that is between their mean values. To get a feeling for the distributions involved, in Figure 2 we show the data distribution along with the posterior distribution of the mean and the posterior predictive distributions for χ_3 of the (3, 1, 3), (3, 2, 3), and (3, 3, 3) rotamers. The posterior distribution for the values of $\mu_{ijk}^{(3)}$ was generated from a simulation of 1000 draws. This was achieved first by drawing values for the variance from its posterior distribution, a scaled inverse χ^2 distribution [Eqs. (16) and (17)] derived from the values of the prior variance and the data variance. With each value of σ^2 a draw was taken from the posterior distribution for the χ_3 angle. This distribution is normal with a variance roughly equal to σ_n^2/κ_n. For large amounts of data, for example the (3, 3, 3) rotamer, this produces a very precise estimate of the value of $\mu_{ijk}^{(3)}$. For small amounts of data, the spread in $\mu_{ijk}^{(3)}$ is larger. We can also predict future data, denoted $\tilde{y}$, from the likelihood and the posterior distribution of the parameters. This is called the *posterior*

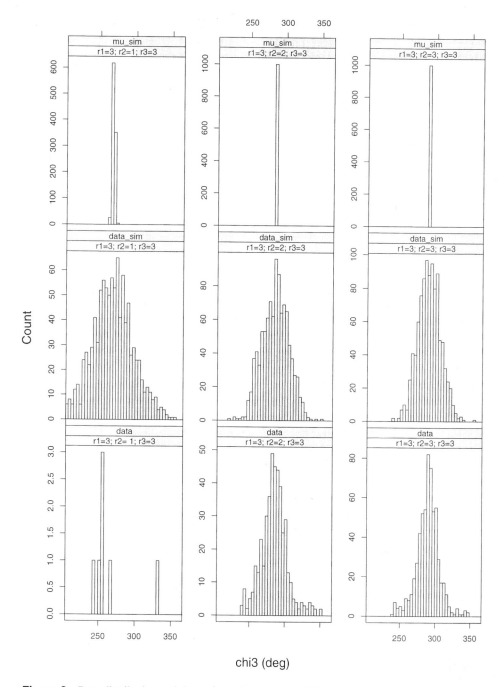

Figure 2 Data distribution and draws from the posterior distribution (mu_sim) and posterior predictive distributions (data_sim) for methionine side chain dihedral angles. The results for three rotamers are shown, ($r_1 = 3$; $r_2 = 1$; $r_3 = 3$), (3, 2, 3), and (3, 3, 3). Each simulation consisted of 1000 draws, calculated and graphed with the program S-PLUS [109].

predictive distribution and can be achieved by making draws from the posterior distribution and from these values, making draws from the likelihood function, i.e.,

$$p(\tilde{y}|y) = \int_\Theta p(\tilde{y}|\theta)p(\theta|y) \, d\theta = \frac{\int_\Theta p(\tilde{y}|\theta)p(y|\theta)p(\theta) \, d\theta}{\int_\Theta p(y|\theta)p(\theta) \, d\theta} \tag{56}$$

This distribution resembles the data closely for rotamer (3, 3, 3) but also forms a very reasonable distribution when there are only seven data points (3, 3, 1). A good posterior predictive distribution for any protein structural feature can be used in simulations of protein folding or structure prediction.

V. CONCLUSION

The field of statistics arose in the eighteenth and nineteenth centuries because of the need to develop good public policy based on demographic and economic data. Applications in the natural sciences were immediate, but generally natural scientists have lagged behind in their knowledge of modern statistics compared to social scientists. This is unfortunate, because many algorithms and methodologies have been developed in the last 20 years or so that make feasible sophisticated analysis of very complex data sets. Bayesian statistics has been used fruitfully in molecular and structural biology in recent years but has enjoyed more applications in genetics and clinical research and in the social sciences. Bayesian methods are particularly useful in modeling complex data, where the distribution of information may be uneven or hierarchical. This is true not only of the sequence and structure databases described in this chapter but also of more recently developed experimental methods such as DNA microarrays for analyzing mRNA expression levels over many thousands of genes [100–106]. The computational challenges for this kind of data are immense [107,108]. Particularly now, when the influx of data in biology is overwhelming, Bayesian statistical analysis promises to be an important tool.

ACKNOWLEDGMENTS

I thank Prof. Marc Sobel of Temple University for many useful discussions on Bayesian statistics. This work was funded in part by an appropriation from the Commonwealth of Pennsylvania and NIH Grant CA06927.

REFERENCES

1. RL Dunbrack Jr. Culling the PDB by resolution and sequence identity. 1999. http://www.fccc.edu/research/labs/dunbrack/culledpdb.html
2. CA Orengo, AD Michie, S Jones, DT Jones, MB Swindells, JM Thornton. CATH—A hierarchic classification of protein domain structures. Structure 5:1093–1108, 1997.
3. L Holm, C Sander. Touring protein fold space with Dali/FSSP. Nucleic Acids Res 26:316–319, 1998.

4. TJ Hubbard, B Ailey, SE Brenner, AG Murzin, C Chothia. SCOP: A structural classification of proteins database. Nucleic Acids Res 27:254–256, 1999.
5. JS Shoemaker, IS Painter, BS Weir. Bayesian statistics in genetics: A guide for the uninitiated. Trends Genet 15:354–358, 1999.
6. S Greenland. Probability logic and probability induction. Epidemiology 9:322–332, 1998.
7. GM Petersen, G Parmigiani, D Thomas. Missense mutations in disease genes: A Bayesian approach to evaluate causality. Am J Hum Genet. 62:1516–1524, 1998.
8. DA Berry, DK Stangl, eds. Bayesian Biostatistics. New York: Marcel Dekker, 1996.
9. G D'Agostini. Bayesian reasoning in high energy physics: Principles and applications. CERN Lectures, 1998.
10. TJ Loredo. In: PF Fougère, ed. From Laplace to Supernova SN 1987A: Bayesian Inference in Astrophysics. Dordrecht, The Netherlands: Kluwer, 1990, pp 81–142.
11. TJ Loredo. In: ED Feigelson, GJ Babu, eds. The Promise of Bayesian Inference for Astrophysics. New York: Springer-Verlag, 1992, pp 275–297.
12. E Parent, P Hubert, B Bobée, J Miquel, eds. Statistical and Bayesian Methods in Hydrological Sciences. Paris: UNESCO Press, 1998.
13. CE Buck, WG Cavanaugh, CD Litton. The Bayesian Approach to Interpreting Archaeological Data. New York: Wiley, 1996.
14. A Zellner. An Introduction to Bayesian Inference in Econometrics. New York: Wiley, 1971.
15. J Zhu, JS Liu, CE Lawrence. Bayesian adaptive sequence alignment algorithms. Bioinformatics 14:25–39, 1998.
16. K Sjölander, K Karplus, M Brown, R Hughey, A Krogh, IS Mian, D Haussler. Dirichlet mixtures: A method for improved detection of weak but significant protein sequence homology. Comput Appl Biosci 12:327–345, 1996.
17. K Karplus, K Sjolander, C Barrett, M Cline, D Haussler, R Hughey, L Holm, C Sander. Predicting protein structure using hidden Markov models. Proteins Suppl: 134–139, 1997.
18. RH Lathrop, TF Smith. Global optimum protein threading with gapped alignment and empirical pair score functions. J Mol Biol 255:641–665, 1996.
19. RH Lathrop, JR Rogers Jr, TF Smith, JV White. A Bayes-optimal sequence–structure theory that unifies protein sequence–structure recognition and alignment. Bull Math Biol 60:1039–1071, 1998.
20. RA Chylla, JL Markley. Improved frequency resolution in multidimensional constant-time experiments by multidimensional Bayesian analysis. J Biomol NMR 3:515–533, 1993.
21. DA d'Avignon, GL Bretthorst, ME Holtzer, A Holtzer. Thermodynamics and kinetics of a folded–folded′ transition at valine-9 of a GCN4-like leucine zipper. Biophys J 76:2752–2759, 1999.
22. JA Lukin, AP Gove, SN Talukdar, C Ho. Automated probabilistic method for assigning backbone resonances of (^{13}C,^{15}N)-labeled proteins. J Biomol NMR 9:151–166, 1997.
23. MT McMahon, E Oldfield. Determination of order parameters and correlation times in proteins: A comparison between Bayesian, Monte Carlo and simple graphical methods. J Biomol NMR 13:133–137, 1999.
24. MF Ochs, RS Stoyanova, F Arias-Mendoza, TR Brown. A new method for spectral decomposition using a bilinear Bayesian approach. J Magn Reson 137:161–176, 1999.
25. TO Yeates. The asymmetric regions of rotation functions between Patterson functions of arbitrarily high symmetry. Acta Crystallogr A 49:138–141, 1993.
26. S Doublie, S Xiang, CJ Gilmore, G Bricogne, CW Carter Jr. Overcoming non-isomorphism by phase permutation and likelihood scoring: Solution of the TrpRS crystal structure. Acta Crystallogr A 50:164–182, 1994.
27. CW Carter Jr. Entropy, likelihood and phase determination. Structure 3:147–150, 1995.
28. RL Dunbrack Jr, FE Cohen. Bayesian statistical analysis of protein sidechain rotamer preferences. Protein Sci 6:1661–1681, 1997.

29. P Baldi, S Brunak. Bioinformatics: The Machine Learning Approach. Cambridge, MA: MIT Press, 1998.

30. ET Jaynes. Probability Theory: The Logic of Science. http://bayes.wustl.edu/etj/prob.html. 1999.

31. M Gardner. The Second Scientific American Book of Mathematical Puzzles and Diversions. New York: Simon and Schuster, 1961.

32. T Bayes. An essay towards solving a problem in the doctrine of chances. Phil Trans Roy Soc Lond 53:370, 1763.

33. PS Laplace. Theorie Analytique des Probabilités. Paris: Courcier, 1812.

34. TM Porter. The Rise of Statistical Thinking. Princeton, NJ: Princeton Univ Press, 1988.

35. JO Berger, M Delampady. Testing precise hypotheses. Stat Sci 2:317–352, 1987.

36. TS Kuhn. Structure of Scientific Revolutions. Chicago: Univ Chicago Press, 1974.

37. DV Lindley. The 1988 Wald Memorial Lecture: The present position of Bayesian statistics. Stat Sci 5:44–89, 1990.

38. H Jeffreys. Theory of Probability. Oxford: Clarendon Press, 1939.

39. LJ Savage. The Foundations of Statistics. New York: Wiley, 1954.

40. WR Gilks, S Richardson, DJ Spiegelhalter, eds. Markov Chain Monte Carlo in Practice. London: Chapman & Hall, 1996.

41. IJ Good. The Bayes/non-Bayes compromise: A brief review. J Am Stat Assoc 87:597–606, 1992.

42. J Cornfield. In: DL Meyer, RO Collier, eds. The Frequency Theory of Probability, Bayes' Theorem, and Sequential Clinical Trials. Bloomington, In: Phi Delta Kappa, 1970, pp 1–28.

43. M Bower, FE Cohen, RL Dunbrack Jr. Prediction of protein sidechain rotamers from a backbone-dependent rotamer library: A new homology modeling tool. J Mol Biol 267:1268–1282, 1997.

44. A Gelman, JB Carlin, HS Stern, DB Rubin. Bayesian Data Analysis. London: Chapman & Hall, 1995.

45. N Metropolis, S Ulam. The Monte Carlo method. J Am Stat Assoc 44:335–341, 1949.

46. N Metropolis, AW Rosenbluth, MN Rosenbluth, AH Teller, E Teller. Equation of state calculations by fast computing machines. J Chem Phys 21:1087–1092, 1953.

47. WK Hastings. Monte Carlo sampling methods using Markov chains and their applications. Biometrika 57:97–109, 1970.

48. CP Robert. In: WR Gilks, S Richardson, DJ Spiegelhalter, eds. Mixtures of Distributions: Inference and estimation. London: Chapman & Hall, 1996, pp 441–464.

49. M Gribskov, AD McLachlan, D Eisenberg. Profile analysis: Detection of distantly related proteins. Proc Natl Acad Sci USA 84:4355–4358, 1987.

50. JU Bowie, ND Clarke, CO Pabo, RT Sauer. Identification of protein folds: Matching hydrophobicity patterns of sequence sets with solvent accessibility patterns of known structures. Proteins Struct Func Genet 7:257–264, 1990.

51. M Brown, R Hughey, A Krogh, IS Mian, K Sjolander, D Haussler. Using Dirichlet mixture priors to derive hidden Markov models for protein families. Intelligent Systems in Molecular Biology 1:47–55, 1993.

52. K Karplus. Evaluating regularizers for estimating distributions of amino acids. Intelligent Systems in Molecular Biology 3:188–196, 1995.

53. RL Tatusov, EV Koonin, DJ Lipman. A genomic perspective on protein families. Science 278:631–637, 1997.

54. TL Bailey, M Gribskov. The megaprior heuristic for discovering protein sequence patterns. Intelligent Systems in Molecular Biology 4:15–24, 1996.

55. S Pietrokovski, JG Henikoff, S Henikoff. The BLOCKS database—A system for protein classification. Nucleic Acids Res 24:197–200, 1996.

56. C Dodge, R Schneider, C Sander. The HSSP database of protein structure–sequence alignments and family profiles. Nucleic Acids Res 26:313–315, 1998.

57. AE Sluder, SW Mathews, D Hough, VP Yin, CV Maina. The nuclear receptor superfamily has undergone extensive proliferation and diversification in nematodes. Genome Res 9:103–120, 1999.

58. MO Dayhoff, WC Barker, PJ McLaughlin. Inferences from protein and nucleic acid sequences: Early molecular evolution, divergence of kingdoms and rates of change. Orig Life 5:311–330, 1974.

59. MO Dayhoff. The origin and evolution of protein superfamilies. Fed Proc 35:2132–2138, 1976.

60. WC Barker, MO Dayhoff. Evolution of homologous physiological mechanisms based on protein sequence data. Comp Biochem Physiol [B] 62:1–5, 1979.

61. JS Liu, CE Lawrence. Bayesian inference on biopolymer models. Bioinformatics 15:38–52, 1999.

62. TF Smith, MS Waterman. Identification of common molecular subsequences. J Mol Biol 147:195–197, 1981.

63. M Hendlich, P Lackner, S Weitckus, H Flöckner, R Froschauer, K Gottsbacher, G Casari, MJ Sippl. Identification of native protein folds amongst a large number of incorrect models. J Mol Biol 216:167–180, 1990.

64. MAS Saqi, PA Bates, MJE Sternberg. Towards an automatic method of predicting protein structure by homology: An evaluation of suboptimal sequence alignments. Protein Eng 5: 305–311, 1992.

65. DT Jones, WR Taylor, JM Thornton. A new approach to protein fold recognition. Nature 358:86–89, 1992.

66. SH Bryant, CE Lawrence. An empirical energy function for threading protein sequence through the folding motif. Proteins Struct Funct Genet 16:92–112, 1993.

67. R Abagyan, D Frishman, P Argos. Recognition of distantly related proteins through energy calculations. Proteins Struct Funct Genet 19:132–140, 1994.

68. TJ Hubbard, J Park. Fold recognition and ab initio structure predictions using hidden Markov models and β-strand pair potentials. Proteins Struct Funct Genet 23:398–402, 1995.

69. NN Alexandrov. SARFing the PDB. Protein Eng 9:727–732, 1996.

70. D Fischer, D Eisenberg. Protein fold recognition using sequence-derived predictions. Protein Sci 5:947–955, 1996.

71. TR Defay, FE Cohen. Multiple sequence information for threading algorithms. J Mol Biol 262:314–323, 1996.

72. B Rost, R Schneider, C Sander. Protein fold recognition by prediction-based threading. J Mol Biol 270:471–480, 1997.

73. WR Taylor. Multiple sequence threading: An analysis of alignment quality and stability. J Mol Biol 269:902–943, 1997.

74. V DiFrancesco, J Garnier, PJ Munson. Protein topology recognition from secondary structure sequences: Application of the hidden Markov models to the alpha class proteins. J Mol Biol 267:446–463, 1997.

75. S Henikoff, JG Henikoff. Performance evaluation of amino acid substitution matrices. Proteins 17:49–61, 1993.

76. JG Henikoff, S Henikoff. BLOCKS database and its applications. Methods Enzymol 266: 88–105, 1996.

77. S Henikoff, JG Henikoff, S Pietrokovski. BLOCKS+: A non-redundant database of protein alignment blocks derived from multiple compilations. Bioinformatics 15:471–479, 1999.

78. M Gerstein, M Levitt. A structural census of the current population of protein sequences. Proc Natl Acad Sci USA 94:11911–11916, 1997.

79. PY Chou, GD Fasman. Prediction of the secondary structure of proteins from their amino acid sequence. Adv Enzymol Relat Areas Mol Biol 47:45–148, 1978.

80. JF Gibrat, J Garnier, B Robson. Further developments of protein secondary structure prediction using information theory. New parameters and consideration of residue pairs. J Mol Biol 198:425–443, 1987.

81. N Qian, TJ Sejnowski. Predicting the secondary structure of globular proteins using neural network models. J Mol Biol 202:865–884, 1988.

82. LH Holley, M Karplus. Protein secondary structure prediction with a neural network. Proc Natl Acad Sci USA 86:152–156, 1989.

83. B Rost, C Sander. Combining evolutionary information and neural networks to predict protein secondary structure. Proteins Struct Funct Genet 19:55–72, 1994.

84. AL Delcher, S Kasif, HR Goldberg, WH Hsu. Protein secondary structure modelling with probabilistic networks. Intelligent Systems in Molecular Biology 1:109–117, 1993.

85. JM Chandonia, M Karplus. Neural networks for secondary structure and structural class predictions. Protein Sci 4:275–285, 1995.

86. JM Chandonia, M Karplus. The importance of larger data sets for protein secondary structure prediction with neural networks. Protein Sci 5:768–774, 1996.

87. GE Arnold, AK Dunker, SJ Johns, RJ Douthart. Use of conditional probabilities for determining relationships between amino acid sequence and protein secondary structure. Proteins 12:382–399, 1992.

88. P Stolorz, A Lapedes, Y Xia. Predicting protein secondary structure using neural net and statistical methods. J Mol Biol 225:363–377, 1992.

89. MJ Thompson, RA Goldstein. Predicting protein secondary structure with probabilistic schemata of evolutionarily derived information. Protein Sci 6:1963–1975, 1997.

90. MJ Thompson, RA Goldstein. Predicting solvent accessibility: Higher accuracy using Bayesian statistics and optimized residue substitution classes. Proteins Struct Funct Genet 25:38–47, 1996.

91. J Janin, S Wodak, M Levitt, B Maigret. Conformations of amino acid side-chains in proteins. J Mol Biol 125:357–386, 1978.

92. E Benedetti, G Morelli, G Nemethy, HA Scheraga. Statistical and energetic analysis of side-chain conformations in oligopeptides. Int J Peptide Protein Res 22:1–15, 1983.

93. JW Ponder, FM Richards. Tertiary templates for proteins: Use of packing criteria in the enumeration of allowed sequences for different structural classes. J Mol Biol 193:775–792, 1987.

94. MJ McGregor, SA Islam, MJE Sternberg. Analysis of the relationship between sidechain conformation and secondary structure in globular proteins. J Mol Biol 198:295–310, 1987.

95. RL Dunbrack Jr, M Karplus. Backbone-dependent rotamer library for proteins: Application to sidechain prediction. J Mol Biol 230:543–571, 1993.

96. RL Dunbrack Jr, M Karplus. Conformational analysis of the backbone-dependent rotamer preferences of protein sidechains. Nature Struct Biol 1:334–340, 1994.

97. H Schrauber, F Eisenhaber, P Argos. Rotamers: To be or not to be? An analysis of amino acid sidechain conformations in globular proteins. J Mol Biol 230:592–612, 1993.

98. J Kuszewski, AM Gronenborn, GM Clore. Improving the quality of NMR and crystallographic protein structures by means of a conformational database potential derived from structure databases. Protein Sci 5:1067–1080, 1996.

99. BI Dahiyat, SL Mayo. Protein design automation. Protein Sci 5:895–903, 1996.

100. M Schena, D Shalon, RW Davis, PO Brown. Quantitative monitoring of gene expression patterns with a complementary DNA microarray. Science 270:467–470, 1995.

101. M Schena, D Shalon, R Heller, A Chai, PO Brown, RW Davis. Parallel human genome analysis: Microarray-based expression monitoring of 1000 genes. Proc Natl Acad Sci USA 93:10614–10619, 1996.

102. D Shalon, SJ Smith, PO Brown. A DNA microarray system for analyzing complex DNA samples using two-color fluorescent probe hybridization. Genome Res 6:639–645, 1996.

103. MB Eisen, PT Spellman, PO Brown, D Botstein. Cluster analysis and display of genome-wide expression patterns. Proc Natl Acad Sci USA 95:14863–14868, 1998.

104. M Wilson, J DeRisi, HH Kristensen, P Imboden, S Rane, PO Brown, GK Schoolnik. Exploring drug-induced alterations in gene expression in Mycobacterium tuberculosis by microarray hybridization. Proc Natl Acad Sci USA 96:12833–12838, 1999.

105. GP Yang, DT Ross, WW Kuang, PO Brown, RJ Weigel. Combining SSH and cDNA microarrays for rapid identification of differentially expressed genes. Nucleic Acids Res 27:1517–1523, 1999.
106. VR Iyer, MB Eisen, DT Ross, G Schuler, T Moore, JCF Lee, JM Trent, LM Staudt, J Hudson Jr, MS Boguski, D Lashkari, D Shalon, D Botstein, PO Brown. The transcriptional program in the response of human fibroblasts to serum. Science 283:83–87, 1999.
107. MQ Zhang. Large-scale gene expression data analysis: A new challenge to computational biologists. Genome Res 9:681–688, 1999.
108. JM Claverie. Computational methods for the identification of differential and coordinated gene expression. Hum Mol Genet 8:1821–1832, 1999.
109. S-PLUS, Version 3.4. Mathsoft Inc., 1996.

16

Computer Aided Drug Design

Alexander Tropsha and Weifan Zheng
University of North Carolina at Chapel Hill, Chapel Hill, North Carolina

I. INTRODUCTION

Computer-aided drug design (CADD) has enjoyed a long and successful history in its development and applications. Even before the computer age, chemists and medicinal chemists used their intuition and experience to design molecules with desired biological profiles. However, as in practically all areas of chemistry, rapid accumulation of vast amounts of experimental information of biologically active molecules naturally led to the development of theoretical and computational tools for the rational analysis and design of bioactive molecules. The power of these tools rests solidly on a variety of well-established scientific disciplines including theoretical and experimental chemistry, biochemistry, biophysics, pharmacology, and computer science. This unique combination of underlying scientific disciplines makes CADD an indispensable complementary tool for most types of experimental research in drug design and development.

The advancement of drug design as a quantitative, analytical methodology should be, perhaps, attributed to Professor Corwin Hanch, who in the early 1960s initiated the development of the quantitative structure–activity relationships (QSAR) method (reviewed in Ref. 1). Early QSAR studies made extensive use of various physicochemical properties (such as the octanol–water distribution coefficient, which serves as a measure of hydrophobicity) and chemical topology based molecular descriptors to arrive at reliable models correlating structure and activity. Beginning in the early 1970s, CADD started to employ computer graphics or, more specifically, molecular graphics. This led to the development of three-dimensional molecular modeling, which can be defined as the generation, manipulation, calculation, and prediction of realistic chemical structures and associated physicochemical and biological properties [adapted from Ref. 2]. The application of molecular modeling approaches in the drug design area culminated in the development of the first three-dimensional (3D) pharmacophore assignment method, the active analog approach (AAA) [3,4]. This was quickly followed by the development of pharmacophore-based database searching methodologies and the establishment of 3D QSAR approaches such as comparative molecular field analysis (CoMFA) [5], which remains one of the most popular methods of CADD today. Finally, beginning in the early 1980s, a growing number of drug design applications started to use experimental structural information about macromolecular drug targets, i.e., proteins and nucleic acids.

Modern methods of computer-aided drug design fall into two major categories: ligand-based and receptor-based methods. The former methods, which include QSAR, various pharmacophore assignment methods, and database searching or mining, are based entirely on experimental structure–activity relationships for receptor ligands or enzyme inhibitors. Their application in the last three or four decades led to several drugs currently on the market (reviewed in Ref. 6). The structure-based design methods, which include docking and advanced molecular simulations, require structural information about the receptor or enzyme that is available from X-ray crystallography, nuclear magnetic resonance (NMR) techniques, or protein homology model building. This strategy became popular only recently owing to rapid advances in structure elucidation methods [7] and has already led to several promising drug candidates and even marketed drugs such as several approved inhibitors of the HIV protease for the treatment of AIDS [8].

It is practically impossible to review all major developments, applications, approaches, and caveats of the various drug design methods in one chapter. Several monographs and numerous publications have been devoted to CADD [9–11], and many specialized reviews addressing various aspects of molecular modeling as applied to drug design have been published in recent years in the ongoing series *Reviews in Computational Chemistry* [12]. We decided to concentrate in this chapter on one outstanding problem of modern chemistry and theoretical biology, that of ligand–receptor recognition, which is the underlying mechanism of action of most drugs. Using this problem as a molecular modeling paradigm, we introduce major molecular modeling concepts and tools that one can apply to drug design. To provide a practical example of a drug design project and make the discussion somewhat didactic, we consider the dopaminergic ligand–receptor system, which has been the focus of our research in recent years [13–15]. We believe that by considering one particular drug design problem we can better illustrate the power and limitations of underlying techniques and at the same time provide a practical introduction to the major concepts of molecular modeling as applied to drug design.

II. RELATIONSHIPS BETWEEN EXPERIMENTAL AND THEORETICAL APPROACHES TO STUDYING DRUG–RECEPTOR INTERACTIONS

The ultimate goal of molecular modeling as a pharmacological and medicinal chemistry tool is to predict, in advance of any laboratory testing, novel biologically active compounds. Molecular modeling research starts from the analysis of experimental observables of drug–receptor interaction. This interaction leads to the formation of the ligand–receptor complex followed by the conformational change of the receptor, which constitutes the putative mechanism of signal transduction. One can obtain either ligand–receptor binding constants from direct measurements in vitro or the biological activity data measured on isolated tissues or on the whole organism. Thus, experimental observables of ligand–receptor systems include the chemical structure of the ligands, biological activity or binding constants, receptor primary sequence, and, in some cases, receptor 3D structure (Fig. 1).

Molecular modeling uses the experimental observations as input data to programs that afford rational analysis of structure–activity relationships. As mentioned above, there are two main approaches to drug design: ligand-based and structure-based methods. First, if the receptor structure has been characterized by either X-ray crystallography or NMR, computer graphics methods can be used to design novel compounds based on complemen-

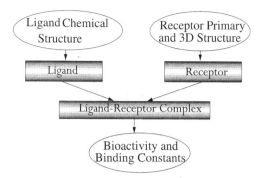

Figure 1   A flow chart of experimental information about drug–receptor interaction. The objects of experimental and theoretical investigation are boxed, and the experimental information is circled.

tarity between the surfaces of the ligand and the receptor binding site. Energy calculations may also be applied to quantitate and refine the graphics design. Second, if the receptor structure is unknown, one is left to infer the shape of the receptor binding site. Several receptor mapping techniques have been developed, such as AAA [3] and CoMFA [5]. The idea of these approaches is to construct a pharmacophore that integrates all the important structural and physicochemical features of the active compounds required for their successful binding to a receptor. Although the pharmacophore itself has significant predictive power as far as rational drug design is concerned, the design can be greatly facilitated if the receptor model is available. In many cases structural information about the receptor can be obtained experimentally by the means of X-rays or NMR. However, in many instances, especially for transmembrane receptors such as G-protein coupled receptors (GPCR), experimental structural determination is difficult if not impossible. Thus, theoretical structure modeling remains the only possible source of structural information about many receptors.

Recent successes in deciphering primary sequences of many receptors and developing several computerized protein-modeling tools enabled researchers to devise putative three-dimensional models of these receptors from their primary sequences. The development of molecular docking methodologies [16,17] (which allow one to bring together the ligand and the receptor and ''dock'' the former into the latter) complete the round of computer representations of natural processes of ligand–receptor interaction (i.e., a pharmacophore-based conformationally constrained ligand is brought into the vicinity of the modeled receptor binding site via docking).

The interaction between ligands and their receptors is clearly a dynamic process. Once the static model of ligand–receptor interaction has been obtained, the stability of ligand–receptor complexes should be evaluated by means of molecular dynamics simulations [18].

One can see that at this point only two types of experimental data are used for the analysis of a ligand–receptor system: ligand chemical structure (from which one generates the 3D pharmacophore conformation) and the primary receptor sequence (which is used to generate the 3D receptor model). The most important property of the receptors, however, is their ability to discriminate ligands on the basis of their chemical structure; we quantify this as ligand binding constants. The ability to reproduce binding constants, or at least their relative order, is the most sensitive test of any putative receptor model. The technique that, in principle, can provide such data is the free energy simulations/thermodynamic

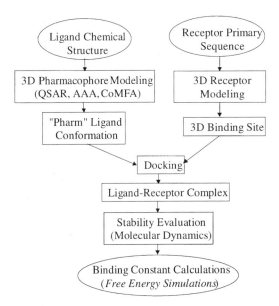

Figure 2 A flow chart of theoretical modeling drug–receptor interaction and relation to experiment. The objects of theoretical investigation are in rectangles, and the experimental information is in the ovals.

cycle approach (see Chapter 9), a method that relies on full atomistic representations of both ligand and receptor in the molecular model [19]. This approach is very robust but requires substantial computational resources, which limits its practical applicability for drug design. Nevertheless, in several reported cases, it was able to reproduce accurately the experimental binding constants [20,21]. The entire modeling process, including tools used for evaluation and refinement of the putative ligand–receptor model, is summarized in Figure 2.

The successful realistic modeling of ligand–receptor interactions can only be based on a combined approach incorporating several computer-assisted modeling tools. As outlined above, these tools may include ligand-based "negative image" receptor modeling, QSAR, primary structure-based "real" receptor modeling, docking, structure refinement (molecular dynamics), and relative binding constant calculations. We now discuss these techniques in the order in which one can proceed while working on a receptor model.

III. COMPUTATIONAL APPROACHES TO MODELING LIGAND–RECEPTOR INTERACTIONS

A. Ligand-Based Approaches

The advent of faster computers and the development of new algorithms led in the late 1970s and early 1980s to the development and active use of 3D negative receptor image modeling tools such as AAA. In this approach, first proposed by Marshall et al. [3], the researcher infers the size, the shape, and some physicochemical parameters of the receptor binding site by modeling receptor ligands. The key assumption of this approach is that all the active receptor ligands adopt, in the vicinity of the binding site, conformations that present common pharmacophoric functional groups to the receptor. Thus, the receptor is thought of as a negative image of the generalized pharmacophore that incorporates all structural and physicochemical features of all active analogs overlapped in their pharmacophoric conformations. The latter are found in the course of a conformation search, starting from the most rigid active analogs and proceeding with more flexible compounds. Conformational restraints imposed by more rigid compounds(s) with respect to the internal geometry of common functional groups are used to facilitate the search for more flexible compounds. The AAA and similar approaches have been successfully applied to negative image modeling of many receptors, yielding important leads for rational drug design (see, e.g., Ref. 14). Most important, these models can be incorporated into a common database providing the source of ligand structures for pharmaceutical lead generation (see, e.g., Ref. 22). Furthermore, the same database can be used to search for potential specific activity or cross-reactivity of independently designed or synthesized ligands by comparing them with known 3D pharmacophores.

For illustration, let us consider the application of AAA to an important pharmacological class of dopaminergic ligands [13]. The analysis of pharmacological activity of D_1 receptor ligands leads to their classification as active or inactive (see Table 1 and Fig. 3). Six compounds were chosen as active on the basis of three criteria: They had affinity for the D_1 receptor ($K_{0.5} < 300$ nM); they could increase cAMP synthesis in rat striatal membranes to the same degree as dopamine; and this increase could be blocked completely by the D_1 antagonist SCH23390. The compounds that met these criteria were dopamine, DHX, SKF89626, SKF82958, A70108, and A77636. As shown in Table 1, all of these compounds caused complete activation of dopamine-sensitive adenylate cyclase in this preparation and had $K_{0.5}$ values ranging from 267 nM (dopamine) to 0.9 nM (A70108).

The molecular modeling studies with this set of dopaminergic ligands involved the following analytical steps:

1. The tentative pharmacophoric elements of the D_1 receptor were determined on the basis of known structure–activity relationships.
2. A rigorous conformational search on the active compounds was performed to determine their lowest energy conformation(s).

Table 1 Pharmacological Analysis of DHX and Related Compounds

Drug	D_1 affinity $K_{0.5}$ (nM)	D_2 affinity $K_{0.5}$ (nM)	Adenylate cyclase EC50 (nM)	Max. stimulation of adenylate cyclase (% vs. DA)
Dopamine	267	36	5000	100
(+)DHX	2.3	43.8	30	120
SKF89626	61	142	700	120
SKF82958	4	73	491	94
A70108	0.9	41	1.95	96
A77636	31.7	1290	5.1	92
cis-DHX	$>10^3$	$>10^3$	$>10^4$	17
N-Propyl-DHX	326	27	$>10^4$	36
N-Allyl-DHX	328	182	$>10^4$	32
Ro 21-7767	477	61	$>10^4$	22
N-Benz-5,6-ADTN	$>10^3$	$>10^3$	$>10^3$	38
N-Benz-6,7-ADTN	$>10^3$	335	$>10^4$	25

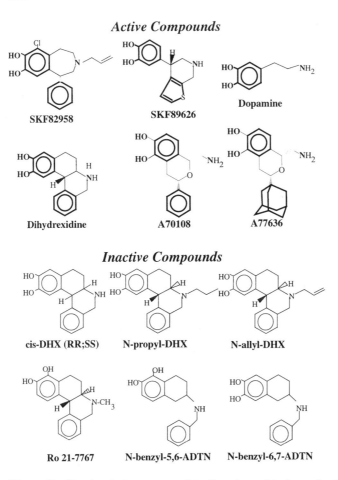

Figure 3 The chemical structures of the ligands used in the molecular modeling study of the D_1 dopamine receptor. The ligands were divided into two groups (active and inactive) based on their pharmacological properties. The hypothesized pharmacophoric elements are shown in bold.

3. Conformationally flexible superimposition of these compounds was done to determine their common (pharmacophoric) conformation.
4. Similar conformational analyses were performed for inactive compounds, and inactive compounds in pharmacophoric conformations were superimposed with the active compounds to determine steric limitations in the active site. Where appropriate, the geometry of each inactive molecule was obtained by modifying the chemical structure of the relevant active analogs followed by the energy minimization of the resulting structure.
5. Finally, an evaluation was made of excluded receptor volume and shape as spatial equivalents of the volume and shape of the pharmacophore.

All molecular modeling studies were performed with the multifaceted molecular modeling software package SYBYL (Tripos Associates Inc., St. Louis, MO).

Based on the experimental SAR data, the following functional groups of agonists were defined as key elements of the D_1 agonist pharmacophore (Fig. 3): the two hydroxyl groups of the catechol ring, the nitrogen, and (except for dopamine) an accessory hydrophobic group (e.g., the aromatic ring in dihydrexidine or SKF82958). Thus, the task of molecular modeling analysis was to identify a pharmacophoric conformation for each compound in which these key pharmacophoric elements were spatially arranged in the same way for all active compounds.

(a) Lowest Energy Conformations of Active Compounds. Construction of the Pharmacophore. The evaluation of the D_1 agonist pharmacophore was based on the following three-step routine.

> *Step 1.* Conduct a conformational search on each of the agonists to identify their lowest energy conformation(s).
>
> *Step 2.* Find common low energy conformations for all of the compounds. The commonality was assessed by comparing the distances between each of the hydroxyl oxygens and the nitrogen and the angle between the planes of the catechol ring and the accessory ring.
>
> *Step 3.* Overlay all agonists in their most common conformations using dihydrexidine as a template compound by superimposition of equivalent pharmacophoric atoms of all the agonists and those of DHX.

(b) D_1 Receptor Mapping and Agonist Pharmacophore Development. The "pharm" configurations of the active molecules also were used to map the volume of the receptor site available for ligand binding. The steric mapping of the D_1 receptor site, using the MVolume routine in SYBYL, involved the construction of a pseudoelectron density map for each of the active analogs superimposed in their pharmacophore conformations. A union of the van der Waals density maps of the active compounds defines the *receptor excluded volume* [3].

The essential feature of the AAA is a comparison of active and inactive molecules. A commonly accepted hypothesis to explain the lack of activity of inactive molecules that possess the pharmacophoric conformation is that their molecular volume, when presenting the pharmacophore, exceeds the receptor excluded volume. This additional volume apparently is filled by the receptor and is unavailable for ligand binding; this volume is termed the *receptor essential volume* [3]. Following this approach, the density maps for each of the inactive compounds (in their "pharm" conformations superimposed with that of active compounds) were constructed; the difference between the combined inactive compound density maps and the receptor excluded volume represents the receptor essential volume. These receptor-mapping techniques supplied detailed topographical data that allowed a steric model of the D_1 receptor site to be proposed.

These modeling efforts relied upon dihydrexidine as a structural template for determining molecular geometry because it was not only a high affinity full agonist, but it also had limited conformational flexibility relative to other more flexible, biologically active agonists. For all full agonists studied (dihydrexidine, SKF89626, SKF82958, A70108, A77636, and dopamine), the energy difference between the lowest energy conformer and those that displayed a common pharmacophore geometry was relatively small (<5 kcal/mol). The pharmacophoric conformations of the full agonists were also used to infer the shape of the receptor binding site. Based on the union of the van der Waals density maps of the active analogs, the excluded receptor volume was calculated. Various inactive analogs (partial agonists with D_1 $K_{0.5} > 300$ nM) subsequently were used to define

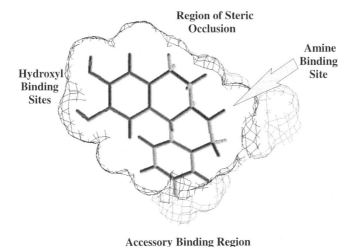

Region of Steric
Occlusion

Amine
Binding
Site

Hydroxyl
Binding
Sites

Accessory Binding Region

Figure 4 Excluded volume for the D_1 agonist pharmacophore. The mesh volume shown by the black lines is a cross section of the excluded volume representing the receptor binding pocket. Dihydrexidine (see text) is shown in the receptor pocket. The gray mesh represents the receptor essential volume of inactive analogs. The hydroxyl binding, amine binding, and accessory regions are labeled, as is the steric occlusion region.

the receptor essential volume (i.e., sterically intolerable receptor regions). These volumes, together with the pharmacophore results, were integrated into a three-dimensional model estimating the D_1 receptor active site topography (see Fig. 4).

Figure 4 represents the typical result of the application of the active analog approach. Based on the steric description of the essential receptor volume, new ligands can be designed that fit geometrical descriptions of the pharmacophore. Indeed, recent research based on the active analog approach led to the design of a novel highly selective dopamine agonist dihydrexidine [14] shown in Figure 4.

B. Quantitative Structure–Activity Relationship Method

The quantitative structure–activity relationship (QSAR) approach was first introduced by Hansch and coworkers [23,24] on the basis of implications from linear free energy relationships (LFERs) in general and the Hammett equation in particular [25]. It is based upon the assumption that the difference in physicochemical properties accounts for the differences in biological activities of compounds. According to this approach, the changes in physicochemical properties that affect the biological activities of a set of congeners are of three major types: electronic, steric, and hydrophobic [26]. These structural properties are often described by Hammett electronic constants [27], Verloop STERIMOL parameters [28], hydrophobic constants [27], etc. The quantitative relationships between biological activity (or chemical property) and the structural parameters could be conventionally obtained by using multiple linear regression (MLR) analysis.

The typical QSAR equation is

$$\log\left(\frac{1}{C}\right) = b_0 + \sum_i b_i D_i$$

where C is drug concentration, b_0 is a constant, b_i is a regression coefficient, and D_i is a molecular descriptor such as the Hansch π parameter (hydrophobicity) or the Hammett σ parameter (electron-donating or accepting properties), or molar refractivity (MR), which describes the volume and electronic polarizability, etc. The fundamentals and applications of this method in chemistry and biology have been summarized by Hansch and Leo [1]. This traditional QSAR approach has generated many useful and, in some cases, predictive QSAR equations and has led to several documented drug discoveries (see, e.g., Ref. 6 and references therein).

Many different approaches to QSAR have been developed since Hansch's seminal work. These include both two-dimensional (2D) and 3D QSAR methods. The major differences among these methods can be analyzed from two viewpoints: (1) the structural parameters that are used to characterize molecular identities and (2) the mathematical procedure that is employed to obtain the quantitative relationship between a biological activity and the structural parameters.

Most of the 2D QSAR methods are based on graph theoretic indices, which have been extensively studied by Randic [29] and Kier and Hall [30,31]. Although these structural indices represent different aspects of molecular structures, their physicochemical meaning is unclear. Successful applications of these topological indices combined with multiple linear regression (MLR) analysis are summarized in Ref. 31. On the other hand, parameters derived from various experiments through chemometric methods have also been used in the study of peptide QSAR, where partial least square (PLS) [32] analysis has been employed [33].

With the development of accurate computational methods for generating 3D conformations of chemical structures, QSAR approaches that employ 3D descriptors have been developed to address the problems of 2D QSAR techniques, e.g., their inability to distinguish stereoisomers. The examples of 3D QSAR include molecular shape analysis (MSA) [34], distance geometry [35,36], and Voronoi techniques [37].

Perhaps the most popular example of 3D QSAR is CoMFA. Developed by Cramer et al. [5], CoMFA elegantly combines the power of molecular graphics and partial least squares (PLS) technique and has found wide applications in medicinal chemistry and toxicity analysis (see Ref. 38 for several excellent reviews). This method is one of the most recent developments in the area of ligand-based receptor modeling. This approach combines traditional QSAR analysis and three-dimensional ligand alignment central to AAA into a powerful 3D QSAR tool. CoMFA correlates 3D electrostatic and van der Waals fields around sample ligands typically overlapped in their pharmacophoric conformations with their biological activity. This approach has been successfully applied to many classes of ligands [38].

CoMFA methodology is based on the assumption that since, in most cases, the drug–receptor interactions are noncovalent, the changes in biological activity or binding constants of sample compounds correlate with changes in electrostatic and van der Waals fields of these molecules. To initiate the CoMFA process, the test molecules should be structurally aligned in their pharmacophoric conformations; the latter are obtained by using, for instance, the AAA described above. After the alignment, steric and electrostatic fields of all molecules are sampled with a probe atom, usually an sp^3 carbon bearing a $+1$ charge, on a rectangular grid that encompasses structurally aligned molecules. The values of both van der Waals and electrostatic interaction between the probe atom and all atoms of each molecule are calculated in every lattice point on the grid using the force field equation described above and entered into the CoMFA QSAR table. This table thus contains thousands of columns, which makes it difficult to analyze; however, application of special multivariate statistical analysis routines such as partial least squares analysis, cross-validation, and bootstrapping ensures the statistical significance of the final CoMFA equation. A cross-validated R^2 (q^2) that is obtained as a result of this analysis serves as a quantitative measure of the quality and predictive ability of the final CoMFA model. The statistical meaning of the q^2 is different from that of the conventional r^2; a q^2 value greater than 0.3 is considered significant.

Recent trends in both 2D and 3D QSAR studies have focused on the development of optimal QSAR models through variable selection. This implies that only a subset of available descriptors of chemical structures that are the most meaningful and statistically significant in terms of correlation with biological activity is selected. The optimum selection of variables is achieved by combining stochastic search methods with correlation methods such as MLR, PLS analysis, or artificial neural networks (ANN) [39–44]. More specifically, these methods employ either generalized simulated annealing [39], genetic algorithms [40], or evolutionary algorithms [41–44] as the stochastic optimization tool. Since the effectiveness and convergence of these algorithms are strongly affected by the choice of a fitting function, several such functions have been applied to improve the performance of the algorithms [41,42]. It has been demonstrated that combinations of these algorithms with various chemometric tools have effectively improved the QSAR models compared to those without variable selection.

The variable selection approaches have also been adopted for region selection in the area of 3D QSAR. For example, GOLPE [45] was developed using chemometric principles. q^2-GRS [46] was developed based on independent CoMFA analyses of small areas (or regions) of near-molecular space to address the issue of optimal region selection in CoMFA analysis. More recently, a genetic algorithm based sampling of 3D regions of CoMFA fields was implemented [47]. These methods have been shown to improve the QSAR models compared to the original CoMFA technique.

Most of the QSAR techniques (both 2D and 3D) assume the existence of a linear relationship between a biological activity and molecular descriptors, which may be an adequate assumption for relatively small datasets (dozens of compounds). However, the fast collection of structural and biological data, owing to recent development of combinatorial chemistry and high throughput screening technologies, has challenged traditional QSAR techniques. First, 3D methods may be computationally too expensive for the analysis of a large volume of data, and in some cases, an automated and unambiguous alignment of molecular structures is not achievable [48], Second, although existing 2D techniques are computationally efficient, the assumption of linearity in the SAR may not hold true, especially when a large number of structurally diverse molecules are included in the analysis. Thus, several nonlinear QSAR methods have been proposed in recent years. Most of these methods are based on either artificial neural network (ANN) [49] or machine learning techniques [50]. Such applications, especially when combined with variable selection, represent fast-growing trends in modern QSAR research.

It is important not to overinterpret the QSAR models in terms of the underlying mechanism of the ligand–receptor interaction or their value for the design of novel compounds with high biological activity. Successful QSAR models present a statistically significant correlation between chemical features of compounds (descriptors) and biological activity, which does not imply any particular mechanism of drug action. In some instances, such as when CoMFA or simple linear regression QSAR models are used with clearcut physicochemical descriptors (e.g., van der Waals and electrostatic molecular fields or hydrophobicity), the models could be formally interpreted in terms of the structural modifications required to increase the biological activity. However, it is essential in all cases to avoid forecasting biological activity for compounds too structurally different from those in the training set.

C. Structure-Based Drug Design

Ligand-based methods of drug design provide only limited information about the stereo-chemical organization of the receptor binding site. Detailed knowledge of the active site geometry can be obtained experimentally by means of X-ray crystallography or NMR; unfortunately, despite rapid progress in practical applications of these techniques, the structures of very few receptors and ligand–receptor complexes are available experimentally. In the absence of any experimental information on the 3D structure of many receptors, predictions from primary sequence remain the only means of generating the receptor structure (cf. Chapter 14). Obviously, the knowledge of the active site geometry significantly facilitates the design of new drugs by providing real spatial and chemical limitations on the structures of newly designed molecules. We briefly review in the following subsections first the methods of molecular simulations applicable to the analysis of known ligand–receptor complexes and then the current state of the art in the area of modeling GPCRs, which include dopamine receptors. Excellent reviews of methods and successes in the area of structure-based drug design were published in the early 1990s [7,51].

1. Ligand–Receptor Docking

In ligand–receptor docking, the ligand is brought (manually or in some automated way) into the vicinity of the binding site and oriented so that electrostatic and van der Walls interactions (which correspond to Coulomb and dispersion terms, respectively, in molecular mechanics energy expressions) between ligand and receptor are optimized. Pharmacophoric conformation of ligands is frequently used in these studies. Earlier docking programs allowed the docking of only a rigid ligand into a rigid receptor, i.e., no conformational change of either receptor or ligand was permitted as the latter approached the former. Clearly, this is not the way the interaction occurs in nature, because both ligand and receptor are relatively flexible molecules that adjust their conformation to each other in the process of binding to maximize steric and chemical complementarity. Flexible docking calculations are very difficult and require an extremely fast computer. Nevertheless, the consideration of both ligand and receptor active site flexibility is becoming part of modern docking approaches [52], Further descriptions of docking algorithms can be found in several publications [16,17,53].

2. Molecular Dynamics

Once the model of a ligand–receptor complex is built, its stability should be evaluated. Simple molecular mechanics optimization of the putative ligand–receptor complex leads only to the identification of the closest local minimum. However, molecular mechanics optimization of molecules lacks two crucial properties of real molecular systems: temperature and, consequently, motion. Molecular dynamics studies the time-dependent evolution of coordinates of complex multimolecular systems as a function of inter- and intramolecular interactions (see Chapter 3). Because simulations are usually performed at normal temperature ($\sim$300 K), relatively low energy barriers, on the order of kT (0.6 kcal), can be easily overcome. Thus if the starting configuration of the whole system (i.e., drug–receptor complex) resulting from docking is separated from the more stable configuration by such a low barrier, molecular dynamics will take the system over the barrier. Molecular simulations may identify more stable, therefore more realistic, conformational states of ligand–receptor complexes. Furthermore, they may provide unique information about conformational changes of the receptor due to ligand binding. They may shed light on the intimate mechanisms of receptor activation that currently cannot be studied by any other technique. Finally, molecular simulations frequently incorporate solvent and thus allow the inclusion of solvent effects in the consideration. Unfortunately, due to the inherently very short elementary simulation step size, ~ 2 fs, this technique is presently limited to relatively short total simulation times, on the order of hundreds of picoseconds to nanoseconds. These limitations are mainly due to the fact that available computer power is still inadequate for significantly longer simulation times.

3. Binding Constant Calculation

The combination of free energy simulation (FES) and the thermodynamic cycle (TC) approach is the most promising modern technique for calculating relative ligand–receptor binding constants by simulating ligand–receptor interaction [19]. Experimentally, equilibrium free energies of binding of two ligands to the same receptor are evaluated in two independent experiments, according to the following scheme.

$$\text{Ligand1 + Receptor} \Leftarrow \Delta G_1^\circ \Rightarrow \text{Ligand1/Receptor}$$
$$\text{Ligand2 + Receptor} \Leftarrow \Delta G_2^\circ \Rightarrow \text{Ligand2/Receptor}$$

To calculate the relative binding constants of the two ligands, Ligand1 and Ligand2, we construct the following cyclic scheme,

$$\text{Ligand1 + Receptor} \Rightarrow \Delta G_1^\circ \Rightarrow \text{Ligand1/Receptor}$$
$$\Downarrow \qquad\qquad\qquad\qquad\qquad \Downarrow$$
$$\Delta G_3^\circ \qquad\qquad\qquad\qquad\qquad \Delta G_4^\circ$$
$$\Downarrow \qquad\qquad\qquad\qquad\qquad \Downarrow$$
$$\text{Ligand2 + Receptor} \Rightarrow \Delta G_2^\circ \Rightarrow \text{Ligand2/Receptor}$$

where ΔG_1° and ΔG_2° correspond to the experimental binding free energies of Ligand1 and Ligand2, respectively, and ΔG_3° and ΔG_4° correspond to the free energy of the formal transformation of the chemical structure of Ligand1 into that of Ligand2 in solution and in the binding site, respectively. From a thermodynamic viewpoint, the foregoing scheme represents a closed thermodynamic cycle that consists of four transformations: the binding of Ligand1 in solution to the receptor (ΔG_1°); the chemical transformation of bound Ligand1 to bound Ligand2 (ΔG_4°); the chemical transformation of Ligand1 in solution to Ligand2 in solution (ΔG_3°); and the binding of Ligand2 in solution to the receptor (ΔG_2°). Using the thermodynamic cycle (TC) relationship, $\Delta\Delta G_{\text{cycle}}^\circ = 0$, one obtains the *difference* in the free energy of binding of Ligand1 and Ligand2, i.e., ($\Delta G_1^\circ - \Delta G_2^\circ$). Thus,

$$\Delta\Delta G_{\text{cycle}}^\circ = \Delta G_1^\circ + \Delta G_4^\circ - \Delta G_3^\circ - \Delta G_2^\circ = 0$$

so

$$\Delta\Delta G_{\text{binding}}^\circ = \Delta G_1^\circ - \Delta G_2^\circ = \Delta G_3^\circ - \Delta G_4^\circ$$

The latter two free energies of chemical transformation are computed in the course of FES. The advantage of this approach is that it avoids calculations of ligand–receptor binding free energy per se, i.e., ΔG_1° and ΔG_2°, which would be extremely computationally intensive, without sacrificing the theoretical rigor of calculation of binding constants from molecular simulations of ligand–receptor interaction. Faster free energy simulation approaches based on the linear response theory have been introduced to improve the computational efficiency of free energy simulations [54–56]. Despite its computational intensity, the FES–TC approach finds important and growing applications in the analysis of ligand–receptor interactions and drug design [57–60].

D. Chemical Informatics and Drug Design

Combinatorial chemical synthesis and high throughput screening have significantly increased the speed of the drug discovery process [61–63]. However, it is still impossible to synthesize all of the library compounds in a reasonably short period of time. As many as 3000^3 (2.7×10^{10}) compounds can be synthesized from a molecular scaffold with three different substitution positions when each of the positions has 3000 different substituents. If a chemist could synthesize 1000 compounds per week, 27 million weeks (~0.5 million years) would be required to synthesize all of these compounds. Furthermore, many of these compounds can be structurally similar to each other, and chemical information contained in the library can be redundant. Thus, there is a need for rational library design (i.e., rational selection of a subset of building blocks for combinatorial chemical synthesis) so that a maximum amount of information can be obtained while a minimum number of compounds are synthesized and tested. Similarly, there is a closely related task in computational database mining, i.e., rational sampling of a subset of compounds from commercially available or proprietary databases for biological testing. Thus, new challenges to computational drug design in the context of combinatorial chemistry include information management, rational library design, and database analysis. These are the research topics of chemoinformatics—a new area of computational chemistry.

Chemoinformatics (or *cheminformatics*) deals with the storage, retrieval, and analysis of chemical and biological data. Specifically, it involves the development and application of software systems for the management of combinatorial chemical projects, rational design of chemical libraries, and analysis of the obtained chemical and biological data. The major research topics of chemoinformatics involve QSAR and diversity analysis. The researchers should address several important issues. First, chemical structures should be characterized by calculable molecular descriptors that provide quantitative representation of chemical structures. Second, special measures should be developed on the basis of these descriptors in order to quantify structural similarities between pairs of molecules. Finally, adequate computational methods should be established for the efficient sampling of the huge combinatorial structural space of chemical libraries.

There are two types of experimental combinatorial chemistry and high throughput screening research directions: targeted screening and broad screening [61,62]. The former approach involves the design and synthesis of chemical libraries with compounds that are either analogs of some active compounds or can specifically interact with the biological target under study. This is desired when a lead optimization (or evolution) program is pursued. On the other hand, a broad screening project involves the design and synthesis of a large array of maximally diverse chemical compounds, leading to diverse (or universal) libraries that are then tested against a variety of biological targets. This design strategy is suited for lead identification programs. Thus, two categories of computational tools should be developed and validated to meet the needs of the two different types of projects.

In a targeted screening project, computational library design involves the selection of a subset of chemical building blocks from an available pool of chemical structures. This subset of selected building blocks affords a limited virtual library with a high content of compounds similar to a lead molecule. Molecular similarity is quantified by using a chosen set of molecular descriptors and similarity metrics [64,65]. Building blocks can also be chosen such that the resulting virtual library could have a high percentage of compounds that are predicted to be active from a preconstructed QSAR model [66]. When the structure of the biological target is known, one can select building blocks such that the resulting library compounds are stereochemically complementary to the binding site structure of the underlying target [67,68]. Other approaches to targeted library design using different criteria have also been considered [69]. Similar approaches have long been used in targeted database mining that were based on the principle of either molecular similarity [70,71] or structure-based drug design [72]. In a broad screening project, computational library design or database mining involves the selection of a subset of compounds that are optimally diverse and representative of available classes of compounds, leading to a nonredundant chemical library or a set of nonredundant compounds for biological testing. Reported methods can be generally classified into several categories.

1. Cluster sampling methods, which first identify a set of compound clusters, followed by the selection of several compounds from each cluster [73].
2. Grid-based sampling, which places all the compounds into a low-dimensional descriptor space divided into many cells and then chooses a few compounds from each cell [74].
3. Direct sampling methods, which try to obtain a subset of optimally diverse compounds from an available pool by directly analyzing the diversity of the selected molecules [75,76].

Many reports have been published that address various aspects of diversity analysis in the context of chemical library design and database mining [77–84].

IV. SUMMARY AND CONCLUSIONS

We have reviewed several methods of computer-aided drug design that are formally divided into ligand-based and receptor-based methods. Ligand-based methods of analysis are used widely because they are not very computationally intensive and afford rapid generation of QSAR models from which the biological activity of newly designed compounds can be predicted. Most of the existing methods require the generation of a 3D pharmacophore hypothesis (i.e., unique 3D arrangements of important functional groups common to all or the majority of the receptor ligands). In many cases, when the receptor ligands are not very structurally diverse and include conformationally rigid compounds, a pharmacophore can be generated in a reasonably unbiased and unique way, using either automated (e.g., DISCO [85]) or semiautomated [3] pharmacophore prediction methods. The pharmacophore hypothesis can help medicinal chemists gain an insight into the key interactions between ligand and receptor when a 3D ligand–receptor complex structure has not been determined. It can be used as the search query for 3D database mining, and this approach has been demonstrated as very productive for lead compound discovery [86]. It can also be used for 3D QSAR analysis, grouping together the compounds that follow the same binding mode and indicating the possible 3D alignment rules [87]. However, this task of unique pharmacophore generation becomes less feasible for more structurally diverse and/or conformationally flexible compounds. In general, in the absence of detailed structural information about the receptor binding site, any pharmacophore inferred from only the ligand structure remains hypothetical.

Structure-based design methods present an appealing alternative to more traditional approaches to CADD. Molecular docking algorithms can generate fairly accurate orientations of known or designed receptor ligands in the active site (e.g., GRID [88], DOCK [16]). However, their application in practice is limited by the availability of macromolecular target structure. The protein and nucleic acid structural databases have been growing exponentially over the past decade, with the size of the Protein Data Bank (PDB) [89] currently exceeding 10,000 entries. Nevertheless, many of these entries are highly homologous proteins, and the structure of many pharmacologically important molecules such as transmembrane receptor proteins cannot be resolved experimentally at the present time. Furthermore, predictions of the binding affinity of receptor ligands are fast but rather inaccurate because the scoring functions used in docking calculations are not robust enough. Alternatively, these predictions can be fairly accurate but very computationally intensive when free energy simulation methods are used [21,90]. Therefore, because of these practical limitations, ligand-based design methods will probably remain the main arena of CADD efforts, supplemented, when possible, by structure-based approaches.

We have shown that successful molecular modeling and the design of new drugs can be achieved by integration of different computational chemistry and molecular modeling techniques. Owing to the rapid increase in computational power due to advances in both hardware and software, molecular modeling has become an important integral part of multidisciplinary efforts in the design and synthesis of new potent pharmaceuticals. Practical knowledge of these techniques and their limitations is a necessary component of modern research in drug discovery.

REFERENCES

1. C Hansch, A Leo. Exploring QSAR: Fundamentals and Applications in Chemistry and Biology. Washington, DC: American Chemical Society, 1995.
2. DB Boyd. In: KB Lipkowitz, DB Boyd, eds. Reviews in Computational Chemistry. New York: VCH, 1991.
3. GR Marshall, CD Barry, HE Bosshard, RA Dammkoehler, DA Dunn. The conformational parameter in drug design: The active analog approach. ACS Symp Ser 112:205–226, 1979.
4. YC Martin. Overview of concepts and methods in computer-assisted rational drug design. Methods Enzymol 203:587–613, 1991.
5. RD Cramer III, DE Patterson, JD Bunce. Comparative molecular field analysis (CoMFA). 1. Effect of shape on binding of steroids to carrier proteins. J Am Chem Soc 110:5959–5967, 1988.
6. DB Boyd. Successes of computer-assisted molecular design. In: KB Lipkowitz, DB Boyd, eds. Reviews in Computational Chemistry I. New York: VCH, 1990, pp 355–372.
7. CE Bugg, WM Carson, JA Montgomery. Drugs by design. Sci Am 269:92–98, 1993.
8. SG Deeks, M Smith, M Holdniy, JO Kahn. HIV-1 protease inhibitors: A review for clinicians. J Am Med Assoc 277:145–153, 1997.

9. GR Marshall, CD Barry, HE Bosshard, RA Dammkoehler, DA Dunn. The conformational parameter in drug design: The active analog approach. ACS Symp Ser 112:205–226, 1979.

10. JL Fauchere, ed. QSAR: Quantitative Structure–Activity Relationships in Drug Design. New York: Alan R Liss, 1989, pp 177–181.

11. H Kubinyi, G Folkers, YC Martin, eds. 3D QSAR in Drug Design: Recent Advances, Vols 2 and 3. Dordrecht, The Netherlands: Kluwer, 1998.

12. KB Lipkowitz, DB Boyd, eds. Reviews in Computational Chemistry, Vols I–XII. New York, VCH, 1991– (ongoing series).

13. DM Mottola, S Laiter, VJ Watts, A Tropsha, SW Wyrick, DE Nichols, P Mailman. Conformational analysis of d_1 dopamine receptor agonists: Pharmacophore assessment and receptor mapping. J Med Chem 39:285–296, 1996.

14. RB Mailman, DE Nichols, A Tropsha. Molecular drug design and dopamine receptors. In: K Neve, R Neve, eds. Dopamine Receptors. Totowa, NJ: Humana Press, 1996, pp 105–133.

15. BT Hoffman, SJ Cho, W Zheng, S Wyrick, DE Nichols, RB Mailman, A Tropsha. QSAR modeling of dopamine d_1 agonists using comparative molecular field analysis, genetic algorithms-partial least squares, and K nearest neighbor methods. J Med Chem 42:3217–3226, 1999.

16. EC Meng, BK Shoichet, ID Kuntz. Automated docking with grid-based energy evaluation. J Comput Chem 13:505–524, 1992.

17. DS Goodsell, AJ Olson. Automated docking of substrates to proteins by simulated annealing. Proteins Struct Funct Genet 8:195–202, 1990.

18. M Karplus, GA Petsko. Molecular dynamics simulations in biology. Nature 347:631–637, 1990.

19. TL Tembe, JA McCammon. Ligand–receptor interactions. Comput Chem 8:281–283, 1984.

20. S Hirono, PA Kollman. Calculation of the relative binding free energy of 2′-GMP and 2′-AMP to ribonuclease T1 using molecular dynamics/free energy perturbation approaches. J Mol Biol 212:197–209, 1990.

21. A Tropsha, J Hermans. Application of free energy simulations to the binding of a transition-state-analogue inhibitor to HIV protease. Protein Eng 51:29–34, 1992.

22. DP Marriott, IG Dougall, P Meghani, Y-J Liu, DR Flower. Lead generation using pharmacophore mapping and three-dimensional database searching: Application to muscarinic M3 receptor antagonists. J Med Chem 42:3210–3216, 1999.

23. C Hansch, RM Muir, T Fujita, PP Maloney, E Geiger, M Streich. The correlation of biological activity of plant growth regulators and chloromycetin derivatives with Hammett constants and partition coefficients. J Am Chem Soc 85:2817–2824, 1963.

24. T Fujita, J Iwasa, C Hansch. A new substituent constant, π, derived from partition coefficients. J Am Chem Soc 86:5175–5180, 1964.

25. LP Hammett. Some relations between reaction rates and equilibrium constants. Chem Rev 17: 125–136, 1935.

26. C Hansch, A Leo. Exploring QSAR: Fundamentals and Applications in Chemistry and Biology (Chapters 1–4). Washington, DC: Am Chem Soc, 1995.

27. C Hansch, A Leo, D Hoekman. Exploring QSAR: Hydrophobic, Electronic, and Steric Constants. Washington, DC: Am Chem Soc, 1995.

28. A Verloop, W Hoogenstraaten, J Tipker. In: EJ Ariens, ed. Drug Design, Vol VII. New York: Academic Press, 1976, p 165.

29. M Randic. On characterization of molecular branching. J Am Chem Soc 97:6609–6615, 1975.

30. LB Kier, LH Hall. Molecular Connectivity in Structure–Activity Analysis. Chichester, England: Research Studies Press, 1986.

31. LH Hall, LB Kier. The molecular connectivity chi indexes and kappa shape indexes in structure–property modeling. In: KB Lipkowitz, DB Boyd, eds. Reviews in Computational Chemistry, Vol. 2. New York: VCH, 1991, pp 367–422.

32. S Wold, A Ruhe, H Wold, WJ Dunn III. The collinearity problem in linear regression. The partial least squares (PLS) approach to generalized inverses. SIAM J Sci Stat Comput 5:735–743, 1984.

33. S Hellberg, M Sjostrom, B Skagerberg, S Wold. Peptide quantitative structure–activity relationships, a multivariate approach. J Med Chem 30:1126–1135, 1987.

34. AJ Hopfinger. A QSAR investigation of dihydrofolate reductase inhibition by Baker triazines based upon molecular shape analysis. J Am Chem Soc 102:7196–7206, 1980.

35. GM Crippen. Distance geometry approach to rationalizing binding data. J Med Chem 22(8):988–997, 1979.

36. GM Crippen. Quantitative structure–activity relationships by distance geometry: Systematic analysis of dihydrofolate reductase inhibitors. J Med Chem 23:599–606, 1980.

37. LG Boulu, GM Crippen. Voronoi binding site models: Calculation of binding models and influence of drug binding data accuracy. J Comput Chem 10(5):673–682, 1989.

38. H Kubinyi, G Folkers, YC Martin, eds. 3D QSAR in Drug Design: Recent Advances, Vols 2 and 3. Dordrecht, The Netherlands: Kluwer Academic, 1998.

39. JM Sutter, SL Dixon, PC Jurs. Automated descriptor selection for quantitative structure–activity relationships using generalized simulated annealing. J Chem Inf Comput Sci 35(1):77–84, 1995.

40. D Rogers, AJ Hopfinger. Application of genetic function approximation to quantitative structure–activity relationships and quantitative structure–property relationships. J Chem Inf Comput Sci 34(4):854–866, 1994.

41. H Kubinyi. Variable selection in QSAR studies. I. An evolutionary algorithm. Quant Struct-Act Relat 13:285–294, 1994.

42. H Kubinyi. Variable selection in QSAR studies. II. A highly efficient combination of systematic search and evolution. Quant Struct-Act Relat 13:393–401, 1994.

43. BT Luke. Evolutionary programming applied to the development of quantitative structure–activity relationships and quantitative structure–property relationships. J Chem Inf Comput Sci 34(6):1279–1287, 1994.

44. SS So, M Karplus. Evolutionary optimization in quantitative structure–activity relationship: An application of genetic neural networks. J Med Chem 39:1521–1530, 1996.

45. M Baroni, G Costantino, G Cruciani, D Riganelli, R Valigi, S Clementi. Generating optimal linear PLS estimations (GOLPE): An advanced chemometric tool for handling 3D-QSAR problems. Quant Struct-Act Relat 12:9–20, 1993.

46. SJ Cho, A Tropsha. Cross-validated R2-guided region selection for comparative molecular field analysis: A simple method to achieve consistent results. J Med Chem 38:1060–1066, 1995.

47. T Kimura, K Hasegawa, K Funatsu. GA strategy for variable selection in QSAR studies: GA-based region selection for CoMFA modeling. J Chem Inf Comput Sci 38:276–282, 1998.

48. SJ Cho, ML Garsia, J Bier, A Tropsha. Structure-based alignment and comparative molecular field analysis of acetylcholinesterase inhibitors. J Med Chem 39(26):5064–5071, 1996.

49. TA Andrea, H Kalayeh. Applications of neural networks in quantitative structure–activity relationships of dihydrofolate reductase inhibitors. J Med Chem 34:2824–2836, 1991.

50. G Bolis, L Pace, F Fabrocini. A machine learning approach to computer-aided molecular design. J Comput Aided Mol Des 5(6):617–628, 1991.

51. PJ Whittle, TL Blundell. Protein structure-based drug design. Annu Rev Biophys Biomol Struct 23:349–375, 1994.

52. DM Lorber, BK Shoichet. Flexible ligand docking using conformational ensembles. Protein Sci 7:938–950, 1998.

53. TJA Ewing, ID Kuntz. Critical evaluation of search algorithms for automated molecular docking and database screening. J Comput Chem 18:1175–1189, 1997.

54. J Aqvist, C Medina, J-E Samuelsson. A new method for predicting binding affinity in computer-aided drug design. Protein Eng 7:385–391, 1994.

55. HA Carlson, WL Jorgensen. An extended linear response method for determining free energies of hydration. J Phys Chem 99:10667, 1995.

56. X Chen, A Tropsha. A generalized linear response method: Application to the hydration free energy calculations. J Comput Chem 20:749–759, 1999.

57. J Apostolakis, A Caflisch. Computational ligand design. Comb Chem High Throughput Screen 2:91–104, 1999.

58. JW Pitera, NR Munagala, CC Wang, PA Kollman. Understanding substrate specificity in human and parasite phosphoribosyltransferases through calculation and experiment. Biochemistry 38:10298–10306, 1999.

59. J Marelius, M Graffner-Nordberg, T Hansson, A Hallberg, J Aqvist. Computation of affinity and selectivity: Binding of 2,4-diaminopteridine and 2,4-diaminoquinazoline inhibitors to dihydrofolate reductases. J Comput-Aided Mol Des 12:119–131, 1998.

60. RH Smith Jr, WL Jorgensen, J Tirado-Rives, ML Lamb, PAJ Janssen, CJ Michejda, MBK Smith. Prediction of binding affinities for TIBO inhibitors of HIV-1 reverse transcriptase using Monte Carlo simulations in a linear response method. J Med Chem 41:5272–5286, 1998.

61. EM Gordon, RW Barret, WJ Dower, SPA Fodor, MA Gallop. Applications of combinatorial technologies to drug discovery. 2. Combinatorial organic synthesis, library screening strategies, and future directions. J Med Chem 37:1385–1401, 1994.

62. MA Gallop, RW Barret, WJ Dower, SPA Fodor, EM Gordon. Applications of combinatorial technologies to drug discovery. 1. Background and peptide combinatorial libraries. J Med Chem 37:1233–1251, 1994.

63. WA Warr. Combinatorial chemistry and molecular diversity. An overview. J Chem Inf Comput Sci 37:134–140, 1997.

64. RP Sheridan, SK Kearsley. Using a genetic algorithm to suggest combinatorial libraries. J Chem Inf Comput Sci 35:310–320, 1995.

65. W Zheng, SJ Cho, A Tropsha. Rational combinatorial library design 1. Focus-2D: A new approach to the design of targeted combinatorial chemical libraries. J Chem Inf Comput Sci 38:251–258, 1998.

66. SJ Cho, W Zheng, A Tropsha. Rational combinatorial library design 2. Rational design of targeted combinatorial peptide libraries using chemical similarity probe and the inverse QSAR approaches. J Chem Inf Comput Sci 38:259–268, 1998.

67. Q Zheng, DJ Kyle. Computational screening of combinatorial libraries. Bioorg Med Chem 4(5):631–638, 1996.

68. EK Kick, DC Roe, AG Skillman, G Liu, TJ Ewing, Y Sun, ID Kuntz, JA Ellman. Structure-based design and combinatorial chemistry yield low nanomolar inhibitors of cathepsin D. Chem Biol 4(4):297–307, 1997.

69. RD Brown, YC Martin. Designing combinatorial library mixtures using a genetic algorithm. J Med Chem 40(15):2304–2313, 1997.

70. P Willett. Algorithms for the calculation of similarity in chemical structure databases. In: MA Johnson, GM Maggiora, eds. Concepts and Applications of Molecular Similarity. New York: Wiley, 1990, pp 43–63.

71. NC Perry, VJ van Geerestein. Database searching on the basis of three-dimensional molecular similarity using the SPERM program. J Chem Inf Comput Sci 32(6):607, 1992.

72. DA Gschwend, AC Good, ID Kuntz. Molecular docking towards drug discovery. J Mol Recogn 9(2):175–186, 1996.

73. P Willett, V Winterman, D Bawden. Implementation of nonhierarchic cluster analysis methods in chemical information systems: Selection of compounds for biological testing and clustering of substructures search output. J Chem Inf Comput Sci 26:109–118, 1986.

74. RS Pearlman. Novel software tools for addressing chemical diversity. Network Sci 1996. http://www.awod.com/netsci/Science/Combichem/feature08.html

75. EJ Martin, JM Blaney, MA Siani, DC Spellmeyer, AK Wong, WH Moos. Measuring diversity:

Experimental design of combinatorial libraries for drug discovery. J Med Chem 38(9):1431–1436, 1995.

76. W Zheng, SJ Cho, CL Waller, A Tropsha. Rational combinatorial library design. 3. Simulated annealing guided evaluation (SAGE) of molecular diversity: A novel computational tool for diverse library design and database mining. J Chem Inf Comput Sci 39:738–746, 1999.

77. H Bauknecht, A Zell, H Bayer, P Levi, M Wagener, J Sadowski, J Gasteiger. Locating biologically active compounds in medium-sized heterogeneous datasets by topological autocorrelation vectors: Dopamine and benzodiazepine agonists. J Chem Inf Comput Sci 36(6):1205–1213, 1996.

78. A Polinski, RD Feinstein, S Shi, A Kuki. LiBrain: Software for automated design of exploratory and targeted combinatorial libraries. In: IM Chaiken, KD Janda, eds. Molecular Diversity and Combinatorial Chemistry: Libraries and Drug Discovery. ACS Conf Proc Ser. Washington, DC: Am Chem Soc, 1996, pp 219–232.

79. DB Turner, SM Tyrell, P Willett. Rapid quantification of molecular diversity for selective database acquisition. J Chem Inf Comput Sci 37:18–22, 1997.

80. DJ Cummins, CW Andrews, JA Bentley, M Cory. Molecular diversity in chemical databases: Comparison of medicinal chemistry knowledge bases and databases of commercially available compounds. J Chem Inf Comput Sci 36:750–763, 1996.

81. S Pickett, JS Mason, IM McLay. Diversity profiling and design using 3D pharmacophores: Pharmacophore-derived queries (PDQ). J Chem Inf Comput Sci 36:1214–1223, 1996.

82. PL Myers, JW Greene, J Saunders, SL Teig. Rapid, reliable drug discovery. Today's Chemist at Work July/August 1997, pp 46–53.

83. AC Good, R Lewis. New methodology for profiling combinatorial libraries and screening sets: Cleaning up the design process with HARPick. J Med Chem 40:3926–3936, 1997.

84. VJ Gillet, P Willett, J Bradshaw. The effectiveness of reactant pools for generating structurally-diverse combinatorial libraries. J Chem Inf Comput Sci 37:731–740, 1997.

85. YC Martin, MG Bures, EA Danaher, J DeLazzer, I Lico, PA Pavlik. A fast new approach to pharmacophore mapping and its application to dopaminergic and benzodiazepine agonists. J Comput Aided Mol Des 7:83–102, 1993.

86. S Wang, DW Zaharevitz, R Sharma, VE Marquez, NE Lewin, L Du, PM Blumberg, GWA Milne. The discovery of novel, structurally diverse protein kinase C agonists through computer 3D-database pharmacophore search. J Med Chem 37:4479–4489, 1994.

87. SA DePriest, D Mayer, CB Naylor, GR Marshall. 3D-QSAR of angiotensin-converting enzyme and thermolysin inhibitors: A comparison of CoMFA models based on deduced and experimentally determined active site geometries. J Am Chem Soc 115:5372–5384, 1993.

88. PJ Goodford. A computational procedure for determining energetically favorable binding sites on biologically important macromolecules. J Med Chem 28:849–957, 1985.

89. FC Bernstein, TF Koetzle, GJB Williams, EF Meyer, MD Brice, JR Rodgers, O Kennard, T Shimanouchi, M Tasumi. Protein Data Bank: Computer based archival file for macromolecular structure. J Mol Biol 112:535–542, 1977.

90. DM Ferguson, RJ Radmer, PA Kollman. Determination of the relative binding free energies of peptide inhibitors to the HIV-1 protease. J Med Chem 34:2654–2659, 1991.

17

Protein Folding: Computational Approaches

Oren M. Becker

Tel Aviv University, Tel Aviv, Israel

I. INTRODUCTION

''Protein folding'' is the term used to describe the complex process in which polypeptide chains adopt their three-dimensional ''native'' conformation. To carry out their functions, proteins must fold rapidly and reliably. They must satisfy a kinetic requirement that folding can be completed within a reasonable time and a thermodynamic requirement that the folded conformation be stable under physiological conditions. Although the folding process in a cell also involves catalytic and control mechanisms, for many, if not all, proteins, the information for folding is contained primarily in the amino acid sequence. Because there are so many possible conformations for any given polypeptide chain, these requirements mean that protein folding must be many orders of magnitude faster than a random search through conformation space [1]. An estimate of this speedup can be obtained by conservatively assuming that each amino acid residue has three possible conformations. If a protein is made up of 100 amino acids, there are about 10^{49} possible conformations for the entire polypeptide chain. Even if the time required to change from one conformation to another is as little as 10 ps (1 picosecond = 10^{-12} s), a random search through all of conformation space would still require 10^{36} s, or about 10^{29} years. This estimate, often referred to as the ''Levinthal paradox,'' clearly indicates that protein folding is not a random search but rather follows a built-in bias toward the native state.

An important feature of protein folding is that the amino acid sequence of the protein uniquely determines its overall structure [2], which is a combination of secondary structure (the regions of α-helix and β-sheet) and tertiary structure (the overall folding pattern). Differences in sequences give rise to differences in secondary and tertiary structure. So far the three-dimensional structures of approximately 6000 proteins have been determined by X-ray crystallography and NMR spectroscopy. The domains in these proteins can be grouped into approximately 350 families of folds, which consist of sequences that have similar structures [3]. It has been estimated that the total number of different folds is only on the order of 1000 [3–5]. This number is much smaller than the total number of different sequences in the human genome, which is on the order of 100,000. Some of these folds are observed in a large number of sequences, whereas others have been found, so far, in only a small number of instances. The frequency with which a fold occurs is probably related to the stability of the fold or to the speed of the folding process.

It should be noted that in almost all cases only one fold exists for any given sequence. The uniqueness of the native state arises from the fact that the interactions that stabilize the native structure significantly destabilize alternate folds of the same amino acid sequence. That is, evolution has selected sequences with a deep energy minimum for the native state, thus eliminating misfolded or partly unfolded structures at physiological temperatures.

The process of protein folding is one of the most fundamental biophysical processes. It is of interest also because of the important role it plays in the mechanisms and controls of a wide range of cellular processes. These include regulation of complex events during the cell cycle and translocation of proteins across membranes to their appropriate organelles [6]. Furthermore, it is known that the failure of proteins to fold correctly is associated with the malfunction of biological systems, leading to a broad range of diseases. Some of these diseases, such as Alzheimer's and Creutzfeldt-Jakob diseases, are associated with the conversion of normal soluble proteins into insoluble aggregated amyloid plaques and fibrils [7]. Others, for example cystic fibrosis, result from mutations that hinder the normal folding and secretion of specific proteins [8].

As with any other chemical reaction, understanding protein folding requires knowledge of the interactions that dominate it as well as an insight into the kinetics and dynamics of the process. Substantial progress has been made in recent decades toward achieving such an understanding for simple chemical reactions [9]; however, our current knowledge is less advanced with regard to the more complex protein folding reaction. Nonetheless, in the last decade or so substantial strides toward a comprehensive understanding of the folding process were achieved through a combination of theoretical and experimental studies. It is important to note that protein folding reactions have very different characteristics from reactions of small molecules. For example, experiments have shown that although the Arrhenius equation can often be applied to protein folding, the preexponential factor has a strong temperature dependence [10]. Furthermore, under physiological conditions the free energy of the native state of a protein is only slightly lower than that of the unfolded state. This is due to a near cancelation of large energetic and entropic contributions. The energetics of protein folding are dominated by the nonbonded van der Waals and electrostatic terms in the potential energy function (see Chapter 2), including both intramolecular interactions between the atoms of the protein and intermolecular interactions between the protein atoms and the solvent [11]. In particular, it was found that the nonpolar (hydrophobic) groups strongly favor the folded state due to the attractive van der Waals interactions in the native structure and to the hydrophobic effect, which favors the burial of nonpolar groups. By contrast, polar groups (the peptide groups and the polar and charged side chains) contribute much less to the stability of the native state due to a balance between the interactions in the interior of the protein and those with the solvent. For example, in lysozyme [12], calculations show that at 25°C the nonpolar groups contribute 450 kcal/mol whereas the polar groups contribute only 87 kcal/mol to the free energy of denaturation. The overall stabilization of the native state due to these energy interactions (about 537 kcal/mol) is counterbalanced by a configuration entropy contribution of about 523 kcal/mol at 25°C. This yields a net free energy of unfolding of only 14 kcal/mol (on the order of 0.1 kcal/mol per residue), which is a typical value for globular proteins. In contrast, the energy or enthalpy difference between the native and unfolded states can be significantly larger; for lysozyme at 25°C, the unfolding enthalpy is 58 kcal/mol [12].

Chemical reactions, including protein folding, are best understood from the vantage point of their underlying "energy landscapes," which are theoretical manifestations of the interactions that contribute to the chemical processes. An energy landscape is a surface defined over conformation space indicating the potential energy of each and every possible conformation of the molecule. Similar to regular topographic landscapes, valleys in an energy landscape indicate stable low energy conformations and mountains indicate unstable high energy conformations. However, although reactions of small molecules can be characterized directly by the potential energy landscape, the high dimensionality of protein conformation spaces often makes a temperature-dependent effective energy landscape (or free energy landscape) the theoretical framework of choice. Such a surface corresponds to a Boltzmann weighted average of the accessible energies along an appropriately chosen reaction coordinate (or progress variable). The latter, which describes the approach to the native state, is obtained by averaging over many nonessential degrees of freedom. Such a reaction coordinate describes the progress of the reaction from the initial to the final state but includes the possibility of many different paths on the original high dimensional energy landscape.

Because protein folding is determined primarily by the amino acid sequence, the difference between foldable sequences and unfoldable sequences should be manifested in their underlying energy landscapes. A folding sequence is expected to have the energy of its conformations proportional to a reaction coordinate Q, with some roughness that is introduced by non-native contacts. This correlation of energy and structure introduces a bias in favor of the native conformation as well as a bias against conformations that are significantly different from the native structure. Such a correlation is responsible for the funnel shape of the landscape (Fig. 1b). A random sequence will not exhibit such a correlation between energy and conformation, and the corresponding energy landscape is expected to be rough (Fig. 1a). Because proteins are finite systems, if they have a single native state there is always a temperature below which the native state is stable. This temperature is called the folding temperature, T_f. On the other hand, due to the roughness of the landscape there is also a temperature below which the kinetics are controlled by nonnative traps and not by the bias toward the native state. This temperature is denoted as T_g, in reference to a similar transition temperature in glasses. For a sequence to fold it is necessary that the folding temperature be higher than the glass temperature, $T_f > T_g$. That is, the competition between the energetic bias toward the native state and the landscape's roughness plays a central role in the folding process, leading to a diversity of folding scenarios [13–15].

Thus, energy landscape theory offers a solution to many of the kinetic and thermodynamic perplexities of protein folding. The kinetic bias toward the native state is explained as an overall bias in the energy landscape itself, where a large depression or "funnel" around the native state biases the folding process toward this structure. An interplay between the bias toward the native conformation, the relative stability of that structure, and the roughness of the landscape gives rise to the non-Arrhenius temperature dependence of the folding process, highlighting the interplay of energy and entropy. Furthermore, the unique topography and the multidimensionality of the landscape allow for multiple folding pathways that pass a multidimensional folding "seam" (rather than a single one-dimensional barrier) that can still be described by an average reaction coordinate.

Naturally, the pivotal role of protein folding in biophysics and biochemistry has yielded a very large body of research. In this chapter we focus primarily on the different theoretical and computational approaches that have contributed to the current understand-

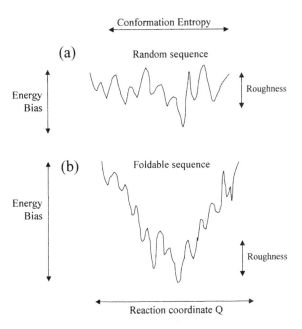

Figure 1 (a) Schematic energy landscape for a random unfoldable heteropolymer. The roughness is on the order of the energy bias, and the sequence is likely to be trapped in low energy states far from its native state. (b) A schematic energy landscape for a foldable proteinlike heteropolymer. The funnel-like topography is characterized by an energy bias toward the native state that is much larger than the roughness of the landscape.

ing of the folding process. Discussion of other aspects of protein folding can be found in many excellent reviews, which address this topic from different points of view. Some of these reviews can be found in Refs. 11, 13, 14, and 16–21.

II. SIMPLE MODELS

Significant theoretical progress in understanding protein folding has been achieved by examining the properties of simple models of energy landscapes. Such models often look at proteins as a special class of heteropolymers. Whereas proteinlike heteropolymers have a well-defined three-dimensional conformation, random heteropolymers with a tendency to collapse do not have such a conformation but rather a collection of different low energy structures. The "minimally frustrated random energy model" introduced by Bryngelson and Wolynes [22] is one of the more successful models for protein folding. The model is based on two assumptions: (1) The energies of non-native contacts may be taken as random variables, and (2) on average, the overall energy of the protein decreases as the protein comes to look more like the native state, regardless of the measure used to gauge its similarity. This second assumption implies that there is an overall energy bias toward the native state.

Representing a heteropolymer, the model tries to capture three contributions to the overall energy E of each conformation: the self-energy ε of each amino acid, a bond energy term $J_{i,i+1}$ between two neighboring residues, and a nonbonded $K_{i,j}$ interaction term, primarily for hydrophobic interactions, that draws the amino acids close to each other,

$$E = -\sum_i \varepsilon_i - \sum_i J_{i,i+1} - \sum_{i,j} K_{i,j} \tag{1}$$

Each of these terms depends on the specific state α_i of each of the N residues and on its position r_i. That is, the energy E is a complex function of $\{\alpha_i\}$ and $\{r_i\}$. Since this problem is too complicated to be solved by standard ensemble statistical mechanics, the researchers replace the above complex Hamiltonian with a stochastic one that shares the same statistical properties. The energy of the protein is thus taken to be a random variable assigned from a distribution that has the same characteristics as the full Hamiltonian, following a technique developed in the study of spin glasses. This generates the so-called random energy model. The bias toward the native state is introduced via the nonbonded interaction term K. In the random energy model, non-native interactions are randomly selected from a distribution of energies with a mean value of $-\langle K \rangle$ and standard deviation ΔK. Only native nonbonded interactions are consistently assigned the value $-K$, where $K > \langle K \rangle$. This proteinlike model is then subjected to an in-depth analysis of its thermodynamics and kinetics, using a single order parameter to describe the distance from the native state. The kinetics of this model were studied for two variants that differ in the kinetic connectivity between different states [23]. In one variant the landscape was "locally connected," meaning that only states with a similar value of the order parameter, which describes similarity to the native state, were kinetically accessible to one another. In the other variant, "global connectivity" between the energy states was assumed.

Kinetic studies such as these use the "master equation" to follow the flow of probability between the states of the model. This equation is a basic loss–gain equation that describes the time evolution of the probability $p_i(t)$ for finding the system in state i [24]. The basic form of this equation is

$$\frac{dp_i(t)}{dt} = \sum_j [W_{ij}p_j(t) - W_{ji}p_i(t)] \tag{2}$$

where W_{ij} is the transition probability from state j to state i. Equation (2) can be rewritten in matrix form by defining the transition matrix elements as

$$\mathbf{W}_{ij} = W_{ij} - \delta_{ij}\left(\sum_k W_{ki}\right) \tag{3}$$

The matrix $\mathbf{W}$ has the properties that $\mathbf{W}_{ij} \geq 0$ for $i \neq j$ and that the sum over each column is zero; i.e., $\sum_i \mathbf{W}_{ij} = 0$ for all j. This last property is required for a closed system so that the flux out of any given state remains within the system (i.e., goes into the other states of the system). In matrix form Eq. (3) becomes

$$\dot{\mathbf{p}}(t) = \mathbf{W}\mathbf{p}(t) \tag{4}$$

which has the formal solution $\mathbf{p}(t) = e^{t\mathbf{W}}\,\mathbf{p}(0)$, where $\mathbf{p}(t)$ is the probability vector at time t.

Solving the master equation for the "minimally frustrated random energy model" showed that the kinetics depend on the connectivity [23]. For the "globally connected" model it was found that the resulting kinetics vary as a function of the energy gap between the folded and unfolded states and the roughness of the energy landscape. The model yielded nonlinear Arrhenius plots that resemble those seen experimentally. It also pointed to the presence of kinetic intermediates that are actually misfolded "traps" and not necessary steps for folding. On the other hand, the "locally connected" model resulted in significantly different kinetics. In one regime of the parameters the overall reaction rate was determined by the rate of going through a bottleneck region (in terms of the order parameter) that corresponds to the state of highest free energy. In other regimes, close to the glass transition, the rate was limited by search through misfolded states.

Another simple model of protein folding kinetics was suggested by Zwanzig [25]. This model assumes that the energy depends solely on the sequence and can be described as a simple function of the distance S between a given conformation and the native state. If N "parameters" (e.g., dihedral angles) characterize the native conformation, then S is the number of parameters in a given conformation that have non-native values. The energy in this model is defined as

$$E_S = SU - \varepsilon \delta_{SU} \tag{5}$$

where $S = 1, 2, \ldots, N$ and both U and ε are assumed to be positive. The positive U ensures a smooth funnel as the energy increases with increasing S, and the positive ε ensures an energy gap between $S = 0$ and $S = 1$. That is, the reaction coordinate is the similarity of a conformation to the native state. The model employs a gap in the energy spectrum, has large configuration entropy, and exhibits a free energy barrier between folded and partially folded states. The folding time in this model was estimated by means of a local thermodynamic equilibrium assumption followed by solving the master equation. It was found that the above set of rules leads to an energy landscape that has two basins, one corresponding to the native state and the other corresponding to an ensemble of partially folded states. Following a short equilibration time the overall kinetics are similar to those of fast-folding two-state systems. The folding time has a maximum near the folding transition temperature and can have a minimum at lower temperatures.

III. LATTICE MODELS

The current understanding of the protein folding process has benefited much from studies that focus on computer simulations of simplified lattice models. These studies try to construct as simple a model as possible that will capture some of the more important properties of the real polypeptide chain. Once such a model is defined it can be explored and studied at a level of detail that is hard to achieve with more realistic (and thus more complex) atomistic models.

In a lattice model the protein is represented as a "string of beads" threaded on a lattice (often denoted as a "self-avoiding walk" on a lattice). Each residue is positioned on a different grid point, and specific nearest-neighbor interactions, which depend on the residues involved, are defined. Once the model is defined the folding process is simulated by local Monte Carlo moves that change the position of the "beads" on the lattice until the chain reaches its lowest energy configuration. In many studies a simple square [20] or a cubic grid was used [26–28], although more complex lattices have also been employed [29,72]. Figure 2 illustrates a simple polypeptide chain with 27 amino acids (27-mer) folded on a $3 \times 3 \times 3$ cubic lattice. All in all there are on the order of 10^{16} conformations of a 27-mer chain on an infinite cubic lattice. Due to an overall attraction between the residues (primarily of hydrophobic nature), the native state of the model protein is "col-

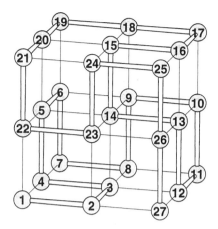

Figure 2 A low energy conformation of a 27-mer lattice model on a 3 × 3 × 3 cubic lattice. (Adapted from Ref. 11.)

lapsed'' and can be fit into a 3 × 3 × 3 cube, which is fully occupied by the polypeptide chain. There are more than 100,000 ways to fit a 27-mer into this cube. The most stable conformation, which corresponds to the native state, is determined by the specific interactions dictated by amino acid sequence. Different sequences are likely to have different native conformations, even in the simplified lattice representation.

As discussed above, folding is driven by nonbonded interactions. In lattice models this is represented by ''contact energies,'' i.e., interactions between residues that are situated on adjacent (or nearest-neighbor) lattice sites but are not covalently bonded to each other. For example, since there are 28 nearest-neighbor contacts in the native structure of a 27-mer in a 3 × 3 × 3 cube, each conformation of this model can be characterized in terms of how many of these native contacts are correctly formed. Indeed, in most lattice models simple contact potentials are thus used to represent the effective energy of a given configuration. The combination of a simple model, which enables extensive enumeration of conformations, together with a simple ''contact'' energy function allows such model studies to determine the thermodynamics and dynamics of the system within a reasonable amount of computer time.

The ''contact'' energy E of a given conformation is typically calculated by summing the values of energies over all nonbonded contacts in the lattice,

$$E = \sum_{\text{neighbors}} \varepsilon(S_i, S_j) \Delta(\vec{r}_i - \vec{r}_j) \tag{6}$$

where $\vec{r}_i$ and $\vec{r}_j$ denote the locations of residues i and j and $\Delta(\vec{r}_i - \vec{r}_j) = 0$ unless residues i and j are on adjacent vertices of the lattice. The term $\varepsilon(s_i, s_j)$ indicates the nonbonded neighboring interaction between a residue of type s_i and a residue of type s_j. These contact interactions are typically on the order of $k_B T$.

Despite their simplicity, certainly compared to the all-atom potentials used in molecular dynamics studies, these contact energy functions enable the exploration of different interaction scenarios. This diversity is achieved by changing the heterogeneity of the sequence, by altering the number N of different types of ''residues'' that are being used. The most elementary lattice model involves only two types of monomers: hydrophobic monomers (H) and polar monomers (P). Such a model is often referred to as an HP model. In HP models, only nearest-neighbor contacts of the type HH have a stabilizing contribution $\varepsilon < 0$ to the overall energy, whereas all other contact types, whether HP or PP, are considered neutral, contributing zero energy [18,30]. It was found that most HP model sequences have rugged energy landscapes with many kinetic traps [20]. In this case, folding kinetics involve at least two stages: a fast collapse to compact non-native conformations followed by a slow barrier-climbing process to escape traps and reattempt to fold [18,30,31].

In this respect, the HP model is unlike many real proteins that appear to have smoother landscapes with few traps, resulting in fast folding and two-state kinetics [11,21]. One way to make the model more proteinlike is to increase its heterogeneity. Another way is to introduce a specific bias toward the native state, resulting in a variant model denoted as the HP+ model [20]. For an HP sequence with a unique native structure, the HP+ energy given by Eq. (6) is defined by a negative ε value for each native HH contact, by $\varepsilon = 0$ for each native HP or PP contact, and by $-\varepsilon$ for all non-native contacts ($-\varepsilon > 0$). As a result the HP and its corresponding HP+ sequences share the same unique native structure, with the only difference being that in the HP+ energy function non-native contacts have unfavorable energies. This extra interaction in the HP+ model is intended to capture, in a very simple way and without additional parameters, more energetic specificity than the original HP model. The HP+ model is similar in principle to the "Go model," which adds an explicit biasing potential to the native structure, ensuring that this structure becomes the global minimum of the whole energy landscape [32,33].

Agreement with the real protein folding process can be obtained by increasing the heterogeneity of the lattice model, using multiple-letter codes and sequence design [26,27,34–37]. A model with 20 different residue types ($N = 20$) is expected to have heterogeneity similar to that of a real protein. In such models the energy is taken from a range of interaction energies, ensuring an overall net attraction. For example, contact energies between adjacent residues may be chosen to have an average of $-2k_BT$ with an effective deviation of about k_BT, ensuring that the stable native contacts are among the most stable nonbonded interactions, with an average energy of about $-3k_BT$ [26]. In other studies the interactions were selected randomly from a continuous range of interactions with special terms to prevent the chain from crossing over itself [34]. Overall, these more complex models show kinetic pathways that converge into folding funnels, guiding the folding to a unique stable native conformation.

A convenient property of all lattice models is the ability to use the "fraction of native contacts" Q as a reaction coordinate or progress variable to describe the folding process. The variable is the ratio between the number of "native contacts" that are observed in any given conformation of the chain and the maximum number of possible native contacts. Thus, Q varies from a value near zero for the highly denatured conformation to unity for the native state. For the 27-mer in a $3 \times 3 \times 3$ cube described above, there are 156 different possible contacts and 28 native contacts. For a 125-mer there are 3782 possible contacts and 176 native contacts in a $5 \times 5 \times 5$ cube [11]. Although there are many more "native contacts" in a real protein, it is expected that even there a smaller subset of contacts can be used to define the native conformation in a way similar to the Q variable in lattice models. The progress variable Q has been very useful for visualizing the average effective energy and the configuration entropy of the polypeptide chain as it folds from the denatured to the native state. The resulting values, which are averaged over many folding simulations, depend as expected on the temperature at which the simulation is performed.

Like real proteins, lattice models have a narrow optimal temperature range in which the folding process is most efficient. At temperatures that are too low, folding may be extremely slow because the chain cannot escape from local minima. At very high temperatures the native state is not stable, and the number of accessible conformations is so large that the folding problem cannot be solved. Indeed, analysis of a low temperature average effective energy/entropy surface calculated for the 27-mer model on a cubic lattice showed that the conformation space accessible to the protein is limited, even at low Q (unfolded conformations) [11]. At such temperatures the polypeptide chain collapses to a misfolded globular state with a Q value near that of the random coil. The change in configuration entropy on collapse is small enough that its destabilizing contribution to the free energy is compensated for by the burial of hydrophobic groups, even in the absence of native contacts. At this temperature the average effective energy surface as a function of Q is ''rough'' due to the presence of energy barriers to reorganization within the collapsed state. The transition region at these temperatures was found to be close to the native state ($Q = 0.7$–0.9).

At high folding temperatures, on the other hand, the average effective energy/entropy surface resulting from lattice simulations indicates a different scenario [11]. Early in folding (e.g., for $Q = 0.2$), the surface is very broad, indicating that most of the unfolded configurations are accessible. As the entropy decreases with the increase of Q to unity for the native structure, the surface becomes narrower, resulting in an overall funnel structure for the average effective energy surface. Thus, regardless of the initial conformation, the molecule moves downward in energy toward the native state as the number of stabilizing contacts increases. Despite the smoothness of the effective energy surface, a transition state barrier in the free energy profile can exist even for the 27-mer at relatively high temperatures. The free energy transition barrier corresponds to an entropy ''bottleneck'' that arises from a reduction of the chain entropy at large Q values (the number of accessible configurations decreases rapidly as Q approaches the native state). In general, it is the balance between the rate of decrease of the energy and that of the entropy that determines whether there is a free energy barrier and where it occurs. A different balance between the two contributions to the free energy could move the transition barrier in the free energy to smaller or larger Q values.

To conclude, although the models used in lattice simulations are very simplified, the results provide general information on possible protein folding scenarios, albeit not on the detailed behavior of specific proteins, which would require more complex models and more accurate potentials. The contribution made by these simulations is that they enable an analysis of the structures, energetics, and dynamics of folding reactions at a level of detail not accessible to experiment.

IV. OFF-LATTICE MINIMALIST MODELS

Despite their contribution to the understanding of protein folding, the correspondence between lattice models and real proteins is still very limited. The first step toward making such models more realistic is to remove the lattice and study off-lattice minimalist models. Simple off-lattice models of proteins can have proteinlike shapes with well-defined secondary structure elements, as in real proteins. In addition, the continuum character of the conformation space allows for the native state to become a basin rather than a single minimum.

An off-lattice minimalist model that has been extensively studied is the 46-mer β-barrel model, which has a native state characterized by a four-stranded β-barrel. The first to introduce this model were Honeycutt and Thirumalai [38], who used a three-letter code to describe the residues. In this model monomers are labeled hydrophobic (H), hydrophilic (P), or neutral (N) and the sequence that was studied is $(H)_9(N)_3(PH)_4(N)_3(H)_9(N)_3(PH)_5P$. That is, two strands are hydrophobic (residues 1–9 and 24–32) and the other two strands contain alternating H and P beads (residues 12–20 and 36–46). The four strands are connected by neutral three-residue bends. Figure 3 depicts the global minimum conformation of the 46-mer β-barrel model. This β-barrel model was studied by several researchers [38–41], and additional off-lattice minimalist models of α-helical [42] and β-sheet proteins [43] were also investigated.

The energy function of the off-lattice three-letter model is much more elaborate than those used in lattice models [Eq. (6)]. Similar to all-atom energy functions, it includes both bonded and nonbonded energy terms. Bond, bond angle, and dihedral angle energy terms give the model flexibility along the bonded structure while a nonbonded van der Waals interaction term is used to mimic the hydrophobic/hydrophilic character of the different monomer types.

$$E = \{\text{bonds}\} + \{\text{angles}\} + \{\text{dihedral}\} + \sum_{i<j-3}\left\{4\varepsilon S_1\left[\left(\frac{\sigma}{R}\right)^{12} - S_2\left(\frac{\sigma}{R}\right)^6\right]\right\} \quad (7)$$

where the bonded energy terms are similar to those used in all-atom models (see Chapter 2), and the parameters S_1 and S_2 in the van der Waals term distinguishes between the different types of beads. There are attractive interactions between all HH residue pairs ($S_1 = 1$ and $S_2 = 1$), repulsion interaction between all PP and PH pairs ($S_1 = 2/3$ and $S_2 = -1$), and only excluded volume interactions between the pairs PN, HN, and NN ($S_1 = 1$ and $S_2 = 0$).

Studies of this model showed that the underlying energy landscape is very rough, probably due to the long-range and nonspecific character of the interactions. To overcome the roughness and smooth the surface, a "Go model"-like variant of the three-letter model was introduced [15]. In this variant the only attractive interactions are those between monomers that form native contacts, i.e., contacts found in the native β-barrel. An analysis of the native β-barrel structure yielded 47 pairs of monomers within a distance of 1.167σ, most of them between hydrophobic monomers. All other pairs have only the repulsive van der Waals term, which accounts for excluded volume. It was shown that this modification removes the roughness that is introduced by the non-native contacts, allowing the sequence to recover a nearly optimal folding behavior.

Recently a different modification of the classic 46-mer β-barrel model was suggested. In this case a single side group, represented by a bead that may be hydrophilic or hydrophobic, was added to the model [44]. Molecular dynamics and quenching simulations showed that the nature and the location of the single side group bead influences both the structure at the global minimum of internal energy and the relaxation process by which the system finds its minima. The most drastic effects occur with a hydrophobic side group in the middle of a sequence of hydrophobic residues.

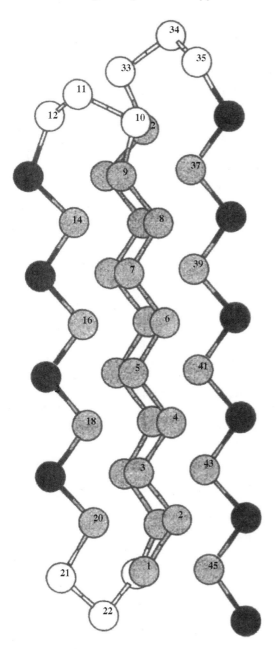

Figure 3 The minimum energy conformation of the off-lattice 46-mer β-barrel model. Hydrophobic residues are in gray, hydrophilic residues in black, and neutral residues are white. (Adapted from Ref. 44.)

V. ATOMISTIC MODELS

The highest level of detail in theoretical studies of protein folding involves the use of detailed atomic models of the protein and the environment. Such models have been discussed in depth in previous chapters of this book. The main limitation of atomic models is that they are computationally much more demanding, a fact that restricts the number of calculations that can be performed with them. In terms of using atomic models for protein folding it is possible to identify two main approaches. The first approach is to study the folding process by performing explicit molecular dynamics simulations of protein unfolding and folding. The other approach is to use conformation sampling techniques to characterize the underlying energy and free energy landscapes.

A. Unfolding/Folding Simulations

The main problem facing the attempt to study room temperature folding by direct molecular dynamics simulations of an all-atom model is that of time scales. Whereas protein folding takes place on the millisecond time scale and up, the time scale accessible to molecular dynamics is on the order of nanoseconds. Recently, using a massively parallel computer, Duan and Kollman [45] performed a 1 μs simulation of the villin headgroup subdomain protein, a 36-residue peptide, in water. Starting from a fully unfolded extended state, including approximately 3000 water molecules, the simulation was able to follow the dynamics of this protein as it adopted a partially folded conformation. Such long-time-scale molecular dynamics (MD) simulations require exceptionally large computational resources. Furthermore, the usefulness of these simulations is limited by the fact that they cannot provide the level of statistics required for studying folding kinetics and thermodynamics. Another problem associated with a direct MD approach to the folding process is that it is unclear how well the MD potential energy functions used fare in the unfolded regime.

Thus, instead of using molecular dynamics to simulate the folding process, many researchers turned their attention to using MD simulations as a tool for studying the inverse process of protein unfolding from the native state. It is hoped, though not proven, that analysis of the unfolding process will contribute to the understanding of the folding process. To speed up the unfolding reaction, which has a significant activation barrier, these studies are typically performed in the high temperature range of 400–600 K. A simple Arrhenius-type calculation shows that the unfolding reaction for a protein that denatures experimentally at 325 K and has an activation barrier for unfolding of 20 kcal/mol is about six orders of magnitude faster at 600 K than at 325 K. Even if the Arrhenius equation is not exact for unfolding reactions, this argument indicates that elevating the temperature reduces the time for unfolding from the experimentally observed millisecond range to the nanosecond time scale, which is accessible to molecular dynamics simulations.

The details of many all-atom unfolding simulation studies have been summarized in several reviews [17,46,47]. These studies include unfolding simulations of α-lactalbumin, lysozyme, bovine pancreatic trypsin inhibitor (BPTI), barnase, apomyoglobin, β-lactamase, and more. The advantage of these simulations is that they provide much more detailed information than is available from experiment. However, it should be stressed that there is still only limited evidence that the pathways and intermediates observed in the nanosecond unfolding simulations correlate with the intermediates observed in the actual experiments.

Of specific interest are the unfolding simulation studies that highlight the role of the solvent in the folding and unfolding process, an insight that is very difficult to obtain experimentally. For example, simulations of the early stages of barnase unfolding at high temperature [47] showed that solvent plays a key role in the denaturation process. It was found that an important element of the helix-unfolding transition is the replacement of an α-helical hydrogen bond (i to $i + 4$, where i is an amino acid residue) by water hydrogen bonds through an intermediate involving a 3_{10} (i to $i + 3$), or reverse turn, hydrogen bond. Denaturation of a β-sheet was also observed to start by the distortion of the β-sheet hydrogen bonds, followed by the insertion of hydrogen-bonding water molecules between the strands. Finally, significant solvent participation was found even in the denaturation of the central stabilizing element of globular proteins—the hydrophobic core. This happens as some water molecules form "cage structures" around hydrophobic groups, often involving hydrogen bonds to water molecules outside the core. There are, however, concerns as to whether the observed water behavior corresponds to the actual denaturation process. The reason is that high temperature unfolding simulations are done either with a room temperature water density [47] or with low water density followed by rapid water penetration when the temperature is set equal to room temperature [48,49]. These procedures create an artificially high pressure, which may force water into protein cavities. Nonetheless, comparisons of unfolding simulations results at different temperatures seem to indicate that this effect is not very great [17].

B. Mapping Atomistic Energy Landscapes

An alternative approach to the study of protein folding on an atomic level is to base the study on conformation sampling rather than on direct simulation of the folding process. Sampling of folded and unfolded conformations allows for reconstructing the underlying energy landscape and for deducing the folding pathway (or pathways) from it.

In principle, energy landscapes are characterized by their local minima, which correspond to locally stable conformations, and by the transition regions (barriers) that connect the minima. In small systems, which have only a few minima, it is possible to use a direct approach to identify all the local minima and thus to describe the entire potential energy surface. Such is the case for small reactive systems [9] and for the alanine dipeptide, which has only two significant degrees of freedom [50,51]. The direct approach becomes impractical, however, for larger systems with many degrees of freedom that are characterized by a multitude of local minima.

A useful procedure for characterizing the multiminimum energy landscape of large systems was introduced by Stillinger and Weber [52]. These researchers investigated the energy landscape of water by quenching (i.e., minimizing) configurations from a molecular dynamics trajectory down to their nearest local minima. Using this procedure a sample of the local minima accessible at a given temperature was obtained, providing a "map" of the underlying landscape. Following the original work this procedure was applied to a variety of systems, including water [52], rare gas clusters [53], and proteins such as myoglobin [54] and bovine pancreatic tripsin inhibitor (BPTI) [55]. The protein studies showed that there are a very large number of local minima in the vicinity of the native state of the protein. Furthermore, the local minima are kinetically clustered into subsets, within which they tend to be connected by low barriers.

Atomic level studies of complex peptide and protein energy landscapes have become more detailed as computers have become faster, allowing for longer sampling simulations and more complicated analysis. A problem that is faced by protein energy landscape cartographers is that of how to represent the resulting conformation sample in a meaningful way that will allow visualization and analysis of the underlying landscape. As far as the folding process is concerned, good results have been obtained by using one (or a few) effective reaction coordinates such as similarity to the native state (Q) or radius of gyration (R_g) [70,71]. These, however, are not very useful in exploring the energy landscape near the native state of a large protein or of peptides. Instead, to reduce the dimensionality of the data and to allow easier analysis of the landscape, it is becoming increasingly popular to use principal component analysis (PCA) (see Chapter 4) for this purpose. PCA is used to project the high-dimensional conformation sample onto a low-dimensional subspace that best represents it. The combination of PCA with long-time molecular dynamics has led to detailed studies of the energy landscape of proteins such as lysozyme [56], CRP:(cAMP)$_2$ [57], cytochrome c [58], and crambin [59]. In all, these systems exhibit complex landscapes with multiple basins. The observed dynamics on these landscapes typically involve long periods of motion within a basin followed by fast transition from one basin to another. These observations led Go and collaborators to suggest a ''jumping among minima'' (JAM) model to help analyze the simulation results [60].

Combining the PCA projection with an energy scale allows for 3D visualization of the underlying landscape. It should be noted, however, that without specific information on the barriers such PCA representations of the landscapes will at most reflect their overall shape, limited by the quality of the projection, and not necessarily their details. Nonetheless, the lack of information on the barriers is somewhat compensated for by the presence of ''empty spaces,'' which correspond to poorly sampled regions associated with high energy [59,61]. A problem associated with generating three-dimensional PCA views of protein energy landscapes is that the other principal coordinates, which are not included in this view, will manifest themselves as ''noise'' or ''roughness'' in the low-dimensional representation. This is because each point in the plain defined by the two main principal coordinates $\{Q_1, Q_2\}$ is associated with many conformations of different energies, separated from each other in the other principal coordinates $\{Q_3, Q_4, \ldots\}$. When the number of sampling points is small this problem can be overcome by a simple smoothing procedure, such as that used in mapping the energy landscape of alanine tetrapeptide [62]. However, when many conformations are included in the conformation sample, the ''minimum energy envelope'' procedure can be used to reduce the roughness [61]. For each value (on a grid) of the two main principal coordinates $\{Q_1, Q_2\}$ this procedure chooses the lowest conformation energy among all conformations that project onto this 2D grid point. The resulting smooth landscape is equivalent to an adiabatic surface, a surface that has been minimized in all coordinates other than Q_1 and Q_2. The resulting 3D view offers a direct visualization of the main basins on the energy landscape. Figure 4 shows the energy landscape of the prion protein (PrP) (residues 124–226) in vacuum [63]. Two large basins are clearly seen. The first is a deep but narrow basin associated with the native PrP^c conformation [7]. The second basin, which is shallower but wider, is associated with a second group of conformations of a partially unfolded protein. These offer a framework for studying the kinetics of protein folding.

Clearly, mapping energy landscapes based only on local minima gives only a partial description of the energy landscape, because the maps do not contain information about the energy barriers that govern the system's kinetics. It is the knowledge of the transition states that allows a detailed exploration of kinetics through the use of the master equation approach [Eqs. (2)–(4)]. One of the first detailed studies of this sort was performed by

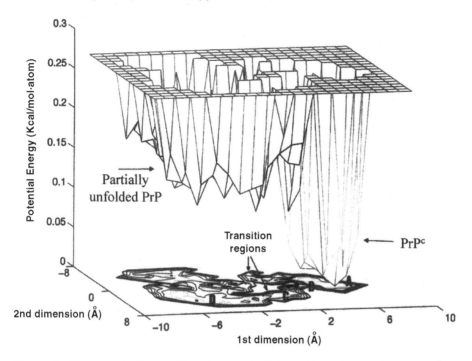

Figure 4 The energy landscape of the prion protein (PrP) (residues 124–226) in vacuum, obtained by principal coordinate analysis followed by the minimal energy envelope procedure. Two large basins are seen. One basin is associated with the native PrP^c conformation; the other is associated with partially unfolded conformations.

Czerminski and Elber [64], who generated an almost complete map of the minima and barriers of an alanine tetrapeptide in vacuum. Using the master equation approach they were able to study aspects of this system's kinetics, which involve the crossing of barriers of different heights.

Obtaining information regarding barriers, which accounts for state-to-state transition states, is a complicated computational task (see Chapter 10). However, even if such data are obtained, their complexity renders it difficult to introduce barrier information into the description of the atomistic energy landscape. In particular, one would like to extract from the raw data information regarding the overall connectivity of the landscape as well as information regarding the global basin-to-basin kinetic transitions. It is the transition from the ensemble of unfolded conformations (''unfolded basin'') to the ensemble of folded conformations (''folded basin'') that is of interest, rather than individual transitions between specific conformations. This type of ''global'' kinetics is in line with the type of observations available experimentally. To address this problem the method of ''topological mapping'' was introduced by Becker and Karplus [65]. Based on barrier information this method partitions conformation space into its component energy basins, thus highlighting the overall basin-to-basin connectivity of the landscape. At any energy level E the molecular conformation space can be partitioned into disconnected regions consisting of local minima that are connected by barriers lower than E. The method of topological mapping follows the way these disconnected regions, or ''basins,'' connect and disconnect as a function of increasing and decreasing energy E. An elementary basin $R(\alpha)$ is defined as a connected set of molecular conformations that, when minimized, map to the same single local minimum. Topological mapping groups these elementary basins according to the barriers between them. At any energy level E (or temperature level T) the multidimensional landscape is thus partitioned into "superbasins," $R^E(\alpha')$, defined as the union of elementary basins $R(\alpha)$ connected by barriers lower than energy E (or T).

$$R^E(\alpha') = \cup R(\alpha) \tag{8}$$

Each such "superbasin" is then mapped to its lowest minimum α' in a way that is analogous to simulated annealing (Fig. 5a). As a result, minima connected by barriers lower than E are grouped together and separated from other minima to which they are connected by higher barriers. A topological "disconnectivity" graph is obtained by following the way these superbasins break up as the system's energy E decreases. Each node on this graph (Fig. 5b) reflects a conformational superbasin on the landscape, and the connecting edges reflect the basin connectivity. The node at the top of the tree-graph corresponds to the ergodic limit, in which all states are connected. As the energy is decreased the graph splits to indicate basins that are becoming disconnected at that energy level. The topological mapping method resembles the Lid method independently developed by Sibani et al. [66] to study the energy landscape of crystals and glasses.

An advantage of topological mapping is that the resulting disconnectivity tree graph reflects, in a straightforward way, the overall topography of the energy landscape. For example, a tree graph reflecting "funnel" topography would be characterized by a single main branch with many small side branches that do not undergo additional splitting. On the other hand, a tree graph that corresponds to a landscape characterized by several large competing basins will exhibit several large branches, each displaying a complex branching pattern of its own. In the case of a completely rough landscape, no dominant branch can be detected in the disconnectivity graph. Application of this analysis method to the energy

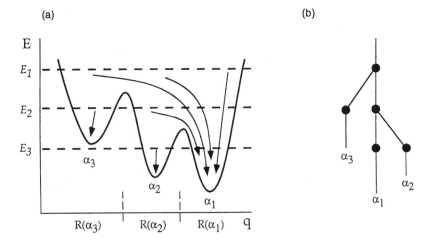

Figure 5 A schematic representation of a "topological mapping" of an energy landscape. (a) The energy landscape is studied at different energies E. Each region of connected conformations, denoted as a "superbasin" $R^E(\alpha')$, is mapped to its lowest minima α'. (b) The corresponding topological "disconnectivity" tree graph reflects the way superbasins become disconnected as the energy is decreased.

landscape of alanine tetrapeptide, based on the data of Czerminski and Elber [64], showed that this all-atom energy landscape is dominated by a ''funnel'' topography although the presence of a large kinetic trap could also be detected [65]. The insight into the connectivity of this landscape was used to study the overall basin-to-basin kinetics of this tetrapeptide, employing the master equation approach [65]. A very clear funnel topography is also seen in the disconnectivity graph of linear alanine hexapeptide (Ala)$_6$ shown in Figure 6 [67]. The method of topological mapping was successfully employed to characterize the energy landscape of different types of atomic and molecular clusters [68].

A different approach for handling barrier information was suggested by Kunz and Berry [69]. In this method conformations are sampled along high temperature dynamical trajectories, with the connectivity, including saddle points, determined for successive coordinate sets along a given trajectory. The minima–barrier–minima triplets are then put together in a way that follows the descent from high energy conformations to low energy structures. This results in linear cross sections through the high-dimensional energy landscape. Applying this method to different types of clusters led to the distinction between ''structure-seeking''clusters, such as the $(KCl)_{32}$ cluster, that exhibit a steep staircase-like

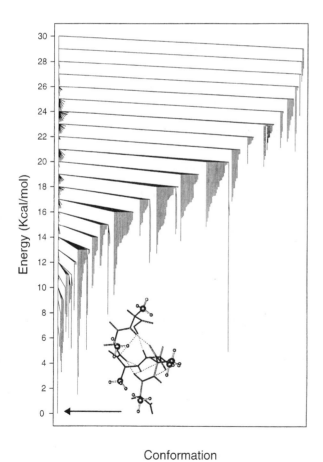

Figure 6 The topological disconnectivity graph of alanine hexapeptide. (Adapted from Ref. 67.)

descent into the native state, and ''glass-forming'' substances, such as Ar_{19} and Ar_{55} clusters, that are characterized by sawtooth-like landscapes [40,41]. Similar to topological mapping this analysis method characterizes the surface by its basin structure, highlighting the connectivity between the basins and following the basin-to-basin transitions. However, whereas topological mapping focuses on global connectivity, the staircase analysis tries to highlight pathways toward the native state.

C. Mapping Atomistic Free Energy Landscapes

The thermodynamics of protein folding, like those of other chemical reactions, are governed not by energy but by free energy, which is a combination of energy (enthalpy) and entropy. The foregoing mapping approaches focus only on the energy component of free energy, mapping energy as a function of conformation space. Although free energy can be inferred from these ''energy landscapes'' by evaluating conformation volumes and relating them to entropy [62], this is not a very accurate estimate of the free energy. Because free energy is not a function of any single conformation but rather of the whole conformation ensemble, ''free energy landscapes'' should be charted as a function of effective reaction coordinates, unlike ''energy landscapes,'' which are a function of conformational coordinates. In lattice studies a convenient reaction coordinate was the discrete enumeration of ''native contacts'' Q. An equivalent, though continuous, reaction coordinate appropriate for an all-atom model must be defined before a detailed free energy map of a protein can be obtained. Once a reaction coordinate is defined, statistical sampling methods can be used to estimate energy and entropy along the reaction coordinate, resulting in the desired chart.

An example of such an effort is the study by Sheinerman and Brooks [70] of the free energy landscape of a small α/β protein, the 56-residue B1 segment of streptococcal protein G. As discussed above, the first step in such an endeavor is to define an appropriate reaction coordinate. To this end the native state of the protein was first characterized through nanosecond time scale molecular dynamics simulations. From these simulations a set of 54 ''native nonadjacent side chain contacts,'' similar to those used in lattice simulations, were identified. These contacts were then employed to define a continuous reaction coordinate ρ, which measures similarity to the native state based on actual distances between the components of these 54 ''native contacts.'' Once the reaction coordinate was defined, high temperature MD simulations were used to sample a large number of protein conformations, which were then divided into groups according on their ρ values. For each group of conformations with a common ρ value, the center of the group was picked and subjected to an importance sampling, using a harmonic biasing potential along the reaction coordinate ρ. The slices of the potential of mean force that were generated in this way were combined to give a free energy map for this protein as a function of two reaction coordinates, the native contacts coordinate ρ and the radius of gyration R_g. The results indicated that this α/β protein undergoes a two-step folding. Folding commences with an overall collapse that is accompanied by the formation of $\sim 35\%$ of native contacts that are not spatially adjacent. Only later do the rest of the contacts gradually form, starting with the α-helix and only later continuing to the β-sheet. Water was present in the protein core up to a late stage of the folding process. A few similar studies have been performed for other proteins, such as the three-helix bundle protein (a 46-residue segment from fragment B of staphylococcal protein A) [71].

VI. SUMMARY

In this chapter we reviewed the main computational approaches that have been used to study the basic biophysical process of protein folding. The current theoretical framework for understanding protein folding is based on understanding the underlying energy and free energy landscapes that govern both the folding kinetics and its thermodynamics. A variety of computational models are being used to study protein folding, ranging from simple theoretical models, through simple lattice and off-lattice models, to atomic level descriptions of the protein and its environment. In recent years the focus of these studies has been gradually shifting toward the more detailed atomic level description of the process, with new computational methods and analytical techniques helping to gain additional insight into this fundamental process.

REFERENCES

1. C Levinthal. In: P Debrunner, JCM Tsibris, E Munck, eds. Mossbauer Spectroscopy in Biological Systems. Proceedings of a meeting held at Allerton House, Monticello, Illinois. Urbana, IL: Univ Illinois Press, 1969, p 22.
2. CB Anfinsen. Principles that govern the folding of protein chains. Science 181:223–230, 1973.
3. L Holm, C Sander. Science 273:595–602, 1996.
4. C Chothia. Nature 360:543–544, 1992.
5. CA Orengo, DT Jones, JM Thornton. Nature 372:631–634, 1994.
6. ML Riley, BA Wallace, SL Flitsch, PJ Booth. Biochemistry 36:192–196, 1997.
7. SB Prusiner. Prion protein biology. Cell 93:337–348, 1998.
8. SC Hyde, P Emsley, MJ Hartshorn, MM Mimmack, U Gileadi, SR Pearce, MP Galagher, DR Gill, RE Hubbard, CF Higgins. Nature 346:362–365, 1990.
9. RD Levine, RB Bernstein. Molecular Reaction Dynamics and Chemical Reactivity. New York: Oxford Univ Press, 1987.
10. M Oliveberg, AR Fersht. Proc Natl Acad Sci USA 92:8926–8929, 1995.
11. CM Dobson, A Sali, M Karplus. Protein folding: A perspective from theory and experiment. Angew Chem Int Ed Engl 37:868–893, 1998.
12. GI Makhatadze, PL Privalov. Adv Protein Chem 47:308–417, 1995.
13. JD Bryngelson, JN Onuchic, N Socci, PG Wolynes. Funnels, pathways, and the energy landscape of protein folding: A synthesis. Proteins 21:167–195, 1995.
14. JN Onuchic, Z Luthey-Schulten, PG Wolynes. Theory of protein folding: The energy landscape perspective. Annu Rev Phys Chem 48:545–600, 1997.
15. H Nymeyer, AE Garcia, JN Onuchic. Folding funnels and frustration in off-lattice minimalistic protein landscapes. Proc Natl Acad Sci USA 95:5921–5928, 1998.
16. TE Creighton, ed. Protein Folding. New York: WH Freeman, 1992.
17. M Karplus, A Sali. Theoretical studies of protein folding and unfolding. Curr Opin Struct Biol 5:58–73, 1995.
18. KA Dill, S Bromberg, K Yue, KM Fiebig, DP Yee, PD Thomas, HS Chan. Principles of protein folding—A perspective from simple exact models. Protein Sci 4:561–602, 1995.
19. EI Shakhnovich. Curr Opin Struct Biol 7:29–40, 1997.
20. HS Chan, KA Dill. Protein folding in the landscape perspective: Chevron plots and non-Arrhenius kinetics. Proteins 30:2–33, 1998.
21. M Karplus. Aspects of protein reaction dynamics: Deviations from simple behavior. J Phys Chem B 104:11–27, 2000.
22. JD Bryngelson, PG Wolynes. Spin glasses and the statistical mechanics of protein folding. Proc Natl Acad Sci USA 84:7524–7528, 1987.

23. JG Saven, J Wang, PG Wolynes. Kinetics of protein folding: The dynamics of globally connected energy landscapes with biases. J Chem Phys 101:11037–11043, 1994.

24. NG van Kampen. Stochastic Processes in Physics and Chemistry. Amsterdam: North-Holland, 1981.

25. R Zwanzig. Simple model of protein folding kinetics. Proc Natl Acad Sci USA 92:9801–9804, 1995.

26. PE Leopold, M Montal, JN Onuchic. Protein folding funnels: A kinetic approach to the sequence–structure relationship. Proc Natl Acad Sci USA 89:8721–8725, 1992.

27. A Sali, E Shakhnovich, M Karplus. How does a protein fold? Nature 369:248–251, 1994.

28. A Dinner, A Sali, M Karplus. Proc Natl Acad Sci USA 93:8356–8361, 1996.

29. A Kolinski, J Skolnick. Monte Carlo simulations of protein folding. I. Lattice model and interaction scheme. Protein 18:338–352, 1994.

30. CJ Camacho, D Thirumalai. Kinetics and thermodynamics of folding in model proteins. Proc Natl Acad Sci USA 90:6369–6372, 1993.

31. D Thirumalai, SA Woodson. Kinetics of folding of proteins and RNA. Acc Chem Res 29:433–439, 1996.

32. H Taketomi, Y Ueda, N Go. Studies on protein folding, unfolding and fluctuations by computer simulation. 1. The effect of specific amino acid sequence represented by specific inter-unit interactions. Int J Peptide Protein Res 7:445–459, 1975.

33. N Go, H Taketomi. Respective roles of short- and long-range interactions in protein folding. Proc Natl Acad Sci USA 75:559–563, 1978.

34. EI Shakhnovich, G Farztdinov, AM Gutin, M Karplus. Protein folding bottlenecks: A lattice Monte Carlo simulation. Phys Rev Lett 67:1665–1668, 1991.

35. AM Gutin, VI Abkevich, EI Shakhnovich. Is burst hydrophobic collapse necessary for protein folding? Biochemistry 34:3066–3076, 1995.

36. A Sali, EI Shakhnovich, M Karplus. Kinetics of protein folding: A lattice model study of the requirements for folding to the native state. J Mol Biol 235:1614–1636, 1994.

37. AR Dinner, M Karplus. A metastable state in folding simulations of a protein model. Nature Struct Biol 5:236–241, 1998.

38. JD Honeycutt, D Thirumalai. The nature of folded states of globular proteins. Biopolymers 32:695–709, 1992.

39. Z Guo, D Thirumalai, JD Honeycutt. Folding kinetics of proteins: A model study. J Chem Phys 97:525–535, 1992.

40. KD Ball, RS Berry, RE Kunz, F-Y Li, A Proykova, DJ Wales. From topographies to dynamics of multidimensional potential energy surfaces of atomic clusters. Science 271:963–966, 1996.

41. RS Berry, N Elmaci, JP Rose, B Vekhter. Linking topography of its potential surface with the dynamics of folding of a protein model. Proc Natl Acad Sci USA 94:9520–9524, 1997.

42. Z Guo, D Thirumalai. J Mol Biol 263:323–343, 1996.

43. MH Hao, H Scheraga. Proc Natl Acad Sci USA 93:4984, 1996.

44. B Vekhter, RS Berry. Simulation of mutation: Influence of a ''side group'' on global minimum structure and dynamics of a protein model. J Chem Phys 111:3753–3760, 1999.

45. Y Duan, PA Kollman. Pathways to a protein folding intermediate observed in a 1-microsecond simulation in aqueous solution. Science 282:740–744, 1998.

46. V Daggett, M Levitt. Protein folding ↔ unfolding dynamics. Curr Opin Struct Biol 4:291–295, 1994.

47. A Caflisch, M Karplus. Molecular dynamics studies of protein and peptide folding and unfolding. In: K Merz Jr, S Le Grand, eds. The Protein Folding Problem and Tertiary Structure Prediction. Boston: Birkhauser, 1994, pp 193–230.

48. V Daggett, M Levitt. A model of the molten globule state from molecular dynamics simulations. Proc Natl Acad Sci USA 89:5142–5146, 1992.

49. A Li, V Daggett. Characterization of the transition state of protein unfolding by use of molecular dynamics: Chymotrypsin inhibitor 2. Proc Natl Acad Sci USA 91:10430–10434, 1994.

50. BM Pettitt, M Karplus. The potential of mean force surface for the alanine dipeptide in aqueous solution: A theoretical approach. Chem Phys Lett 121:194–201, 1985.
51. AG Anderson, J Hermans. Microfolding: Conformational probability map for the alanine dipeptide in water from molecular dynamics simulations. Proteins 3:262–265, 1988.
52. FH Stillinger, TA Weber. Packing structures and transitions in liquids and solids. Science 225:983–989, 1984.
53. RS Berry. Potential surfaces and dynamics: What clusters tell us. Chem Rev 93:2379–2394, 1993.
54. R Elber, M Karplus. Multiple conformational states of proteins: A molecular dynamics analysis of myoglobin. Science 235:318–321, 1987.
55. T Noguti, N Go. Structural basis of hierarchical multiple substates of a protein. I–V. Proteins 5:97, 104, 113, 125, 132, 1989.
56. A Amadei, ABM Linssen, HJC Berendsen. Essential dynamics of proteins. Proteins 17:412–425, 1993.
57. AE Garcia, JG Harman. Simulations of CRP:(cAMP)2 in noncrystalline environments show a subunit transition from the open to the closed conformation. Protein Sci 5:62–71, 1996.
58. AE Garcia, G Hummer. Conformational dynamics of cytochrome c: Correlation to hydrogen exchange. Proteins 36:175–191, 1999.
59. LSD Caves, JD Evanseck, M Karplus. Locally accessible conformations of proteins: Multiple molecular dynamics simulations of crambin. Protein Sci 7:649–666, 1998.
60. A Kitao, S Hayward, N Go. Energy landscape of a native protein: Jumping-among-minima model. Proteins 33:496–517, 1998.
61. OM Becker. Principal coordinate maps of molecular potential energy surfaces. J Comput Chem 19:1255–1267, 1998.
62. OM Becker. Quantitative visualization of a macromolecular potential energy 'funnel.' J Mol Struct (THEOCHEM) 398–399:507–516, 1997.
63. Y Levy, OM Becker. Wild-type and mutant prion proteins: Insights from energy landscape analysis. In: E Katzir, B Solomon, A Taraboulos, eds. Conformational Diseases. In press.
64. R Czerminski, R Elber. Reaction path study of conformational transitions in flexible systems: Application to peptides. J Chem Phys 92:5580–5601, 1990.
65. OM Becker, M Karplus. The topology of multidimensional potential energy surfaces: Theory and application to peptide structure and kinetics. J Chem Phys 106:1495–1517, 1997.
66. P Sibani, JC Schon, P Salamon, JO Andersson. Emergent hierarchical structures in complex-system dynamics. Europhys Lett 22:479–485, 1993.
67. Y Levy, OM Becker. Effect of conformational constraints on the topography of complex potential energy surfaces. Phys Rev Lett 81:1126–1129, 1998.
68. DJ Wales, MA Miller, TR Walsh. Archetypal energy landscapes. Nature 394:758–760, 1998.
69. RE Kunz, RS Berry. Statistical interpretation of topographies and dynamics of multidimensional potentials. J Chem Phys 103:1904–1912, 1995.
70. FB Sheinerman, CL Brooks III. Molecular picture of folding of a small α/β protein. Proc Natl Acad Sci USA 95:1562–1567, 1998.
71. Z Guo, EM Boczko, CL Brooks III. Exploring the folding free energy surface of a three-helix bundle protein. Proc Natl Acad Sci USA 94:10161–10166, 1997.
72. J Skolnick, A Kolinski, AR Ortiz. Application of reduced models to protein structure prediction. In: J Leszczynski, ed. Computational Molecular Biology. Theor Comput Chem Ser New York: Elsevier Science, 1999, pp 397–440.

18

Simulations of Electron Transfer Proteins

Toshiko Ichiye
Washington State University, Pullman, Washington

I. INTRODUCTION

Electron transfer proteins serve key roles as electron carriers in a wide variety of processes in all organisms, including the major energy-transducing functions of photosynthesis and respiration, other metabolic functions such as nitrogen fixation, and biosynthesis [1]. One of the most intriguing questions about these proteins is that of how the protein itself influences the electron transfer properties of its redox site. In this chapter, computational methods used to understand the donor–acceptor energetic interactions of electron transfer proteins at a molecular level are described. The focus is on the electron transfer metalloproteins, which are the blue copper proteins, the iron-sulfur proteins, and the cytochromes. However, many of the issues discussed are important for redox-active proteins and/or metalloproteins in general.

II. ELECTRON TRANSFER PROPERTIES

The properties of electron transfer proteins that are discussed here specifically affect the electron transfer reaction and not the association or binding of the reactants. A brief overview of these properties is given here; more detailed discussions may be found elsewhere (e.g., Ref. 1). The process of electron transfer is a very simple chemical reaction, i.e., the transfer of an electron from the donor redox site to the acceptor redox site.

$$D^- + A \xrightarrow{k} D + A^-$$
(R1)

Thus, it is characterized by $\Delta G°$, the overall free energy of the reaction, and k, the reaction rate.

An electron transfer reaction may be separated into two half-reactions or redox couples so that the free energy, $\Delta G°$, can be separated into $\Delta G_A°$ and $\Delta G_D°$ the free energies of reduction of the donor (D) and the acceptor (A), respectively, by

$$\Delta G° = \Delta G_A° - \Delta G_D°$$
(1)

Therefore, ΔG° is related to the redox potentials E_i° of the donor and acceptor via the Nernst equation,

$$\Delta G_i = -nFE_i \tag{2}$$

where n is the number of electrons transferred and $F = 23.06$ kcal/(mol · V) is Faraday's constant. Thus, an understanding of the redox potentials of electron transfer proteins in an electron transport chain is essential, because the redox potentials determine the direction of favorable flow of the electrons. One caution: There may be differences between the redox potentials of the isolated reactants and those of the reactants in an electron transfer complex [2].

The electron transfer rates in biological systems differ from those between small transition metal complexes in solution because the electron transfer is generally long-range, often greater than 10 Å [1]. For long-range transfer (the nonadiabatic limit), the rate constant is

$$k = \frac{H_{AB}^2}{\hbar} \left(\frac{\pi}{\lambda RT}\right) \exp\left[\frac{-\Delta G^\ddagger}{RT}\right] \tag{3}$$

where H_{AB} is the electronic (or tunneling) matrix element, λ is the environmental reorganization energy, and $\Delta G^\ddagger$ is the activation energy for the reaction. H_{AB} is a measure of the electronic coupling between the reactants and the products at the transition state, and λ is a measure of the energy required to change the *environment* from the equilibrium state around the reactants to what would be the equilibrium state around the products while the reactants remain in their initial state. According to Marcus theory [3], $\Delta G^\ddagger$ is given by

$$\Delta G^\ddagger = (\lambda + \Delta G^\circ)^2/4\lambda \tag{4}$$

where ΔG° is the free energy described above. Thus, understanding λ is important, because it entails understanding how the protein keeps the activation energy $\Delta G^\ddagger$ low enough to promote the long-range transfer.

The calculation of E_i° and λ from computational methods is the focus here. Generally, the energetics of these quantities are separated into contributions from the inner and outer shells. For transfer between small molecules, the inner shell generally is defined as the entire solutes A and D, and the outer shell is generally defined as only the solvent. However, in a more practical approach for proteins, the inner shell is defined as only the redox site, which consists of the metal plus its ligands no further than atoms of the side chains that are directly coordinated to the metal, and the outer shell is defined as the rest of the protein plus the surrounding solvent. Thus

$$\Delta G_i^\circ = -nF(\text{IP} + \text{SHE}) + \Delta G_i^{\text{in}} + \Delta G_i^{\text{out}} = \Delta G_i^{\text{site}} + \Delta G_i^{\text{out}} \tag{5}$$

where IP, the ionization potential, is the negative of the Franck–Condon energy required to add an electron to the species i [4]; ΔSHE, the standard hydrogen electrode correction, is -4.43 V [5]; ΔG_i^{in}, the inner shell relaxation energy, is the change in energy for species i when the metal ligand geometry (i.e., the internal coordinates) has the oxidized versus reduced values; and ΔG_i^{out}, the solvent or outer shell relaxation energy, is the change upon reduction of the solvation polarization energy [3] of species i. In the second equality, ΔG_i^{site} is the change upon reduction in the intrinsic energy of the redox site i. Similarly,

$$\lambda = \lambda^{\text{in}} + \lambda^{\text{out}} \tag{6}$$

where λ^{in}, the inner shell reorganization energy, is the energy required to change the metal ligand geometry of the reactants from the initial to the final values while the charges of the reactants remain in the initial state and λ^{out}, the outer shell reorganization energy, is the energy required to change the solvent polarization from an equilibrium configuration around the reactants to one around the products, again while the charges of the reactants remain in the initial state.

The simplest approach to calculating both ΔG_i and λ is to assume that the energetics of the redox site and the outer shell can be determined independently (i.e., the energies are uncoupled). Thus, as a first approximation, the change in the energy of the redox site can be calculated from quantum mechanical calculations of analogs, and the change in the outer shell energy can be calculated from classical calculations. The coupling between the redox site and outer shell energies is through the potential energy parameters, especially the partial charges, of the redox site used in the classical calculations. Because environmental effects due to the protein and/or solvent may influence the electronic structure, a higher order approximation is to use mixed quantum–classical methods (see Chapter 11) such as are used in a study of electronic tunneling pathways in ruthenium-modified myoglobin [6]; however, such calculations are not yet routine.

III. CALCULATION TECHNIQUES FOR ELECTRON TRANSFER PROTEINS

Computer simulations of electron transfer proteins often entail a variety of calculation techniques: electronic structure calculations, molecular mechanics, and electrostatic calculations. In this section, general considerations for calculations of metalloproteins are outlined; in subsequent sections, details for studying specific redox properties are given. Quantum chemistry electronic structure calculations of the redox site are important in the calculation of the energetics of the redox site and in obtaining parameters and are discussed in Sections III.A and III.B. Both molecular mechanics and electrostatic calculations of the protein are important in understanding the outer shell energetics and are discussed in Section III.C, with a focus on molecular mechanics.

A. Quantum Chemistry of the Redox Site

Quantum mechanical electronic structure calculations of the redox site are often necessary for even classical calculations of electron transfer proteins, such as when parameters are lacking for a metal redox site. Although semiempirical methods have been used for these types of studies, here the focus is on ab initio methods. Currently, density functional theory (DFT) methods are the ab initio quality electronic structure calculations of choice for metal sites because Hartree–Fock (HF) methods are insufficient and the inclusion of configuration interactions (CI), which are important for these systems, with high level basis sets leads to intractable calculations (i.e., the computational dependence on the number of electrons n is n^3 for DFT, n^4 for MP2, and n^5 for CI techniques). In addition, DFT gives excellent results for transition metals, including bond dissociation energies, bond lengths, conformational analysis, ionization potentials, and electron affinities [7,8]. An excellent review covers the application of these methods to transition metal complexes, including redox sites within proteins and analogs of protein redox sites [9].

Whether DFT or HF methods are used, there are several issues regarding the details of the calculation that will affect various aspects of the quality of the calculation; these issues are mentioned here and are discussed more fully with respect to the specific calculation later. Comparisons throughout this chapter refer to unrestricted HF calculations of the [1Fe] or rubredoxin-type site by our group, which use a full core on the iron and geometry optimization unless otherwise specified, and to spin-unrestricted DFT calculations by Noodleman and coworkers [9], which use the local density approximation for exchange and correlation with nonlocal corrections as a perturbation or as part of the SCF procedure, frozen core orbitals, and experimental geometries unless otherwise specified.

One issue concerns whether or not the geometry of the redox site is optimized. For the iron-sulfur sites, geometry optimization has given longer Fe—S bond lengths than are found in experiment—both an HF calculation of a [1Fe] site (0.1–0.2 Å) [10] and a DFT calculation of a [2Fe-2S] site (0.5–0.2 Å) [11]. For the blue copper sites, geometry optimization has given metal–ligand distances that differ by up to 0.2 Å from experimental averages. However, the latter come from crystal structures of the blue copper proteins, which are at much lower resolution than analog structures and have a large range of metal–ligand bonds (0.2–0.5 Å). Another approach is to use experimental geometries from high resolution crystal structures of analogs of the redox site in the proper oxidation state. Less preferably crystal structures of the protein may be used, although they are lower resolution and may exist in only a single oxidation state [12]. The use of analogs is a good approximation for iron-sulfur proteins, because the redox site structures are relatively invariant even for proteins with very different folds and/or redox potentials [13]. However, care must be taken for blue copper proteins, because the redox site structure can vary significantly in different proteins [4].

Another issue is whether the calculation is of the redox site in vacuum or with environmental effects. If the energetics of the metal site are independent of its environment, the calculation of an analog in vacuum is a good approximation. The influence of the protein on the electronic structure of the redox site apparently is rather small for Fe-S proteins, from both experimental [13] and theoretical [10] studies. However, the protein plays a much larger role in the blue copper proteins, as is seen in both experiment [4] and theory [14]. A presumably more accurate but more computationally intensive process entails adding a reaction field that represents the environment. In one approach, a dielectric continuum reaction field (possibly including fixed charges and a low dielectric region representing a protein environment) is incorporated into the self-consistent field (SCF) of the entire system to give a self-consistent reaction field. Examples include DFT calculations of the manganese superoxide dismutase (Mn-SOD) active site using oxidized geometries only in aqueous solution [12] and the [2Fe-2S] site in two different proteins [11]. Other approaches include using QM/MM methods (see Chapter 11).

B. Potential Energy Parameters

The first major obstacle in studying electron transfer and/or metalloproteins is often the lack of potential energy parameters for metal sites in proteins. Although parameters for hemes existed in some of the earliest parameter sets because of the numerous studies of myoglobin [15], hemoglobin [16], and cytochrome [17], there is a dearth of parameters for other metal sites. Parameters for iron-sulfur sites have been recently developed [18–21] based on spectroscopic data for the force constants, crystallographic data for the equilibrium values, and quantum mechanical calculations for the partial charges and for the van der Waals parameters (see Chapter 2). Parameters for other sites have also been developed [22–25].

The force constants and equilibrium values for the internal coordinates should be obtained from experimental data given the computational intensity of obtaining electronic structure calculations of sufficient quality for these quantities. If possible, the equilibrium values should be obtained from analogs of the redox site, because generally the structures are at much higher resolution than those of proteins and it is possible that the protein strains the analog away from its equilibrium internal coordinates.

The partial charges for the metal sites are difficult to obtain via experiment, so quantum mechanical calculations provide the best source. Because the electrostatics of the redox site are potentially a crucial factor in calculating electron transfer properties, it is important to understand the sensitivity of the results to the values of the partial charges. The energetics of fixed structures are relatively insensitive to the values of the partial charges as long as the total charge is correct [26,27]. However, the partial charges become important for energy minimization and molecular dynamics simulations because perturbations of the protein may occur [28]. Either DFT or more traditional ab initio calculations can be used to provide wave functions, which are then used to generate single-point ''partial charges.'' Because Mulliken population analyses tend to be dependent on the basis set, the preferred method of generating the partial charges is to fit the molecular electrostatic potential determined by the nuclear interactions and the electron charge density while constraining the total charge and desired higher moments to match those determined from the electronic wave function using programs such as CHELPG [29]. However, difficulties may be encountered in obtaining good values for a coordinated metal site, because such fitting methods are relatively insensitive to the values of the partial charges of buried atoms such as the metal and can thus be more conformation-dependent than the Mulliken population analysis indicates [10]. Thus, Mulliken charges may be a better indicator of the conformational dependence of the partial charges, whereas fitted charges may be a better indicator of the magnitude of the partial charges. To obtain the best charges, fitted charges should be determined for conformations with a high degree of symmetry, which are likely to be the most accurate. Mulliken population analyses should then be compared for the high symmetry and any other desired conformation to determine if the desired conformation is likely to have the same charge distribution. The details of the electronic structure calculation may also have a significant effect on the values of the partial charges. For instance, the use of an effective core versus a full core on the metal in unrestricted HF calculations of the [1Fe] site lead to 10–15% changes in the CHELPG charge of the Fe [10]. Moreover, the use of geometry for the wrong oxidation state in the same study can lead to 5–10% changes in the CHELPG charge of the Fe. This study also indicated that overestimation of the bond lengths tends to reduce the CHELPG charge on the Fe but increase the Mulliken charge. For partial charges, the treatment of the environment becomes an issue, because the condensed phase environment tends to enhance polarization so that approaches such as using a dielectric continuum to represent either a solvent or protein environment should be considered. The effect of adding environmental effects via a dielectric continuum plus protein charges in a DFT calculation indicates less than a 1% change in charge on the Fe in the [2Fe-2S] site [11], but up to an almost 20% change in CHELPG charge on the Mn of the Mn-SOD site [12]. In determining partial charges of other organic molecules, the tendency for HF to overestimate polarization for the molecule in vacuum had been viewed as an approximate means of determining the increase in polarization as the molecule is transferred to the condensed phase [30]. Experimental verification is essential, because in studies of the [1Fe] site of rubredoxin, DFT calculations [27] and unrestricted HF calculations [10] differ by almost 50% on the Fe even though the identical fitting method was used to obtain the charges.

The van der Waals parameters of the metal in a liganded site may be significantly different from that of the free ion in solution. Experimental determination of these parameters has not been forthcoming, so quantum chemical calculations again provide the best means. In cases in which there are several ligands such as when there is octahedral coordination, the metal may be inaccessible due to the van der Waals spheres of the ligands, so it is not necessary to determine separate van der Waals parameters for the metal as has been assumed for the Fe in the heme group in the CHARMM potential [31,32]. However, for the tetrahedral coordination found in the Fe-S sites, the blue copper sites, and the zinc sites, the metal is very accessible, and thus good metal van der Waals parameters are essential. Although DFT methods are generally preferable for metal sites, they are less well developed for treating the interactions between molecules, so HF methods have been used [21,33].

C. Molecular Mechanics and Electrostatics Calculations of the Protein

Molecular mechanics and electrostatics calculations have both played an important role in studying electron transfer proteins. Molecular mechanics calculations of these proteins use the same techniques (molecular dynamics, energy minimization) as for other proteins, although special consideration must be made in simulation conditions.

The treatment of long-range electrostatics and dielectric effects is an important issue in calculations of the outer shell energetics because of the importance of the electrostatic contribution to the redox properties. Two basic approaches have been used in atomistic studies of biological molecules: (1) treating the protein as a collection of charges embedded in a low dielectric continuum and the solvent as a high dielectric continuum, possibly with an ionic strength associated with it, and solving the Poisson or Poisson–Boltzmann equation as appropriate and (2) treating both the protein and solvent atoms explicitly as a collection of partial charges, with both allowed to reorient to mimic the orientational dielectric response, in a standard molecular mechanics calculation. The molecular mechanics techniques discussed here are energy minimization and molecular dynamics in standard implementations such as in CHARMM and DISCOVER; a review of the Protein Dipoles Langevin Dipoles (PDLD) method in the program POLARIS as applied to iron-sulfur proteins was published in 1996 [34].

Poisson and Poisson–Boltzmann calculations involve calculating the electrostatics due to partial charges of atoms in the protein, generally assuming that the protein itself has a low dielectric constant ε_p and the surrounding solvent has a high dielectric constant ε_w (~80 for water) such as in the programs DelPhi [35,36], UHBD [37,38], and MEAD. Poisson–Boltzmann calculations include nonzero salt concentration, whereas Poisson calculations are for the special case of zero salt concentration. Among the strengths of this approach are that the long-range contributions (i.e., the Born solvation energy) and salt effects are calculated accurately. One of the weaknesses is that a single ε_p is assumed for the entire protein, whereas dielectric relaxation is a molecular phenomenon so the dielectric response varies within a protein. It is not even clear if there is a good average value for proteins, and values ranging from $\varepsilon_p \approx 2$ to $\varepsilon_p \approx 10$ are used. Another potential problem arises if specific interactions of the solvent with the redox site are important, as appears to be the case for rubredoxin from calculations [19] and resonance Raman experiments (J. Sanders-Loehr, personal communication) and also for ferredoxin from resonance Raman experiments [13]. This is an important consideration in general, because the redox sites are close to the surface in many electron transfer proteins. In addition, most Poisson-type calculations do not allow relaxation of the protein or dynamic effects.

The treatment of electrostatics and dielectric effects in molecular mechanics calculations necessary for redox property calculations can be divided into two issues: electronic polarization contributions to the dielectric response and reorientational polarization contributions to the dielectric response. Without reorientation, the electronic polarization contribution to ε is ~2 for the types of atoms found in biological systems. The reorientational contribution is due to the reorientation of polar groups by charges. In the protein, the reorientation is restricted by the bonding between the polar groups, whereas in water the reorientation is enhanced owing to cooperative effects of the freely rotating solvent molecules.

Electronic polarization is included in simulations in two different ways. In many of the potential energy functions used for biological molecules such as the CHARMM and AMBER potentials and also those used for water in simulations of biological systems such as SPC [39], SPC/E [40], and TIP3P and TIP4P [41], electronic polarization effects are included implicitly in the potential energy parameters. In other words, these potential energy functions have been parametrized to give good structure and energetics in an "average" environment without including electronic polarization explicitly [42]. Thus, the approximation will tend to break down in mixed environments. For instance, although electronic polarization plays a relatively small role around singly charged ions, it plays a significant role for charges of magnitude greater than 2 [43], so it may be an issue for metal centers. The PDLD method [44,45] has electronically polarizable dipoles representing the polar groups of the protein surrounded by a grid of Langevin dipoles representing the solvent. However, some other features of the PDLD approach that warrant caution are that the Langevin dipoles, although including some reorientational dielectric effects, do not accurately reflect the structure of water because of the cubic grid to which they are confined and that charged side chains are often neutralized, which is reasonable only for those at the surface far from the point of interest. Several of the standard water models have been modified to include polarizability [46], but as yet they have not been integrated with a polarizable potential function for proteins.

The reorientational polarization contribution to the dielectric response comes from including the interactions of all polar groups, including those of both the protein and the solvent, in the calculation of the electrostatic component. As generally recommended for simulations of proteins, electron transfer proteins must be simulated with explicit water and counterions. However, if the interaction energy of the redox site with the rest of the protein needed for calculating redox properties is to be calculated explicitly from the simulation, then the use of methods such as droplet boundary conditions and/or spherical continuous or discontinuous cutoffs becomes questionable in calculating the energetics of a protein in solution because the contribution of the Born solvation energy of an ion in water is significant even at large distances [47].

For instance, the contribution of water beyond 12 Å from a singly charged ion is 13.7 kcal/mol to the solvation free energy or 27.3 kcal/mol to the solvation energy of that ion. The optimal treatment is to use Ewald sums, and the development of fast methods for biological systems is a valuable addition (see Chapter 4). However, proper account must be made for the finite size of the system in free energy calculations [48].

IV. REDOX POTENTIALS

Calculations of redox potentials may have two somewhat different purposes: (1) to calculate the redox potential of a given protein or (2) to calculate the *differences* in redox potential between two proteins, most often the wild-type and a mutant. The calculation of the absolute redox potential E_i (or ΔG_i) of a given protein is important for determining the relative importance of various contributions to the redox potential. Electronic structure calculations are necessary to obtain the change upon reduction in the intrinsic energy of the redox site and are discussed in Section IV.A. Both molecular mechanics and electrostatic energy calculations are useful for calculating the change upon reduction of the energy of the protein and solvent (the outer shell) and are discussed in Section IV.B. Although calculation of differences in redox potentials between two proteins ΔE_i (or $\Delta\Delta G_i$) would seem to necessitate calculation of ΔG_i, the assumption that most of the contributions to the redox potential will remain constant allows the use of much simpler calculation techniques and are discussed in Section V.

A. Calculation of the Energy Change of the Redox Site

The change upon reduction in the intrinsic energy of the redox site (ΔG_i^{site}) is composed of the negative of the Frank–Condon ionization potential (IP) and the energy change due to changes in the internal coordinates of the redox site (ΔG_i^{in}) [Eq. (5)]. Thus, it is the difference in the absolute energy of the oxidized species in its equilibrium (or relaxed) conformation versus the reduced species in its equilibrium (or relaxed) conformation, which can be calculated via quantum chemistry calculations of the oxidized and reduced species that include geometric optimization in the presence of the reaction field due to the environment. These absolute energies are often given in atomic units (au), where 1 au = 27.21 eV.

The preferred method for calculating ΔG_i^{site} is to use DFT, for the reasons just described. The difference is significant for the [1Fe] analog in vacuum, because the DFT calculation gives a value of 1.79 eV whereas the HF calculation gives a value of 2.13 eV. The HF calculation is clearly sensitive to the lack of CI, because calculations using an effective core potential on the Fe give values of 0.251 and 1.18 eV at the HF and MP2 (second-order Møller–Plesset) levels, respectively. The values are also sensitive to the use of geometric optimization, lowering ΔG_i^{site} by 0.150–0.130 eV for the [2Fe-2S] site relative to geometries of the oxidized and reduced species [11]. Assuming a single geometry for both oxidation states can lead to considerable errors in ΔG_i^{site}, because ΔG_i^{in} is about 0.5 eV in the HF calculation of the [1Fe] site. The effects of the environment on ΔG_i^{site} alone are variable, less than ± 0.1 eV in the Mn-SOD site but 0.2–0.3 eV in the [2Fe-2S] clusters.

B. Calculation of the Energy Changes of the Protein

The calculation of ΔG_i^{out}, the energy change upon reduction of the energy of the outer shell—the rest of the protein plus solvent—along with the quantum chemical value for ΔG_i^{site} allows comparisons to experimental redox potentials through Eq. (5). In addition, it allows the decomposition of the various factors contributing to ΔG_i^{out}. Calculation of ΔG_i^{out} from molecular mechanical techniques entails a free energy simulation between the oxidized and reduced states (see Chapter 9). However, it is useful to perform molecular dynamical simulations or even energy minimizations of the oxidized and reduced states prior to such a calculation because information can be gained about the structures of both states when there is an experimental structure of only one state. In addition, simulations of both states, especially when higher net charges concentrated at the redox site are involved, are important in assessing the simulation condition for free energy simulations. Moreover, these simulations can then be used to calculate ΔE_i^{out}, which gives the enthalpic contribution minus the PV term to ΔG_i^{out}. The decomposition of the contributing factors to ΔE_i^{out} is generally simpler than for ΔG_i^{out}. The calculation of ΔE_i^{out} from molecular mechanics is described first, followed by the calculation of ΔG_i^{out} from free energy simulations and finally the calculation of ΔG_i^{out} from Poisson–Boltzmann calculations.

1. Molecular Mechanics Calculations of ΔE_i^{out}

Calculation of $\Delta E_i^{\text{out}} = E_{i,\text{red}}^{\text{out}} - E_{i,\text{oxd}}^{\text{out}}$ (where the subscripts red and oxd denote the reduced and oxidized states, respectively, of species i) from molecular mechanics entails calculating the total classical potential energy of the system when species i is in both the reduced and oxidized states, where the state is determined by the partial charges of the redox site. Since the total potential energy is dependent on the conditions (temperature, pressure, number of particles) of the system, it is thus necessary to have the same conditions in the oxidized and reduced states to calculate the total ΔE_i^{out}. This type of calculation is useful in determining various contributions to the energetics of a redox reaction such as the contributions of structural relaxation of the protein versus change in charge of the redox site. In addition, because the energies are simply sums of various terms, electrostatics versus van der Waals versus internal coordinate contributions and various components of the protein such as backbone versus side chain can be calculated by summing over only the appropriate terms. Thus, specific contributions may be focused on with the caveat that there may be compensating changes in other components.

When only the energetics of the protein are of interest, a simple approximation is to calculate the energetics of a single structure of the oxidized state and a single structure of the reduced state. The structures may simply be from experimental crystal or NMR studies of the protein in both states as in a study of cytochrome c [26]. If structures do not exist for both states, the structures of each state may be obtained by energy minimization using the potential energy parameters of that state, as in studies of rubredoxin [18,49]; however, the structures must each be carefully minimized to a local minimum. The structures can also be obtained from average structures from molecular dynamics simulations of each state. It is important that if the structures are from energy minimizations or molecular dynamics calculations that have used electrostatic cutoffs, the calculation of the energy itself should not use cutoffs. A more accurate calculation is to calculate the average energies from the molecular dynamics simulations. If the energetics of the solvent are also important, then it is crucial to do molecular dynamics simulations of each state with periodic boundary conditions so that either the pressure or the volume can be fixed.

The calculation of ΔE_i^{out} requires calculations of the *total* energies of each state, not just the interaction energies of the redox site with its environment. This means that interactions between atoms of the environment (protein and solvent) must also be included; the protein contribution may be over 20 kcal/mol [50]. One simplification is to examine only the energetics due to the protein itself (i.e., redox site with protein atoms and between protein atoms), which is useful when the solvent has not been treated by way of molecular dynamics. This restricts the analysis to the backbone plus the nonpolar, polar, and buried charged side chains, because it is necessary to include solvent and counterions to calculate the surface charged side chain contributions correctly. The contribution of the charged side chains at the surface is actually quite small owing to cancellation by the solvent and counterions. Interestingly, calculations of structures from molecular dynamics of rubredoxin [19] indicate that most of the large positive electrostatic potential from polar groups is due to the backbone within 8 Å of the iron, which indicates that the local back-bone structure around the iron provides an electrostatic environment that sets the redox potential to a certain range (i.e., rubredoxins have a redox potential of 0.8 V). The lack of polar side chain contributions indicates that much of the sequence variability in the homologous rubredoxins does not affect the redox potential significantly, which is consistent with the homogeneity of the redox potentials found for these proteins.

One important question about the energetics of a redox reaction is how much of the energy change is due to structural relaxation of the protein in response to the change in charge versus that due to simply the different interaction energy with the new charge. The contribution of structural relaxation of the outer shell in response to the change in charge of species i for the protein P can be determined by assuming that the reaction can be broken down into a change in charge followed by structural relaxation as shown in Scheme 1.

In Scheme 1 the state of the protein is indicated by (q,r), where q denotes the oxidized or reduced charge state of the redox site and r denotes the coordinates of the outer shell in equilibrium with the oxidized or reduced redox site. The energy of the reaction, ΔE, for the reduction ($o \rightarrow r$) and the oxidation ($r \rightarrow o$) reactions is broken down as

$$\Delta E = \Delta E^q + \Delta E^r \tag{7}$$

where ΔE^q is the energy change when the charge of the redox site changes but the environment does not relax, and ΔE^r is the energy change when the environment relaxes after the charge change. Thus, the energy of the protein must be calculated for the four states defined by the values of q and r, as indicated in Scheme 1, which requires the structure of the protein in the oxidized and reduced states. Simple estimates can be obtained from a single structure of the oxidized state and a single structure of the reduced states such as can be obtained from experimental structures.

It is also important to assess the individual contributions of the non-bonded internal coordinate energies. For rubredoxin, it has been determined that the electrostatic energy accounts for all but 2 kcal/mol of the approximately -60 kcal/mol change in the energy of the protein upon reduction, even though the equilibrium internal coordinates of the redox site are different [18]. This means that the protein accommodates the change in charge by moving atoms while maintaining the internal coordinates and without creating bad van der Waals contacts. Although this has not been shown for all electron transfer proteins, it is consistent with the idea that there is very little strain due to the protein

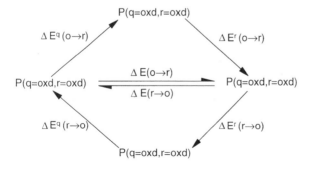

Scheme 1

because proteins are so flexible. Thus, an important simplification for many problems is to assume that when the relaxation energy ΔE^r [Eq. (7)] is mainly due to the electrostatic contribution, i.e.,

$$\Delta E^r = \Delta E^{r,\,\text{el}} + \Delta E^{r,\,\text{nonel}} \approx \Delta E^{r,\,\text{el}} \tag{8}$$

where $\Delta E^{r,\,\text{el}}$ is the electrostatic energy change when the environment relaxes after the charge change and ΔE^{nonel} is the nonelectrostatic energy change when the environment relaxes after the charge change. Of course, ΔE^q [Eq. (7)] is purely electrostatic.

2. Free Energy Simulations of ΔG_i^{out}

The free energy changes of the outer shell upon reduction, ΔG_i^{out}, are important, because the Nernst equation relates the redox potential to ΔG. Free energy simulation methods are discussed in Chapter 9. Here, the free energy change of interest is for the reaction

$$P^{\text{oxd}} \xrightarrow{\;\;\Delta G_{\text{oxd}\to\text{red}}\;\;} P^{\text{red}} \tag{R2}$$

When applying free energy methods, the length of simulations and λ increment needed for the reduction free energy calculation are of some concern, because changes in charge upon reduction result in large changes in energy. However, these changes do not generally involve large changes in structure or configurations; i.e., the primary structural change is the reorientation of dipoles [51]. Indeed, fast convergence (40 ps) was found for the calculation of the free energy difference of hydration between uncharged Ne and Na$^+$ in water using the slow growth (SG) method [52]. This is particularly encouraging, because the change from a neutral species to a singly charged species, which involves breaking the symmetry of the disordered environment around the neutral species to impose an orientation of the solvent dipoles, is a greater perturbation than increasing the charge for charged species, which simply involves increasing the existing orientation. Both thermodynamic perturbation (TP) and thermodynamic integration (TI) can be used. A possible advantage of the TI method is that the contributions to the total $\Delta\Delta G$ are additive and thus decomposable, although the validity of this decomposition is controversial [53–55].

3. Poisson–Boltzmann Calculations of ΔE_i^{out}

Calculations of ΔG_i^{out} by Poisson–Boltzmann methods have also been carried in which ΔG_i^{out} is assumed to be the sum of (1) the electrostatic interactions of the heme charges with the protein charges, screened by the polarizability of the water and protein, and (2) the reaction field energy from the polarization of solvent and protein [56]. Thus, both dielectric screening and entropic contributions are included in an approximate fashion, although no relaxation effects are included, because only a single structure is used. For the four hemes in the photosynthetic reaction center of *Rhodopseudomonas viridis* [56], major contributing factors include the axial ligands and the propionic acids. The reaction field energy is similar for all four hemes, indicating that this is not likely to be a major contributor to free energy differences between sites. However, the calculations for the protein assumed that the charge change was localized on the iron, which is likely to influence the results, especially for the highly delocalized heme system.

4. Summary

Recently, calculations of the photosynthetic reaction center by three different groups have been reviewed [57], which used DelPhi [58], PDLD [59], and CHARMM [60]. The unusual situation of having studies using all three different methods on the same system provided the opportunity for comparison. One point made was that each method has specific strengths, which are summarized in the review. MD methods have the advantages of allowing reorganization of the atoms, including dynamic information, and including explicit solvent, and FEMD has the added advantage that the connection to formal redox potentials is direct, because free energies rather than enthalpies are calculated. In addition, all three groups came to similar conclusions, although with different magnitudes of energetics, although the only common electrostatic contribution was the polar groups of the backbone and polar side chains.

V. DIFFERENCES IN REDOX POTENTIALS

Electron transfer proteins can modulate electron transfer processes by varying the outer shell energy. For instance, differences in redox potential of up to a few hundred millivolts are seen between homologous proteins with the same redox site, and even larger differences occur between nonhomologous electron transfer proteins with the same redox site [61]. Despite the rapidly growing number of crystal structures of electron transfer proteins, the structural origins of differences in redox potentials for a given redox site remain unclear. The differences may be due to intrinsic changes in the ionization potential of the redox site or to extrinsic changes in the energetics of the environment surrounding the redox site. By assuming that only the energetics of a few degrees of freedom lead to the redox potential differences, the calculation of differences in the energy change upon reduction for different proteins with the same redox site can be greatly simplified. Electronic structure calculations are necessary to examine differences that arise in the intrinsic energy of the redox site and are discussed in Section V.A. Both molecular mechanics and electrostatic energy calculations can be used for calculating the differences in the reduction energy of the outer shell and are discussed in Section V.B.

A. Calculation of Differences in the Energy Change of the Redox Site

Changes in the electronic structure of the redox site can lead to changes in the redox potential, mainly by changing the ionization potential or, more subtly, because geometry or charge distribution changes may alter the energetics of the environment surrounding the redox site. Three possible ways that the electronic structure of the redox site can be altered are by mutations of the ligands of the metal, by mutations that alter ligand geometry, and by substitution of the metal ion by another metal ion. The effects of these changes on the ionization potential and partial charges of the redox site can be determined simply by electronic structure calculations of the appropriate analogs of the wild-type and mutant using the methods of the previous section. Other mutations of the outer shell are thought not to influence the ionization potential but rather the electrostatic interaction energy between the redox site and the outer shell.

B. Calculation of Differences in the Energy Change of the Protein

Changes in the environment of the redox site can lead to changes in the redox potential via alteration of the interaction energy of the redox site with the outer shell. In many proteins such as the cytochromes [62] and the iron-sulfur proteins [63], the major way that nature influences the redox potentials is apparently through the protein environment of the cluster. One important contribution comes from the electrostatic environment created by the protein and solvent [34,64]. However, mutations involving surface charged residues often show very little effect [65,66] or unpredictable effects [67] on the redox potential. Moreover, our studies show surface charged residues are screened by solvent and counterions [19,68]. Buried charged side chains have been shown to be important in some proteins [69,70] but do not occur naturally in all proteins. Theoretical calculations indicate that a *combination* of the electrostatic contributions from the polar backbone, polar side chains, and solvent can impact the redox potential [68,71–73]. These polar interactions are complex because they depend on both orientation and distance and may vary dynamically. Moreover, large redox potential differences appear to be composed of many such small interactions as opposed to a few key interactions. Because of this, it has been almost impossible to sort out the contributions by simple inspection of structural data or by mutations guided by physical intuition alone. Therefore, molecular mechanics calculations provide a powerful means to sort out important effects in these molecules.

Calculation of how changes in the protein environment influence the redox potential depend in part on the magnitude of the changes involved. The rationale is that the overall backbone fold of the protein determines the general range of the potential and thus the general function, while specific sequences tune this potential (or are involved in binding specificity or are nonfunctional). Generally, most single-site mutations that affect neither the metal site nor the folding result in changes in redox potential of less than 100 mV. At one extreme are two proteins with the same redox site but completely different folds and redox potentials. For instance, the bacterial ferredoxins and the high potential iron-sulfur proteins (HiPIPs) both have the [4Fe-4S] site, but the redox potential for the $2-/3-$ couple is so much lower for the HiPIPs that only the $1-/2-$ couple has been seen experimentally whereas the $2-/3-$ couple is seen experimentally for the ferredoxins. For such cases, the approach should be to perform calculations of each couple according to the methods of Section V.A. On the other hand, proteins with similar folds but different sequences can be compared by using much simpler strategies, which are discussed here.

1. Structure/Sequence–Function Analysis

The simplest approach to understanding differences in redox potentials between homologous proteins is to analyze the experimental structures of multiple homologous proteins plus additional sequences of multiple homologous proteins in conjunction with available experimental redox potentials, in what might be termed a structure/sequence–function analysis. This analysis is based on the idea that a redox potential difference between homologous proteins can be identified by differences in the electrostatic potential contribution of specific residues as calculated from experimental structures of representative proteins. Furthermore, similar redox potential differences can be introduced into a given protein by introducing the appropriate amino acid mutations identified from the homologous protein study. Both MD and FEMD provide means of testing the latter assumption prior to experimental studies. The major advantage of examining multiple experimental X-ray crystal or NMR solution structures rather than molecular dynamics simulations is that changes in energy are often small (less than 2 kcal/mol) and may involve structural changes that are small (distances less than 0.5 Å). Thus, fluctuations in the simulations or inaccuracies due to the potential may be on the order of the changes that are seen in different experimental structures. However, a disadvantage of examining these experimental structures is that specific protein–solvent interactions and dynamic effects are difficult to examine.

The first step in the analysis is to examine the energetics of experimental structures of multiple homologous proteins, preferably with measured redox potentials. An assumption that is generally made is that most of any redox potential differences between homologous proteins are not due to entropic effects, so that $\Delta G_2 - \Delta G_1 = \Delta\Delta G \approx \Delta\Delta E$, where the subscripts refer to proteins 1 and 2. Furthermore, it is assumed that most of the differences lie in the electrostatic energy [18]. In most cases experimental structures are determined only for one oxidation state of the protein. An approximation to overcome this limitation is to assume that the structural relaxation of the homologous proteins is similar so that $\Delta\Delta E^r \approx 0$, and the electrostatics of only a single state can be examined. Thus, using Eq. (7),

$$\Delta\Delta E \approx \Delta\Delta E^q = -neF(\phi_2 - \phi_2) \tag{9}$$

where ϕ_i is the electrostatic potential of protein i, n is the number of electrons added, and e is the magnitude of an electron charge. If the change in charge is delocalized over several atoms, such as in the case of the Fe-S redox sites, a delocalized electrostatic potential can be defined as

$$\phi = \frac{\displaystyle\sum_{\substack{i=\text{redox}\\ \text{site atoms}}} \sum_{j \neq i}^{N} \left[\frac{\Delta q_i q_j}{r_{ij}} \right]}{\displaystyle\sum_{\substack{i=\text{redox}\\ \text{site atoms}}} \Delta q_i} \tag{10}$$

where Δq_i is the change in charge of atom i of the redox site upon reduction, q_j is the charge of any non-redox site atom j, and r_{ij} is the distance from atom i to atom j. The first sum is over all atoms of the redox site (i.e., those atoms that change charge upon reduction), and the second summation is over all atoms excluding the redox site. However, this definition is dependent on the partial charge parameters of the redox site, which are often uncertain or even unknown. Thus, a more approximate definition of ϕ, the electrostatic potential at a specific point 0, may be also be used:

$$\phi = \sum_{\substack{j \neq \text{redox}\\ \text{site atom}}} \frac{q_j}{r_{0j}} \tag{11}$$

where r_{0j} is the distance of the jth atom from the point 0 and the summation is over all atoms excluding the redox site. This relationship implies that all of the charge change is localized at a single point, which would be exact for a point charge description of a simple monatomic ion and is reasonable for the iron in the [1Fe] site. This latter definition of ϕ may be used as a parameter-independent estimate when there is uncertainty in the parameters. In either case, a dielectric constant of 1.0 with no cutoffs is used. The dielectric screening of the protein and solvent may be accounted for in an approximate manner using Poisson or Poisson–Boltzmann calculations as was done in a calculation of the differences between plastocyanin and rusticyanin [74]. However, it may be preferable not to, both for speed of calculation and because of the uncertainty in the local screening by the protein. If dielectric screening has not been included, caution should be exercised. In particular, the polar and charged side chain contributions should be calculated separately from each other, because the charged side chain contributions are likely to be greatly exaggerated. However, polar contributions have been the most important contribution in many of the cases studied [68,74].

An effective method for localizing causes of redox potentials is to plot the total backbone and side chain contributions to ϕ per residue for homologous proteins as functions of the residue number using a consensus sequence, with insertions treated by summing the contribution of the entire insertion as one "residue." The results for homologous proteins should be examined for differences in the contributions to ϕ per residue that correlate with observed redox potential differences. These differences can then be correlated with any other sequence–redox potential data for proteins that lack crystal or NMR structures. In addition, any sequences of homologous proteins that lack both redox potentials and structures should be examined, because residues important in defining the redox potential are likely to have semi-sequence conservation of a few key amino acid types.

One example of a sequence determinant of redox potentials that has been identified in this manner is an Ala-to-Val mutation at residue 44, which causes a 50 mV decrease in redox potential (and vice versa) in the rubredoxins [68]. The mutation was identified because the sum of the backbone contributions to ϕ of residues 43 and 44 change by 40 mV due to an ~0.5 Å backbone shift away from the redox site. This example points out the importance of examining the backbone contributions. The corresponding site-specific mutants have confirmed both the redox potential shift [75] and the structural shift [75].

A second example is that of an Ala-to-Cys mutation, which causes the formation of a rare SH $\cdots$ S hydrogen bond between the cysteine and a redox site sulfur and a 50 mV decrease in redox potential (and vice versa) in the bacterial ferredoxins [73]. Here, the side chain contribution of the cysteine is significant; however, a backbone shift can also contribute depending on whether the nearby residues allow it to happen. Site-specific mutants have confirmed the redox potential shift [76,77] and the side chain conformation of cysteine but not the backbone shift in the case with crystal structures of both the native and mutant species [78]; the latter can be attributed to the specific sequence of the ferredoxin studied [73].

2. Molecular Dynamics Simulations

Molecular dynamics simulations are useful in understanding whether the electrostatic energy shifts seen between homologous proteins can be translated into site-specific mutations that shift the redox potential. The molecular dynamics simulations of the wild-type and proposed mutant can address whether the mutant will have the same structural and energetic shifts as occur between the homologous proteins with different redox potentials. The use of MD simulation allows for greater sampling of conformational space than energy minimization, thus enhancing the probability of properly modeling structural changes.

3. Free Energy Simulations

Free energy simulations are a useful means of quantitating whether the free energy and not simply the energy is shifting in the predicted manner for the mutant (see Chapter 9). The difference in the free energy changes upon reduction between a wild-type and a mutant, $\Delta\Delta G = \Delta G^* - \Delta G$, where the asterisk indicates the mutant, can be calculated in two ways via the thermodynamic cycle shown in Scheme 2,

$$\begin{array}{ccc}
\mathrm{P^{oxd}} & \xrightarrow{\;\Delta G_{\mathrm{oxd}\to\mathrm{red}}\;} & \mathrm{P^{red}} \\
\Delta G_{\mathrm{oxd}\to\mathrm{red*}}\downarrow & & \downarrow \Delta G_{\mathrm{red}\to\mathrm{red*}} \\
\mathrm{P^{*oxd}} & \xrightarrow[\;\Delta G_{\mathrm{oxd*}\to\mathrm{red*}}\;]{} & \mathrm{P^{*red}}
\end{array}$$

Scheme 2

where P and P* denote the native and mutated protein, respectively, and oxd and red denote the oxidized and reduced forms, respectively. The free energy difference is given by

$$\Delta\Delta G = \Delta G_{\mathrm{oxd*}\to\mathrm{red*}} - \Delta G_{\mathrm{oxd}\to\mathrm{red}} \tag{12}$$

$$\Delta\Delta G = \Delta G_{\mathrm{red}\to\mathrm{red*}} - \Delta G_{\mathrm{oxd}\to\mathrm{oxd*}} \tag{13}$$

where Eq. (12) is simply the definition of $\Delta\Delta G$ as the difference between the reduction free energies of the wild-type and mutant, and Eq. (13), which comes from the thermodynamic cycle, gives $\Delta\Delta G$ as the difference between the free energies of mutation of the oxidized and reduced states. Calculation of $\Delta\Delta G$ by Eqs. (12) and (13) will be referred to as reduction and mutation, respectively, free energy calculations.

VI. ELECTRON TRANSFER RATES

The environmental (i.e., solvent and/or protein) free energy curves for electron transfer reactions can be generated from histograms of the polarization energies, as in the works of Warshel and coworkers [79,80].

A. Theory

This section contains a brief review of the molecular version of Marcus theory, as developed by Warshel [81]. The free energy surface for an electron transfer reaction is shown schematically in Figure 1, where R represents the reactants D^- and A, P represents the products D and A^-, and the reaction coordinate X is the degree of polarization of the solvent. The subscript o for R and P denotes the equilibrium values of R and P, while P′ is the Franck–Condon state on the P-surface. The activation free energy, $\Delta G^{\ddagger}$, can be calculated from Marcus theory by Eq. (4). This relation is based on the assumption that the free energy is a parabolic function of the polarization coordinate. For self-exchange transfer reactions, we need only λ to calculate $\Delta G^{\ddagger}$, because $\Delta G^{\circ} = 0$. Moreover, we can write

$$\begin{aligned}
\lambda &\equiv \Delta G(\mathbf{r}_{\mathrm{D,red}}, \mathbf{q}_{\mathrm{D,oxd}}; \mathbf{r}_{\mathrm{A,oxd}}, \mathbf{q}_{\mathrm{A,red}}) - \Delta G(\mathbf{r}_{\mathrm{D,oxd}}, \mathbf{q}_{\mathrm{D,oxd}}; \mathbf{r}_{\mathrm{A,red}}, \mathbf{q}_{\mathrm{A,red}}) \\
&= \Delta G(\mathbf{r}_{\mathrm{D,red}}, \mathbf{q}_{\mathrm{D,oxd}}; \mathbf{r}_{\mathrm{A,oxd}}, \mathbf{q}_{\mathrm{A,red}}) - \Delta G(\mathbf{r}_{\mathrm{D,red}}, \mathbf{q}_{\mathrm{D,red}}; \mathbf{r}_{\mathrm{A,oxd}}, \mathbf{q}_{\mathrm{A,oxd}})
\end{aligned} \tag{14}$$

where the difference lies in the second term. Here, $\mathbf{r}_D$ and $\mathbf{r}_A$ represent the nuclear configurations of all atoms near D and A, respectively; $\mathbf{q}_D$ and $\mathbf{q}_A$ represent the charged states (via the partial charges of the appropriate atoms) of D and A, respectively; and oxd and red denote oxidized and reduced, respectively. Thus, λ is the energy of the Franck–Condon transition from the R (i.e., $\mathbf{q}_D = \mathbf{q}_{\mathrm{D,red}}$ and $\mathbf{q}_A = \mathbf{q}_{\mathrm{A,oxd}}$) to the P (i.e., $\mathbf{q}_D = \mathbf{q}_{\mathrm{D,oxd}}$ and $\mathbf{q}_A = \mathbf{q}_{\mathrm{A,red}}$) surface at $X = R_0$. Note that by examining λ, one does not have to explicitly define the polarization coordinate.

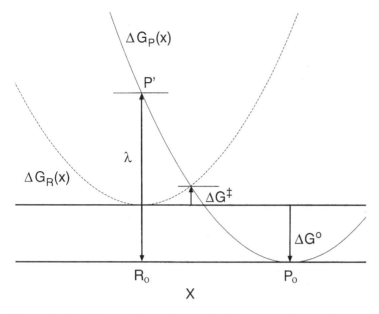

Figure 1 Reaction coordinate diagram for electron transfer reactions.

More generally, the connection between the free energy surface and the simulation data can be made by the relation [81]

$$\Delta G(X) = -k_B T \ln[P(X)/P(X_{min})] \tag{15}$$

where $P(X)$ is the probability of a value of X, k_B is Boltzmann's constant, and T is the temperature. $P(X)$ can be calculated by making a histogram of X obtained from a simulation. The activation free energy can also be calculated from the ratio of the probability of being at the transition state ($X = 0$) versus the probability of being in the equilibrium state ($X = X_{min}$),

$$\Delta G^{\ddagger} = -k_B T \ln[P(0)/P(X_{min})] \tag{16}$$

This definition of $\Delta G^{\ddagger}$ is thus dependent on the exact definition of X. Here the polarization coordinate X is defined to be the difference in the energy between when the excess electron is on A [curve $\Delta G_P(X)$ of Fig. 1] and when the excess electron is on D [curve $\Delta G_R(X)$ of Fig. 1] [79], i.e.,

$$
\begin{aligned}
X \equiv \Delta V &= V(\mathbf{r}_D, \mathbf{q}_{D,oxd}; \mathbf{r}_A, \mathbf{q}_{A,red}) - V(\mathbf{r}_D, \mathbf{q}_{D,red}; \mathbf{r}_A, \mathbf{q}_{A,oxd}) \\
&= \Delta G(\mathbf{r}_D, \mathbf{q}_{D,oxd}; \mathbf{r}_A, \mathbf{q}_{A,red}) - \Delta G(\mathbf{r}_D, \mathbf{q}_{D,red}; \mathbf{r}_A, \mathbf{q}_{A,oxd})
\end{aligned} \tag{17}
$$

where the second equality holds because there is no entropy change upon making a vertical transition between the two curves of Figure 1, plus some small terms corresponding to the change in geometry of the redox site (i.e., the potential energy parameters) on going from the oxidized to the reduced state and vice versa. Therefore, at the left-hand minimum, R_0, $X_{min} = \lambda$. For a self-exchange reaction, the transition state corresponds to when the energy of the excess electron on either protein is the same; i.e., $\Delta V = 0$.

If the distribution of X is assumed to be a Gaussian about X_{min}, the minimum of either the left- or the right-hand side, the activation free energy [Eq. (4)] becomes [82]

$$\Delta G^{\ddagger} = k_B T \lambda^2 / 2\sigma^2 \qquad (18)$$

where $\sigma^2 = \langle (X - X_{min})^2 \rangle$, i.e., the mean-square fluctuations of X, which can also easily be calculated from the simulation. This will be referred to as the Gaussian fluctuation approximation.

As indicated above, the free energy curve predicted by simulation is obtained by making a histogram of values of X in the trajectory [Eq. (15)]. However, because thermal fluctuations in a molecular dynamics simulation generally are not sufficient to allow sampling far away from $P(X_{min})$, two methods are used to extend the free energy surface. First, non-Boltzmann or ''umbrella'' sampling [79,83,84] can be used to obtain the free energy surface away from $X = X_{min}$ (see Chapter 10). Here, the nuclear rearrangement due to intermediate values of charge on the two redox centers is considered, meaning that the charge density of the transferring electron is $-(1 - z)e$ on one center and $-ze$ on the other. More specifically, for a protein, all parameters (partial charges and equilibrium bond lengths and angles) of the redox site should be scaled according to

$$U_{oxd,z} = (1 - z)U_{oxd} + zU_{red} \qquad (19a)$$

and

$$U_{red,z} = (1 - z)U_{red} + zU_{oxd} \qquad (19b)$$

The expression for the probability with $X = \Delta V$ is

$$P(\Delta V) = c^{(z)} e^{\beta z \Delta V} P^{(z)}(\Delta V) \qquad (20)$$

where $c^{(z)}$ is a normalization constant and $P^{(z)}$ indicates the probability when the system has the charge distribution characterized by z. The second method is actually a special case of umbrella sampling when $z = 1$, so that $c^{(1)} = c^{(0)}$ [81,85]. Thus, the left-hand side of the P curve, ΔG_P, is given by $\Delta G_R + \Delta V$ and the right-hand side of the R curve is given by $\Delta G_P + \Delta V$ (Fig. 1).

B. Application

Free energy curves for the self-exchange reaction between two rubredoxins (Rd_1 and Rd_2) were generated from MD simulations [86,87].

$$Rd_1^{2-} + Rd_2^{1-} \rightarrow Rd_1^{1-} + Rd_2^{2-} \qquad (R3)$$

The study of self-transfer reactions is a great simplification for theoretical studies, although not for experimental studies, and is thus a useful starting point. The free energy for electron transfer reactions in a variety of small molecule systems has been studied using MD methods [79–81,84,85,88–90]. The applications to proteins have been more limited. Several applications have been made by the Warshel group, including studies of the crystal structures of oxidized and reduced cytochrome c [91]. The simulations used to evaluate $\Delta G^{\ddagger}$ for the self-exchange reaction of Rd were actually separate simulations of Rd in the oxidized and reduced form [19], which implies that the two rubredoxins are separated by a distance great enough that the polarization of solvent around one does not affect the other, i.e., the infinite separation limit. Thus, the polarization of the entire system is estimated by summing that of separate simulations of the protein in the oxidized and reduced forms,

$$\Delta V = V_1(\mathbf{r}_1, \mathbf{q}_{1,\mathrm{oxd}}) + V_2(\mathbf{r}_2, \mathbf{q}_{2,\mathrm{red}}) - V_1(\mathbf{r}_1, \mathbf{q}_{1,\mathrm{red}}) - V_2(\mathbf{r}_2, \mathbf{q}_{2,\mathrm{oxd}}) \tag{21}$$

The first and third terms on the right-hand side of Eq. (21) can be evaluated from the reduced simulation by using the oxidized and reduced partial charges, respectively, and the second and fourth terms can be evaluated from the oxidized simulation by using the reduced and oxidized partial charges, respectively. Using Eqs. (17) and (21), the solvent reorganization energy becomes

$$\lambda = X_{\mathrm{min}} = V_1(\mathbf{r}_{1,\mathrm{red}}, \mathbf{q}_{1,\mathrm{oxd}}) + V_2(\mathbf{r}_{2,\mathrm{oxd}}, \mathbf{q}_{2,\mathrm{red}}) \\ - V_1(\mathbf{r}_{1,\mathrm{red}}, \mathbf{q}_{1,\mathrm{red}}) - V_2(\mathbf{r}_{2,\mathrm{oxd}}, \mathbf{q}_{2,\mathrm{oxd}}) \tag{22}$$

Thus, the $z = 0$ surface (and equivalently the $z = 1$ surface) is generated from the original oxidized and reduced simulations. Additional simulations were performed to sample values of the polarization coordinate away from the minima, using parameters scaled according to Eq. (19).

REFERENCES

1. HB Gray, WR Ellis Jr. Electron transfer. In: I Bertini, HB Gray, SJ Lippard, JS Valentine, eds. Bioinorganic Chemistry. Sausalito, CA: University Science Books, 1994, pp 315–363.
2. KA Gray, VL Davidson, DB Knaff. Complex formation between methylamine dehydrogenase and amicyanin from Paracocuus denitrificans. J Biol Chem 263:13987–13990, 1988.
3. RA Marcus, N Sutin. Electron transfer in chemistry and biology. Biochim Biophys Acta 811: 265–322, 1985.
4. RH Holm, P Kennepohl, EI Solomon. Structural and functional aspects of metal sites in biology. Chem Rev 96:2239–2341, 1996.
5. H Reiss, A Heller. J Phys Chem 89:4207–4213, 1985.
6. A Kuki, PG Wolynes. Electron tunneling paths in proteins. Science 236:1647–1652, 1987.
7. T Ziegler. Approximate density functional theory as a practical tool in molecular energetics and dynamics. Chem Rev 91:651–667, 1991.
8. W Kohn. Density functional theory of electronic structure. J Phys Chem 100:12974–12980, 1996.
9. J Li, L Noodleman, DA Case. Electronic structure calculations: Density functional methods with applications to transition metal complexes. In: EIS Lever, ABP Lever, eds. Inorganic Electronic Structure and Spectroscopy, Vol. 1. Methodology. New York: Wiley, 1999, pp 661–724.
10. JB Koerner, T Ichiye. Conformational dependence of the electronic properties of $[Fe(SCH_3)_4]^{-,2-}$. J Phys Chem B 101:3633–3643, 1997.
11. J Li, MR Nelson, CY Peng, D Bashford, L Noodleman. Incorporating protein environments in density functional theory: A self-consistent reaction field calculation of redox potentials of [2Fe2S] clusters in ferredoxin and phthalate dioxygenase reductase. J Phys Chem A 102:6311–6324, 1998.
12. CL Fisher, J-L Chen, J Li, D Bashford, L Noodleman. Density-functional and electrostatic calculations for a model of a manganese superoxide dismutase active site in aqueous solution. J Phys Chem 100:13498–13505, 1996.
13. G Backes, Y Mino, TM Loehr, TE Meyer, MA Cusanovich, WV Sweeny, ET Adman, J Sanders-Loehr. The environment of Fe_4S_4 clusters in ferredoxins and high-potential iron proteins. New information from X-ray crystallography and resonance Raman spectroscopy. J Am Chem Soc 113:2055–2064, 1991.

14. MHM Olsson, U Ryde, BO Roos, K Pierloot. On the relative stability of tetragonal and trigonal Cu(II) complexes with relevance to the blue copper proteins. J Biol Inorg Chem 3:109–125, 1998.

15. DA Case, M Karplus. Dynamics of ligand binding to heme proteins. J Mol Biol 132:343–368, 1979.

16. BR Gelin, M Karplus. Mechanism of tertiary structural change in hemoglobin. Proc Natl Acad Sci USA 74:801–805, 1977.

17. SH Northrup, MR Pear, JA McCammon, M Karplus. Molecular dynamics of ferrocytochrome c. Nature 286:304–305, 1980.

18. VS Shenoy, T Ichiye. Influence of protein flexibility on the redox potential of rubredoxin: Energy minimization studies. Proteins 17:152–160, 1993.

19. RB Yelle, N-S Park, T Ichiye. Molecular dynamics simulations of rubredoxin from Clostridium pasteurianum: Changes in structure and electrostatic potential during redox reactions. Proteins 22:154–167, 1995.

20. BW Beck. Theoretical Investigations of Iron-Sulfur Proteins. Ph.D. thesis. Pullman, WA: Washington State University, 1997.

21. BW Beck, JB Koerner, RB Yelle, T Ichiye. Unusual hydrogen bonding ability of sulfurs in Fe-S redox sites: Ab initio quantum and classical mechanical studies. J Phys Chem B, submitted.

22. A Vedani, DW Huhta. A new force field for modeling metalloproteins. J Am Chem Soc 112: 4759–4767, 1990.

23. SC Hoops, KW Anderson, KM Merz Jr. Force field design for metalloproteins. J Am Chem Soc 113:8262–8270, 1991.

24. J Shen, CF Wong, S Subramaniam, TA Albright, JA McCammon. Partial electrostatic charges for the active center of Cu,Zn superoxide dismutase. J Comput Chem 11:346–350, 1990.

25. U Ryde. Molecular dynamics simulations of alcohol dehydrogenase with a four- or five-coordinate catalytic zinc ion. Proteins 21:40–56, 1995.

26. AK Churg, A Warshel. Control of the redox potential of cytochrome c and microscopic dielectric effects in proteins. Biochemistry 25:1675, 1986.

27. J-M Mouesca, JL Chen, L Noodleman, D Bashford, DA Case. Density functional/Poisson-Boltzmann calculations of redox potentials for iron-sulfur clusters. J Am Chem Soc 116: 11898–11914, 1994.

28. RB Yelle, BW Beck, JB Koerner, CA Sacksteder, T Ichiye. Influence of the metal site on the structure and solvation of rubredoxin and its analogs: A molecular dynamics study. Proteins accepted.

29. CM Breneman, KB Wiberg. Determining atom-centered monopoles from molecular electrostatic potentials. The need for high sampling density in formamide conformational analysis. J Comput Chem 11:361–373, 1990.

30. WL Jorgensen, CJ Swenson. Optimized intermolecular potential functions for amides and peptides. Structure and properties of liquid amides. J Am Chem Soc 107:569–578, 1985.

31. BR Brooks, RE Bruccoleri, BD Olafson, DJ States, S Swaminathan, M Karplus. CHARMM: A program for macromolecular energy, minimization, and dynamics calculations. J Comput Chem 4:187–217, 1983.

32. AD MacKerell Jr, D Bashford, M Bellot, RL Dunbrack Jr, MJ Field, S Fischer, J Gao, H Guo, S Ha, D Joseph, K Kuchnir, K Kuczera, FTK Lau, M Mattos, S Michnick, DT Nguyen, T Ngo, B Prodhom, B Roux, M Schlenkrich, J Smith, R Stote, J Straub, J Wiorkiewicz-Kuczera, M Karplus. All-atom empirical potential for molecular modeling and dynamics studies of proteins. J Phys Chem B 102:3586–3616, 1998.

33. JB Koerner, T Ichiye. Interactions of the rubredoxin redox site analogue [Fe(SCH$_3$)$_4$]$^{2-}$ with water: An ab initio quantum chemistry study. J Phys Chem B 104:2424–2431, 2000.

34. PJ Stephens, DR Jollie, A Warshel. Protein control of redox potentials of iron-sulfur proteins. Chem Rev 96:2491–2513, 1996.

35. MK Gilson, KA Sharp, B Honig. Calculating the electrostatic potential of molecules in solution: Method and error assessment. J Comput Chem 9:327–335, 1988.

36. MK Gilson, B Honig. Calculation of the total electrostatic energy of a macromolecular system: Solution energies, binding energies, and conformational analysis. Proteins 4:7–18, 1988.

37. ME Davis, JD Madura, BA Luty, JA McCammon. Electrostatics and diffusion of molecules in solution: Simulations with the University of Houston Brownian dynamics program. Comput Phys Commun 62:187–197, 1991.

38. JD Madura, JM Briggs, RC Wade, ME Davis, BA Luty, A Ilin, J Antosiewicz, MK Gilson, B Bagheri, LR Scott, JA McCammon. Electrostatics and diffusion of molecules in solution: Simulations with the University of Houston Brownian Dynamics Program. Comput Phys Commun 91:57–95, 1995.

39. HJC Berendsen, JPM Postma, WF van Gunsteren, J Hermans. Interaction models of water in relation to protein hydration. In: B Pullman, ed. Intermolecular Forces. Dordrecht, Holland: Reidel, 1981, pp 331–341.

40. HJC Berendsen, JR Grigera, TP Straatsma. J Phys Chem 91:6269, 1987.

41. WL Jorgensen, J Chandrasekhar, JD Madura, RW Impey, ML Klein. Comparison of simple potential functions for simulating liquid water. J Chem Phys 79:926–935, 1983.

42. JA McCammon, SC Harvey. Dynamics of Proteins and Nucleic Acids. New York: Cambridge Univ Press, 1987.

43. DE Smith, LX Dang. Computer simulations of NaCl association in polarizable water. J Chem Phys 100:3757–3766, 1994.

44. ST Russell, A Warshel. Calculations of electrostatic energies in proteins: The energetics of ionized groups in bovine pancreatic trypsin inhibitor. J Mol Biol 185:389–404, 1985.

45. A Warshel, ST Russell. Calculations of electrostatic interactions in biological systems and in solutions. Quart Rev Biophys 17:283–422, 1984.

46. PE Smith, BM Pettitt. Modeling solvent in biomolecular systems. J Phys Chem 98:9700–9711, 1994.

47. J-K Hyun, CS Babu, T Ichiye. Apparent local dielectric response around ions in water: A method for its determination and its applications. J Phys Chem 99:5187–5195, 1995.

48. G Hummer, LR Pratt, AE García. Free energy of ionic hydration. J Phys Chem 100:1206–1215, 1996.

49. PD Swartz, T Ichiye. Protein contributions to redox potentials of iron-sulfur proteins: An energy minimization study. Biophys J 73:2733–2741, 1997.

50. VS Shenoy. Contribution of Protein Environment to Redox Potentials of Rubredoxin and Cytochrome c. M.S. Thesis. Pullman, WA: Washington State University, 1992.

51. P Kollman. Free energy calculations: Applications to chemical and biochemical phenomena. Chem Rev 93:2395–2417, 1993.

52. TP Straatsma, HJC Berendsen. Free energy of ionic hydration: Analysis of a thermodynamic integration technique to evaluate free energy differences by molecular dynamics simulations. J Chem Phys 89:5876–5886, 1988.

53. WF van Gunsteren. Molecular dynamics studies of proteins. Curr Opinion Struct Biol 3:277–281, 1993.

54. AE Mark, WF van Gunsteren. Decomposition of the free energy of a system in terms of specific interactions. J Mol Biol 240:167–176, 1994.

55. WL Jorgensen, JK Buckner, S Boudon, J Tirado-Rives. Efficient computation of absolute free energies of binding by computer simulations. Application to the methane dimer in water. J Chem Phys 89:3742–3746, 1988.

56. MR Gunner, B Honig. Electrostatic control of midpoint potentials in the cytochrome subunit of the Rhodopseudomonas viridis reaction center. Proc Natl Acad Sci USA 88:9151–9155, 1991.

57. T Ichiye. A dawning light: The beginnings of an understanding of the photosynthetic reaction center. Structure 4:1009–1012, 1996.

58. MR Gunner, A Nicholls, B Honig. Electrostatic potentials in Rhodopseudomonas viridis reaction centers: Implications for the driving force and directionality of electron transfer. J Phys Chem 100:4277–4291, 1996.

59. WW Parson, Z-T Chu, A Warshel. Electrostatic control of charge separation in bacterial photosynthesis. Biochim Biophys Acta 1017:251–272, 1990.

60. M Marchi, JN Gehlen, D Chandler, M Newton. Diabatic surfaces and the pathway for primary electron transfer in a photosynthetic reaction center. J Am Chem Soc 115:4178–4190, 1993.

61. R Cammack. Iron-sulfur cluster in enzymes: Themes and variations. Adv Inorg Chem 38: 281–322, 1992.

62. J-C Marchon, T Mashiko, CA Reed. How does nature control cytochrome redox potentials? In: C Ho, ed. Electron Transport and Oxygen Utilization. New York: Elsevier North-Holland, 1982, pp 67–72.

63. WV Sweeney, JC Rabinowitz. Proteins containing 4Fe-4S clusters: An overview. Annu Rev Biochem 49:139–161, 1980.

64. A Schejter, I Aviram, T Goldkorn. The contribution of electrostatic factors to the oxidation-reduction potentials of c-type cytochromes. In: C Ho, ed. Electron Transport and Oxygen Utilization. New York: Elsevier North-Holland, 1982, pp 95–109.

65. B Shen, DR Jollie, CD Stout, TC Diller, FA Armstrong, CM Gorst, GN La Mar, PJ Stephens, BK Burgess. Azotobacter vinelandii ferredoxin I: Alteration of individual surface charges and the $[4Fe\text{-}4S]^{2+/+}$ cluster reduction potential. J Biol Chem 269:8564–8575, 1994.

66. I Quinkal, V Davasse, J Gaillard, J-M Moulis. On the role of conserved proline residues in the structure and function of Clostridium pasteurianum 2[4Fe-4S] ferredoxin. Protein Eng 7: 681–687, 1994.

67. Q Zeng, ET Smith, DM Kurtz, RA Scott. Protein determinants of metal site reduction potentials. Site directed mutagenesis studies of Clostridium pasteurianum: rubredoxin. Inorg Chim Acta 242:245–251, 1996.

68. PD Swartz, BW Beck, T Ichiye. Structural origins of redox potential in iron-sulfur proteins: Electrostatic potentials of crystal structures. Biophys J 71:2958–2969, 1996.

69. RL Cutler, AM Davies, S Creighton, A Warshel, GR Moore, M Smith, AG Mauk. Role of arginine-38 in regulation of the cytochrome c oxidation-reduction equilibrium. Biochemistry 28:3188–3197, 1989.

70. R Varadarajan, TE Zewert, HB Gray, SG Boxer. Effects of buried ionizable amino acids on the reduction potential of recombinant myoglobin. Science 243:69–72, 1989.

71. R Langen, GM Jensen, U Jacob, PJ Stephens, A Warshel. Protein control of iron-sulfur cluster redox potentials. J Biol Chem 267:25625–25627, 1992.

72. T Ichiye, RB Yelle, JB Koerner, PD Swartz, BW Beck. Molecular dynamics simulation studies of electron transfer properties of Fe-S proteins. Biomacromolecules: From 3-D Structure to Applications. Hanford Symposium on Health and the Environment 34, Pasco, WA, 1995, pp 203–213.

73. BW Beck, Q Xie, T Ichiye. Computational study of S—H···S hydrogen bonds in [4Fe-4S]-type ferredoxin x-ray and NMR structures: Characterization and implications for redox potentials. Protein Sci, submitted.

74. MV Botuyan, A Toy-Palmer, J Chung, RC Blake II, P Beroza, DA Case. NMR solution structure of Cu(I) rusticyanin from Thiobacillus ferrooxidans: Structural basis of the extreme acid stability and redox potential. J Mol Biol 263:752–767, 1996.

75. MK Eidsness, AK Burden, KA Richie, DMJ Kurtz, RA Scott, ET Smith, T Ichiye, C Kang. Modulation of the redox potential of the $Fe(SCys)_4$ site in rubredoxin by the orientation of a peptide dipole. Biochemistry 38:14803–14809, 1999.

76. SE Iismaa, AE Vázquez, GM Jensen, PJ Stephens, JN Butt, FA Armstrong, BK Burgess. Site-

directed mutagenesis of Azotobacter vinelandii ferredoxin I. J Biol Chem 266:21563–21571, 1991.

77. PS Brereton, FJM Verhagen, ZH Zhou, MWW Adams. Effect of iron-sulfur cluster environment in modulating the thermodynamic properties and biological function of ferredoxin from Pyrococcus furiosus. Biochemistry 37:7351–7362, 1998.

78. J Soman, S Iismaa, CD Stout. Crystallographic analysis of two site-directed mutants of Azotobacter vinelandii ferredoxin. J Biol Chem 266:21558–21562, 1991.

79. J-K Hwang, A Warshel. Microscopic examination of free-energy relationships for electron transfer in polar solvents. J Am Chem Soc 109:715–720, 1987.

80. G King, A Warshel. Investigation of the free energy functions for electron transfer reactions. J Chem Phys 93:8682–8692, 1990.

81. A Warshel. Dynamics of reactions in polar solvents. Semiclassical trajectory studies of electron-transfer and proton-transfer reactions. J Phys Chem 86:2218–2224, 1982.

82. T Ichiye. Solvent free energy curves for electron transfer: A non-linear solvent response model. J Chem Phys 104:7561–7571, 1996.

83. JP Valleau, GM Torrie. A guide to Monte Carlo for statistical mechanics. 2. Byways. In: BJ Berne, ed. Modern Theoretical Chemistry, Vol. 5. New York: Plenum, 1977, p 137.

84. RA Kuharski, JS Bader, D Chandler, M Sprik, ML Klein, RW Impey. Molecular model for aqueous ferrous-ferric electron transfer. J Chem Phys 89:3248–3257, 1988.

85. M Tachiya. Relation between the electron-transfer rate and the free energy change of reaction. J Phys Chem 93:7050–7052, 1989.

86. RB Yelle. Theoretical studies of the electron transfer properties of rubredoxin. Ph.D. Thesis. Washington State University, Pullman, WA, 1996.

87. RB Yelle, T Ichiye. Unpublished results.

88. T Kakitani, N Mataga. Comprehensive study on the role of coordinated solvent mode played in electron-transfer reactions in polar solutions. J Phys Chem 91:6277–6285, 1987.

89. EA Carter, JT Hynes. Solute-dependent solvent force constants for ion pairs and neutral pairs in a polar solvent. J Phys Chem 93:2184–2187, 1989.

90. RB Yelle, T Ichiye. Solvation free energy reaction curves for electron transfer: Theory and simulation. J Phys Chem B 101:4127–4135, 1997.

91. AK Churg, RM Weiss, A Warshel, T Takano. On the action of cytochrome c: Correlating geometry changes upon oxidation with energies of electron transfer. J Phys Chem 87:1683–1694, 1983.

19

The RISM-SCF/MCSCF Approach for Chemical Processes in Solutions

Fumio Hirata and Hirofumi Sato
Institute for Molecular Science, Okazaki National Research Institutes, Okazaki, Japan

Seiichiro Ten-no
Nagoya University, Nagoya, Japan

Shigeki Kato
Kyoto University, Kyoto, Japan

I. SOLVENT EFFECT ON CHEMICAL PROCESSES

Most chemical reactions in nature as well as in the laboratory take place in liquid solutions. Chemical reactions of molecules in living systems take place exclusively in the solution phase. One of the most important tactics in organic chemistry is to choose an appropriate solvent for their reactions or to get higher yields. The reaction yield and rate are controlled by the solvent through changes in the free energy difference between reactants and products and thus in the activation free energy. In living cells, chemical reactions are controlled by the solvent with an additional complication—conformational fluctuations in biomolecules. In all those reactions, the solvent effect manifests itself not only through solute–solvent interactions and solvent reorganizations but also through intramolecular processes that are always associated with changes in electronic structure. In this regard, theoretical investigations of chemical processes in solutions are inevitably coupled with studies of quantum and statistical mechanics.

One of the most straightforward realizations of such couplings is the ab initio molecular dynamics (MD) approach originated by Car and Parrinello (CP) and published by Laasoninen et al. [1]. The method, which consists of solving the Kohn–Sham density functional (DF) equation by simulated annealing, has been proven to be a capable theory in many applications. However, the treatment has serious limitations or difficulties in terms of system size and the handling of excited states. In a sense, the approaches based on the path integral technique share similar difficulties with the CP theory for chemical processes in solution. Combined with molecular simulations and the statistical mechanics of liquids, the method could have revealed many interesting physical aspects of quantum processes in solution. But, again, it is largely limited to a simple and/or small system such as a harmonic oscillator or a solvated electron. In this respect, it is highly desirable to exploit the molecular orbital (MO) theory for the electronic structure, which has been proven to be the most powerful tool for exploring molecular processes. This method does not have the serious limitations characteristic of the CP and path integral approaches.

However, theories that are based on a basis set expansion do have a serious limitation with respect to the number of electrons. Even if one considers the rapid development of computer technology, it will be virtually impossible to treat by the MO method a small system of a size typical of classical molecular simulation, say ~1000 water molecules. A logical solution to such a problem would be to employ a hybrid approach in which a chemical species of interest is handled by quantum chemistry while the solvent is treated classically.

Roughly speaking, three types of hybrid approaches have been proposed depending on the level of coarse graining with respect to solvent coordinates. Methods based on continuum models smear all solvent coordinates and represent solvent characteristics with a single parameter, namely, a dielectric constant. Molecular simulations, on the other hand, take all solvent coordinates into account explicitly and sample many solvent configurations in order to realize meaningful statistics for the physical quantities of concern. The approach based on the statistical mechanics of molecular liquids coarse-grains solvent coordinates in a level of the (density) pair correlation functions. In what follows, we briefly outline these three methods.

A. Continuum Model

The continuum model, in which solvent is regarded as a continuum dielectric, has been used to study solvent effects for a long time [2,3]. Because the electrostatic interaction in a polar system dominates over other forces such as van der Waals interactions, solvation energies can be approximated by a reaction field due to polarization of the dielectric continuum as solvent. Other contributions such as dispersion interactions, which must be explicitly considered for nonpolar solvent systems, have usually been treated with empirical quantity such as macroscopic surface tension of solvent.

A variety of methodologies have been implemented for the reaction field. The basic equation for the dielectric continuum model is the Poisson–Laplace equation, by which the electrostatic field in a cavity with an arbitrary shape and size is calculated, although some methods do not satisfy the equation. Because the solute's electronic structure and the reaction field depend on each other, a nonlinear equation (modified Schrödinger equation) has to be solved in an iterative manner. In practice this is achieved by modifying the electronic Hamiltonian or Fock operator, which is defined through the shape and size of the cavity and the description of the solute's electronic distribution. If one takes a dipole moment approximation for the solute's electronic distribution and a spherical cavity (Onsager's reaction field), the interaction can be derived rather easily and an analytical expression of the Fock operator is obtained. However, such an expression is not feasible for an arbitrary electronic distribution in an arbitrary cavity fitted to the molecular shape. In this case the Fock operator is very complicated and has to be prepared by a numerical procedure.

Numerous attempts have been made to develop hybrid methodologies along these lines. An obvious advantage of the method is its handiness, while its disadvantage is an artifact introduced at the boundary between the solute and solvent. You may obtain agreement between experiments and theory as close as you desire by introducing many adjustable parameters associated with the *boundary conditions*. However, the more adjustable parameters are introduced, the more the physical significance of the parameter is obscured.

B. Simulations

Molecular simulation techniques, namely Monte Carlo and molecular dynamics methods, in which the liquid is regarded as an assembly of interacting particles, are the most popular approach. Various properties related to the macroscopic and microscopic quantities are computed by these methods.

A typical hybrid approach is the QM/MM (quantum mechanical–molecular mechanical) simulation method [4]. In this method, the solute molecule is treated quantum mechanically, whereas surrounding solvent molecules are approximated by molecular mechanical potentials. This idea is also used in biological systems by regarding a part of the system, e.g., the activation site region of an enzyme, as a quantum "solute," which is embedded in the rest of the molecule, which is represented by molecular mechanics. The actual procedure used in this method is very simple: The total energy of the liquid system (or part of a protein) at an instantaneous configuration, generated by a Monte Carlo or molecular dynamics procedure, is evaluated, and the modified Schrödinger equations are solved repeatedly until sufficient sampling is accumulated. Since millions of electronic structure calculations are needed for sufficient sampling, the ab initio MO method is usually too slow to be practical in the simulation of chemical or biological systems in solution. Hence a semiempirical theory for electronic structure has been used in these types of simulations.

The QM/MM methods have their own disadvantages, the obvious one being the computational load added to the already complex calculation of the electronic structure.

C. Reference Interaction Site Model

The integral equation method is free of the disadvantages of the continuum model and simulation techniques mentioned in the foregoing, and it gives a microscopic picture of the solvent effect within a reasonable computational time. Since details of the RISM-SCF/MCSCF method are discussed in the following section we here briefly sketch the reference interaction site model (RISM) theory.

The statistical mechanics of liquids and liquid mixtures has its own long history, but the major breakthrough toward the theory in chemistry was made by Chandler and Andersen in 1971 with the reference interaction site model (RISM) [5]. This theory can be regarded as a natural extension of the Ornstein–Zernike (OZ) equation for simple atomic liquids to a mixture of atoms with chemical bonds represented by intramolecular correlation functions. Introducing this correlation function enables us to take into account the geometry of molecules. However, it cannot handle electrostatics in its original form, though the charge distribution in a molecule plays an essential role in determining the chemical specificity of the molecular system. The next important development in the theory was made in 1981 with the extended RISM theory [6–8]. The extended RISM theory takes into account not only the geometry but also the charge distribution of a molecule, which completes the chemical characterization of a species for the statistical mechanics of a molecular liquid.

A general expression of the RISM equation for a system consisting of several molecular species can be written as

$$\rho \mathbf{h} \rho = \omega \mathbf{c} * \omega + \omega * \mathbf{c} * \rho \mathbf{h} \rho \tag{1}$$

where the asterisk indicates convolution integrals and matrix products, and ρ denotes a diagonal matrix consisting of the density of molecular species. $\mathbf{c}$ and $\mathbf{h}$ are the direct and total correlation matrices with matrix elements of $c_{\alpha\beta}(r)$ and $h_{\alpha\beta}(r)$, respectively. These functions, $c_{\alpha\beta}$ and $h_{\alpha\beta}$, represent the intermolecular correlation functions between the sites (atoms or atomic groups) α and β. ω is the intramolecular correlation function, which embodies information about a molecular geometry. In the case of rigid molecules, ω is expressed by

$$\omega_{\alpha\beta}(r) = \rho_\alpha \delta_{\alpha\beta} \delta(r) + (1 - \delta_{\alpha\beta}) \frac{1}{4\pi L_{\alpha\beta}^2} \delta(r - L_{\alpha\beta}) \qquad (2)$$

where $L_{\alpha\beta}$ indicates the rigid constraint (bond length) representing the chemical bond between sites α and β.

The general equation can be further reduced to the case of infinite dilution limit, a binary mixture, ionic solutions, and so on. These equations are supplemented by closure relations such as the Percus–Yevick (PY) and hypernetted chain (HNC) approximations.

$$c_{\alpha\beta}(r) = \exp[-\beta u_{\alpha\beta}(r)]\{1 + h_{\alpha\beta}(r) - c_{\alpha\beta}(r)\} - 1 \qquad \text{(PY)} \qquad (3)$$

and

$$c_{\alpha\beta}(r) = \exp[-\beta u_{\alpha\beta}(r) h_{\alpha\beta}(r) - c_{\alpha\beta}(r)] - \{h_{\alpha\beta}(r) - c_{\alpha\beta}(r)\} - 1 \qquad \text{(HNC)} \qquad (4)$$

where $u_{\alpha\beta}(r)$ is the intermolecular interaction potential and $\beta = 1/k_B T$. All the thermodynamic functions can be calculated from the correlation functions. The excess chemical potential or the solvation energy of a molecule has particular importance and is calculated from the correlation functions [9].

$$\Delta\mu = -\frac{\rho}{\beta} \sum_{\alpha s} \int d\mathbf{r} \left[c_{\alpha s}(r) - \frac{1}{2} h_{\alpha s}^2(r) + \frac{1}{2} h_{\alpha s}(r) c_{\alpha s}(r) \right] \qquad (5)$$

Essentially, the RISM and extended RISM theories can provide information equivalent to that obtained from simulation techniques, namely, thermodynamic properties, microscopic liquid structure, and so on. But it is noteworthy that the computational cost is dramatically reduced by this analytical treatment, which can be combined with the computationally expensive ab initio MO theory. Another aspect of such treatment is the transparent logic that enables phenomena to be understood in terms of statistical mechanics. Many applications have been based on the RISM and extended RISM theories [10,11].

II. OUTLINE OF THE RISM-SCF/MCSCF METHOD

We recently proposed a new method referred to as RISM-SCF/MCSCF based on the ab initio electronic structure theory and the integral equation theory of molecular liquids (RISM). Ten-no et al. [12,13] proposed the original RISM-SCF method in 1993. The basic idea of the method is to replace the reaction field in the continuum models with a microscopic expression in terms of the site–site radial distribution functions between solute and solvent, which can be calculated from the RISM theory. Exploiting the microscopic reaction field, the Fock operator of a molecule in solution can be expressed by

$$F_i^{\text{solv}} = F_i - f_i \sum_{\lambda \in \text{solute}} b_\lambda V_\lambda \qquad (6)$$

where the first term on the right-hand side is the operator for an isolated molecule and the second term represents the solvation effect. b_λ is a population operator of solute atoms, and V_λ represents the electrostatic potential of the reaction field at solute atom λ produced by solvent molecules. The potential V_λ takes a microscopic expression in terms of the site–site radial distribution functions between solute and solvent:

$$V_{\lambda \in \text{solute}} = \rho \sum_{\alpha \in \text{solvent}} \int \frac{q_\alpha}{r} g_{\lambda\alpha}(r) \, d\mathbf{r} \tag{7}$$

where q_α is the partial charge on site α, ρ is the bulk density of the solvent, and $g_{\lambda\alpha}$ is the site–site radial distribution function (RDF) or pair correlation function (PCF).

In the RISM-SCF theory, the statistical solvent distribution around the solute is determined by the electronic structure of the solute, whereas the electronic structure of the solute is influenced by the surrounding solvent distribution. Therefore, the ab initio MO calculation and the RISM equation must be solved in a self-consistent manner. It is noted that "SCF" (self-consistent field) applies not only to the electronic structure calculation but to the whole system, e.g., a self-consistent treatment of electronic structure and solvent distribution. The MO part of the method can be readily extended to the more sophisticated levels beyond Hartree–Fock (HF), such as configuration interaction (CI) and coupled cluster (CC).

The solvated Fock operator can be naturally derived from the variational principles [14] defining the Helmholtz free energy of the system ($\mathcal{A}$) by

$$\mathcal{A} = \langle \Psi | \hat{H} + \hat{\Delta\mu} | \Psi \rangle + E_{\text{nuc}} \tag{8}$$

Here, E_{nuc} is the nuclear repulsion energy and

$$\langle \Psi | \hat{\Delta\mu} | \Psi \rangle = \Delta\mu \tag{9}$$

is the modified version of the solvation free energy originally defined by Singer and Chandler [9] [Eq. (5)]. This is a functional of the total correlation function $h_{\alpha s}(r)$, the direct correlation function $c_{\alpha s}(r)$, and the solute wave function $|\Psi\rangle$. Energy of the solute molecule E_{solute} is defined as follows:

$$E_{\text{solute}} = \langle \Psi | \hat{H} | \Psi \rangle + E_{\text{nuc}} \tag{10}$$

Note that this is also a functional of $h_{\alpha s}(r)$, $c_{\alpha s}(r)$, and $|\Psi\rangle$. Imposing constraints concerning the orthonormality of the configuration state function (C) and one-particle orbitals (ϕ_i) on the equation, one can derive the Fock operator from $\mathcal{A}$ based on the variational principle:

$$\delta(\mathcal{A}[\mathbf{c}, \mathbf{h}, \mathbf{t}, \mathbf{v}, \mathbf{C}] - [\text{constraints to orthonormality}]) = 0 \tag{11}$$

and

$$F_{ij}^{\text{solv}} = F_{ij} - \gamma_{ij} \sum_{\lambda \in \text{solute}} b_\lambda \frac{\partial}{\partial q_\lambda} \left(-\frac{\rho}{\beta} \sum_{\alpha s} \int d\mathbf{r} \, \exp[-\beta u_{\alpha s}(r) + h_{\alpha s}(r) - c_{\alpha s}(r)] \right) \tag{12}$$

If classical Coulombic interactions are assumed among point charges for electrostatic interactions between solute and solvent, and the term for the CI coefficients (C) is omitted, the solvated Fock operator is reduced to Eq. (6). The significance of this definition of the Fock operator from a variational principle is that it enables us to express the analytical first derivative of the free energy with respect to the nuclear coordinate of the solute molecule R_α,

$$\frac{\partial \mathscr{A}}{\partial R_a} = \frac{\partial E_{\text{nuc}}}{\partial R_a} - \frac{1}{2(2\pi)^3\beta} \sum_{\alpha,\gamma,s,s'} \int d\mathbf{k} \; \hat{c}_{\alpha s}(k)\hat{c}_{\gamma s'}(k) \frac{\partial\hat{\omega}_{\alpha\gamma}(k)}{\partial R_a} \hat{\chi}_{ss'}(k)$$

$$+ \sum_{i,j} \gamma_{ij} h_{ij}^\alpha + \frac{1}{2}\sum_{i,j,k,l} \Gamma_{ijkl}(\phi_i\phi_j|\phi_k\phi_l)^a - \mathbf{V}^t\mathbf{q}^a - \sum_i \sum_j \varepsilon_{ij}S_{ij}^a \tag{13}$$

The second term of the right-hand side of Eq. (13) corresponds to the change of the solute–solvent distribution function due to the modification of the intramolecular correlation function ω. Other notations used here have the usual meanings. It has been well recognized that the energy gradient technique in the ab initio electronic structure theory is a powerful tool for investigating the mechanism of chemical reactions of polyatomic systems, and it opens up a variety of applications to the actual chemical processes in solution: carrying out the geometric optimization of reactant, transition state, and product in the solvated molecular system; constructing the free energy surfaces along the proper reaction coordinates; computing the vibrational frequencies and modes; and so on.

In analyzing the computational results, the following quantities are very important:

$$E_{\text{reorg}} = E_{\text{solute}} - E_{\text{isolate}} \tag{14}$$

where E_{isolate} is the total energy of the solute molecule in an isolated condition and E_{solute} is the energy of the solute molecule defined above. The quantity E_{reorg} represents the reorganization energy associated with the relaxation or distortion of the electronic cloud and molecular geometry in solution.

Now we have the tools in hand to tackle various problems in solvated molecules. In the following sections, we present our recent efforts to explore such phenomena by means of the RISM-SCF/MCSCF method.

III. SOLVATION EFFECT ON A VARIETY OF CHEMICAL PROCESSES IN SOLUTION

A. Molecular Polarization in Neat Water*

The molecular and liquid properties of water have been subjects of intensive research in the field of molecular science. Most theoretical approaches, including molecular simulation and integral equation methods, have relied on the effective potential, which was determined empirically or semiempirically with the aid of ab initio MO calculations for isolated molecules. The potential parameters so determined from the ab initio MO in vacuum should have been readjusted so as to reproduce experimental observables in solutions. An obvious problem in such a way of determining molecular parameters is that it requires the reevaluation of the parameters whenever the thermodynamic conditions such as temperature and pressure are changed, because the effective potentials are state properties.

One of the most efficient ways to treat this problem is to combine the ab initio MO method and the RISM theory, and this has been achieved by a slight modification of the original RISM-SCF method. Effective atomic charges in liquid water are determined such that the electronic structure and the liquid properties become self-consistent, and along the route of convergence the polarization effect can be naturally incorporated.

The temperature dependence of the effective charges and dipole moment of water

---

* This discussion is based on Ref. 15.

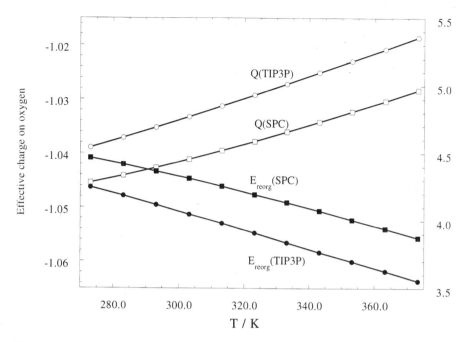

Figure 1   Temperature dependence of the reorganization energy (E_{reorg}) and effective charges on oxygen atom based on (□, ■) SPC and (○, ●) TIP3P models.

are plotted in Figure 1. The parameters associated with the short-range part of the interaction and geometry are borrowed from two typical models of water, SPC and TIP3P. In both models, the magnitudes of the effective charges and dipole moment monotonically decrease with increasing temperature. The results can be explained in terms of the increase in molecular motion, especially rotational motion, with increasing temperature. As the motion of a molecule (molecule A) increases, the average electrostatic field produced by the surrounding water molecules becomes less anisotropic, which decreases the polarity of molecule A. Conversely, the reaction field from the water (molecule A) become more isotropic, which decreases the polarity of other molecules.

The pair correlation function of water has a marked feature that distinguishes water from other liquids (Fig. 2). One of the important features characterizing the liquid water structure is a peak around $r = 1.8$ Å observed in the oxygen–hydrogen (O–H) pair, which is a direct manifestation of the hydrogen bond between a pair of water molecules. Another feature is the position of the second peak in the oxygen–oxygen (O–O) PCF, which is caused by the tetrahedral icelike coordination. Since the icelike structure becomes less pronounced as temperature increases because of the thermal disruption of the hydrogen-bonded network, those features in PCF become less prominent.

B. Autoionization of Water*

A water molecule has amphoteric character. This means it can act as both an acid and a base. The autoionization equilibrium process in water,

---

* This discussion is based on Ref. 16.

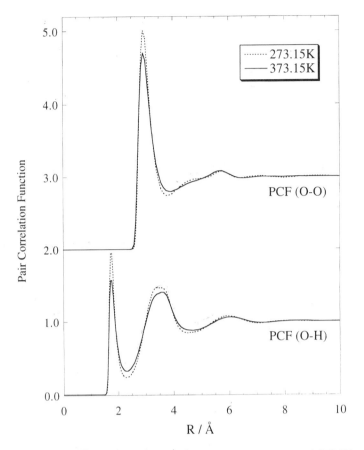

Figure 2 Pair correlation functions of O–O and O–H at (· · ·) 273.15 and (——) 375.15K computed with the parameters of the SPC water model.

$$H_2O + H_2O \rightleftharpoons H_3O^+ + OH^- \tag{15}$$

is one of the most important and fundamental reactions in a variety of fields in chemistry, biology, and biochemistry. The ionic product (K_w) and its logarithm defined by

$$K_w = [H_3O^+][OH^-], \qquad pK_w = -\log K_w \tag{16}$$

are measures of the autoionization. The quantity can be related to the free energy change (ΔG^{aq}) associated with the reaction of Eq. (15) by the standard thermodynamic relation

$$\Delta G^{aq} = 2.303 \, RT \, pK_w \tag{17}$$

It is experimentally known that the pK_w value shows significant temperature dependence, i.e., it decreases with increasing temperature [17]. However, there is no easy explanation for this phenomenon even from the phenomenological point of view. The free energy change consists of various contributions, including changes in the electronic energy and solvation free energy of the molecular species taking part in the reaction, which are related to each other. Therefore, a theory that accounts for both the electronic and liquid structures of water with a microscopic description of the reaction is required.

The free energy change associated with the reaction in Eq. (15) can be written in terms of the energy change associated with the reaction in vacuo (ΔG^{vac}) and the free energy change of the reacting species due to solvation as

$$\Delta G^{aq} = \Delta G^{vac} + \delta G(H_3O^+) + \delta G(OH^-) - 2\delta G(H_2O) \qquad (18)$$

where $\delta G(H_3O^+)$, $\delta G(OH^-)$, and $\delta G(H_2O)$ are, respectively, the free energy changes of H_3O^+, HO^-, and H_2O upon solvation. It is also possible to decompose ΔG^{aq} into intra- and intermolecular contributions as

$$\Delta G^{aq} = \Delta E^{vac}_{elec} + \Delta \delta G_{kin} + \Delta \delta G_{reorg} + \Delta \delta \mu \qquad (19)$$

where ΔE^{vac}_{elec}, $\Delta \delta G_{kin}$, and $\Delta \delta G_{reorg}$ are electronic energy in vacuo, kinetic free energy, and electronic reorganization energy, respectively, which are intramolecular contributions. $\Delta \delta \mu$ is the solvation free energy change. (We use Δ for changes of quantities associated with the chemical reaction and δ for changes due to solvation.)

The value of pK_w at temperature T relative to that at $T = 273.15$ K, given by

$$\Delta_T\, pK_w(T) = pK_w(T) - pK_w (273.15) \qquad (20)$$

is further decomposed into four contributions corresponding to the free energy components:

$$\begin{aligned}\Delta_T\, pK_w(T) &= \Delta_T\, pK^{vac}_{w,elec} + \Delta_T\, pK_{w,kin}(T) + \Delta_T\, pK_{w,reorg}(T) \\ &\quad + \Delta_T\, pK_{w,\delta\mu}(T)\end{aligned} \qquad (21)$$

The resultant $\Delta_T\, pK_w(T)$ values and their components are plotted in Figure 3. As shown

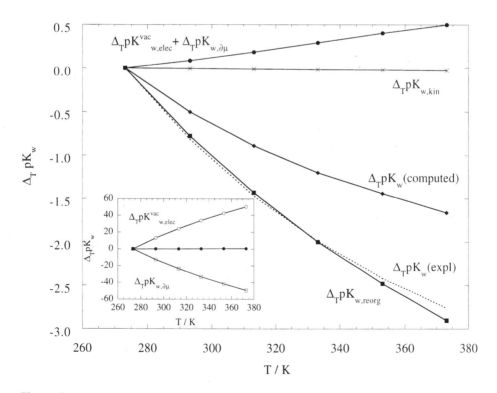

Figure 3 Temperature dependence of calculated pK_w. Dashed line indicates experimental values.

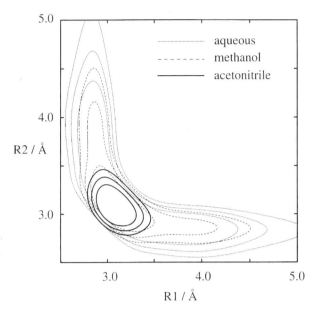

Figure 4 Free energy surfaces of triiodide ion in various solutions. Contours correspond to isoenergy lines of $1k_BT$, $2k_BT$, and $3k_BT$, respectively.

in the figure, contributions from $\Delta_T\,pK_{w,\delta\mu}$ and $\Delta_T\,pK_{w,\text{elec}}^{\text{vac}}$ are very large, but they compensate for each other. The final temperature dependence of pK_w is determined by an interplay of several contributions with different physical origins. It is also interesting that the temperature dependence is dominated by $\Delta_T\,pK_{w,\text{reorg}}$ after the compensation for the largest contributions. The theoretical results for temperature dependence of the ionic product show fairly good agreement with experiments and also demonstrate the importance of polarization effects.

C. Solvatochromism*

The molecular properties of the triiodide ion (I_3^-) in polar liquids have been studied by many techniques, motivated in part by the expected strong coupling between the solute I_3^- electronic structure and the environment. Interestingly, Raman and resonance Raman spectra in several solvents show a weak band corresponding to the antisymmetrical stretch mode, which is expected to be symmetry-forbidden [19], whereas the infrared spectra of many triiodide complexes with cations were also reported to show a band corresponding to the symmetrical stretching mode, which again should be symmetry-forbidden. Ab initio MO calculations for the solvated triiodide ion had been impracticable because of the difficulties in dealing with the character of the solvation, which is strongly coupled with solute electronic structure.

Computed free energy surfaces of the triiodide ion in its ground state in acetonitrile, methanol, and aqueous solution are presented in Figure 4, in which the two I—I bond

---

* This discussion is based on Ref. 18.

lengths ($R1$ and $R2$) are taken as coordinates. Note that the free energy surface in solution does not correspond to the potential energy surface but governs the relative population of different structures in solution. It would be misleading to estimate vibrational frequencies from the curvature of the surfaces. Contours in the figures represent isoenergy lines of 1 $k_B T$ at room temperature, 298.15 K, showing how large a population of triiodide ions exists in the different structures. The free energy profiles in solution strongly depend on the solvent. The profiles in acetonitrile solution are very localized and similar to those in the gas phase, consistent with resonance Raman experimental results in which the symmetry-forbidden band does not appear. The free energy surface in aqueous solution is markedly different and indicates a dramatically enhanced probability of structures with lower symmetry. The observation of nominally symmetry-forbidden bands in vibrational spectra is attributed to these species.

D. Conformational Equilibrium*

The acidity or basicity of organic acids and bases such as carboxylic acid and amines is governed by many factors: solvent, substitution, conformation, and so forth. Among those factors, the effect of conformational change is of special interest in terms of its significance in biological systems. In bimolecular systems such as protein, the acidity or basicity of related functional groups depends sensitively upon the molecular conformations; due to such sensitivity, the property is sometimes exploited to detect the conformational change of protein [21].

A prototype of such phenomena can be seen in even the simplest carboxylic acid, acetic acid (CH_3CHOOH). Acidity is determined by the energy or free energy difference between the dissociated and nondissociated forms, whose energetics usually depend significantly on their conformation, e.g., the *syn/anti* conformational change of the carboxylate group in the compound substantially affects the acid–base equilibrium. The coupled conformation and solvent effects on acidity is treated in Ref. 20.

Potential and potential of mean force curves along the torsional angle θ ($H-O-C-C$) is illustrated in Figure 5. In the gas phase the *syn*-acetic acid ($\theta = \pm 180°$) is more stable than the *anti* conformer by 6.9 kcal/mol, and the barrier height of rotation between these conformers is estimated as 13.2 kcal/mol. In aqueous solution, the calculated free energy difference is significantly reduced to 1.7 kcal/mol. The rotational barrier also becomes lower than that in the gas phase, 10.3 kcal/mol. The reduction of the free energy gap indicates that the pK_a difference between the two conformers is drastically changed from 5.1 to 1.2 on transferring from the gas phase to aqueous solution at room temperature.

Stabilization of the *syn* conformer in the gas phase is explained rather intuitively in terms of the extra stabilization due to increased interactions between the H atom in the OH group and the O atom in $C=O$ group. As one can see in Figure 5, the extra stabilization in the *anti* conformer in aqueous solution arises from the solvation energy, especially at the carbonyl oxygen site.

The change in the electronic redistribution on transferring the molecule from the gas phase to aqueous solution is another interesting issue. Analysis of the computed Mulliken charge population demonstrates a substantial change on the hydrogen and oxygen in

---

* This discussion is based on Ref. 20.

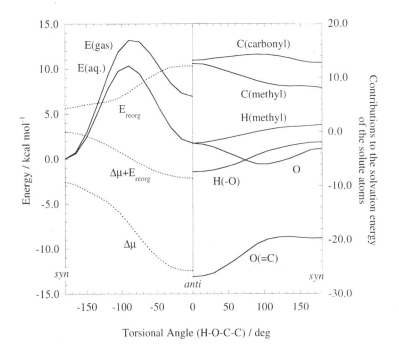

Figure 5 Calculated potential and potential of mean force of acetic acid along the torsional angle of θ(H—O—C—C). The left-hand side shows the total energies and the components, and the right-hand side shows the decomposed Δμ into the site.

the OH group: for the *anti*-conformer, the partial charges are altered from 0.355 (H) and −0.428 (O) to 0.377 (H) and −0.434 (O) upon transfering from the gas to the aqueous solution phases, and for the *syn* conformer, from 0.373 (H) and −0.463 (O) to 0.395 (H) and −0.476 (O), respectively. One can notice that the character of the proton increases on the hydrogen in the OH group when it is immersed into aqueous solution. (Note that these values are evaluated at the optimized geometry in aqueous solution. Geometrical changes from the gas to aqueous phases should also contribute to the modification of the electron density.)

E. Acid–Base Equilibrium

In this section, we review three studies on the coupled substitution and solvent effects on basicity and acidity.

1. Basicity of the Methylamines in Aqueous Solution*

Basicity and acidity are fundamental and familiar concepts in chemistry and biochemistry. Quantum chemistry has provided a theoretical understanding of the phenomena as far as the gas phase in concerned. However, it is known that in solution reactivity is seriously affected by solvents. One example of such a well-known phenomenon is that the basicity

---

* This discussion is based on Ref. 22 and Ref. 23.

of the methylamines increases monotonically with successive methyl substitutions in the gas phase,

$$NH_3 < (CH_3)NH_2 < (CH_3)_2NH < (CH_3)_3N$$

while the order reverses at the trimethylamine in an aqueous environment [24],

$$NH_3 < (CH_3)NH_2 < (CH_3)_2NH > (CH_3)_3N$$

The monotonic increase in basicity in the gas phase has been explained in terms of the "negative induction" or the polarization effect due to the methyl groups. Essentially two important factors are considered responsible for the solvent effect on the proton affinity: the solvation free energy and the energy change associated with the electron reorganization upon solvation. The solvation free energy in turn consists of the solute–solvent interaction energy and the free energy change associated with the solvent reorganization.

Let us define the respective basicity by $-\Delta G^g$ in the gas phase and $-\Delta G^s$ in aqueous solution. For discussions concerning the relative strength in basicity of a series of methyl-amines, only the relative magnitudes of these quantities are needed. Thus the free energy changes associated with the protonation of the methylamines relative to those of ammonia are defined as

$$\Delta\Delta G^i_{298}[(CH_3)_nNH_{3-n}] = \Delta G^i_{298}[(CH_3)_nNH_{3-n}] - \Delta G^i_{298}[NH_3], \quad n = 1, 2, 3 \quad (22)$$

Computed values are plotted in Figure 6 against the number of methyl groups. Note that these components include the electronic contributions (net contribution for isolated mole-

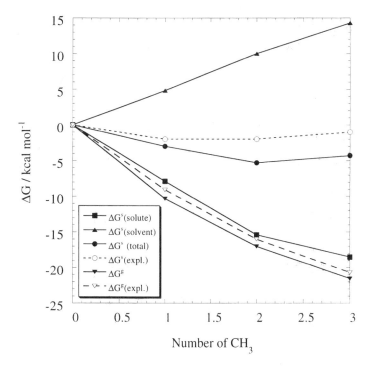

Figure 6 Free energy changes of methylamines in aqueous solution upon protonation referred to NH_3.

cule and E_{reorg}), solvation energy ($\Delta\mu$) described above, and kinetic contributions evaluated from the elementary statistical mechanics of ideal systems. The contribution from solute itself, $\Delta\Delta G^s_{298}$(solute), exhibits similar monotonic behavior with the gas phase result ($\Delta\Delta G^g_{298}$), which is in good agreement with the experimental data. The difference between $\Delta\Delta G^s_{298}$(solute) and $\Delta\Delta G^g_{298}$ is due essentially to the electron reorganization energy. The solvation free energy $\Delta\Delta G^s_{298}$(solvent) shows the monotonic increase with successive methyl substitution. The sum of the two contributions produces an inversion in the overall free energy change $\Delta\Delta G^s_{298}$, which is in qualitative accord with the experimental result.

2.   Acidities of Haloacetic Acids in Aqueous Solution*

Another example is the acidities of a series of carboxylic acids. It is known that the substitution effect on these compounds also depends on the environment. The behavior of the halo-substituted acetic acids is one of the prototype problems for the solvent effect on acidity: The order in strength of the haloacetic acids in the gas phase is

$$CH_3COOH < CH_2FCOOH < CH_2ClCOOH < CH_2BrCOOH$$

whereas in aqueous solution it is drastically altered [27] to

$$CH_3COOH < CH_2FCOOH > CH_2ClCOOH > CH_2BrCOOH$$

The observed acidities in the gas phase are interpreted in terms of the negative induction effect of the halo substituents; however, the microscopic picture of the solvent effects in addition to such induction effects of the solute have not been clarified.

Procedures to compute acidities are essentially similar to those for the basicities discussed in the previous section. The acidities in the gas phase and in solution can be calculated as the free energy changes ΔG^g and ΔG^s upon proton release of the isolated and solvated molecules, respectively. To discuss the relative strengths of acidity in the gas and aqueous solution phases, we only need the magnitude of $-\Delta G^g$ and $-\Delta G^s$ for haloacetic acids relative to those for acetic acids. Thus the free energy calculations for acetic acid, haloacetic acids, and each conjugate base are carried out in the gas phase and in aqueous solution.

In Figure 7, ΔG^s and its components, ΔG^s(solute) and ΔG^s(solvent), are plotted. ΔG^s(solute) contributes to the increase in the net free energy upon proton release or to the acidity, which is similar to ΔG^g in the gas phase. ΔG^s(solvent) is positive relative to acetic acid and contributes to hinder the proton release and to decrease the acidity. The net increase in ΔG^s(solvent) from the fluoro derivative to the chloro derivative is caused by the greater destabilization of the negative ions compared to that of the neutral molecules. The essential difference between the ionic and neutral species lies in the electrostatic interactions, which play an important role in determining the order of the acidities in solution. The inversion in the acidities from the fluoro substitution to the chloro substitution is due to the greater increase in the electrostatic contribution in the solvation free energy compared to the decrease in the contribution from the change in the solute electronic structure. As shown in Figure 7, agreement with respect to the relative order in acidity is obtained in aqueous solution as well as in the gas phase.

---

* This discussion is based on Ref. 25 and Ref. 26.

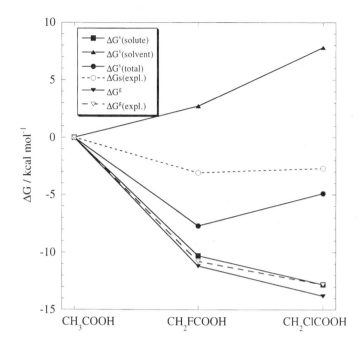

Figure 7 Free energy changes of halo-substituted carboxyl acid in aqueous solution upon deprotonation referred to acetic acid.

3. Acid Strength of the Hydrogen Halides in Aqueous Solution*

It is known that the order of acidity of hydrogen halides (HX, where X = F, Cl, Br, I) in the gas phase can be successfully predicted by quantum chemical considerations, namely, F < Cl < Br < I. However, in aqueous solution, whereas hydrogen chloride, bromide, and iodide completely dissociate in aqueous solutions, hydrogen fluoride shows a small dissociation constant. This phenomenon is explained by studying free energy changes associated with the chemical equilibrium $HX + H_2O \rightleftharpoons X^- + H_3O^+$ in the solution phase for a series of hydrogen halides. In this study, the species in the equilibrium reaction, HX, X^-, H_2O, and H_3O^+, are regarded as "solute" in the infinitely dilute solution. Thus the free energy difference in aqueous solution can be obtained in terms of the free energy difference associated with the reaction in vacuo and solvation free energy.

Figure 8 shows the PCF between the halogen site in HX and the hydrogen site in solvent water. The fluoride shows a distinct peak at 1.82 Å. There is no corresponding peak in the other X—H correlation functions. From other PCFs and geometrical considerations, we can conclude that hydrogen bonds between solute hydrogen and solvent oxygen are strong and are found in all the hydrogen halides and that only hydrogen fluoride forms a distinct F—H (solvent water) hydrogen bond. The liquid structure around HF is expected to be markedly different from those around the other hydrogen halides.

The free energy difference is mainly governed by the subtle balance of the two energetic components, the formation energies of hydrogen halides and the solvation ener-

---

* This discussion is based on Ref. 28.

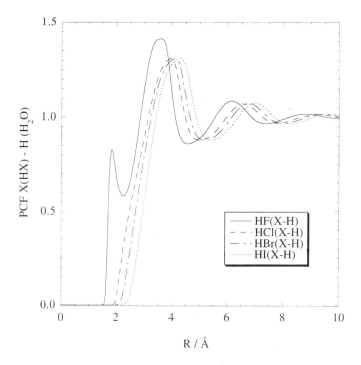

Figure 8 Pair correlation functions between X (in HX) and H (in solvent water) for a series of hydrogen halides.

gies of the halide anions. The characteristic behavior of hydrogen fluoride as a weak acid is explained in terms of the enhanced stability of the nondissociated form of the molecule in aqueous solution due to the extra hydrogen bonding with solvent. From these modern theoretical considerations, one can say that Pauling's heuristic argument seems to be correct in a qualitative sense.

F.   Tautomerization in Formamide*

Solvent effects on chemical equilibria and reactions have been an important issue in physical organic chemistry. Several empirical relationships have been proposed to characterize systematically the various types of properties in protic and aprotic solvents. One of the simplest models is the continuum reaction field characterized by the dielectric constant, ε, of the solvent, which is still widely used. Taft and coworkers [30] presented more sophisticated solvent parameters that can take solute–solvent hydrogen bonding and polarity into account. Although this parameter has been successfully applied to rationalize experimentally observed solvent effects, it seems still far from satisfactory to interpret solvent effects on the basis of microscopic information of the solute–solvent interaction and solvation free energy.

Among many examples of the solvent effects on chemical equilibria and reactions, the solvent effect on tautomerization has been one of the most extensively studied. Experi-

---

* This discussion is based on Ref. 29.

mentally, it is known that such equilibrium constants depend sensitively on the solvent polarity in solution. Tautomerization in formamide has been studied to obtain the microscopic solvation structure around the solute and to find a relationship between the empirical relation (Taft's parameter) and microscopic information given by ab initio theory.

The empirical parameters derived by Taft and coworkers are well known as a measure of the hydrogen-bonding abilities of solvents. For aprotic solvents, β parameters indicate the strength of hydrogen bonding. On the other hand, the height of the first peak in PCFs around solute hydrogen, which are obtained from the RISM-SCF calculation, also represent the strength of hydrogen bonding. As shown in Figure 9, Taft's β parameters are actually well correlated to the calculated well depth of the hydrogen bonding corresponding to the logarithm of the height of the first peak in the PCFs. The figures give a microscopic explanation for the origin of the empirical solvent parameters representing the ability of a solvent to form hydrogen bonds.

Figure 10 presents a correlation between the solvation energy computed by the RISM-SCF method and the Onsager–Kirkwood parameter, $(\varepsilon - 1)/(2\varepsilon + 1)$, which is a typical parameter empirically derived from the macroscopic parameter ε. If the parameters are good, all the solvation energies must lie on a straight line. One can see that the irregularity of the solvation free energy in acetonitrile is remarkable for both tautomers. Considering the situation that the present theory reproduces well the experimentally derived hydrogen bonding strengths, the irregularity observed here clearly demonstrates the breakdown of the continuum solvation model.

G. The S_N2 Reaction*

Chemical reactions are undoubtedly the most important issue in theoretical chemistry, where electronic structure plays an essential role. However, as will be demonstrated in this section, solvent effects also often play a crucial role in the mechanism of a chemical reaction in solution.

The Menshutkin-type S_N2 reaction in aqueous solution,

$$NH_3 + CH_3Cl \rightarrow NH_3CH_3^+ + Cl^-$$

is a prototype reaction in quantum chemistry that requires treatment of the solvent effect, and substantial computational studies have been reported based on the dielectric continuum model and QM/MM method.

A free energy profile along the reaction path is constructed by taking the difference of the C—Cl and C—N distances as the reaction coordinate (see Fig. 11). Although this reaction is found by the Hartree-Fock method to be endothermic in the gas phase by 106.3 kcal/mol, it becomes exothermic in aqueous solution by 27.8 kcal/mol with a barrier height of 17.7 kcal/mol. The barrier height in aqueous solution computed by the MP2 method (not shown) is slightly higher, 20.9 kcal/mol, but the global feature of the energy surfaces is not affected by taking the electron correlation into account.

It is interesting that the molecular structure in the transition state is also subject to a solvent effect. Compared to the gas phase, the solute molecular geometry at the transition state shifts toward the reactant side in aqueous solution; the C—N and C—Cl distances

---

* This discussion is based on Ref. 31.

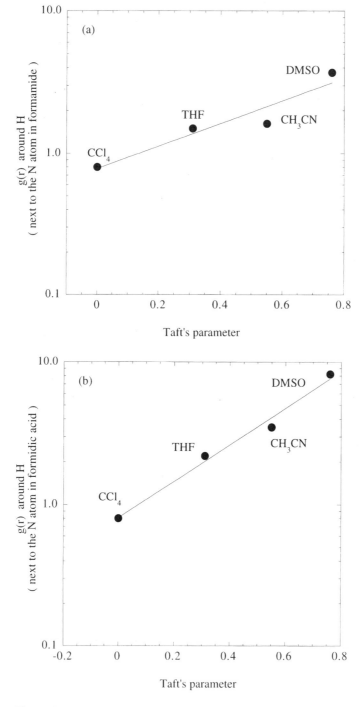

Figure 9 Logarithmic plots of the heights of the first peak in the PCF against Taft's β parameters: (a) formamide; (b) formamidic acid.

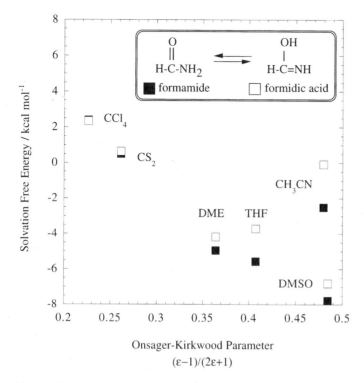

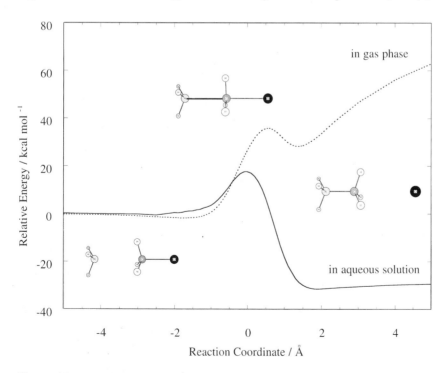

Figure 10 Solvation free energy versus the Onsager–Kirkwood parameter $(\varepsilon - 1)/(2\varepsilon + 1)$.

Figure 11 Potential energy and potential of mean force of the Menshutkin reaction. The dashed line is for reaction in the gas phase, and the solid line for reaction in aqueous solution.

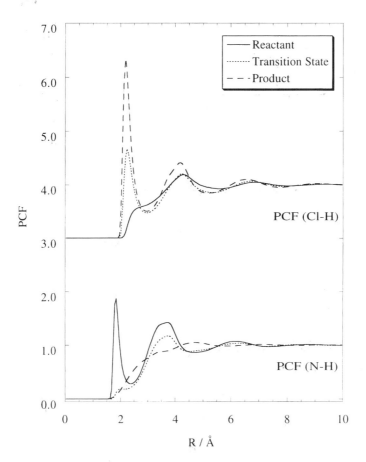

Figure 12 Pair correlation functions of Cl—H and N—H along the reaction path.

are almost the same (2.258 Å) in aqueous solution, whereas those in the gas phase are 1.890 and 2.490 Å, respectively, which is consistent with Hammond's postulate [32].

The great stabilization of the product in aqueous solution is explained from changes in PCFs along the reaction path, which are shown in Figure 12. The PCF of Cl—H obviously illustrates the progress of hydrogen bonding along the reaction coordinate: For the reactant no peak corresponding to the Cl—H interaction is observed, but a distinct peak around 2.2 Å gradually appears with slight inward shifts as the reaction proceeds. In contrast, the hydrogen bond between the ammonia N and water H is observed to break. The first peak around 2.0 Å corresponds to the hydrogen bonding of a water hydrogen to the lone pair electron in nitrogen. The lone pair electron participates in a new chemical bond with C of CH_3Cl, and the peak disappears as the reaction proceeds. The formation of Cl—H hydrogen bonds (solvation) and breaking of N—H bonds are key features in the understanding of solvent effects on the reaction mechanism.

IV. SUMMARY AND PROSPECTS

In this chapter, we have reviewed the RISM-SCF/MCSCF method, which combines electronic structure and liquid-state theories to deal with the chemistry of solutions. The ability of the method to treat solvent effects on chemical processes has been demonstrated. Electronic structure plays a primary role in determining the structure of a molecule. However, changes in the electronic energy associated with a chemical process are comparable, in many cases, with those due to solvation in solution. This subtle balance between changes in electronic energy and in the solvation free energy sometimes causes drastic changes in the stability of chemical species, as we have seen in several examples. The solvent effect even reverses the equilibrium between a reactant and a product.

As has been repeatedly stated throughout the chapter, the theory provides good qualitative descriptions to solution chemistry in most cases, but quantitative agreement between theoretical and experimental results is only moderate. It is always possible to improve the results numberwise by tuning the molecular parameters and/or by introducing empirical parameters into either or both elements of the theory, MO and RISM. This direction, in fact, has been pursued by Gao and coworkers [33], who have shown almost perfect quantitative agreement between the theory and experiments by replacing the ab initio method by the semiempirical approach for the MO part and by adjusting the Lennard-Jones parameters of atoms. The effort should be greatly appreciated, because it could have demonstrated the capability of the combined MO and RISM method to account for experimental data at least at the same quantitative level as the continuum model descriptions. However, it will not be an ultimate goal of the combined quantum and statistical mechanics theory. The real strength of the theory lies in the fact that *it does not require in principle any adjustable or empirical parameter* to describe complicated solution chemistry. The theory could be or should be naturally improved by theoretical development of either or both elements of the method and by coupling them, not by adjusting or introducing empirical parameters. Since considerable efforts have been continuously devoted to improvement in both theoretical fields in chemical physics, there is no doubt that the RISM-SCF/MCSCF approach will find greater application in the future.

There are several directions conceivable to extend further the horizon of the theory. One such direction is to seek an experimental method to prove electron distributions in a molecule in solution: the partial atomic charges are an *effective* representation of the electron distribution. As has been described in the preceding sections, the RISM-SCF/MCSCF method provides information on the electron distribution that is more detailed than information on the dipole moment. Therefore, if we could find a means to observe the electron distribution, it will provide more detailed information on molecular structure in solution. It will also provide a reliable tool for testing theory experimentally. One possible candidate among experimental methods to observe the electron distribution may be the NMR chemical shift, because the chemical shift is a manifestation of changes in the screening of nuclear magnetic fields due to electron clouds. It is highly desirable to establish a theory to bridge the NMR chemical shift and the RISM-SCF/MCSCF method.

In the first example of applications of the theory in this chapter, we made a point with respect to the polarizability of molecules and showed how the problem could have been handled by the RISM-SCF/MCSCF theory. However, the current level of our method has a serious limitation in this respect. The method can handle the polarizability of molecules in neat liquids or that of a single molecule in solution in a reasonable manner. But in order to be able to treat the polarizability of both solute and solvent molecules in solution, considerable generalization of the RISM side of the theory is required. When solvent molecules are situated within the influence of solute molecules, the solvent molecules are polarized differently depending on the distance from the solute molecules, and the solvent can no longer be ''neat.'' Therefore, the polarizable model developed for neat liquids is not valid. In such a case, solvent–solvent PCF should be treated under the solute field, which is typical of in-homogeneous liquids. The density functional theory in classical theory may be the best choice for extending the theory to such a problem [34].

Nonequilibrium processes including chemical reactions present a most challenging problem in theoretical chemistry. There are two aspects to chemical reactions: the reactivity or chemical equilibrium and the reaction dynamics. The chemical equilibrium of molecules is a synonym for the free energy difference between reactant and product. Two important factors determining the chemical equilibrium in solution are the changes in electronic structure and the solvation free energy. Those quantities can be evaluated by the coupled quantum and extended RISM equations, or RISM-SCF theory. Exploration of the reaction dynamics is much more demanding. Two elements of reaction dynamics in solution must be considered: the determination of reaction paths and the time evolution along the reaction path. The reaction path can be determined most naively by calculating the free energy map of reacting species. The RISM-SCF procedure can be employed for such calculations. If the rate-determining step of the reaction is an equilibrium between the reactant and the transition state, the reaction rate can be determined from the free energy difference of the two states based on transition state theory. On the other hand, for such a reaction in which the dynamics of solvent reorganization determines the reaction rate, the time evolution along the reaction path may be described by coupling RISM and the generalized Langevin equation (GLE) in the same spirit as the Kramers theory: The time evolution along a reaction path can be viewed as a stochastic barrier crossing driven by thermal fluctuations and damped by friction. Our treatment features the microscopic treatment of solvent structure on the level of the density pair correlation functions, which distinguishes it from earlier attempts that used phenomenological solvent models. One of the prerequisites for developing such a treatment is a theory to describe liquid dynamics on the molecular level. We recently proposed a new theory based on the interaction site model in which liquid dynamics is decoupled into the collective modes of density fluctuation: the acoustic and optical modes corresponding, respectively, to transnational and rotational motion of molecules [35]. From this point of view, transport coefficients such as the friction coefficient can be realized as a response of the collective modes of the solvent to perturbations due to solute. It is the first step in developing a theory of reaction dynamics to describe stochastic barrier crossing in terms of the collective fluctuations of solvent to reacting species along a properly defined reaction coordinate.

ACKNOWLEDGMENTS

We thank all our collaborators, especially Drs. Masaaki Kawata, Kazunari Naka, Tateki Ishida, and Akihiro Morita. We are also grateful to Profs. Masaru Nakahara, Yasuhiko Kondo, and Tadashi Okuyama, who have given invaluable advice and encouragement.

REFERENCES

1. K Laasonen, M Sprik, M Parrinello, R Car. J Chem Phys 99:9080–9089, 1993.
2. CJF Böttcher. Theory of Electric Polarization. Amsterdam: Elsevier Scientific, 1973.
3. J Tomasi, M Persico. Chem Rev 94:2027–2094, 1994.
4. J Gao. Acc Chem Res 29:298–305, 1996.
5. D Chandler, HC Andersen. J Chem Phys 57:1930–1937, 1972.

6. F Hirata, PJ Rossky. Chem Phys Lett 83:329–334, 1981.
7. F Hirata, BM Pettitt, PJ Rossky. J Chem Phys 77:509–520, 1982.
8. F Hirata, PJ Rossky, BM Pettitt. J Chem Phys 78:4133–4144, 1983.
9. SJ Singer, D Chandler. Mol Phys 55:621–625, 1985.
10. F Hirata. Bull Chem Soc Jpn 71:1483–1499, 1998.
11. PJ Rossky. Annu Rev Phys Chem 36:321–346, 1985.
12. S Ten-no, F Hirata, S Kato. J Chem Phys 100:7443–7453, 1994.
13. S Ten-no, F Hirata, S Kato. Chem Phys Lett 214:391–396, 1993.
14. H Sato, F Hirata, S Kato. J Chem Phys 105:1546–1551, 1996.
15. SA Maw, H Sato, S Ten-no, F Hirata. Chem Phys Lett 276:20–25, 1997.
16. H Sato, F Hirata. J Phys Chem A 102:2603–2608, 1998.
17. D Dobos. Electrochemical Data: A Handbook for Electrochemists in Industry and Universities. Amsterdam: Elsevier Scientific, 1975.
18. H Sato, F Hirata, AB Myers. J Phys Chem A 102:2065–2071, 1998.
19. AE Johnson, AB Myers. J Chem Phys 102:3519–3533, 1995.
20. H Sato, F Hirata. J Mol Struct (THEOCHEM) 461–462:113–120, 1999.
21. CR Cantor, PR Schimmel. Biophysical Chemistry. New York: WH Freeman, 1980.
22. M Kawata, S Ten-no, S Kato, F Hirata. Chem Phys 203:53–67, 1996.
23. M Kawata, S Ten-no, S Kato, F Hirata. J Am Chem Soc 117:1638–1640, 1995.
24. EJ King. Acid Base Equilibria. New York: Pergamon Press, 1965.
25. M Kawata, S Ten-no, S Kato, F Hirata. Chem Phys Lett 240:199–204, 1995.
26. M Kawata, S Ten-no, S Kato, F Hirata. J Phys Chem 100:1111–1117, 1996.
27. K Hiraoka, R Yamdagni, P Kebarle. J Am Chem Soc 95:6833–6835, 1973; P Haberfield, AK Rakshit. J Am Chem Soc 98:4393–4394, 1976.
28. H Sato, F Hirata. J Am Chem Soc 121:3460–3467, 1999.
29. T Ishida, F Hirta, H Sato, S Kato. J Phys Chem B 102:2045–2050, 1998.
30. MJ Kamlet, J-LM Abboud, MH Abraham, RW Taft. J Org Chem 48:2877–2887, 1983.
31. K Naka, H Sato, A Morita, F Hirata, S Kato. Theor Chem Acc 102:165–169, 1999.
32. GS Hammond. J Am Chem Soc 77:334–338, 1955.
33. L Shao, H-A Yu, J Gao. J Phys Chem A 102:10366–10373, 1998.
34. D Chandler, JD McCoy, SJ Singer. J Chem Phys 85:5971–5982, 1986.
35. S-H Chong, F Hirata, Phys Rev E 57:1691–1701, 1998.

20

Nucleic Acid Simulations

Alexander D. MacKerell, Jr.
University of Maryland, Baltimore, Maryland

Lennart Nilsson
Karolinska Institutet, Huddinge, Sweden

I. INTRODUCTION

The biological functions of DNA and RNA were initially assumed to involve only their primary sequence as required for storage of the genetic code. Consistent with this view was the helical B structure of DNA initially proposed by Watson and Crick [1]. While initial experimental work based on fiber diffraction indicated heterogeneity in DNA structure, such as the A and B forms of DNA, it propagated the idea that the structure of DNA was that of a regular helix [2]. This view started to change when the first structures of DNA based on single-molecule X-ray crystallography were obtained, which showed local conformational heterogeneity to be present in DNA while the overall structures still assumed canonical forms. Later, structural studies of RNA, particularly transfer RNA (tRNA), revealed the structure of RNA to have significant tertiary characteristics beyond the helical structures dominating DNA. More recently, X-ray crystallographic studies of DNA–protein complexes revealed DNA structures that are significantly distorted from the helical conformations traditionally envisioned for DNA. Furthermore, it has become evident that the structural distortion of DNA and the wide variety of tertiary structures of RNA are essential for their biological activity [2,3].

Although experimental studies of DNA and RNA structure have revealed the significant structural diversity of oligonucleotides, there are limitations to these approaches. X-ray crystallographic structures are limited to relatively small DNA duplexes, and the crystal lattice can impact the three-dimensional conformation [4]. NMR-based structural studies allow for the determination of structures in solution; however, the limited amount of nuclear overhauser effect (NOE) data between nonadjacent stacked basepairs makes the determination of the overall structure of DNA difficult [5]. In addition, nanotechnology-based experiments, such as the use of optical tweezers and atomic force microscopy [6], have revealed that the forces required to distort DNA are relatively small, consistent with the structural heterogeneity observed in both DNA and RNA.

Computational studies of nucleic acids offer the possibility to enhance and extend the information available from experimental work. Computational approaches can facilitate the experimental determination of DNA and RNA structures. Dynamic information, although often isotropic in nature from experimental studies, can be obtained from computations at an atomic level of detail. Of particular interest is a detailed knowledge of the influence of base sequence, base composition, and environment on DNA structure. Finally, computational approaches can reveal the subtle relationship between structure and energetics, yielding an understanding of the properties of oligonucleotides that allow for the conformational changes required for their biological function.

In Section II we provide an overview of the current status of nucleic acid simulations, including studies on small oligonucleotides, DNA, RNA, and their complexes with proteins. This is followed a presentation of computational methods that are currently being applied for the study of nucleic acids. The final section of the chapter includes a number of practical considerations that may be useful in preparing, performing, and analyzing MD simulation based studies of nucleic acids.

II. OVERVIEW OF COMPUTATIONAL STUDIES ON OLIGONUCLEOTIDES

A. DNA

Computational studies of nucleic acids initially lagged behind protein-based calculations. Nucleic acids, being extended polyanions, require a more rigorous treatment of the solvent environment, whereas the globular structure of many proteins allows for greater tolerance for the vacuum environment applied in early MD simulations due to computational limitations. Early attempts at simulating nucleic acids in vacuum involved decreasing the phosphate charges or setting them to zero [7,8]. Alternative approaches included the inclusion of ''hydrated'' counterions, or solvatons, that mimicked a sodium ion hydrated by six water molecules and the use of distance-dependent dielectrics for buffering the electrostatic interactions [9,10]. Whereas results from these calculations produced some insights into the properties of DNA, the quality of the results were generally in poor agreement with experiment, emphasizing the need for better treatment of the solvent environment.

Inclusion of explicit solvent in calculations on DNA involved simulations in which the DNA was both held rigid and allowed to evolve along with the solvent molecules. Application of the former approach allowed for a better understanding of the solvation of DNA to be obtained. For example, the hydration of AT and polyA–polyT B-form tracts was studied via Monte Carlo calculations [11], and MD simulations were used to investigate differences in hydration of the B and Z forms of DNA [12]. Although these works contributed to the understanding of the solvation of oligonucleotides, the local conformational heterogeneity of DNA structure observed in crystal structures of DNA emphasized the need to include both the DNA and solvent as flexible degrees of freedom in the simulations. One of the earlier calculations on DNA with an explicit solvent representation was performed on a d(CGCGA) duplex in a sphere of water that included neutralizing counterions [13]. The structure resulting from this simulation was shown to be similar to the B form of DNA; however, the total simulation time was only 114 ps, not long enough to allow for significant relaxation of the DNA, which has been more recently shown to require 1 ns or longer. Though limited, this work strongly indicated that MD simulations of DNA duplexes with an explicit solvent representation were both feasible and a useful method to better our understanding of DNA structure.

Over the next decade a number of efforts were made to apply MD simulations using explicit solvent representations to DNA. A number of these calculations were performed on the d(CGCGAATTCGCG) or "Drew" dodecamer [14], making this sequence the benchmark for DNA duplex calculations. Calculations performed on this molecule using explicit solvent included a 140 ps simulation with harmonic constraints on the hydrogens involved in Watson–Crick basepairs [15] and a 1 ns simulation with reduced charges on the phosphates [16]. In both cases relatively stable structures were obtained, although significant distortions of the structures from the canonical B form were evident. Similar results were obtained on the Drew dodecamer in a 150 ps simulation that included a 9 Å solvation shell of explicit water molecules, where the structure continued to drift from the canonical B form over the course of the simulation [17]. Although these and other efforts produced reasonably stable simulations, calculations up to approximately 1995 generally yielded distorted DNA structures if performed long enough to allow for significant relaxation of the system.

During 1995 several groups showed that stable simulations were possible. A simulation on the Drew dodecamer in the crystal environment was performed for 2.2 ns, yielding root-mean-square (RMS) differences in the range of 1.0–1.5 Å from the experimental structure [18]. Similarly good results were obtained on a crystal simulation of a Z-DNA d(CGCGCG) duplex [19]. These successes were reproduced for DNA in solution, with simulations of the d(CCAACGTTGG) duplex [20] and a DNA triplex [21] both performed for 1 ns. Initially, these successes were attributed to the use of the Ewald method [22], in some cases via the particle mesh Ewald (PME) approach [23]. This assertion seemed appropriate given the highly charged nature of oligonucleotides and the ability of the Ewald methods to accurately treat the long-range electrostatic interactions that should dominate in these systems. Subsequent studies, however, using atom-based truncation methods for simulations of the Drew dodecamer [24] and deoxy and ribo GCGCGCGCGCGC duplexes [25] in solution showed that stable structures could be obtained without the use of Ewald-based methods. Furthermore, simulations of the d(CGCGCG) DNA duplex in aqueous solution with a nonbond cutoff of 12 Å with smooth shifting to zero at the cutoff distance [26] of just the electrostatic energy or both the electrostatic energy and force gives simulation results indistinguishable from those obtained using Ewald summation in terms of the RMS difference from the initial structure and RMS fluctuations around the final average [27]. These results indicated that improvements in the AMBER [28] and CHARMM [29] force fields used in the calculations contributed significantly to the ability to perform stable simulations.

Since those initial successful simulations, several simulations of DNA in solution of more than 1 ns have been performed. Two different simulations have been performed on the Drew dodecamer. One simulation, extended for 3.5 ns using the CHARMM force field, showed a change in the overall conformation from the canonical B to the A form of DNA [30]. The second simulation, using AMBER [28], was initially performed for 5 ns, over which time the structure was shown to fluctuate approximately 2.5–3.5 Å from the canonical B form of DNA [31]; that simulation has since been extended to 14 ns [32]. Two 10 ns simulations have been performed on the DNA duplex d(C_5T_5) using the AMBER and CHARMM force fields. This system was selected because of the suggestion that it assumes a B-type structure in the AT region and a A-like structure in the GC region [33]. Results revealed that both force fields yield reasonable results, although disagreements in both force fields with experiment were identified. Of note are the relaxation times of the overall structures indicated from these simulations. Initial relaxation times of 1 ns or more are reported, with significant conformational fluctuations occurring for the remainder of the simulations.

The methodological advances just presented have brought the field of nucleic acid force field calculations to a point where results from the calculations can be used with reasonable confidence to aid in the interpretation of experimental data as well as to be used for scientific investigations that are not accessible to experiment. Accordingly, a number of studies based on MD simulations, as well as other methods, have been undertaken to study a wide array of biologically relevant events associated with DNA. A brief overview of some of these efforts follows.

1. Environmental and Base Sequence Influences on Duplex DNA

Alterations of DNA structure associated with changes in water activity have long been known [2], although the exact mechanisms associated with these phenomena are still in question. Further, it is known that base sequence as well as base composition also have a central role in dictating both the local and overall structure of duplex DNA. Transitions from the A to B [34] and from the B to A [30] forms of DNA have been observed, indicating the lack of significant energetic barriers between these two forms of DNA. The presence of ethanol [35,36], hexamminecobalt (III) [37], and 4 M NaCl [38] have been shown to stabilize the A form of DNA, consistent with experiment. Studies combining results from MD simulations with entropy estimates from harmonic analysis and continuum models to estimate the free energies of solvation indicate that internal energies favor the B form and solvation contributions favor the A form; however, the approach does not fully account for the switch to the A form of DNA at high salt concentrations. This discrepancy is consistent with calculations indicating that changes in the hydration of the phosphodiester backbone of DNA lead to changes in the conformational preferences of the backbone that influence the equilibrium between the A and B forms [39]. A similar combined MD–continuum study, however, properly predicted stabilization of the A and B forms in low and high water activity, respectively, when ethanol is used to alter the water activity [40].

Concerning the influence of sequence on the structure and dynamics of DNA, a number of interesting studies have been performed on the TATA box. The TATA box is a consensus 7mer that is essential for the initiation of transcription in eukaryotes. Crystal structures have been determined for the TATA box DNA bound to the TATA box binding protein (TBP) [41]. In these structures the DNA is observed to be significantly distorted from the B form to a form closer to the A form of DNA that has been dubbed the TA form [42]. On the basis of these results it was suggested that the inherent conformational preference of TATA box DNA may be similar to the TA form, thereby facilitating binding with the TBP. Two separate MD studies of the TATA box, using different force fields and simulation methodologies, both indicated that sequence to indeed assume a more A-like conformation than other forms of DNA [43,44]. These results are an example of how MD simulations can provide information that is difficult to obtain or inaccessible via experimental approaches.

2. DNA–Protein Interactions

Protein–DNA complexes present demanding challenges to computational biophysics: The delicate balance of forces within and between the protein, DNA, and solvent has to be faithfully reproduced by the force field, and the systems are generally very large owing to the use of explicit solvation, which so far seems to be necessary for detailed simulations. Simulations of such systems, however, are feasible on a nanosecond time scale and yield structural, dynamic, and thermodynamic results that agree well with available experimental data. Some aspects common to the various systems are briefly summarized in the following paragraphs.

Only a handful of MD simulations of protein–DNA complexes have been reported. All but one, a model of the chromosomal HMG-D system [45], deal with sequence-dependent DNA binding either to a restriction endonuclease (EcoRI) [46,47] or to transcription factors. Studies on transcription factors include the repressors [48,49], the antennapedia homeodomain [50], the TATA box binding protein [51,52], and the DNA binding domains (DBDs) of hormone receptors [53–59]. This set of systems contains representative proteins of several of the known DNA-binding structural motifs: helix–turn–helix (e.g., homeodomain and the lac repressor), Zn finger proteins (e.g., hormone receptors), ribbon proteins (TBPs), and the HMG box (HMG-D, SRY). These examples contain DNA conformers of the canonical B type as well as DNA with bends and kinks present. All of the listed studies are on solvated systems that contain several thousand atoms. For example, the first simulation of the lac repressor headpiece with 51 amino acids bound to a 14 basepair DNA duplex in water contained 12,889 atoms and lasted for 0.125 ns, in part with NOE restraints. Simulations up to 2 ns [50] and of systems as large as 36,000 atoms [54,60] have been performed. Comparisons of similar systems, e.g., wild-type versus mutants or cognate versus noncognate complexes, have also been made in some cases.

Some proteins bind to DNA of any sequence as part of their biological function, such as in the tight packing of DNA in chromosomes. The structures of at least two HMG-box-containing proteins that are important for chromatin structure have been experimentally determined in complexes with the proteins specifically bound to DNA, but no complex between a sequence-independent HMG box protein–DNA complex has been determined. In the study by Balaeff et al. [45], three models of a complex between the HMG-D protein (a nonspecific HMG protein) and DNA were constructed, and the model complexes were subjected to 160 ps of MD simulation. The quality of these models was assessed on the basis of a number of criteria, including the stability of the structure and the geometry of the protein and the DNA. The model based on docking HMG-D to a DNA model similar to the bent DNA conformer observed in TBP–DNA complexes was chosen for a final 60 ps MD simulation that indicated that the protein adapted its conformation slightly to better fit the DNA. In addition to a number of contacts between basic amino acid residues with the DNA phosphodiester backbone, there were many hydrophobic interactions in the DNA minor groove formed by hydrophobic residues on the surface of the HMG box. Comparisons with the sequence-dependent HMG box protein–DNA complexes showed how nonspecific HMG domain proteins can bind in a similar way to many different DNA sequences by using nonpolar interactions instead of the polar interactions found at key sites in the specific complexes.

For the proteins that bind to a specific DNA sequence it is quite natural to compare cognate complexes with complexes in which either the protein or the DNA has been altered. Such comparisons are also undertaken in simulations of hormone receptors, including variations of both the protein and DNA, [56,61,62], variation of the DNA bound to EcoRI [63], the SRY–DNA complex [64], and the TBP in which TBPs bound in different orientations to the same DNA sequence are compared [51]. Here the nuclear hormone receptors have attracted the most attention, with several simulations presented for both the glucocorticoid and estrogen receptors. The overall picture emerging from these simulations is that the systems are well behaved; the DNA adapts its conformation to the protein, which holds on to the DNA with a number of well-defined hydrogen bonds to the phosphate backbone, allowing specific recognition by rather complex, and dynamic, networks of both direct and water-mediated hydrogen bonds between the protein and DNA. Point mutations can, to some degree, be accommodated by this network through side chain rearrangements and moving water molecules, but there can also be larger changes. In the glucocorticoid receptor DBD changes of a single amino acid at the protein/DNA interface lead not only to a slight change in the orientation of the protein on the DNA but also to significant conformational changes some 15 Å away, in the part of the protein where contacts are made with the other protein in the dimer that is bound to the DNA. Other recent simulation studies of protein–DNA complexes were performed on the TRP operator [65] and the ZIF268–DNA complex [66,67]. From all these studies it is clear that calculations of DNA or RNA in complexes with protein will greatly facilitate our understanding of a wide variety of processes associated with growth, differentiation, and signal transduction at an atomic level of detail.

3. DNA–Drug Interactions

Numerous drugs, including many antibiotics, function via direct interactions with DNA. In addition, a number of anticancer agents, including cisplatin, function through alkylation of DNA. Computational approaches offer the means to better understand the nature of the interactions between drugs and DNA as well as a rational approach for the optimization as well as identification of lead compounds (see Chapter 16). The mode of interaction of two antibiotics, Esperamicin and Dynamicin, both of which lead to the cleavage of DNA, were investigated via MD simulations [68,69]. These studies yielded information on the mechanisms and cleavage patterns of DNA. In another study, the relative binding of daunomycin and 9-aminoacridine to B-DNA were studied via free energy perturbation calculations [70]. Although the calculations reproduced the experimental trends, the agreement may have been fortuitous considering that the calculations were performed in vacuum. An interesting study was the application of QM/MM methods to investigate the cross-linking of guanine bases in DNA by nitrous acid [71]. Although not a study of a drug per se, the work strongly indicates that details of the reactions of alkylating agents with DNA can be investigated.

B. RNA

RNA structures, compared to the helical motifs that dominate DNA, are quite diverse, assuming various loop conformations in addition to helical structures. This diversity allows RNA molecules to assume a wide variety of tertiary structures with many biological functions beyond the storage and propagation of the genetic code. Examples include transfer RNA, which is involved in the translation of mRNA into proteins, the RNA components of ribosomes, the translation machinery, and catalytic RNA molecules. In addition, it is now known that secondary and tertiary elements of mRNA can act to regulate the translation of its own primary sequence. Such diversity makes RNA a prime area for the study of structure–function relationships to which computational approaches can make a significant contribution.

To date, RNA calculations have been performed on a variety of systems of different topologies including helical duplexes, hairpin loops, and single strands from tRNA, rRNA, and ribozymes. In a simulation of an RNA tetraloop of the GRNA type, which is very common and known to be remarkably stable, it was found that without imposing any external information the simulation found the right conformation even when it started from the wrong one [72]. Studies have used Ewald summation methods to handle the long-range electrostatic interactions in several simulations of tRNAAsp; both the whole molecule [73] and the anticodon arm have been simulated [74,75]. These simulations basically find that RNA molecules maintain the experimentally observed structures, and the authors proceed to analyze hydrogen bonding and hydration patterns; the latter, being more difficult to observe directly in experiments, provide new information from the simulations. In particular, several C—H···O "hydrogen bonds" are found. They are shown to be important for stabilizing the preferred nucleotide conformation in RNA through base–backbone interactions and also to stabilize the anticodon loop conformation. Results from the anticodon studies are similar to what is found in the simulation of the whole tRNA, indicating that no serious artifacts were introduced in the fragment simulations. Simulation studies have also been performed on a ribozyme [76]; once again the molecule remained structurally stable, and hydrogen bonding and hydration as well as specific interactions involving Mg ions were analyzed. The structural stability observed in all these simulations is attributed to the use of Ewald summation for long-range electrostatic interactions. As discussed earlier, this structural stability is not unique to MD simulations using Ewald summation. In quite a large number of studies, standard spherical truncation schemes were successfully used for DNA as well as RNA systems [24,25,39,77,78]. Some spherical truncation schemes are known to cause problems [26], especially for charged systems, e.g., "neutral group" switching or truncation, and these should thus be avoided. Furthermore, it is not clear that stable results can be obtained using a "mixed" force field, in which charges for different portions of the system (e.g., DNA versus protein) are obtained from different sources without reparametrization.

C. Dynamics and Energetics of Oligonucleotides

One of the most powerful attributes of computational studies is the ability to obtain direct relationships between energetics and structure. Chapters 9, 10 and 11 of this book address different approaches for the determination of free energies associated with conformational and chemical alterations. As discussed earlier, the structures of both DNA and RNA are extremely sensitive to environmental conditions. In essence, alteration in the environment leads to changes in the conformational free energy surface of the molecules. Moreover, experimental studies have shown that DNA can be significantly distorted from the canonical forms by using very small forces [6,79]. Such plasticity presumably allows for the opening of DNA required for transcription and replication, for formation of nucleosomes and other processes of central biological importance. Detailed knowledge of the phenomena that allow for this plasticity will, in addition to furthering our knowledge of biological processes, facilitate the use of oligonucleotides as mechanical devices as the field of nanotechnology develops [80,81].

Initial applications of computational techniques have involved the use of potentials of mean force (PMFs) or umbrella sampling (see Chapters 9 and 10) to investigate the energetics of ribo- and deoxyribodinucleotides. From a series of PMF calculations on the 16 combinations of the dinucleotide X_pY (X,Y = A,C,G,U) and their deoxyribose counterparts, it was found, in accordance with the experimental data, that purine–purine pairs stack best, pyrimidine–pyrimidine pairs not at all, and the purine–pyrimidine heterodimers were in between [82]. It is quite clear from these studies that the relative free energies are not dominated by direct base–base interactions, but that the driving force for stacking is of an enthalpic character [83]. Differences between DNA and RNA were small, with the methyl group in thymine stabilizing stacking and the 2'-hydroxyl group of RNA in some cases stabilizing and in other cases destabilizing [84]. Reduced stabilization of the stacked conformation in nonaqueous solvents (DMSO, chloroform, methanol) was also observed [85].

PMF calculations have also been used to investigate the end-to-end extension of a duplex DNA 16mer, with the calculations designed to reproduce the conditions used in atomic force microscopy (AFM) experiments [86]. Results from these calculations were consistent with the barrierless extension of DNA observed in the AFM experiments. Detailed analysis of the simulated results yielded a model in which the unfavorable intramolecular mechanical energy of DNA associated with extension is compensated for by DNA–solvent interactions to yield the barrierless extension of the DNA. Alternative computational studies of the extension of DNA have been performed in vacuo with both atom-based and internal coordinate based methods. Results from the internal coordinate based calculations, which have been developed to implicitly include solvent effects, yielded a qualitative picture of the structural changes in DNA as end-to-end extension is observed from which a new form of DNA, the S form, was identified [79]. Atom-based calculations of the stretching DNA in vacuum indicated that Watson–Crick hydrogen bonds remained intact throughout the extension of the DNA; however, the omission of solvent could significantly influence this conclusion [87]. In combination, the calculations to date on the energetics of oligonucleotides associated with structural perturbations strongly indicate that much information is to be gained from these types of computational studies.

D. DNA Phase Transitions

Molecular dynamics simulations have also been used to interpret phase behavior of DNA as a function of temperature. From a series of simulations on a fully solvated DNA hexamer duplex at temperatures ranging from 20 to 340 K, a ''glass transition'' was observed at 220–230 K in the dynamics of the DNA, as reflected in the RMS positional fluctuations of all the DNA atoms [88]. The effect was correlated with the number of hydrogen bonds between DNA and solvent, which had its maximum at the glass transition. Similar transitions have also been found in proteins.

E. Modified Oligonucleotides

Interest in chemically altered oligonucleotides has been generated by the possibility of using antisense technology in drug therapy [89] as well as to exploit chemically modified species in oligonucleotide structure–function studies, including nucleic acid–protein interactions [90]. Modifications studied to date via computational approaches include modifications of the phosphates, sugars, and bases. Modified phosphodiester backbones subjected to computational studies include phosphoramidate-modified species, peptide nucleic acids (PNAs), and 2′–5′ phosphodiester linkages [91–95]. Results from these studies yield insights into the influence of the backbone on the overall conformation of the oligonucleotide, including stabilization of triple helical structures. In another study, calculations were performed on DNA duplexes and triplexes, with guanidine groups replacing the normal phosphodiester linkages, yielding a polycation [96]. Results showed the triplex to be more stable than a polyanionic triplex. Examples of the study of modified sugar include calculations on oligonucleotides containing hexitol instead of the furanose rings [97]. These calculations indicated that oligonucleotides with hexitol sugars may form more stable duplexes with RNA than DNA due to alterations in solvation of the minor groove. Some of the calculations involving modified bases included free energy perturbation calculations investigating the influence of bromination and methylation on the stability of Z-form oligonucleotides [98,99]. Other studies of modified oligonucleotides have been performed on the thymine dimers caused by UV photodamage of DNA showing the presence of bent structures consistent with experiment along with information on possible mechanisms by which repair enzymes may identify damaged sites [100]. Although these studies yielded useful insights, it should be emphasized that the quality of the results will be strongly influenced by the force field parameters and methods used in the calculations. Special care is required to ensure that the new parameters introduced into a force field to treat the novel chemical structures are (1) consistent with the force field and (2) modeling the modified structures correctly (see Chapter 2).

F. Alternative Secondary and Tertiary Motifs of Oligonucleotides

Beyond the helical structures and the RNA folded structures already discussed, both DNA and RNA are known to assume alternative structures, such as quadraplexes, that are of biological relevance. In addition, the nature of the hydrogen bonding patterns in oligonucleotides allows the design and construction of structures with the potential to significantly advance the field of nanotechnology [80,81]. To date the number of calculations on alternative oligonucleotide structures is not large; however, some good examples exist. MD simulations have been performed on four-stranded iDNA, which involves intercalated cytosines and have been shown to yield stable structures [101]. Quadraplexes are structures suspected to exist at the end of telomeres, stabilizing the terminal DNA in chromosomes. The quadraplexes themselves are stabilized by the presence of cations. Free energy perturbation calculations have been applied to investigate the influence of cation type on the extent of stabilization [102]. MD simulations have also been used to facilitate the determination of the structure and dynamics of quadraplexes based on NMR data [103]. Although only a few studies on alternative oligonucleotide folds have been studied to date, the ability of computational methods to investigate the stability of alternative structural motifs should facilitate the use of oligonucleotides in nanotechnology.

III. METHODOLOGICAL CONSIDERATIONS

Biological functions of nucleic acids occur at the level of single nucleosides up to chromosomes, with the majority of the functions intimately involving proteins as well as other biological molecules. Accordingly, computational methods for the study of nucleic acids must be able to access these variously sized systems. In this volume, the focus is on calculations at the atomistic level, and the present chapter remains consistent with this goal. Such a limitation, however, confines us to a discussion of computational studies of oligonucleotides that contain less than 50 basepairs. To overcome this limitation, a brief section on the different computational methods for the study of oligonucleotides, including approaches to the study of larger structures, follows.

A. Atomistic Models

Initial atomistic calculations on nucleic acids were performed in the absence of an explicit solvent representation, as discussed earlier. To compensate for this omission, various changes in the energy functions were employed, including reduced charges in the phosphate groups, the inclusion of hydrated sodiums, and altered dielectric constants. These approaches did yield results that furthered our knowledge of the structure–function relationships of nucleic acids in a qualitative fashion. However, as computational methods matured, the demand for more quantitative results increased, requiring more realistic models in oligonucleotide calculations. These demands have been met by way of both improved algorithms for simulations and improved potential energy functions, not to mention the incredible increase in computational power that has occurred over the last two decades. These aspects are discussed in this section.

Central to the quality of any computational study is the mathematical model used to relate the structure of a system to its energy. General details of the empirical force fields used in the study of biologically relevant molecules are covered in Chapter 2, and only particular information relevant to nucleic acids is discussed in this chapter.

Initial applications of computational techniques to study nucleic acids used models adapted from protein force fields [104]. Other parameter work of note included early studies on the properties of the furanose ring in nucleic acids [105,106], which showed the importance of the proper treatment of the pseudorotation profile in potential energy functions. The first widely used force fields for the study of nucleic acids were associated with the programs CHARMM [107] and AMBER [108,109]. These force fields were developed for both extended atom and all-atom representations. Extended atom models, which have been used extensively for simulation studies of proteins, are of lesser value for calculations on nucleic acids because the smaller number of nonpolar hydrogens in nucleic acids makes the gain in computer efficiency for an extended atom model smaller for them than for proteins. Both force fields were also developed primarily for modeling and simulation studies in vacuum, based on the use of distance-dependent dielectric constants in the calculation of electrostatic interactions. Such an approach allows short-range electrostatic interactions to dominate while longer range interactions are damped in accord with the damping due to the dielectric constant of water. Both of these force fields were quite useful, and a number of the studies cited earlier were based on these works.

Computational studies of nucleic acids via atomistic models in the absence of solvent are expected to yield poor representations owing to the polyanionic nature of oligonucleotides. Although a number of efforts were made to circumvent the use of explicit solvent models, the results were generally unsatisfactory (see above). These failures motivated the inclusion of solvent models in MD calculations. Initial efforts, primarily based on the TIP3P water model [110], used the parameter sets developed for vacuum calculations; the only difference was the use of a dielectric constant of 1 for the electrostatic calculations rather than the distance-dependent dielectric. Results from these studies, along with work based on an earlier version of the GROMOS force field [111], generally led to improved agreement with experiment compared to the vacuum calculations. However, as MD simulations were significantly extending beyond 100 ps, it was observed that the structures deviated significantly from the canonical forms of DNA. It was not until the "second generation" force fields were developed that stable simulations of oligonucleotides in solution were achieved.

The second generation force fields for nucleic acids were designed to be used with an explicit solvent representation along with inclusion of the appropriate ions [28,29]. In addition, efforts were made to improve the representation of the conformational energetics of selected model compounds. For example, the availability of high level ab initio calculations on the conformational energetics of the model compound dimethylphosphate yielded significant improvements in the models for the phosphodiester backbone. Similarly, new data on the interaction strengths for the Watson–Crick and Hoogsteen basepair interactions allowed for improvements in the nonbonded parameters. In the context of the nonbonded parameters is the need to properly balance the solute–solute interactions (e.g., Watson–Crick basepair interactions) with solute–solvent and solvent–solvent interactions. In the development of the CHARMM force field, special emphasis was placed on this balance [29]. On the basis of these force fields it was shown that stable MD simulations could be performed on a variety of sequences and structures; however, the results still had limitations [33].

Motivated by the need for further improvements, the AMBER [112] and CHARMM [36,113] force fields were optimized further. In addition, the BMS force field was developed, which incorporates features from the second generation CHARMM and AMBER force fields along with additional optimization of the parameters [38]. In all of these force fields, emphasis was placed on further optimization of the conformational energetics, yielding improved agreement with data on the canonical structures of DNA, surveys of dihedral distributions, and helical parameters of crystal structures from the Nucleic Acid Database (NAD) [114], along with a variety of other experimental and ab initio data. With the CHARMM force field, emphasis was placed on balancing local contributions, based on potential energy data for a series of model compounds, with global properties for duplexes in solution. This approach was designed to ensure that the proper contributions in the force field were combining to yield the desired properties of DNA and RNA in solution. It is expected that these improved energy functions will further facilitate the ability of MD simulations on nucleic acids to yield better quantitative agreement with experimental data.

Further extension of the atomistic models of nucleic acids will be achieved through additional optimization of the force fields using the present forms of the potential energy functions and extending the form of the potential energy function to include electronic polarizability along with other terms. Current work indicates that improvements in the revised CHARMM force field with respect to the treatment of sugar puckering are possible [113]. Concerning electronic polarizability, the polyanionic nature of nucleic acids and the influence of salt on their structure strongly indicate that gains in the quality of results from MD simulations will be made via its inclusion. To date, no calculations on nucleic acids using models that include electronic polarizability have been performed; however, QM studies have suggested that the polarization contribution to the solvation free energy of DNA is only 1–3% [115]. Thus, conclusions concerning the true gains to be made via the inclusion of electronic polarizability in nucleic acid simulations must wait until more detailed studies are performed. At this time one may speculate that some calculations addressing specific problems (e.g., energetics of phosphate–counterion interactions) will require polarizable models whereas other phenomena (e.g., conformational sampling of DNA and RNA duplexes) will see little improvement.

B. Alternative Models

Although the discussion thus far has concentrated on atomistic models of oligonucleotides in which the solvent is included explicitly, alternative models exist that allow for computational studies of larger oligonucleotides. For example, supercoiling of circular DNA has been studied using ribbonlike models [116]. This approach allows for the generation of trajectories using time steps of 100 fs, from which folding from the circular state into supercoiled forms was observed. Another approach is the use of internal coordinates combined with the implicit treatment of solvent, as in the program JUMNA [117]. This method is basically atomistic, but the movements of the system are described entirely in internal coordinates, greatly facilitating the locating of minima and the sampling of conformational space via Monte Carlo methods [118]. The internal coordinate method has also been used with a minimal hydration model where a 5 Å shell of explicit water molecules was used to hydrate the DNA, allowing for an integration time step of 10 fs [119]. Results showed the structures to be close to the B form of DNA; however, variations in RMS differences were smaller than occur in simulations using full solvent representations with periodic boundary conditions. Another "low resolution" method involves treating individual basepairs as three-point representations. The degrees of freedom between the individual "basepairs" can then be sampled to investigate the structural properties of extended DNA or RNA duplexes [120]. This approach can be combined with atomistic models to allow for both the overall fold of the oligonucleotide and specific interactions in small portions of the structure to be modeled, an approach that has been used to study portions of the 16S rRNA. A method based on the use of a segmented rod model along with Brownian dynamics allows for studies of DNA molecules hundreds of basepairs in length [121,122]. Although these methods sacrifice varying levels of detail, they extended computational approaches to significantly larger oligonucleotides, allowing for access to a wide variety of biological processes, such as the winding of DNA into supercoils and mechanisms associated with nucleosome and, ultimately, chromatin formation.

IV. PRACTICAL CONSIDERATIONS

Performing successful calculations on nucleic acids requires selection of the appropriate models for the goals of the calculations followed by determination of the proper starting configuration. When designing a computational study one should carefully consider the type of information desired from the calculations along with the available resources. In many instances, atomic details of interactions between oligonucleotides and the environment or with a bound protein are desired, making the use of atomistic models appropriate. These methods, however, require significant computational resources for the generation, storage, and analysis of the MD simulations. Continuing increases in computational power with the simultaneous decrease in computer costs makes the required facilities accessible to most laboratories. An alternative is the use of supercomputing centers. For systems larger than about 50 basepairs or where atomistic details of interactions between the nucleic acid and the solvent are not required, the methods discussed in the preceding section are appropriate.

The remainder of this chapter focuses on practical aspects of the preparation and implementation of atomistically based computations of nucleic acids. A flow diagram of the steps involved in system preparation and the performance of MD studies of nucleic acids is presented in Figure 1. Additional details on many of the procedures described here may be found in books by Allen and Tildesly [123] and Frenkel and Smit [124].

A. Starting Structures

A significant advantage of computational studies on nucleic acids is that reasonable guesses of the starting geometries can be made. When studying duplexes, these are typi-

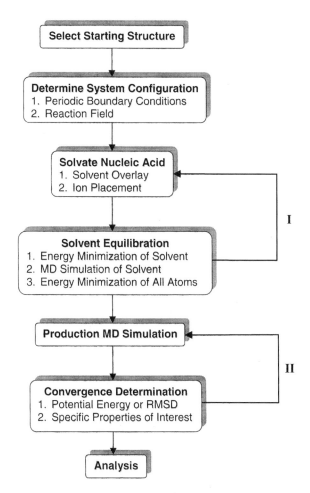

Figure 1 Flow diagram of the parameter optimization process. Loops I–II represent iterative stages of the optimization process as discussed in the text.

cally based on the canonical forms of DNA and RNA [2,125–130]. A number of available modeling and graphics packages have the ability to generate canonical structures for a given sequence. Alternatively, experimental structures from crystal or NMR studies, obtained from the nucleic acid [114] or protein databanks [131], can be used. While with DNA and RNA duplexes, crystal and NMR structures generally do not differ significantly from canonical structures, in cases where there are loops, bulges, hairpins, or unstacked bases, as in tRNA, the use of experimental structures is helpful. Alternatively, if the helical and nonhelical regions are known, reasonable guesses for a starting geometry, followed by relaxation of the structure via MD simulations, can be applied. This approach is useful when low resolution data on a nucleic acid structure are available [132–136]. A useful alternative is the program NAB, which generates structures of both helical and nonhelical regions of oligonucleotides [137] and is accessible via the Internet. When creating starting models of RNA or DNA, efforts should be made to check that the model is consistent with available biophysical and biochemical experimental data.

B. System Configuration, Solvation, and Ion Placement

Essential for MD simulations of nucleic acids is a proper representation of the solvent environment. This typically requires the use of an explicit solvent representation that includes counterions. Examples exist of DNA simulations performed in the absence of counterions [24], but these are rare. In most cases neutralizing salt concentrations, in which only the number of counterions required to create an electrically neutral system are included, are used. In other cases excess salt is used, and both counterions and co-ions are included [30]. Though this approach should allow for systematic studies of the influence of salt concentration on the properties of oligonucleotides, calculations have indicated that the time required for ion distributions around DNA to properly converge are on the order of 5 ns or more [31]. This requires that preparation of nucleic acid MD simulation systems include careful consideration of both solvent placement and the addition of ions.

As a first step in setting up an MD study of nucleic acids in solution, the overall configuration of the system must be considered. This configuration is defined by the boundary conditions to be used in the solvent simulation. Boundary conditions are required to maintain the proper density of the system as well as to minimize edge effects if the system is set up so that the condensed phase environment is finite, thereby interacting directly with vacuum. The most commonly used and most rigorously correct are periodic boundary conditions (PBCs). In this approach the system (nucleic acid, surrounding solvent, and ions) is created in, typically, a cubic or rectangular shape. The edges of the system, however, do not see a vacuum, but the edge on the opposite side of the cube or rectangle, allowing for interaction of the solvent on each edge of the cube or rectangle with that on the opposing edge. In addition to cubic or rectangular systems, PBC simulations may also be performed using octahedral or rhombic dodecahedral symmetries, which are appropriate for spherical molecules and minimize the total number of solvent molecules required to properly solvate the spherical macromolecule [138,139]. An advantage with the PBC approach is that it can be used with Ewald methods [22,123], which are currently considered the most rigorous methods for treating long-range electrostatic interactions (see Chapter 5). Alternative approaches for the treatment of boundaries are reaction field based methods [140,141], which include a potential energy barrier that keeps the solvent molecules from diffusing away from the simulation system and reaction field terms that account for the absence of water (or the presence of vacuum) outside the barrier. If a reaction field is not present, then the water molecules at the surface will tend to interact to a greater extent with each other than with interior water molecules, leading to problems with the solvent density and solvent transport problems at the surface of the system that may adversely effect the properties of the entire simulation system. These approaches are typically used on spherical systems, although cylinders and planes have also been used. For treatment of long-range electrostatics, reaction field methods can use atom truncation [26], extended electrostatic [142], or fast multipole methods [143]. Reaction field methods are most useful for systems too large to treat via PBC, such as protein–nucleic acid complexes. In all cases it is important that care be taken to ensure that the boundary conditions being used do not adversely affect the properties of the systems under study. This can typically be checked by performing simulations on the system and comparing properties calculated from the system with available experimental data, the most accessible being structural properties from X-ray or NMR experiments.

When preparing a PBC or reaction field calculation, the total size of the system is important. In general, the larger the system, as judged by the amount of solvent, the better, in that there are less likely to be adverse contributions from the boundary condition on the nucleic acid. The minimum amount of solvent surrounding the nucleic acid is dictated by the treatment of long-range electrostatics. With PBC the solvent box should be of such a size that the nucleic acid molecules in adjacent cells are further apart than (1) the real space cutoff for Ewald-based methods or (2) the atom truncation interaction cutoff distance. For reaction field based calculations, the distance from the nucleic acid to the edge of the solvation shell should be greater than the atom truncation distance. Concerning these distances, for Ewald methods a real space cutoff of 10 Å or greater and for atom truncation a cutoff distance of 12 Å or more are suggested. In all cases the simulator should perform tests to ensure that the applied boundary methods and treatment of long-range electrostatic interactions do not adversely affect the calculated result.

Once the geometry and size of the system to be studied are determined, a pure solvent system (i.e., no DNA or RNA) of those dimensions should be built. This can typically be done via standard procedures included with the various modeling packages. These systems should then be subjected to MD simulations using the identical methods for treatment of the nonbonded interactions to be used in the final calculation. This will (1) allow the solvent to properly equilibrate with respect to itself and any ions included at this stage and (2) offer a test of the proposed methodology by ensuring that water density and transport properties are in satisfactory agreement with experiment. Once the solvent is equilibrated, it is overlaid onto the nucleic acid molecule, and all solvent molecules with nonhydrogen atoms within a given distance of solute nonhydrogen atoms (typically 1.8 Å) are then deleted. At this stage ions can be added to the system as required.

Ion placement in simulations of nucleic acids can have a significant impact on the computed results, consistent with the role of water activity on the structure of DNA. A comparison of the influence of ion placement on MD simulations was reported by Young et al. [31]. Methods applied included a Monte Carlo based method that places counterions at low energy positions around DNA using a sigmoidal distance-dependent dielectric function for calculation of the interaction energy of the ion with the environment. The second method used was based on calculation of the electrostatic potential around the DNA followed by the placement of ions at the most favorable locations in the potential. This is performed in an iterative fashion such that subsequent ions take into account previously placed ions. The final method used involved the placement of sodium counterions "6 Å from the P atoms along the bisector of the backbone OPO groups." All three methods yielded similar results when the electrostatic interactions were treated with the Ewald method. Additional methods include the replacement of water oxygens in a previously solvated system with counterions, with the selection criteria based on the interaction with the surrounding water molecules and the oligonucleotide [144,145]. This can also be done in an iterative fashion that allows for the positions of ions to be sensitive to the presence of other ions. In this approach, the omission of water hydrogens from the interaction energy calculation eliminates orientational problems that would require energy minimization or dynamics, thereby significantly decreasing the computational requirements. Other simulators simply replaced randomly selected water molecules with ions [146]. A final approach is to initially overlay the DNA or RNA with a solvent box or sphere that already contains ions at the desired concentration and has been previously equilibrated [24]. In this method all ions and solvent molecules that overlap the solute are removed and additional ions are added at random positions or deleted, based on those furthermost from the DNA or RNA, to obtain electrical neutrality. The last two methods are particularly well suited for the placement of excess salt (counter- and co-ions beyond those needed to achieve electrical neutrality). In all cases, after the addition of ions it is essential that the solvent and ions be adequately equilibrated around the nucleic acid, as discussed in the following paragraph. The simulator should also be aware that ions strongly interacting at specific sites on the nucleic acid can strongly influence both the structural and dynamic properties.

Once the DNA or RNA has been overlaid with solvent and ions, the energy of the system is minimized, and the system is subjected to an equilibration MD simulation. These calculations should be performed with the nucleic acid atoms fixed or harmonically constrained to allow for the solvent to relax around the nucleic acid. Properties such as the potential energy and the solvent–solute interaction energy should be monitored as a function of simulation time to ensure that the system has relaxed to a satisfactory extent. In cases where the MD studies will be performed at constant volume (i.e., NVT or NVE simulations) it is suggested that the pressure of the system be monitored to ensure that it is in the vicinity of 1 atm. If the pressure differs significantly from 1 atm, the distances used for the deletion of solvent molecules can be increased or decreased accordingly (loop I of Fig. 1) [146]. Note that in simulations containing fixed or harmonically constrained atoms the calculated pressures may be incorrect, requiring that short MD simulations without constraints be performed to obtain accurate pressures. Once the solvent has adequately equilibrated around the solute, the entire system, including the nucleic acid, should be energy minimized. At this stage of the system preparation it is important that checks be performed to ensure that the system is still properly solvated. If problems in the simulation system are evident, the system should be resolved in appropriate fashion to correct the problem and the equilibration redone, as shown in loop I of Figure 1.

C. Production MD Simulation

At this stage the production MD simulation can be initiated. It is generally preferable to perform simulations in the isobaric, isothermal (NPT) ensemble, which yields thermodynamic properties that correspond to the experimentally accessible Gibbs free energy [147]. In some cases initiation of the MD simulations can involve gradual heating of the system, although initiation of the simulation at the final desired temperature is often sufficient. Several algorithms for NPT simulations are available, including Berendsen's method [148], the Langevin piston [149], and an extended method from Klein and coworkers [150]. It should be noted that Berendsen's method does not correspond to a true ensemble, and problems associated with systematic oscillations in the system volume can occur [149]. When the simulation is initiated it is important to closely monitor both structural and energetic properties to ensure that significant perturbations of the solute do not initially occur due to the applied methodology. If such perturbations are present, the system preparation and equilibration approach should be evaluated for potential problems.

D. Convergence of MD Simulations

Rigorous proof of convergence of MD simulations cannot be performed [31]. Efforts can be made, however, to ensure that various properties calculated from a simulation have reached satisfactory levels of convergence. First, the simulations should be analyzed to determine if global (1) energetic and (2) structural properties have stabilized. Energetic stability is typically investigated by monitoring the potential energy versus time. During most simulations the potential energy initially relaxes, after which it fluctuates around a constant value for the remainder of the simulation. In NVE simulations the ratio of RMS fluctuations of the potential energy and kinetic energy can be monitored, with $RMS_{PE}/RMS_{KE} = 0.01$ indicating proper energy conservation. Analysis of structural properties versus time (e.g., RMS difference with respect to the starting structure) will also typically show an initial relaxation followed by fluctuations around an average value for the remainder of the simulation. In certain cases these fluctuations may be relatively large, indicating the sampling of alternative conformations that may be biologically relevant [16]. From the analysis of structural and energetic properties versus time, the initial portion of the production simulation during which relaxation of these properties occurs is discarded. Additional convergence tests are performed only on the remainder of the production simulation.

Tests of convergence on the remaining portion of the simulation can most readily be performed by calculating the desired property (i.e., the property that is of most interest to the simulator) over different simulation time lengths and monitoring the change in the average value as a function of simulation time. Convergence is indicated by the lack of significant change in the average value of the property as the simulation time increases. Another method involves separating the total trajectory into independent blocks (e.g., a 500 ps production simulation may be separated into five 100 ps blocks), calculating average values for each block, and comparing the average values from the individual blocks. These block averages can also be used to calculate overall averages and standard errors for individual properties [151]. This is an excellent method for obtaining the statistical significance of results from a simulation.

For an entire MD simulation, as well as for separate blocks from a simulation, determination of convergence of a property can be most rigorously carried out by calculating the time series of a property and determining its autocorrelation function and accompanying relaxation time. For adequate convergence the total MD simulation time or the block time should be approximately four times as long as the relaxation time. This test is also appropriate for determining the length of blocks that can be considered independent. It should be emphasized that the relaxation time and, accordingly, the amount of simulation time required for convergence are dependent on the property being investigated. Thus, the total required simulation time is dependent on the type of information that is to be obtained from a simulation. If it is determined that the properties of interest have not converged, then the production simulation should be extended, as indicated by loop II of Figure 1.

E. Analysis of MD Simulations

Molecular dynamics simulations at an atomic level of detail contain enormous amounts of information, making rational analysis of simulations essential. In cases where studies have been designed in close collaboration with experiment, the information to be obtained is that which corresponds to the experimental data. Alternatively, systematic analysis of the MD trajectory may be performed to identify specific properties of interest. Properties determined from simulations of nucleic acids can be separated into those associated with structure, hydration, and energetics. Structural properties include the dihedrals associated with the phosphodiester linkage, the sugar and the glycosidic linkage, Watson–Crick base pairing, and a variety of intramolecular distances such as the minor groove width based typically on interstrand P distances. Nucleic acid sugar puckering is typically not analyzed on the basis of the individual dihedrals, but rather with respect to the concept of pseudorotation, where the sugar conformation on is defined in terms of two variables, the pseudorotation angle and the amplitude [2,152]. An important class of structural properties comprises the helical or helicoidal parameters [153–155]. These terms define the orientations of the bases with respect to a global helical axis or locally with respect to adjacent bases or basepairs. Both methods have been used extensively, and a number of programs have been developed to perform this analysis, including CURVES [156], FREEHELIX [157], and a program by Babcock and Olson [158]. The CURVES method has been incorporated into an analysis and plotting package, DIALS and WINDOWS, which analyzes MD trajectories as well as individual coordinate files and presents data, including time series, in a highly compact fashion [159]. When using these programs the user should be aware of both the global and local definitions of the helicoidal parameters. Whereas in idealized helices the two definitions yield similar or identical results, the global method is preferable for standard helices and the local definition is more suited for distorted helices, such as those found in DNA–protein complexes, where a global helical axis cannot be rigorously defined. When performing structural analysis it is often best to initially calculate overall averages, use that information to identify terms that may be of interest, and then analyze those terms via time series or probability distributions. Probability distributions of DNA dihedrals and helicoidal parameters calculated via CURVES have been published [31], and those based on FREEHELIX are available (N Banavali, AD MacKerell Jr, unpublished).

Hydration properties, including interactions with ions, are strong determinants of DNA structure [2,160–162]. Hydration numbers can be determined on the basis of a variety of criteria [29,163,164] for comparison with individual crystal structures or with data from surveys of the NDB [165,166]. This type of analysis is often performed on different portions of the DNA such as the minor groove and is typically limited to first shell hydration. Complicating such analysis is the overlap of waters hydrating different sites. For example, a water molecule hydrating the sugar O4′ atom may also be hydrating atoms in the minor groove, and it has been shown that counterions can replace water molecules in the first shell of hydration [167]. Hydration can also be studied with respect to the dynamic properties of the water molecules. Examples include changes in water dynamics around DNA leading to local alterations of the dielectric environment [32] and results indicating that decreased water activity due to increased salt concentration is associated with changes in water mobility rather than hydration number [24]. Dynamics properties that have been analyzed include diffusion constants and rotational correlation times. Rotational correlation times may be analyzed on the basis of the water dipole axis, the axis perpendicular to the plane of the water, and the H—H axis. These motions are associated with molecular twisting (dipole axis), rocking (perpendicular), and wagging (H—H vector), respectively [168]. Difficulties in determining correlation times of individual waters occur with water molecules that undergo little motion on the time scale of the simulation, leading in some cases to undefined correlation times. A method of analysis based on the initial 4 ps of the decay of the autocorrelation function [169] can overcome some, but not all, of these problems.

Supplementing direct analysis of structural and hydration properties is the use of energetic analysis. For example, instead of monitoring individual Watson–Crick hydrogen bonds, interaction energies between basepairs can be monitored. Similarly, interaction energies between water and the nucleic acid can be determined to supplement the information discussed in the preceding paragraph. The advantage of energetics over direct structural or hydration information is the ability to more readily take into account all possible contributions, not just those envisioned by the simulator. Furthermore, it may be possible to better gauge the relative impact of different types of interactions on the trajectory obtained from an MD simulation. Such analysis has been used to identify energetic and, subsequently, structural phenomena allowing for the barrierless extension of DNA beyond its canonical B form and contributions to the barrier to strand separation [86]. Typically, energetic analysis should be performed using the same truncation scheme as that used in the MD simulation. This can be problematic in Ewald simulations, as it is currently not possible to use Ewald summations to determine interaction energies. One suggestion is to calculate interaction energies between atoms with the real-space truncation distance used in the simulation. An interesting alternative to direct analysis of interaction energies is the recalculation of solvation free energies using continuum models on individual nucleic acid structures from a solvated MD simulation. This approach has been used to understand the equilibrium between the A and B forms of DNA as a function of water activity [40,91]. Although the method successfully ordered the equilibrium when water activity was modified with ethanol, it was not successful at predicting the salt effect, which may be associated with atomic detail interactions not modeled in continuum approaches. Overall, energetic analysis of MD simulations offers an additional method to analyze simulations, often allowing for identification of structural contributions that are difficult to identify via direct structural analysis.

V. CONCLUSION

In this chapter we have attempted to give an overview of the types of nucleic acid systems that are accessible to computational study. These vary from nucleosides, through duplex DNA and RNA, up to nucleic acid–protein complexes. Accessibility to both longer time scales and larger systems is expected to increase as advances in both computational power and methods continue to occur. Although the work cited cannot be considered complete, it should allow the reader to access information required for moving into and obtaining a more general background of the field of nucleic acid simulations.

WEB SITES OF INTEREST

> CHARMM program: yuri.harvard.edu
> CHARMM force field: https://rxsecure.umaryland.edu/research/amackere/research.html
> GROMOS: igc.ethz.ch/gromos/welcome.html
> AMBER: www.amber.ucsf.edu/amber/amber.html
> NAB: www.scripps.edu/case
> Dials and Windows (via D. Beverige):   www.wesleyan.edu/chem/faculty/beveridge
> CURVES (also JUMNA):   www.ibpc.fr/UPR9080/Curindex.html
> Protein Data Bank: www.rcsb.org/pdb
> Nucleic Acid Data Bank (also FREEHELIX):   ndbserver.rutgers.edu
> National Partnership for Advanced Computational Infrastructure (NPACI): www.npaci.edu

REFERENCES

1. JD Watson, FHC Crick. Nature 171:737–738, 1953.
2. W Saenger. Principles of Nucleic Acid Structure. New York: Springer-Verlag, 1984.

3. CR Calladine, HR Drew. Understanding DNA: The Molecule and How It Works. New York: Academic Press, 1997.

4. RE Dickerson, DS Goodsell, S Neidle. Proc Natl Acad Sci USA 91:3579–3583, 1994.

5. Nuclear Magnetic Resonance, Part C, TL James, NJ Oppenheimer, eds. New York: Academic Press, 1994. Vol. 239.

6. GU Lee, LA Chrisey, RJ Colton. Science 266:771–773, 1994.

7. M Levitt. Cold Spring Harbor Symp Quant Biol 47:271–275, 1983.

8. B Tidor, KK Irikura, BR Brooks, M Karplus. J Biomol Struct Dyn 1:231–252, 1983.

9. UC Singh. Proc Natl Acad Sci USA 82:755–759, 1985.

10. J Srinivasan, JM Withka, DL Beveridge. Biophys J 58:533–547, 1990.

11. F Eisenhaber, VG Tumanyan, RA Abagyan. Biopolymers 30:563–581, 1990.

12. KN Swamy, E Clementi. Biopolymers 26:1901–1927, 1987.

13. GL Seibel, UC Singh, PA Kollman. Proc Natl Acad Sci USA 82:6537–6540, 1985.

14. HR Drew, RM Wing, T Takano, C Broka, S Tanaka, K Itakura, RS Dickerson. Proc Natl Acad Sci USA 78:2179–2183, 1981.

15. S Swaminathan, G Ravishanker, DL Beveridge. J Am Chem Soc 113:5027–5040, 1991.

16. KJ McConnell, R Nirmala, MA Young, G Ravishanker, DL Beveridge. J Am Chem Soc 116:4461–4462, 1994.

17. K Miaskiewicz, R Osman, H Weinstein. J Am Chem Soc 115:1526–1537, 1993.

18. DM York, W Yang, H Lee, T Darden, LG Pedersen. J Am Chem Soc 117:5001–5002, 1995.

19. H Lee, T Darden, LG Pedersen. J Chem Phys 102:3830–3834, 1995.

20. TE Cheatham III, JL Miller, T Fox, TA Darden, PA Kollman. J Am Chem Soc 117:4193–4194, 1995.

21. S Weerasinghe, PE Smith, V Mohan, Y-K Cheng, BM Pettitt. J Am Chem Soc 117:2147–2158, 1995.

22. PP Ewald. Ann Phys 64:253–287, 1921.

23. TA Darden, D York, LG Pedersen. J Chem Phys 98:10089–10092, 1993.

24. AD MacKerell Jr. J Phys Chem B 101:646–650, 1997.

25. J Norberg, L Nilsson. J Chem Phys 104:6052–6057, 1996.

26. PJ Steinbach, BR Brooks. J Comp Chem 15:667–683, 1994.

27. J Norberg, L Nilsson. Biophysical J 79, 2000.

28. WD Cornell, P Cieplak, CI Bayly, IR Gould, J Merz, DM Ferguson, DC Spellmeyer, T Fox, JW Caldwell, PA Kollman. J Am Chem Soc 117:5179–5197, 1995.

29. AD MacKerell Jr, J Wiórkiewicz-Kuczera, M Karplus. J Am Chem Soc 117:11946–11975, 1995.

30. L Yang, BM Pettitt. J Phys Chem 100:2550–2566, 1996.

31. MA Young, G Ravishanker, DL Beveridge. Biophys J 73:2313–2336, 1997.

32. MA Young, B Jayaram, DL Beveridge. J Phys Chem B 102:7666–7669, 1998.

33. M Feig, BM Pettitt. Biophys J 75:134–149, 1998.

34. TE Cheatham III, PA Kollman. J Mol Biol 259:434–444, 1996.

35. TE Cheatham III, MF Crowley, T Fox, PA Kollman. Proc Natl Acad Sci USA 94:9626–9630, 1997.

36. AD MacKerell Jr, N Banavali. J Comput Chem 21:105–120, 2000.

37. TE Cheatham III, PA Kollman. Structure 5:1297–1311, 1997.

38. DR Langley. J Biomol Struct Dyn 16:487–509, 1998.

39. AD MacKerell Jr. J Chim Phys 94:1436–1447, 1997.

40. B Jayaram, D Sprous, MA Young, DL Beveridge. J Am Chem Soc 120:10629–10633, 1998.

41. SK Burley, RG Roeder. Ann Rev Biochem 65:769–799, 1996.

42. G Guzikevich-Guerstein, Z Shakked. Nature Struct Biol 3:32–37, 1996.

43. D Flatters, M Young, DL Beveridge, R Lavery. J Biomol Struct Dyn 14:757–765, 1997.

44. N Pastor, L Pardo, H Weinstein. Biophys J 73:640–652, 1997.

45. A Balaeff, ME Churchill, K Schulten. Proteins 30:113–135, 1998.

46. Y Duan, P Wilkosz, JMJ Rosenberg. J Mol Biol 264:546–555, 1996.

47. Y Duan, P Wilkosz, M Crowley, JM Rosenberg. J Mol Biol 272:553–572, 1997.
48. DL Beveridge, KJ McConnell, MA Young, S Vijayakumar, G Ravishanker. Mol Eng 5:255–269, 1995.
49. J de Vlieg, HJ Berendsen, WF van Gunsteren. Proteins 6:104–127, 1989.
50. M Billeter, P Guntert, P Luginbuhl, K Wuthrich. Cell 85:1057–1065, 1996.
51. K Miaskiewicz, RL Ornstein. J Biomol Struct Dyn 13:593–600, 1996.
52. L Pardo, N Pastor, H Weinstein. Biophys J 75:2411–2421, 1998.
53. TC Bishop, K Schulten. Proteins 24:115–133, 1996.
54. TC Bishop, D Kosztin, K Schulten. Biophys J 72:2056–2067, 1997.
55. MAL Eriksson, T Hard, L Nilsson. Biophys J 68:402–426, 1995.
56. MAL Eriksson, L Nilsson. Protein Eng 11:589–600, 1998.
57. LF Harris, MR Sullivan, PD Popken-Harris. J Biomol Struct Dyn 15:407–430, 1997.
58. LF Harris, MR Sullivan, PD Popken-Harris, DF Hickok. J Biomol Struct Dyn 13:423–440, 1995.
59. LF Harris, MR Sullivan, PD Popken-Harris, DF Hickok. J Biomol Struct Dyn 12:249–270, 1994.
60. D Kosztin, et al. Biophys J 73:557–570, 1997.
61. MAL Eriksson, L Nilsson. J Mol Biol 253:453–472, 1995.
62. MAL Eriksson, L Nilsson. Eur Biophys J 28:102–111, 1999.
63. S Sen, L Nilsson. Biophys J 77:1801–1810, 1999.
64. Y Tang, L Nilsson. Proteins 35:101–113, 1999.
65. AMJJ Bonvin, M Sunnerhagen, G Ottin, WF vanGunsteren. J Mol Biol 282:859–873, 1998.
66. G Roxstrom, I Velazquez, M Paulino, O Tapia. J Biomol Struct Dyn 16:301–312, 1998.
67. G Roxstrom, I Velazquez, M Paulino, O Tapia. J Phys Chem B 102:1828–1832, 1998.
68. DR Langley, TW Doyle, DL Beveridge. J Am Chem Soc 113:4395–4403, 1991.
69. DR Langley, J Golik, B Krishman, TW Doyle, DL Beveridge. J Am Chem Soc 116:15–29, 1994.
70. P Cieplak, SN Rao, PDJ Grootenhuis, PA Kollman. Biopolymers 29:717–727, 1990.
71. AH Elcock, PD Lyne, AJ Mulholland, A Nandra, WG Richards. J Am Chem Soc 117:4706–4707, 1995.
72. J Miller, PA Kollman. J Mol Biol 270:436–450, 1997.
73. P Auffinger, S Louise-May, E Westhof. Biophys J 76:50–64, 1999.
74. P Auffinger, E Westhof. Biophys J 71:940–954, 1996.
75. P Auffinger, E Westhof. J Mol Biol 269:326–341, 1997.
76. TP Hermann, E Westhof. Eur Biophys J 27:153–165, 1998.
77. J Norberg, L Nilsson. J Biomol NMR 7:305–314, 1996.
78. Y Tang, L Nilsson. Biophys J 77:1284–1305, 1999.
79. P Cluzel, A Lebrun, C Heller, R Lavery, J-L Viovy, D Chatenay, F Caron. Science 271:792–794, 1996.
80. C Mao, W Sun, Z Shen, NC Seeman. Nature 397:144–146, 1999.
81. E Winfree, F Liu, LA Winzler, NC Seeman. Nature 394:539–544, 1998.
82. J Norberg, L Nilsson. J Am Chem Soc 117:10832–10840, 1995.
83. J Norberg, L Nilsson. J Phys Chem 99:13056–13058, 1995.
84. J Norberg, L Nilsson. Biophys J 69:2277–2285, 1995.
85. J Norberg, L Nilsson. Biophys J 74:394–402, 1998.
86. AD MacKerell Jr, GU Lee. Eur J Biophys 28:415–426, 1999.
87. MW Konrad, JI Bolonick. J Am Chem Soc 118:10989–10994, 1996.
88. J Norberg, L Nilsson. Proc Natl Acad Sci USA 93:10173–10176, 1996.
89. ST Crooke. Antisense Nucleic Acid Drug Dev 8:115–122, 1998.
90. S Verma, F Eckstein. Annu Rev Biochem 67:99–134, 1998.
91. J Srinivasan, TE Cheatham III, P Ceiplak, PA Kollman, DA Case. J Am Chem Soc 120:9401–9409, 1998.
92. AR Srinivasan, WK Olsen. J Am Chem Soc 120:492–499, 1998.

93. P Cieplak, TE Cheatham III, PA Kollman. J Am Chem Soc 119:6722–6730, 1997.

94. GC Shields, CA Lauthton, M Orozco. J Am Chem Soc 120:5895–5904, 1998.

95. S Sen, L Nilsson. J Am Chem Soc 120:619–631, 1998.

96. J Luo, TC Bruice. J Am Chem Soc 120:1115–1123, 1998.

97. H De Winter, E Lescrinier, A Van Aerschot, P Herdewijn. J Am Chem Soc 120:5381–5394, 1998.

98. WS Ross, CC Hardin, I Tinoco Jr, SN Rao, DA Pearlman, PA Kollman. Biopolymers 28: 1939–1957, 1989.

99. DA Pearlman, PA Kollman. Biopolymers 29:1193–1209, 1990.

100. K Miaskiewicz, J Miller, M Cooney, R Osman. J Am Chem Soc 118:9156–9163, 1996.

101. N Spackova, I Berger, M Egli, J Sponer. J Am Chem Soc 120:6147–6151, 1998.

102. WS Ross, CC Hardin. J Am Chem Soc 116:6070–6080, 1994.

103. GD Strahan, MA Keniry, RH Shafer. Biophys J 75:968–981, 1998.

104. M Levitt. Proc Natl Acad Sci USA 75:640–644, 1978.

105. WK Olson, JL Sussman. J Am Chem Soc 104:270–278, 1982.

106. WK Olson. J Am Chem Soc 104:278–286, 1982.

107. L Nilsson, M Karplus. J Comput Chem 7:591–616, 1986.

108. SP Weiner, PA Kollman, DA Case, UC Singh, C Ghio, G Alagona, S Profeta, P Weiner. J Am Chem Soc 106:765–784, 1984.

109. SJ Weiner, PA Kollman, DT Nguyen, DA Case. J Comput Chem 7:230–252, 1986.

110. WL Jorgensen, J Chandrasekhar, JD Madura, RW Impey, ML Klein. J Chem Phys 79:926–935, 1983.

111. WF van Gunsteren, HJC Berendsen. GROMOS 86: Groningen Molecular Simulation Program Package. University of Groningen, The Netherlands, 1986.

112. TE Cheatham III, P Cieplak, PA Kollman. J Biomol Struct Dyn 16:845–861, 1999.

113. N Foloppe, AD MacKerell Jr. J Comput Chem 21:86–104, 2000.

114. HM Berman, WK Olson, DL Beveridge, J Westbrook, A Gelbin, T Demeny, S-H Hsieh, AR Srinivasan, B Schneider. Biophys J 63:751–759, 1992.

115. DM York, TS Lee, W Yang. J Am Chem Soc 118:10940–10941, 1996.

116. T Schlick, WK Olsen. J Mol Biol 223:1089–1119, 1992.

117. D Flatters, K Zakrzewska, R Lavery. J Comput Chem 18:1043–1055, 1997.

118. HA Gabb, C Prevost, G Bertucat, CH Robert, R Lavery. J Comput Chem 18:2001–2011, 1997.

119. AK Mazur. J Am Chem Soc 120:10928–10937, 1998.

120. SC Harvey, MS VanLoock, TR Rasterwood, RK-Z Tan. Mol Modeling Nucleic Acids 682: 369–378, 1998.

121. H Merlitz, K Rippe, KV Klenin, J Langowski. Biophys J 74:773–779, 1998.

122. K Klenin, H Merlitz, J Langowski. Biophys J 74:780–788, 1998.

123. MP Allen, DJ Tildesley. Computer Simulation of Liquids. New York: Oxford Univ Press, 1989.

124. D Frenkel, B Smit. Understanding Molecular Simulation: From Algorithms to Applications. New York: Academic Press, 1996.

125. S Arnott, DWL Hukins, SD Dover, W Fuller, AR Hodgson. J Mol Biol 81:102–122, 1973.

126. S Arnott, DWL Hukins. J Mol Biol 81:93–105, 1973.

127. S Arnott, E Selsing. J Mol Biol 88:551–552, 1974.

128. S Arnott, R Chandrasekaran, DWL Hukins, PJC Smith, L Watts. J Mol Biol 88:523–533, 1974.

129. S Arnott, E Selsing. J Mol Biol 88:509–521, 1974.

130. S Arnott, R Chandrasekaran, DL Birdsall, AGW Leslie, RL Ratcliff. Nature 282:743–745, 1980.

131. FC Bernstein, TF Koetzle, GJB Williams, DF Meyer Jr, MD Brice, JR Rodgers, O Kennard, T Shimanouchi, M Tasumi. J Mol Biol 112:535–542, 1977.

132. AR Banerjee, A Berzal-Herranz, J Bond, S Butcher, JA Esteban, JE Heckman, B Sargueil, N Walter, JM Burke. Mol Modeling Nucleic Acids 682:360–368, 1998.
133. T Elgavish, MS VanLoock, SC Harvey. J Mol Biol 285:449–453, 1999.
134. MS VanLoock, TR Easterwood, SC Harvey. J Mol Biol 285:2069–2078, 1999.
135. S Lemieux, P Chartrand, R Cedergren, F Major. RNA 4:739–749, 1998.
136. C Zwieb, K Gowda, N Larsen, F Müller. Mol Modeling Nucleic Acids 682:405–413, 1998.
137. TJ Macke, DA Case. Mol Modeling Nucleic Acids 682:379–392, 1998.
138. BR Brooks, RE Bruccoleri, BD Olafson, DJ States, S Swaminathan, M Karplus. J Comput Chem 4:187–217, 1983.
139. AD MacKerell Jr, B Brooks, CL Brooks III, L Nilsson, B Roux, Y Won, M Karplus. CHARMM: The energy function and its paramerization with an overview of the program. In: PvR Schleyer, NL Allinger, T Clark, J Gasteiger, PA Kollman, HF Schaefer III, PR Schreiner, eds. Encyclopedia of Computational Chemistry, Vol 1. Chichester, UK: Wiley, 1998, pp 271–277.
140. CL Brooks, M Karplus. J Chem Phys 79:6312, 1983.
141. D Beglov, B Roux. J Chem Phys 100:9050–9063, 1994.
142. RH Stote, DJ States, M Karplus. J Chim Phys 88:2419–2433, 1991.
143. TC Bishop, RD Skeel, K Schulten. J Comput Chem 18:1785–1791, 1997.
144. S-H Jung. Simulation of DNA and Its Interactions with Ligands. PhD Dissertation, Harvard University, 1989.
145. AD MacKerell Jr. J Phys Chem 99:1846–1855, 1995.
146. V Mohan, PE Smith, BM Pettitt. J Phys Chem 97:12984–12990, 1993.
147. DA McQuarrie. Statistical Mechanics. New York: Harper & Row, 1976.
148. HJC Berendsen, JPM Postma, WF van Gunsteren, A DiNola, JR Haak. J Chem Phys 81: 3684–3690, 1984.
149. SE Feller, Y Zhang, RW Pastor, RW Brooks. J Chem Phys 103:4613–4621, 1995.
150. GJ Martyna, DJ Tobias, ML Klein. J Chem Phys 101:4177–4199, 1994.
151. RJ Loncharich, BR Brooks, RW Pastor. Biopolymers 32:523, 1992.
152. C Altona, M Sundaralingam. J Am Chem Soc 94:8205–8212, 1972.
153. B Hartmann, R Lavery. Quart Rev Biophys 29:309–368, 1996.
154. RE Dickerson. Methods Enzymol 211:67–111, 1992.
155. RE Dickerson. J Biomol Struct Dyn 6:627–634, 1989.
156. R Lavery, H Sklenar. J Biomol Struct Dyn 6:63–91, 1988.
157. RE Dickerson. Nucleic Acids Res 26:1906–1926, 1998.
158. MS Babcock, WK Olson. J Mol Biol 237:98–124, 1994.
159. G Ravishanker, S Swaminathan, DL Beveridge, R Lavery, H Sklenar. J Biomol Struct Dyn 6:669–699, 1989.
160. W Saenger. Annu Rev Biophys Biophys Chem 16:93–114, 1987.
161. E Westhof. Annu Rev Biophys Biophys Chem 17:125–144, 1988.
162. B Jayaram, G Ravishanker, DL Beveridge. J Phys Chem 92:1032–1034, 1988.
163. B Schneider, DM Cohen, L Schleifer, AR Srinivasan, WK Olsen, HM Berman. Biophys J 65:2291–2303, 1993.
164. M Mezei, DL Beveridge. Methods Enzymol 127:22–47, 1986.
165. B Schneider, HM Berman. Biophys J 69:2661–2669, 1995.
166. B Schneider, K Patel, HM Berman. Biophys J 75:2422–2434, 1998.
167. MA Young, B Jayaram, DL Beveridge. J Am Chem Soc 119:59–69, 1997.
168. H Jóhannesson, B Halle. J Am Chem Soc 120:6859–6870, 1998.
169. A Wallqvist, BJ Berne. J Phys Chem 97:13841–13851, 1993.

21

Membrane Simulations

Douglas J. Tobias
University of California at Irvine, Irvine, California

I. INTRODUCTION

Biological membranes hold cells together and divide them into compartments. They are also home to the lipid-soluble proteins that comprise roughly one-third of the genome and perform a variety of functions, including energy production and storage, signal transduction, and the formation of channels for the transport of substances into and out of cells and between their compartments. In 1972, Singer and Nicolson [1] assimilated available data into a model for the molecular organization of biological membranes that is now taken for granted. According to this "fluid mosaic" model, a lipid bilayer forms the membrane matrix in which proteins are embedded. This matrix is fluid at physiological temperatures, and both lipids and proteins are free to diffuse in the plane of the membrane.

Biological membranes contain a complex mixture of several different types of lipids. Membrane lipids are amphiphilic molecules, with both polar and nonpolar substituents. A common membrane lipid that has been the subject of numerous experimental and theoretical studies, dipalmitoylphosphatidylcholine (DPPC), is diagrammed in Figure 1. Like other membrane lipids containing two chains, DPPC is classified as an insoluble, swelling amphiphile [2]. These lipids exhibit a variety of liquid crystalline phases that may be lamellar, cubic, or hexagonal, depending on the temperature and water content. Lamellar phases consist of stacks of lipid bilayers separated by layers of water (Fig. 2). The structures and mechanical properties of lipid bilayers depend on the amount of water present. As water is added to a lamellar phase, the spacing between the bilayers increases until the "fully hydrated" state is reached. At this point, which in DPPC is about 28 water molecules per lipid, additional water molecules do not go between the bilayers but rather go into a bulk water phase that is distinct from the lamellar bilayer water phase.

Hydrated bilayers containing one or more lipid components are commonly employed as models for biological membranes. These model systems exhibit a multiplicity of structural "phases" that are not observed in biological membranes. In the state that is analogous to fluid biological membranes, the "liquid crystal" or L_α bilayer phase present above the main bilayer phase transition temperature, T_m, the lipid hydrocarbon chains are conformationally disordered and fluid ("melted"), and the lipids diffuse in the plane of the bilayer. At temperatures well below T_m, hydrated bilayers exist in the "gel," or L_β, state in which the mostly all-trans chains are collectively tilted and pack in a regular two-dimensional

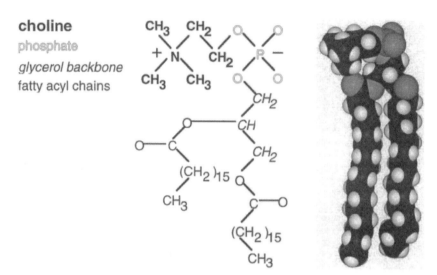

choline
phosphate
glycerol backbone
fatty acyl chains

Figure 1 Chemical structure and space-filling representation of a phosphatidylcholine, DPPC. Different parts of the molecule are referred to by the labels at the left; together the choline and phosphate are referred to as the headgroup, which is zwitterionic. In the space-filling model, H atoms are white, O and P gray, and C black. (From Ref. 55.)

array. Because the biologically relevant liquid crystalline phase is highly disordered, it is not amenable to atomic resolution structure determination. However, electron density and scattering length density profiles as a function of depth in the membrane have been mapped out by X-ray and neutron diffraction [3–6], and deuterium NMR has been used to characterize the average order in the hydrocarbon chains [7]. The dynamics of lipids and water molecules in membranes on time scales presently accessible to simulations have been characterized by various experimental techniques, including neutron scattering [8,9] and NMR [8,10].

In recent years molecular dynamics (MD) simulations have emerged as a useful tool for filling in some of the details on the molecular structure and dynamics of membranes that are not available from experiments and for providing feedback on models used to interpret experimental data. Computer hardware and simulation methodology have matured to the point where stable simulations of membranes can now be routinely performed for at least a few nanoseconds, and many of the results of membrane simulations agree remarkably well with available experimental data [11]. With the help of simulations, Singer and Nicolson's cartoon of a membrane drawn as a flat bilayer of lollypop-like lipids is evolving into images such as the one shown in Figure 2.

This chapter describes some of the technical aspects of simulating membranes, presents results that illustrate the novel insight into membrane structure and dynamics that can be provided by simulation, and discusses the correspondence of the emerging atomic scale picture with the results of NMR and X-ray and neutron scattering experiments. We restrict our attention to pure lipid bilayers, which have largely been the focus of the field of membrane simulations to date (recent reviews can be found in Refs. 11 and 12), with an emphasis on dynamics, which has been somewhat neglected compared to structure. However, we should mention that although the machinery for simulating membranes has been developed and tested on pure lipid bilayers, the number of applications to more

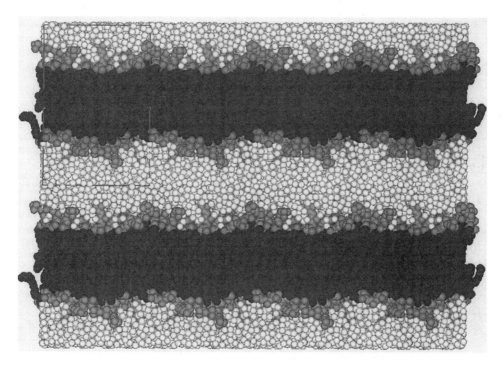

Figure 2 Snapshot from an MD simulation of a multilamellar liquid crystalline phase DPPC bilayer. Water molecules are colored white, lipid polar groups gray, and lipid hydrocarbon chains black. The central simulation cell containing 64 DPPC and 1792 water molecules, outlined in the upper left portion of the figure, is shown along with seven replicas generated by the periodic boundary conditions. (From Ref. 55.)

complicated systems has been growing rapidly. A by no means exhaustive list includes the transport of small molecules (water [13], benzene [14], anesthetics [15,16]) and ions [17] within and across bilayers, cholesterol effects on bilayers [18], and membrane-bound peptides and proteins (surface-bound tripeptides [19,20], transmembrane polyalanine [21], an amphipathic helical peptide [22], an amphipathic segment from the human cortico-trophin releasing factor [23], the lytic peptide mellitin [24], the antimicrobial peptide dermaseptin B [25], bacteriophage Pf1 coat protein [26], the transmembrane domain of ErbB-2 tyrosine kinase receptor [27], and bacteriorhodopsin [28–30]), including pores and ion channels (gramicidin [31–34], *E. coli* OmpF porin [35], alamethicin [36], and the influenza A M2 channel [37]).

II. MOLECULAR DYNAMICS SIMULATIONS OF MEMBRANES

A. System Size and Construction

Algorithms and computer hardware have limited the size of most of the all-atom membrane simulations performed to date to 50–100 lipids plus water. However, with the increasing availability of parallel computers, larger systems containing up to 1000 lipids are starting to be simulated [75], and systems containing more than 100 lipids will soon be routine.

In the typical setup, the lipids are arranged in a bilayer, with water molecules on both sides, in a central simulation cell, or ''box,'' which is then replicated by using three-dimensional periodic boundary conditions to produce an infinite multilamellar system (Fig. 2). It is important to note that the size of the central cell places an upper bound on the wavelength of fluctuations that can be supported by the system.

Because there are no crystal structures of fully hydrated lipid bilayers to start from, the initial configurations for the first membrane simulations had to be constructed ''by hand.'' Many of the early simulations were based on the crystal structure of dimyristoyl-phosphatidylcholine dihydrate [38], in which the DMPC molecules form a bilayer with slightly tilted all-trans hydrocarbon chains. The unit cell was replicated to give the desired number of molecules, and the lipid center-of-mass positions were scaled to give the desired area per lipid. Then the system was hydrated and equilibrated to allow hydrocarbon chains to melt and the water molecules to penetrate through the headgroup regions. A less used approach that reduced the equilibration time was to pack together lipids randomly chosen from a library of configurations generated by simulating a single lipid in a mean field [39]. Woolf and Roux extended this approach by using prehydrated lipids [31]. Presently, most simulations are initiated from the end of a previous simulation, and lipid bilayer coordinates are generally provided upon request by authors of simulation papers. Incorporating solutes into bilayers can be tedious. The simplest approach, which consists of simply deleting enough lipids to create ample space, works well in some cases (e.g., for inserting cholesterol molecules or transmembrane helices), but more sophisticated approaches are more generally useful. The latter include using radial forces to create spherical or cylindrical holes in the bilayer [36], and growing groups of soft spheres at locations that create a cavity in the membrane with the desired size and shape [26].

B. Force Fields

The force fields used in atomistic membrane simulations have the same form as the analytical, empirical molecular mechanics potentials used in classical simulations of proteins and nucleic acids (see Chapter 2). These potential functions contain harmonic terms for deforming bonds and bond angles, periodic and harmonic potentials for torsions, and van der Waals and Coulomb nonbonded interactions. The atomic charges are often obtained from quantum chemical calculations on lipid fragments. Most of the other potential parameters are taken from existing force fields for proteins and nucleic acids, which are generally built up from parametrizations of model compounds. Details on biomolecular force field parametrization are given in Refs. 40 and 41, and in Chapter 2. The quality of lipid force fields may be evaluated with simulations of crystals of lipid fragments [40,42,43] and by checking their ability to reproduce well-established experimental results in simulations of hydrated lipid bilayers (see Section III.A). The level of accuracy that we have been able to achieve is typified by the densities of phospholipid and cholesterol crystals from constant-pressure MD simulations with fully flexible unit cells, plotted in Figure 3 versus the experimental values.

Calculation of the energies and forces due to the long-range Coulomb interactions between charged atoms is a major problem in simulations of biological molecules (see Chapter 5). In an isolated system the number of these interactions is proportional to N^2, where N is the number of charged atoms, and the evaluation of the electrostatic interactions quickly becomes intractable as the system size is increased. Moreover, when periodic

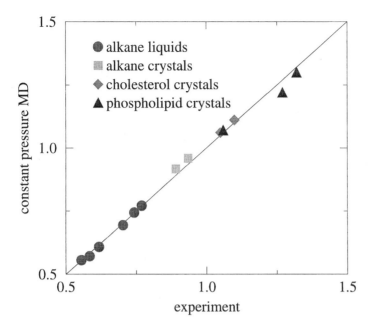

Figure 3 Comparison of the densities (in g/cm^3) of model compounds for membrane lipids computed from constant-pressure MD simulations with the corresponding experimental values. The model compounds include solid octane and tricosane, liquid butane, octane, tetradecane, and eicosane, and the glycerylphosphorylcholine, cyclopentylphosphorylcholine monohydrate, dilaurolyglycerol, anhydrous cholesterol, cholesterol monohydrate, and cholesterol acetate crystals. (Models from Refs. 18, 42, and 43).

boundary conditions are used, the direct sum over all the Coulomb interactions in the periodic system is conditionally convergent. There are two common solutions to these problems. The first, which was used in most biomolecular simulations until recently, is to employ a spherical truncation (''cutoff'') scheme where the electrostatic interactions are smoothly switched off at a cutoff distance (typically around 10 Å). When used in conjunction with a neighbor list, the computational work is proportional to N. An alternative approach, which is essentially exact for crystals, is to include all of the electrostatic interactions by using the Ewald method (or a related technique) for summing long-range interactions in a rapidly convergent fashion in an infinite periodic system. The disadvantages of the Ewald method are that it might enhance the artificial periodicity of a finite system replicated by periodic boundary conditions and that more computational effort is necessary compared to cutoffs (the straightforward implementation scales as $N^{3/2}$). The latter deterred most biomolecular simulators from using the Ewald method until recently, when the particle mesh Ewald method (PME) [44] was introduced. In PME, the reciprocal space part of the Ewald sum is evaluated using fast Fourier transforms, resulting in an overall $N \log N$ scaling. The potential deleterious effects of the strict periodicity that results when lattice sums are used have not been assessed in membrane simulations. However, it has been explicitly demonstrated that the use of spherical truncation introduces serious artifacts into a variety of structural, thermodynamic, and transport properties of interfacial systems and lipid bilayers [45], even when large cutoff radii are used.

C. Ensembles

A molecular dynamics simulation consists of numerically solving the equations of motion of a set of particles (atoms), given the forces on the particles. Classical MD simulations that solve Newton's equations of motion generate trajectories belonging to the microcanonical [constant number of particles, volume, and energy (NVE)] statistical mechanical ensemble. It is generally desirable to perform simulations in other ensembles such as the isobaric-isothermal ensemble (constant NPT). In addition to being the natural choice for correspondence with typical experimental conditions, the NPT ensemble is useful for validating force fields by checking their ability to reproduce important structural parameters known from experimental measurements, such as the surface area per lipid, the interlamellar spacing, and the membrane thickness, and for predicting these quantities when they are not known (e.g., in membrane–protein systems). Constant pressure and temperature are enforced in simulations by controlling the fluctuations of the particle kinetic energy and system volume, respectively, and there are various ways to do this. The best algorithms, in terms of their ability to rigorously generate the NPT ensemble, are based on the "extended system" approach [46], in which additional dynamic variables are introduced, for example a time-dependent friction coefficient ("thermostat") to control the temperature and a piston to control the pressure. The equations of motion and a conserved energy are consistently formulated so that the microcanonical distribution function for the extended phase space gives the isobaric-isothermal distribution function for the particles after integration over the additional dynamic variables [47].

It is now widely accepted that it is best to simulate membranes at constant pressure, but there is some disagreement concerning the assumed form of the pressure tensor, i.e., whether the pressure should be the same in all directions (isotropic) or whether the pressure in the plane of the membrane should be different from the pressure perpendicular to the membrane [11]. The latter is appropriate for a true interfacial system, such as the air/water and air/hydrocarbon interfaces, with a significant surface tension. Because membrane/water interfaces have either a vanishing or very small surface tension [48], in principle membranes should be simulated with an isotropic pressure tensor. However, some simulators have argued that to correct for finite size effects in simulations of small membrane patches, a modest surface tension should be imposed [49]. As larger and larger bilayer patches are being simulated ($\approx$1000 lipids), it is becoming evident that there are finite-size effects in systems containing $\approx$100 lipids. For example, in a recent comparison of bilayers simulated at constant isotropic pressure, the area per lipid in a system of 100 lipids was about 3% less than in a system of 1000 lipids [75]. However, the explanation of the origin of this contraction and the best remedy for avoiding it in small systems (1000 lipids is still out of reach for most simulators) are debatable. On the one hand, one could argue that a surface tension should be applied to stretch a small membrane patch to counteract the shrinkage due to finite-size effects. However, an imposed surface tension can be expected to change the spectrum of fluctuations of the interface (in effect, stiffening the interface). On the other hand, one could impose constant isotropic pressure, keeping in mind the systematic error that results from finite-size effects. The results presented in this chapter were obtained by constant isotropic pressure calculations on bilayers containing 64 lipids [50,51], and hence the areas per lipid reported here are likely a couple of percent below the large system limit for the potential function employed.

D. Time Scales

Until recently, membrane simulations were generally limited to a couple of nanoseconds duration at most, and many of the simulations reported in the literature are less than a nanosecond long. As a result of this time scale limit, along with the size limit of roughly 100 lipids, many interesting phenomena occurring in membranes, such as lateral diffusion and complete rotation of lipids, collective undulations, bilayer phase transitions, and lateral phase separation in multicomponent membranes, to name a few, were well beyond the reach of the first generation of membrane simulations. Nonetheless, the early membrane simulations were useful for studying interactions between lipids, water, and membrane proteins at the atomic level, the complicated rearrangements of individual lipids, and the diffusion of water and other small molecules in and near membranes [11,12]. In addition to permitting larger systems to be studied, parallel computing is presently extending the accessible time scale by an order of magnitude. Indeed, simulations of 10 ns duration are appearing, and these are enabling additional phenomena (e.g., the early events in long-range lateral diffusion, undulations) to be characterized in unprecedented detail [10,52,75].

III. LIPID BILAYER STRUCTURE

A. Overall Bilayer Structure

We compare experimental results on DPPC bilayer dimensions and organization with our simulation results in Table 1 and the average locations of individual methyl and methylene groups along the bilayer normal in Figure 4, to demonstrate the level of accuracy that we have been able to achieve, which is considered quite high by present standards. Our simulations, which contain 64 lipids and numbers of water molecules corresponding to full hydration under the specified conditions, are described in greater detail in Refs. 50 and 51.

B. Density Profiles

The snapshot from a fluid bilayer simulation shown in Figure 2 reveals that the bilayer/ water interface is quite rough and broad on the scale of the diameter of a water molecule.

Table 1 Comparison of MD and X-Ray Diffraction Results for Structural Parameters of Fully Hydrated DPPC Bilayers

Quantity	Gel phase (19°C, 12 water molecules/lipid)		Liquid crystal phase (50°C, 28 water molecules/lipid)	
	MD [51]	X-ray [70,71]	MD [50]	X-ray [5]
Area/lipid (Å^2)	45.8	47.2	61.8	62.9
Interlamellar spacing (Å)	65.2	63.4	67.3	67.2
Bilayer thickness (Å)	45.6	45.0	37.2	36.4
Chain tilt angle (°)	33.6	32.0		
Chain lattice parameters a, b (Å)	8.6, 5.5	8.5, 5.6		

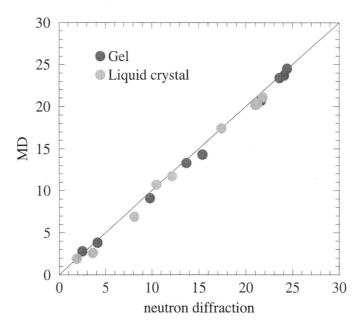

Figure 4 Comparison of average distances from the bilayer center along the bilayer normal for deuterated methyl and methylene groups distributed throughout the DPPC molecule computed from constant-pressure MD calculations and neutron diffraction measurements on gel and liquid crystalline phase DPPC bilayers.

The first atomic scale picture of the average structure of the interface was produced by measurement of the density distributions of different types of atoms along the bilayer normal by a combined neutron and X-ray diffraction study of a phosphatidylcholine bilayer at low hydration [6]. The corresponding picture for fully hydrated bilayers has been provided by MD simulations and is exemplified by our results for the electron density profiles of the liquid crystalline phase of DPPC shown in Figure 5. Defining the bilayer thickness as the distance between the peaks in the total electron density, we obtain 37.2 Å, which is in reasonable agreement with the values determined by X-ray diffraction analysis [5]. In Figure 5b we show the decomposition of the electron density profile into contributions from the lipid polar groups. The chemical heterogeneity of the membrane/water interface is clearly evident in the overlapping distributions of the lipid polar groups and water molecules in the broad interfacial region connecting the bulk water in the middle of the interlamellar space to the hydrocarbon in the middle of the bilayer. In Figure 5a the total contributions from the water and the lipid show that the water density decays smoothly from the bulk value roughly 30 Å from the bilayer center and penetrates deeply into the bilayer. This is in contrast to the water density oscillations observed next to flat hydrophobic surfaces [53] and the relatively narrow, noninterpenetrating air/water and hydrocarbon/water interfaces [54]. Defining the interface as the range over which the water density goes from 90% to 10% of its bulk value, we find that the two interfaces occupy 23 Å, which is more than half of the total bilayer thickness. Thus, the membrane should be thought of as a broad hydrophilic interface, with only a thin slab of pure fluid hydrocarbon in the middle.

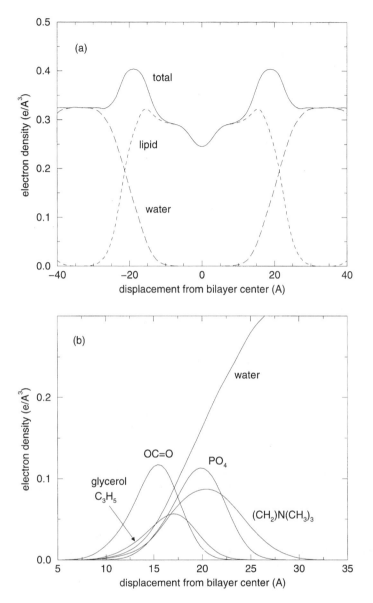

Figure 5 Electron density distributions along the bilayer normal from an MD simulation of a fully hydrated liquid crystalline phase DPPC bilayer. (a) Total, lipid, and water contributions; (b) contributions of lipid components in the interfacial region.

C. Solvation of the Lipid Polar Groups

To discuss the interactions between water molecules and specific lipid polar groups, we consider radial distribution functions for the water oxygen atoms surrounding the phosphate P, choline N, and carbonyl C atoms (Fig. 6). The P—O $g(r)$ has a sharp first peak at 3.8 Å, indicating tight solvation of the negatively charged phosphate by an average of four water molecules [obtained by integrating $g(r)$ to the first minimum]. Inspection of

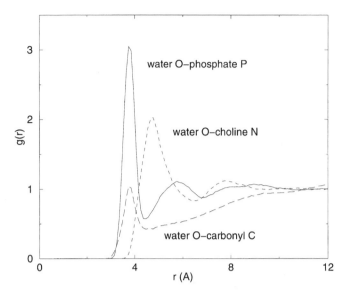

Figure 6   Radial distributions of water oxygen atoms around sites in the polar groups in a DPPC bilayer.

the individual phosphate O–water O $g(r)$ functions (not shown; see Ref. 55) reveals that the water molecules are almost exclusively associated with the unesterified phosphate oxygens. The N—O $g(r)$ displays a broader first peak at 4.8 Å corresponding to 15 water molecules in the first solvation shell of the positively charged choline ammonium group. The relatively inaccessible carbonyl groups at the glycerol–fatty acyl ester linkages are solvated by an average of only 1.5 water molecules each.

The radial distributions of water around the lipid polar groups all contain well-defined first peaks signaling the existence of well-defined solvation shells. In subsequent analyses, we distinguish the water molecules that are closely associated with lipids by defining "bound" waters as those that fall within the first peaks of the water–phosphate, water–choline, or water–carbonyl $g(r)$ functions [55]. Thus, on average, roughly four waters are bound to the phosphate ("P-bound"), 15 to the choline ("N-bound"), and 1.5 to each carbonyl group ("CO-bound"). Some of these bound waters are shared between the polar groups, and overall roughly half of the 28 waters per lipid in the fully hydrated fluid phase bilayer are considered bound to lipids by our definition. We distinguish these bound waters from the "bulk" waters, a slab of water molecules roughly two to three water molecules thick in the middle of the interlamellar space, which we define as being more than 30 Å from the bilayer center. Of course, our definitions of bound and bulk water molecules are somewhat arbitrary, as are alternative definitions based on interaction energies, residence times, etc. However, we see in the next section that these definitions are useful for showing that the dynamics of water molecules near membrane surfaces depend on the strength of their interactions with specific lipid polar groups.

D.   Water Orientational Polarization and the Membrane Dipole Potential

In membranes containing phospholipids such as DPPC, the negatively charged phosphate groups exert a strong influence on the structure of the water molecules. As the unesterified

P—O bonds are preferentially oriented, on average pointing away from the membrane and into the water, the P-bound water molecules tend to orient their dipoles with their positive ends pointing toward the negative phosphates, resulting in a net orientational "polarization" [55]. The orienting power of the phosphate becomes clear when one observes that the orientational polarization is much less pronounced at water interfaces with lipid components that do not contain phosphate groups, e.g., glycerol, decane, and decyl-β-glycoside [54].

These orientationally polarized water molecules appear to have a profound influence on the electrostatic properties of membranes. Experimental measurements on lipid monolayers and bilayers have demonstrated that there is an electric potential difference across lipid/water interfaces, typically a few hundred millielectronvolts, negative on the water side relative to the hydrocarbon [56,57]. Thus, for the purpose of describing the electrostatics, the membrane can be thought of as a planar array of dipoles whose negative ends point toward the water [58]. The molecular origins of the resulting "dipole potential" are of interest because the electrical properties of the bilayer surface influence the binding and passive transport of charged species. Experiments on phospholipid bilayers suggest that the primary negative contributions to the dipole potential arise from oriented water molecules [57] and, to a lesser extent, from the carbonyl groups in the acyl ester linkages [58].

The dipole potential can be easily calculated from a simulation as a double integral of the average charge density (for example, see Ref. 59). The total dipole potential profile from our DPPC bilayer simulation plotted in Figure 7 monotonically decreases in the membrane/water interface to a value in the bulk water of about −500 mV. The lipid contribution is nonzero only in the interfacial region. The negative lipid contribution on the hydrocarbon side of the interface is canceled by a positive contribution on the water side, while the water contribution is monotonically decreasing throughout the interface. Thus, we conclude that the net dipole potential is due primarily to an excess of water

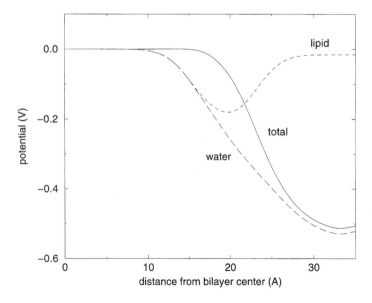

Figure 7 The electric potential relative to the hydrocarbon ("dipole" potential) as a function of distance from the center of a fully hydrated DPPC bilayer.

molecules oriented with their dipoles pointing toward the membrane surface. Qualitatively similar results have been obtained from other simulations of DPPC bilayers [52,54], but different simulations do not agree on the extent of the lipid contribution.

IV. MOLECULAR DYNAMICS IN MEMBRANES

A. Overview of Dynamic Processes in Membranes

As in other biomolecular assemblies, functionally relevant motions in membranes span a wide range of length scales and time scales [9,60]. Motions that have been detected experimentally include isomerizations in the hydrocarbon chains (10–100 ps), single lipid protrusions (10 ps to 1 ns), reorientation of the headgroups (1 ns), rotations of the lipid molecules about their long axes (10 ns), collective bilayer undulations (>10 ns), and long-range lipid diffusion in the plane of the bilayer (10 ns to 1 s, depending on the length scale). Some of these are presently accessible to MD simulations, and more will become accessible as membrane simulations are extended to longer times. Detailed analyses of dynamics in membrane simulations have been relatively scarce compared to structural analyses. Pastor and Feller [61] used short simulations and model calculations to analyze motions over a range of time scales, including chain, headgroup, and glycerol isomerizations, lateral diffusion, and whole lipid wobbling motion. More recently, Essmann and Berkowitz [52] studied lipid center-of-mass diffusion and rotational dynamics using a 10 ns MD simulation, and Feller et al. [10] used a 10 ns simulation to elucidate NOESY cross-relaxation observed in two-dimensional NMR experiments on lipid bilayers. In the remainder of this chapter, we analyze lipid and water motions on time scales up to 100 ps from our simulation of a liquid crystalline phase DPPC bilayer. Our discussion emphasizes a detailed comparison with experimental results, including the first close comparison with, and interpretation of, neutron scattering experiments that probe motions on the same time scale as the simulation.

B. Qualitative Picture on the 100 ps Time Scale

A qualitative picture of lipid and water dynamics in a membrane is given in Figure 8, where we show configurations of water and lipid molecules at intervals spanning a period of 100 ps. On this time scale it is clear that the lipid centers of mass are not freely diffusing laterally, nor are they undergoing large excursions in the direction normal to the bilayer due to long-wavelength undulations; rather, they are "rattling in a cage" formed by their neighbors, with roughly equal amplitudes parallel and perpendicular to the bilayer plane. In addition to the isotropic center-of-mass motion, there is a considerable amount of motion in the internal degrees of freedom, especially in the acyl chains and, to a lesser degree, in the choline groups, involving the formation and disappearance of gauche bonds, gauche–trans–gauche kinks, and other chain defects. Although the details vary considerably among the individual water molecules, it is evident from Figure 8 that, overall, the "bound" waters display less motion than the "bulk" waters and that the waters bound to the lipid phosphate and carbonyl groups are less mobile than those bound to the choline groups. Moreover, the trajectory of the bulk water molecule shown suggests that the motion of the bulk waters is anisotropic. In the remainder of this chapter, we expose these qualitative observations concerning the lipid and water dynamics on the 100 ps time scale in greater detail and discuss our results in relation to neutron scattering and NMR experiments.

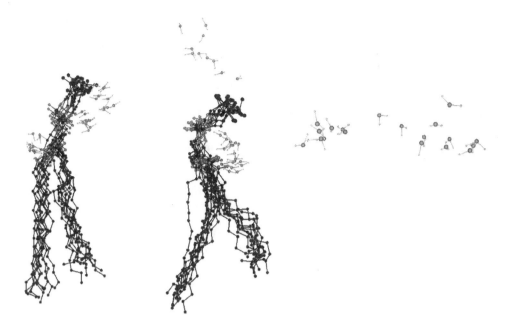

Figure 8 Configurations of lipid and water molecules spanning a 100 ps interval during an MD simulation of a DPPC bilayer. The two left-hand panels show 10 configurations of two different lipids and three of their associated water molecules (one N-bound, one P-bound, and one CO-bound). The right-hand panel shows 20 configurations of a "bulk" water molecule in the interlamellar space of a bilayer stack. (From Ref. 55.)

C. Incoherent Neutron Scattering Measurements of Lipid Dynamics

Neutrons produced by present research-oriented sources typically have wavelengths on the order of angstroms and energies of a few millielectronvolts (1 meV $\approx$ 8 cm^{-1}) and hence are well suited for probing the structure and dynamics of molecules. Neutrons scatter from nuclei, and the total scattering is the sum of "coherent" and "incoherent" contributions. Coherent scattering depends on the relative positions and motions of the nuclei and thus conveys information on the structure and collective dynamics, whereas incoherent scattering reflects the motion of individual nuclei, i.e., self-correlations or single-particle dynamics. The incoherent scattering cross section of hydrogen is much larger than all the other cross sections in organic molecules. Therefore, because hydrogen atoms are uniformly distributed in biological molecules, incoherent neutron scattering (INS) is particularly useful for studying the global molecular dynamics in biological systems. Moreover, selective deuteration provides a powerful mask for isolating the dynamics of selected parts of the system (e.g., to separate water and biomolecule contributions). An energy-resolved spectrum typically has three regions: an elastic peak arising from atoms that move slowly compared to the resolution of the experiment, inelastic peaks due to scattering by normal modes, and quasielastic scattering, which appears as a broadening of the elastic peak and is due to diffusive motions. The range of energy and momentum transfers accessible on presently available neutron spectrometers overlaps well with the duration and size of MD simulations that are presently routine for biological molecules. Thus, INS data are a valuable resource for testing the dynamics produced by simulations, and simulations constitute a potentially valuable tool for interpreting INS data.

Most neutron spectroscopic experiments essentially measure the total dynamic structure factor, $S_{tot}^{meas}(\mathbf{Q}, \omega)$, in which $\mathbf{Q}$ and $\hbar\omega$ are the momentum and energy transfers, respectively. The measured structure factor is the sum of coherent and incoherent contributions. However, because the incoherent scattering length of hydrogen is an order of magnitude larger than the scattering lengths of all the other atoms in lipids and proteins, we will assume that the coherent contribution is negligible, so that $S_{tot}^{meas}(\mathbf{Q}, \omega) = S_{inc}^{meas}(\mathbf{Q}, \omega)$. In practice, the unavoidable spread in energies of the neutrons incident on the sample results in a finite energy resolution, and the measured spectrum is a convolution of the true spectrum, $S_{inc}(\mathbf{Q}, \omega)$, and the instrumental resolution function, $R(\omega)$:

$$S_{inc}^{meas}(\mathbf{Q}, \omega) = S_{inc}(\mathbf{Q}, \omega) \otimes R(\omega) \tag{1}$$

where $\otimes$ denotes a convolution product. The width of the resolution function determines the time scale of the dynamics probed by the instrument in a nontrivial way, with narrower widths (higher resolution) corresponding to longer observation times.

From the theory of neutron scattering [62], $S_{inc}(\mathbf{Q}, \omega)$ may be written as the Fourier transform of a time correlation function, the "intermediate scattering function," $I_{inc}(\mathbf{Q}, t)$:

$$S_{inc}(\mathbf{Q}, \omega) = \frac{1}{2\pi} \int_{-\infty}^{\infty} I_{inc}(\mathbf{Q}, t)e^{-i\omega t}dt \tag{2}$$

$$I_{inc}(\mathbf{Q}, t) = \frac{1}{N}\sum_{j=1}^{N} \langle e^{i\mathbf{Q}\cdot\mathbf{r}_j(t)}e^{-i\mathbf{Q}\cdot\mathbf{r}_j(0)} \rangle \tag{3}$$

Here $\mathbf{r}_j$ is the position operator of atom j, or, if the correlation function is calculated classically as in an MD simulation, $\mathbf{r}_j$ is a position vector; N is the number of scatterers (i.e., H atoms); and the angular brackets denote an ensemble average. Note that in Eq. (3) we left out a factor equal to the square of the scattering length. This is convenient in the case of a single dominant scatterer because it gives $I(\mathbf{Q}, 0) = 1$ and $S_{inc}(\mathbf{Q}, \omega)$ normalized to unity.

The intermediate scattering function, $I_{inc}(\mathbf{Q}, t)$, is readily computed from an MD trajectory by using Eq. (3), and the result may be numerically Fourier transformed to give $S_{inc}(\mathbf{Q}, \omega)$. For the purpose of quantitatively comparing the result to neutron scattering data obtained on a given spectrometer, the instrumental resolution should be taken into account. Instrumental resolution functions are generally represented by a Gaussian or other peaked function centered at $\omega = 0$, with width $\Delta\omega$ (or $\Delta E = \hbar\Delta\omega$). Noting that a convolution in the frequency domain is equivalent to a product in the time domain, a resolution-broadened spectrum, $S_{inc}^{meas}(\mathbf{Q}, \omega)$, is obtained by Fourier transforming the product, $I_{inc}(\mathbf{Q}, t)R(t)$, where $R(t)$ is the Fourier transform of $R(\omega)$. It is instructive to look at $R(t)$ itself because it provides a direct measure of the longest time scale probed by a measurement at a given resolution. For example, the time-of-flight spectrometer IN5 at the Institut Laue Langevin in Grenoble has a resolution function that is well represented by a Gaussian with a full width at half-maximum (FWHM) ΔE value of 0.050 meV. Thus, $R(t)$ is also a Gaussian, with an FWHM of approximately 30 ps, and the product $I_{inc}(\mathbf{Q}, t)R(t)$ does not become negligible until $t \approx 100$ ps. Because $R(t)$ is generally a peaked function, there is no well-defined averaging time corresponding to a particular energy resolution, and motions with short correlation times will be weighted more heavily that those with long correlation times in the product $I_{inc}(\mathbf{Q}, t)R(t)$.

The **Q** and ω dependence of neutron scattering structure factors contains information on the geometry, amplitudes, and time scales of all the motions in which the scatterers participate that are resolved by the instrument. Motions that are slow relative to the time scale of the measurement give rise to a δ-function "elastic" peak at ω = 0, whereas diffusive motions lead to "quasielastic" broadening of the central peak and vibrational motions attenuate the intensity of the spectrum. It is useful to express the structure factors in a form that permits the contributions from vibrational and diffusive motions to be isolated. Assuming that vibrational and diffusive motions are decoupled, we can write the measured structure factor as

$$S_{\text{inc}}^{\text{meas}}(Q, \omega) = \exp[-Q^2 \langle u^2 \rangle] S_{\text{inc}}^{\text{diff}}(Q, \omega) \otimes R(\omega) + B \tag{4}$$

where the Debye–Waller exponential factor represents the attenuation of the intensity due to vibrational motions with mean-square amplitudes $\langle u^2 \rangle$, $S_{\text{inc}}^{\text{diff}}(Q, \omega)$ is the structure factor corresponding to the diffusive motions, and B is a constant background that may arise from coherent scattering and other, extraneous, scattering particular to the experimental setup. In Eq. (4) and the remainder of this chapter, we write the structure factor as a function of the magnitude of the momentum transfer Q, because we are interested in experiments in which the spectra are "powder averaged" (as opposed to single crystals).

Specification of $S_{\text{inc}}^{\text{diff}}(Q, \omega)$ requires models for the diffusive motions. Neutron scattering experiments on lipid bilayers and other disordered, condensed phase systems are often interpreted in terms of diffusive motions that give rise to an elastic line with a Q-dependent amplitude and a series of Lorentzian quasielastic lines with Q-dependent amplitudes and widths, i.e.,

$$S_{\text{inc}}^{\text{diff}}(Q, \omega) = A_0(Q)\delta(\omega) + \sum_{i=1}^{n} A_i(Q)L_i(\Gamma_i(Q), \omega) \tag{5}$$

where $L_i(\Gamma_i(Q), \omega)$ is a Lorentzian centered at $\omega = 0$ with half-width at half-maximum $\Gamma_i(Q)$:

$$L_i(\Gamma_i(Q), \omega) = \frac{1}{\pi}\left(\frac{\Gamma_i(Q)}{\Gamma_i(Q)^2 + \omega^2}\right) \tag{6}$$

The amplitude of the elastic scattering, $A_0(Q)$, is called the elastic incoherent structure factor (EISF) and is determined experimentally as the ratio of the elastic intensity to the total integrated intensity. The EISF provides information on the geometry of the motions, and the linewidths are related to the time scales (broader lines correspond to shorter times). The Q and ω dependences of these spectral parameters are commonly fitted to dynamic models for which analytical expressions for $S_{\text{inc}}^{\text{diff}}(Q, \omega)$ have been derived, affording diffusion constants, jump lengths, residence times, and so on that characterize the motion described by the models [62].

D. Comparison of MD and Neutron Scattering Results on Lipid Dynamics

Clearly, the best way to assess the ability of MD simulations to reproduce neutron scattering results is to compare measured and computed spectra directly, one on top of the other.

However, this is generally not possible because measured spectra are not available to simulators. The spectral parameters (EISFs and line widths) along with the parameters of the dynamic models used to fit their Q and ω dependence are what are generally reported in the literature. Quantities such as diffusion constants are often computed from MD simulations and their values compared to those extracted from neutron data. For the comparison to be meaningful, the same models must be used to analyze the simulations and the experiments. However, as we discuss in more detail later, this is often not the case. It is actually easier, and certainly more appropriate, to directly compare the spectral parameters when they are available. Then, once the ability of the simulations to reproduce the data has been assessed, the simulations can be used to discuss the validity of the dynamic models used to fit the spectra and/or to inspire new models for interpreting experimental data. In the remainder of this section we show how the spectral parameters can be derived from simulations, and we compare our simulations to the results of neutron scattering experiments published by König et al. [63,64] on DPPC bilayers. In subsequent sections we discuss models used to describe lipid dynamics in membranes on time scales up to 100 ps.

For the comparison to be meaningful, it is essential to compute spectra from the MD trajectory that are broadened by the resolution of the experiment to which they will be compared and to process them by the same procedure as that used in the experimental data reduction. The processing consists of determining $\langle u^2 \rangle$, dividing the resolution-broadened spectra by the Debye–Waller factor, and fitting the result to an elastic line plus a sum of Lorentzians. Following König et al., we use two Lorentzians in our fits. The quality of the fits is illustrated in Figure 9, where we show a typical spectrum computed for the lipid H atoms in our simulation of a fluid-phase DPPC bilayer. The value of $\langle u^2 \rangle$ is deter-

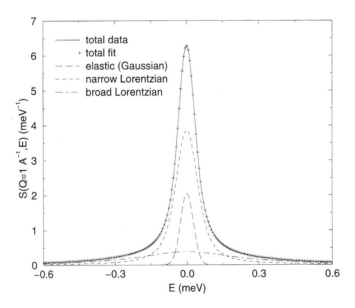

Figure 9 Fit of an incoherent neutron scattering structure factor, $S(Q, \omega)$, computed for lipid H atom motion in the plane of the bilayer in a simulation of a DPPC bilayer, by the sum of an elastic line, a narrow Lorentzian with width Γ_1, and a broad Lorentzian with width Γ_2, convoluted with a Gaussian resolution function with $\Delta E = 0.050$ meV.

mined, as in the experimental analysis, as the slope of a plot of integrated intensity versus Q^2, according to the relationship

$$\int_{\omega_{min}}^{\omega_{max}} S_{inc}^{diff}(Q, \omega)d\omega = \exp[-Q^2\langle u^2 \rangle] \tag{7}$$

For lipid H atom motions in the plane of the bilayer, using an integration range -1.5 meV $< E < 1.5$ meV, we obtain $\langle u^2 \rangle = 0.24$ Å^2, which is about twice as large as the 0.11 Å^2 reported by König [65].

In the remainder of this section, we compare EISFs and Lorentzian line widths from our simulation of a fully hydrated liquid crystalline phase DPPC bilayer at 50°C with experiments by König et al. on oriented bilayers that, in order to achieve high degrees of orientation, were not fully hydrated. We consider two sets of measurements at 60°C on the IN5 time-of-flight spectrometer at the ILL: one in which the bilayer preparations contained 23% (w/w) pure D_2O and another in which bilayer orientation was preserved at 30% D_2O by adding NaCl. The measurements were made on samples with two different orientations with respect to the incident neutron beam to probe motions either in the plane of the bilayers or perpendicular to that plane.

The EISFs for motions in the plane and perpendicular to the plane of the bilayer are compared in Figures 10a and 10b, respectively. The EISFs from the simulations at 50°C and 43% hydration agree well with those measured at 60°C and 30% hydration. The experimental EISFs at the lower hydration decay more slowly with Q, indicating more restricted motion. This is reasonable, as we expect the surface area per lipid, and hence the range of lipid mobility, to decrease with dehydration. Carefully comparing Figures 10a and 10b, we can see that the experimental EISFs at a given hydration level are very similar for motions in and perpendicular to the plane of the bilayer, indicating that the motion is isotropic. In contrast, the MD EISF for in-plane motion decays slightly faster with Q than that for motion perpendicular to the bilayer, suggesting slightly more mobility in the plane of the bilayer. Unfortunately, we cannot tell at this time if this is a minor fault of the simulation or if it is a correct prediction that there is a slight anisotropy to the motion at full hydration.

The widths of the narrow Lorentzians representing slow motions in the plane and perpendicular to the plane of the bilayer are compared in Figures 11a and 11b, respectively. For the in-plane motion, the MD values for $Q = 0.5$ Å^{-1} agree well with the experimental results, but the increase with Q is significantly overestimated in the simulation compared to the experimental values. This suggests that the slower component of the in-plane motion in the simulation is too fast at short distances. On the other hand, the MD line widths for the slower component of the out-of-plane motion agree well with the experimental results at 30% hydration. As in the case of the EISF, the simulation predicts a slight anisotropy not seen in the experimental data.

The widths of the broad Lorentzians representing fast motions in the plane and perpendicular to the plane of the bilayer are compared in Figures 12a and 12b, respectively. Only data at the lower hydration (23%) are available for comparison, and these agree well with the MD results, which show a slow, monotonic increase with Q. Although we expect the fast process to be at most only weakly dependent on hydration, it is not clear to what extent the comparison validates the simulation.

Overall the MD results at 50°C and 43% hydration agree well with the neutron

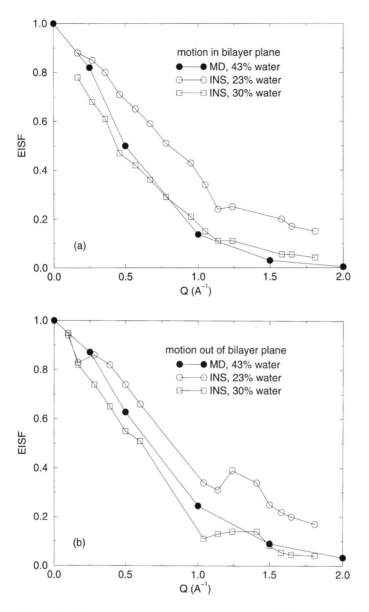

Figure 10 Elastic incoherent structure factors for lipid H atoms obtained from an MD simulation of a fully hydrated DPPC bilayer, and quasielastic neutron scattering experiments on DPPC bilayers at two hydration levels for (a) motion in the plane of the bilayer and (b) motion in the direction of the bilayer normal.

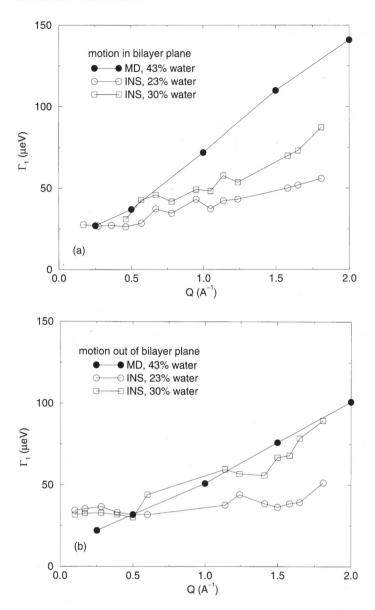

Figure 11 Widths of the narrow Lorentzian components fit to structure factors for lipid H atoms obtained from an MD simulation of a fully hydrated DPPC bilayer and quasielastic neutron scattering experiments on DPPC bilayers at two hydration levels for (a) motion in the plane of the bilayer and (b) motion in the direction of the bilayer normal. The error bars on the experimental points are approximately ±5 μeV.

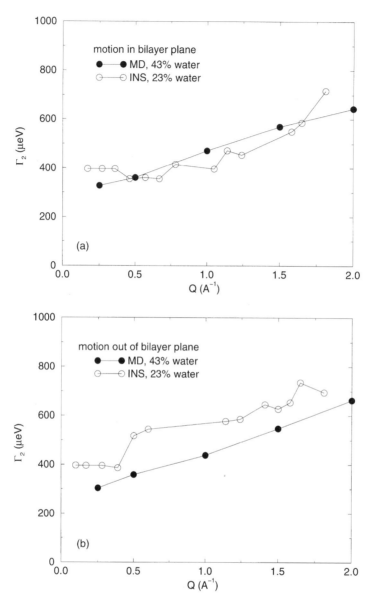

Figure 12 Widths of the broad Lorentzian components fit to structure factors for lipid H atoms obtained from an MD simulation of a fully hydrated DPPC bilayer and quasielastic neutron scattering experiments on DPPC bilayers containing 23% water for (a) motion in the plane of the bilayer and (b) motion in the direction of the bilayer normal. The error bars on the experimental points are approximately ± 150 μeV.

results at 60°C and 30% hydration. Although it is desirable to compare simulation and experimental results under identical conditions, this is the most direct comparison we can make given the available data. We conclude from this comparison that the lipid dynamics reflected in H atom motions on time scales up to 100 ps are reasonably well reproduced by the simulation. The most significant discrepancy is that the time scale of the slower diffusive process decreases too quickly with Q for lipid motions in the plane of the bilayer.

In the stroboscopic picture in Fig. 8, the lipid dynamics appears to consist of two predominant dynamical processes on the 100 ps time scale: (1) intramolecular motions (librations, conformational transitions) superimposed upon (2) a rattling motion of the whole molecules in a confined space ("cage"). The center-of-mass motion appears to be essentially isotropic, with similar amplitudes in and perpendicular to the plane of the bilayer. The intramolecular motion primarily involves the creation and disappearance of a variety of conformational defects (gauche conformers, kinks, etc.) in the hydrocarbon chains. It is clear that on the 100 ps time scale there is no long-range diffusion and no significant rotational motion of the lipids around their long axes. This qualitative description of the lipid dynamics had been proposed, before simulation pictures such as those in Fig. 8 were available, by attempting to fit four dynamic models to neutron time-of-flight data [63]: (I) rotation of the whole molecule plus out-of-plane diffusion inside a box (i.e., with a fixed maximum amplitude); (II) diffusion of each proton inside a sphere (with different protons having different sphere volumes) superimposed on diffusion of the center of mass inside a cylinder; (III) diffusion of kinks in the hydrocarbon chains superimposed on diffusion of the whole molecule inside a cylinder; (IV) same as (III), except that rather than diffusing along the chains, the kinks stochastically appear and disappear. Model I was ruled out, but models II, III, and IV were all consistent with the data. Thus, the picture of lipid dynamics on the 100 ps time scale that has emerged from the neutron time-of-flight measurements is somewhat ambiguous, in the sense that physically distinct dynamic models fit the data equally well.

Having demonstrated that our simulation reproduces the neutron data reasonably well, we may critically evaluate the models used to interpret the data. For the models to be analytically tractable, it is generally assumed that the center-of-mass and internal motions are decoupled so that the total intermediate scattering function can be written as a product of the expression for the center-of-mass motion and that for the internal motions. We have confirmed the validity of the decoupling assumption over a wide range of Q (data not shown). In the next two sections we take a closer look at our simulation to see to what extent the dynamics is consistent with models used to describe the dynamics. We discuss the motion of the center of mass in the next section and the internal dynamics of the hydrocarbon chains in Section IV.F.

E. Lipid Center-of-Mass "Diffusion"

Center-of-mass translational motion in MD simulations is often quantified in terms of diffusion constants, D, computed from the Einstein relation,

$$\langle r^2(t) \rangle = 2dDt \tag{8}$$

where $\langle r^2(t) \rangle$ is the mean-square displacement (MSD) in d dimensions. The application of this relation is certainly valid for liquids in which, at long times, the molecules undergo Brownian dynamics and the MSDs display the required linear increase in time, but it is questionable for lipids on time scales of at least hundreds of picoseconds (likely much

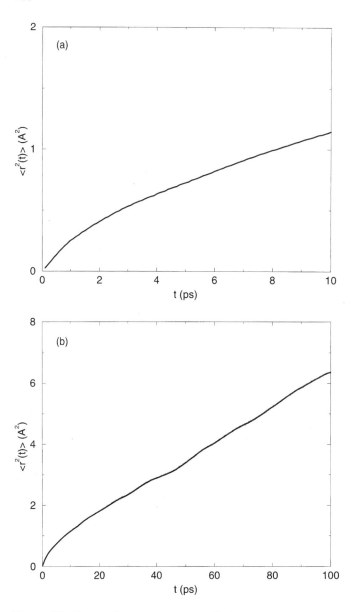

Figure 13 Center-of-mass mean-square displacements computed from MD simulations at 323 K. (a) DPPC motion in the plane of a lipid bilayer averaged over 10 ps; (b) DPPC motion in the plane of a lipid bilayer averaged over 100 ps; (c) comparison of the DPPC in-plane mean-square displacement to linear and power law functions of time; (d) comparison of the center-of-mass mean-square displacement from an MD simulation of liquid tetradecane to a linear function of time.

longer), where it has been deduced from neutron scattering measurements that the center-of-mass motion is confined [63]. The anomalous (non-Brownian) diffusion of the lipid centers of mass is evident in the MSDs plotted in Figure 13. At first glance it appears that the MSDs shown in Figures 13a and 13b approach a linear time dependence on the 10 ps and 100 ps time scales, respectively. However, upon close inspection it is evident

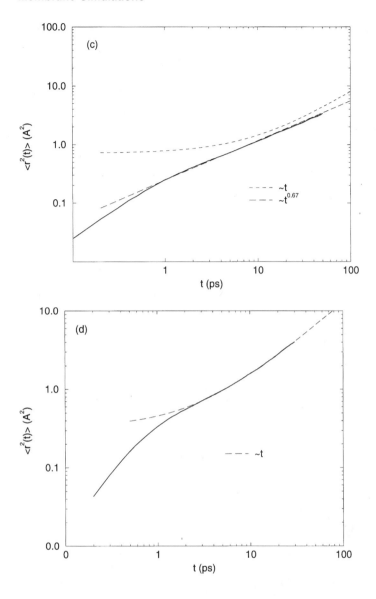

that the slopes of the MSDs are changing in time. In fact, it is clear from Figure 13c that the MSD shows a power law, or fractal, time dependence:

$$\langle r^2(t) \rangle \sim t^\alpha \tag{9}$$

where for lipid center-of-mass motion in the plane of the bilayer, we obtain $\alpha = 0.67$. This is in contrast to the situation in liquid tetradecane, in which the molecules undergo normal diffusion, with a linear time dependence of the MSD setting in after about 3 ps (see Fig. 13d). Application of the Einstein relation to the DPPC MSD in Figure 13c would give a diffusion constant that depends on the length of the MSD used to calculate it and hence is ill-defined. The anomalous lipid diffusion is reminiscent of the cage effect that is universal in supercooled liquids and glasses [66]. In light of the observation that lipids begin to change place only after several nanoseconds in fluid phase bilayers [52], it is likely that Brownian lateral motion of lipids will not be observed until well beyond 10 ns.

Analysis of neutron data in terms of models that include lipid center-of-mass diffusion in a cylinder has led to estimates of the amplitudes of the lateral and out-of-plane motion and their corresponding diffusion constants. It is important to keep in mind that these diffusion constants are not derived from a Brownian dynamics model and are therefore not comparable to diffusion constants computed from simulations via the Einstein relation. Our comparison in the previous section of the Lorentzian line widths from simulation and neutron data has provided a direct, model-independent assessment of the integrity of the time scales of the dynamic processes predicted by the simulation. We estimate the amplitudes within the cylindrical diffusion model, i.e., the length (twice the out-of-plane amplitude) L and the radius (in-plane amplitude) R of the cylinder, respectively, as follows:

$$L = 2\langle \Delta z^2 \rangle \tag{10}$$

$$R = (\langle \Delta x^2 \rangle + \langle \Delta y^2 \rangle)^{1/2} \tag{11}$$

where $\langle \Delta x^2 \rangle$, $\langle \Delta y^2 \rangle$, and $\langle \Delta z^2 \rangle$ are the mean-square fluctuations in the x, y, and z directions, respectively. Averaging over 100 ps, we find $L = 1.3$ Å and $R = 0.96$ Å. These may be compared to the values $L = 2.3$ Å and $R = 1.1$ Å obtained in the model III/IV fits to neutron time-of-flight data taken on IN5 by König et al. [64] on a DPPC bilayer at 60°C and 23% water. The cylinder radius derived from the simulation is slightly smaller than that derived from the neutron data, and the cylinder length is substantially shorter. This in contrast to our expectation that the simulation values should, if anything, be larger because the simulation was carried out at full hydration. Moreover, the simulation predicts an appreciable anisotropy that it absent from the experimental data. There are two possible explanations for the discrepancies. The first is that simulation models the lipid center-of-mass motion poorly. The second, which is probably more correct, is that the combined diffusion in a cylinder-and-chain defect model used to determine the L and R values quoted here attributes too little motion to the chains and too much to the center of mass. We examine models for the motion of the lipid internal degrees of freedom in the next section.

F. Hydrocarbon Chain Dynamics

Two physically reasonable but quite different models have been used to describe the internal motions of lipid molecules observed by neutron scattering. In the first the protons are assumed to undergo diffusion in a sphere [63]. The radius of the sphere is allowed to be different for different protons. Although the results do not seem to be sensitive to the details of the variation in the sphere radii, it is necessary to have a range of sphere volumes, with the largest volume for methylene groups near the ends of the hydrocarbon chains in the middle of the bilayer and the smallest for the methylenes at the tops of the chains, closest to the bilayer surface. This is consistent with the behavior of the carbon–deuterium order parameters, S_{CD}, measured by deuterium NMR:

$$S_{CD} = \frac{1}{2} \langle 3 \cos^2\theta - 1 \rangle \tag{12}$$

where θ is the angle between the C—D bond vector and the bilayer normal [7]. We show the negative of the usual order parameters, which may be thought of as "disorder parameters," averaged over both chains in all the molecules in our bilayer simulation, in

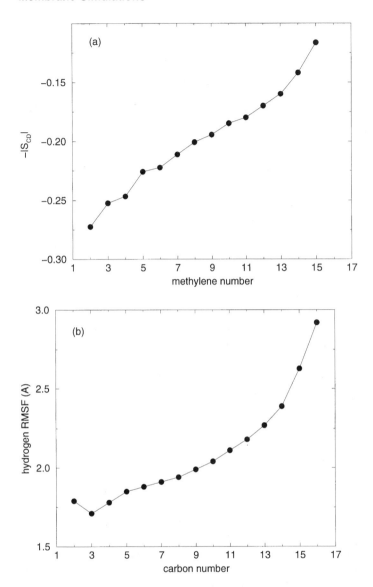

Figure 14 Measures of disorder in the acyl chains from an MD simulation of a fluid phase DPPC bilayer, (a) Order parameter profile of the C—H bonds; (b) root-mean-square fluctuation of the H atoms averaged over 100 ps.

Figure 14a. Although our values are not in quantitative agreement with the experimental values (see Ref. 50 for a discussion), they faithfully reproduce the well-known observation that the disorder increases from the tops of the chains toward the middle of the bilayer, with a more pronounced increase in the last few carbons of the chain. The S_{CD} represent different types of disorder, including chain conformational defects and whole molecule "wobbling" [67].

To make contact with the diffusion-in-a-sphere model, we have defined the spherical radius as the root-mean-square fluctuation of the protons averaged over 100 ps. The variation of the spherical radius as a function of position in the chains is plotted in Figure 14b. As expected, the radius increases monotonically from a minimum value of ≈ 1.8 Å at the tops of the chains to a maximum of ≈ 3 Å in the middle of the bilayer, with a profile closely resembling that of the disorder parameter. Our range of spherical radii is significantly narrower than the ranges reported by König et al. [63] for the bilayer at 30% hydration, assuming a linear increase from the minimum to maximum values: 0.8–7.6 Å from measurements with **Q** in the plane of the bilayer, and 1.1–5.8 Å from measurements with **Q** along the bilayer normal. The origin of this discrepancy is difficult to ascertain without knowing the sensitivity of the fits to the data on the range and shape of the radius profile. To gain more insight it would be useful to see if the neutron data could be fit well using a variation of spherical radii resembling the profile in Figure 14b.

Although the diffusion-in-a-sphere model is illuminating in the sense that it conveys the notion of increasing dynamic disorder moving toward the bilayer interior, as a model for hydrocarbon chain dynamics it is not completely satisfying because it assumes independent motion for each proton and therefore lacks an explicit connection to conformational transitions. In this regard perhaps a more appealing model is the chain defect model introduced subsequently by König et al. [64] as an alternative to the diffusion-in-a-sphere model. In this model it is assumed that the primary mechanism for the proton jumps in the hydrocarbon chains detected by neutron time-of-flight measurements is the creation/annihilation of a $gtg'(g^-tg^+$ or $g^+tg^-)$ kink from/to an all-trans conformation. Two types of kink dynamics were considered. In the first, the kink is assumed to diffuse along the chains, whereas in the second a kink forms and disappears randomly at different points in the chains. Both models fit the data equally well, but the stochastic kink model is more consistent with the picture from MD simulations (for example, Fig. 8). The resulting model contains two parameters: the average number of gtg' kinks per chain, 0.9, and the transition rate, $r_2 = 7.5 \times 10^{-10}$ s^{-1}, which corresponds to a kink lifetime of $1/r_2 = 13$ ps.

The chain defect model is useful because it incorporates a realistic mechanism for proton jumps into a model that leads to analytical expressions that can be used to fit neutron data. However, as pointed out by König et al., this model is an oversimplification, because clearly there are other types of chain defects (g, gg, g^-tg^-, and g^+tg^+) that are forming and disappearing on the time scale of the experiment. Indeed, we find in our simulation that gtg' kinks account for less than 20% of the gauche bonds in a fluid phase DPPC bilayer (Table 2). Moreover, we find that there are as many gtg (g^-tg^- or g^+tg^+) kinks as gtg' kinks. This is remarkable because a gtg kink produces a chain defect that significantly alters the path of the chain, whereas a gtg' kink leaves the chain path largely unaltered. Consequently, a ttt to gtg transition is expected to lead to a larger displacement of more protons than a ttt to gtg' transition. Thus, if our simulation results are trustworthy, we may conclude that the chain defect model including only gtg' kinks significantly under-

Table 2 Conformational Defects in the Hydrocarbon Chains in Fluid Phase Lipid Bilayers

Type of defect	Average number per chain	
	MD	Expt
g	3.6	3.7 [72], 3.6–4.2 [73]
gtg	0.35	1.2 [72]
$gtg + gtg'$	0.61	1.0 [74]

estimates the proton motions arising from conformational transitions in the chains. The average total number of *gauche* bonds per chain in our simulation is in good agreement with estimates from FTIR spectra, but the numbers of kinks are significantly lower than the experimental estimates (Table 2). The seriousness of the discrepancy in the number of kinks is hard to assess because of uncertainties in assigning the spectra and the considerable disagreement between independent experimental measurements. We finish this section by noting that our simulation predicts a kink lifetime of 6 ps, which is about half that assumed in the chain defect model [64].

G. Water Dynamics

In contrast to the lipids, which can exhibit a wide variety of motions over a wide range of time scales, water dynamics in membranes on the time scale of tens of picoseconds are relatively uncomplicated, consisting of (ignoring biologically uninteresting high frequency vibrations) rigid-body translation and rotation. In this section we analyze the translational and rotational motion of water in multilamellar bilayers, highlighting the distinction between the bound and bulk waters identified earlier in Section III.C.

The center-of-mass mean-square displacements of the bulk and the three different classes of bound water molecules are plotted in Figure 15a. In all cases the MSDs display a linear time dependence beyond about 5 ps. Thus, the water translation at times longer than a few picoseconds can be described as Brownian motion, and the mobility can be quantified by a diffusion constant, D, proportional to the slope of the linear part of the MSD. The values obtained for motion in the plane and perpendicular to the plane of the bilayer, from the average of five 20 ps MSD increments, with the slopes calculated from 10 ps to 20 ps, are listed in Table 3. As expected, the bulk water molecules have the greatest translational mobility, followed by the choline-bound, phosphate-bound, and carbonyl-bound water molecules. The tightly associated P-bound and CO-bound water molecules have roughly equal diffusion constants for in-plane and out-of-plane translational motion. The structural organization of the "bulk" waters that occupy a thin slab in the middle of the interlamellar space is quite similar to that of pure water at the same temperature, but the dynamics are different. As one might expect for water confined between two slabs (Fig. 2), we find that the diffusion constant in the plane of the bilayer is significantly larger (~50%) than in the direction normal to the bilayer.

We discuss the rotational dynamics of water molecules in terms of the time correlation functions, $C_l(t) = \langle P_l[\cos \theta_l(t)] \rangle$ ($l = 1, 2$), where P_l is the lth Legendre polynomial, $\cos \theta_l(t) = \mathbf{u}_l(0) \cdot \mathbf{u}_l(t)$, $\mathbf{u}_1$, is a unit vector along the water dipole (HOH bisector), and $\mathbf{u}_2$ is a unit vector along an OH bond. Infrared spectroscopy probes $C_1(t)$, and deuterium NMR probes $C_2(t)$. According to the Debye model (Brownian rotational motion), both correlation functions are exponential, and $C_2(t)$ decays three times as fast as $C_1(t)$. The $C_1(t)$ for the different classes of water molecules, plotted in Figure 15b, display a rapid initial decay (<10 ps) decay followed by a slower relaxation. The $C_2(t)$ (not shown) look similar but show a more rapid initial decay, as expected. Although the $C_l(t)$ are better described by multiexponential or stretched exponential functions, we have estimated rotational correlation times by fitting the short-time (first 10 ps) decays to single exponentials, $\exp(-t/\tau_l)$ (Table 3). The trend in the rotational rates ($1/\tau_l$) is the same as that in the translational diffusion constants, i.e., bulk > bound, and the ratios τ_1/τ_2 are approximately 1.5 for all classes of water molecules, suggesting that the Debye model is not appropriate for describing the rotational motion of water molecules in the vicinity of membranes.

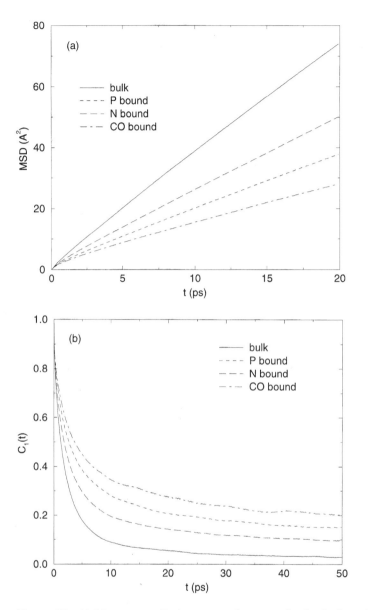

Figure 15 (a) Mean-square displacements of water molecules in three dimensions (see text for definitions of bound and bulk waters). (b) Time correlation functions for reorientation of the water O—H bonds.

We finish this section by comparing our results with NMR and incoherent neutron scattering experiments on water dynamics. Self-diffusion constants on the millisecond time scale have been measured by NMR with the pulsed field gradient spin echo (PFGSE) method. Applying this technique to oriented egg phosphatidylcholine bilayers, Wassall [68] demonstrated that the water motion was highly anisotropic, with diffusion in the plane of the bilayers hundreds of times greater than out of the plane. The anisotropy of

Table 3 Diffusion Constants and Rotational Correlation Times of Water Molecules from an MD Simulation of a Fully Hydrated Fluid Phase DPPC Bilayer[a]

Water class	In-plane D $(10^{-5}$ cm^2/s)	Out-of-plane D $(10^{-5}$ cm^2/s)	Dipole τ_1 (ps)	OH bond τ_2 (ps)
Bulk	6.6	4.5	3.6	2.2
N-bound	4.4	3.6	5.6	3.7
P-bound	3.2	2.7	7.1	5.0
CO-bound	2.3	2.0	9.1	5.9

[a] Note that in Ref. 55 the τ values for the P-bound and N-bound water molecules were erroneously interchanged.
Source: Ref. 55.

the long-range diffusion observed on the microsecond time scale is consistent with our results but much more pronounced, as expected, given the disparity in time and length scales probed in the experiment and simulation. Wassall also observed that the diffusion constant increased with the number of water molecules per lipid molecule according to the Finer model of hydration shells [69] and consistent with the differences we found in the diffusion constants between the different classes of bound and bulk water [55].

An INS study of water dynamics in oriented DPPC bilayers was reported by König et al. [8]. Neutron time-of-flight measurements were performed on oriented samples at 44°C and two hydration levels, using a spectrometer with an energy resolution (0.015 meV) that corresponds to a maximum observable correlation time of a few hundred picoseconds. The spectral parameters are not available for comparison, so we compare parameters of the model used to fit the data, which includes bound water molecules that are assumed to undergo rotation diffusion only, and quasi-free water molecules that undergo both rotational and translational diffusion. At low hydration (three to four water molecules per lipid molecule) the data could be modeled as rotation only, with a rotational correlation time $\tau_R = 60$ ps. Apparently, this $\tau_R = 1/D_R$, where D_R is the rotational diffusion constant, should be divided by 6 for comparison with the simulation results in Table 2. The resulting 10 ps is close to our bound water values. To fit the data at higher hydration (11 waters per lipid), the parameters for the rotational motion were kept fixed and the translational motion was modeled as jump diffusion, giving a residence time of 2 ps and a diffusion constant $D = 1.6 \times 10^{-5}$ cm^2/s. Considering the 6°C difference in temperature, the diffusion constant extracted from the scattering data is close to the values we obtained for the most tightly bound waters (P- and CO-bound) in our simulation.

V. SUMMARY AND CONCLUSIONS

This chapter has given an overview of the structure and dynamics of lipid and water molecules in membrane systems, viewed with atomic resolution by molecular dynamics simulations of fully hydrated phospholipid bilayers. The calculations have permitted a detailed picture of the solvation of the lipid polar groups to be developed, and this picture has been used to elucidate the molecular origins of the dipole potential. The solvation structure has been discussed in terms of a somewhat arbitrary, but useful, definition of bound and bulk water molecules.

The majority of the chapter was focused on an analysis of dynamics, which to date have been neglected relative to structure in simulation studies of membranes. We compared results from our simulation of a fluid phase bilayer with incoherent neutron scattering experiments that probed the motions of lipid H atoms on time scales of up to 100 ps. For the most part the simulation and experimental results were in good agreement, but the simulation predicted a slight anisotropy in the lipid motion that was not detected experimentally and overestimated the spatial dependence of the time scale of the slower of two dynamic processes resolved by the experiment. Having established a reasonable level of agreement between the simulation and neutron data, we examined the correspondence between the motions observed in the simulation and the dynamic models used to fit the experimental data. The center-of-mass motion and internal rearrangements are decoupled, and the former is well described as diffusion in a confined space (cylinder) on the 100 ps time scale, but not as Brownian motion. There were some significant discrepancies between the picture that emerged from the simulation and the models used to describe the internal motions. In particular, the simulation predicted a much weaker variation in the radii of the diffusion-in-a-sphere model and the involvement of more conformations in the acyl chain dynamics than the single kink assumed in the chain defect model.

Our simulation suggested that both the translational and rotational dynamics of water molecules in a fully hydrated, multilamellar lipid bilayer system depend on where the water molecules are located. As expected, both the translational and rotational mobilities of the ''bulk'' water molecules located in the middle of the interlamellar space are significantly greater than those of the ''bound'' water molecules located in the first solvation shell of the lipid polar groups (carbonyl, phosphate, and choline). The translational diffusion constants and the rotational rates of the bound water molecules increase in the order carbonyl-bound < phosphate-bound < choline-bound. On the time scale of tens of picoseconds, the bound water molecules exhibit largely isotropic translational motion, whereas the bulk water molecules diffuse approximately 50% faster in the plane of the membrane than out of the plane. None of the water molecules in the membrane–water system obey the Debye rotational diffusion model. The detailed picture of water dynamics near membranes derived from the simulation is in qualitative agreement with available data from NMR and neutron scattering experiments, but a more quantitative, model-independent comparison with neutron scattering would be useful for a better assessment.

ACKNOWLEDGMENTS

I am pleased to acknowledge that the simulation results presented in this chapter were obtained from calculations carried out in collaboration with Kechuan Tu, Mike Klein, and Kent Blasie. The calculations and fitting of the neutron scattering spectra benefited from discussions with Mounir Tarek. Financial support was provided by the School of Physical Sciences at the University of California at Irvine and a grant from the donors of The Petroleum Research Fund, administered by the American Chemical Society (ACS-PRF 33247-G7).

REFERENCES

1. SJ Singer, GL Nicolson. Science 175:720–731, 1972.
2. DM Small. The Physical Chemistry of Lipids. New York: Plenum Press, 1986.

3. G Büldt, HU Gally, J Seelig, G Zaccai. J Mol Biol 134:673–691, 1979.

4. G Zaccai, G Büldt, A Seelig, J Seelig. J Mol Biol 134:693–706, 1979.

5. JF Nagle, R Zhang, S Tristram-Nagle, W Sun, HI Petrache, RM Suter. Biophys J 70:1419–1431, 1996.

6. MC Wiener, SH White. Biophys J 61:434–447, 1992.

7. A Seelig, J Seelig. Biochemistry 13:4839–4845, 1974.

8. S König, E Sackmann, D Richter, R Zorn, C Carlile, TM Bayerl. J Chem Phys 100:3307–3316, 1994.

9. S König, E Sackmann. Curr Opin Colloid Int Sci 1:78–82, 1996.

10. SE Feller, D Huster, K Gawrisch. J Am Chem Soc 121:8963–8964, 1999.

11. DJ Tobias, K Tu, ML Klein. Curr Opin Colloid Int Sci 2:15–26, 1997.

12. DP Tieleman, SJ Marrink, HJC Berendsen. Biochim Biophys Acta 1331:235–270, 1997.

13. SJ Marrink, HJC Berendsen. J Phys Chem 98:4155–4168, 1994.

14. D Bassolino-Klimas, HE Alper, TR Stouch. Biochemistry 32:12624–12637, 1993.

15. A Pohorille, P Cieplak, MA Wilson. Chem Phys 204:337–345, 1996.

16. K Tu, M Tarek, ML Klein, D Scharf. Biophys J 75:2123–2134, 1998.

17. MA Wilson, A Pohorille. J Am Chem Soc 118:6580–6587, 1996.

18. K Tu, ML Klein, DJ Tobias. Biophys J 75:2147–2156, 1998.

19. KV Damodaran, KM Merz Jr. Biophys J 69:1299–1308, 1995.

20. KV Damodaran, KM Merz Jr. J Am Chem Soc 117:6561–6571, 1995.

21. LY Shen, D Bassolino, T Stouch. Biophys J 73:3–20, 1997.

22. K Belohorcova, JH Davis, TB Woolf, B Roux. Biophys J 73:3039–3055, 1997.

23. P Huang, GH Loew. J Biomol Struct Dyn 12:937–956, 1995.

24. S Berneche, M Nina, B Roux. Biophys J 75:1603–1618, 1998.

25. P LaRocca, Y Shai, MSP Sansom. Biophys Chem 76: 145–159, 1999.

26. B Roux, TB Woolf. Molecular dynamics of Pf1 coat protein in a phospholipid bilayer. In: KM Merz Jr, B Roux, eds. Biological Membranes: A Molecular Perspective from Computation and Experiment. Boston: Birkhauser, 1996, pp 555–587.

27. JP Duneau, S Crouzy, N Garnier, Y Chapron, M Genest. Biophys Chem 76: 35–53, 1999.

28. O Edholm, O Berger, F Jähnig. J Mol Biol 250:94–111, 1995.

29. TB Woolf. Biophys J 73:2376–2392, 1997.

30. TB Woolf. Biophys J 74:115–131, 1998.

31. TB Woolf, B Roux, Proc Natl Acad Sci USA 91:11631–11635, 1994.

32. TB Woolf, B Roux. Proteins: Struct, Funct, Genetics 24:92–114, 1996.

33. SW Chiu, S Subramanian, E Jakobsson. Biophys J 76:1929–1938, 1999.

34. SW Chiu, S Subramanian, E Jakobsson. Biophys J 76:1939–1950, 1999.

35. DP Tieleman, HJC Berendsen. Biophys J 74:2786–2801, 1998.

36. DP Tieleman, HJC Berendsen, MSP Sanson. Biophys J 76:1757–1769, 1998.

37. T Husslein, PB Moore, QF Zhong, DM Newns, PC Pattnaik, ML Klein. Faraday Disc 111: 201–208, 1998.

38. RH Pearson, I Pascher. Nature 281:499–501, 1979.

39. RM Venable, Y Zhang, BJ Hardy, RW Pastor. Science 262:223–226, 1993.

40. TR Stouch, KB Ward, A Altieri, A Hagler. J Comput Chem 12:1033–1046, 1991.

41. MJ Schlenkrich, J Brickmann, AD MacKerell Jr, M Karplus. Criteria for parameters optimization and applications. In: KM Merz Jr, B Roux, eds. Biological Membranes: A Molecular Perspective from Computation and Experiment. Boston: Birkhauser, 1996, pp 31–81.

42. K Tu, DJ Tobias, ML Klein. J Phys Chem 99:10035–10042, 1995.

43. DJ Tobias, K Tu, ML Klein. J Chim Phys Phys-Chim Biol 94:1482–1502, 1997.

44. U Essmann, L Perera, ML Berkowitz, T Darden, LG Pedersen. J Chem Phys 103:8577–8593, 1995.

45. SE Feller, RW Pastor, A Rajnuckarin, S Bogusz, BR Brooks. J Phys Chem 100:17011–17020, 1996.

46. GJ Martyna, DJ Tobias, ML Klein. J Chem Phys 101:4177–4189, 1994.
47. ME Tuckerman, CJ Mundy, GJ Martyna. Europhys Lett 49:149–155, 1999.
48. F Jähnig. Biophys J 71:1348–1349, 1996.
49. SE Feller, RW Pastor. Biophys J 71:1350–1355, 1996.
50. K Tu, DJ Tobias, ML Klein. Biophys J 69:2558–2562, 1995.
51. K Tu, DJ Tobias, ML Klein. Biophys J 70:595–608, 1996.
52. U Essmann, ML Berkowitz. Biophys J 76:2081–2089, 1999.
53. CY Lee, JA McCammon, PJ Rossky. J Chem Phys 80:4448–4455, 1984.
54. S-J Marrink, DP Tieleman, AR van Buuren, HJC Berendsen. Faraday Disc 103:191–201, 1996.
55. DJ Tobias. Water and membranes: Molecular details from MD simulations. In: M-C Bellissent-Funel, ed. Hydration Processes in Biology: Theoretical and Experimental Approaches. Amsterdam: IOS Press, 1999, pp 293–310.
56. GL Gaines. Insoluble Monolayers at Liquid–Gas Interfaces. New York: Interscience, 1966.
57. KD Gawrisch, D Ruston, J Zimmerberg, VA Parsegian, RP Rand, N Fuller. Biophys J 61: 541–552, 1992.
58. RF Flewelling, WL Hubbell. Biophys J 49:541–552, 1986.
59. SW Chiu, M Clark, V Balaji, S Subramanian, HL Scott, E Jakobsson. Biophys J 69:1230–1245, 1995.
60. W Pfeiffer, T Henkel, E Sackmann, W Knoll, D Richter. Europhys Lett 8:201–206, 1989.
61. RW Pastor, SE Feller. Time scales of lipid dynamics and molecular dynamics. In: KM Merz Jr, B Roux, eds. Biological Membranes: A Molecular Perspective from Computation and Experiment. Boston: Birkhauser, 1996, pp 3–30.
62. M Beé. Quasielastic Neutron Scattering: Principles and Applications in Solid State Chemistry, Biology, and Materials Science. Bristol: Adam Hilger, 1988.
63. S König, W Pfeiffer, T Bayerl, D Richter, E Sackmann. J Phys II France 2:1589–1615, 1992.
64. S König, TM Bayerl, G Coddens, D Richter, E Sackmann. Biophys J 69:1871–1880, 1995.
65. S König. Untersuchungen zur Molekularen Bewegung und Diffusion von Membranen Mittles Inkohärenter, Quasielastischer Neutronenstreung und Deuterium-NMR. PhD Dissertation, TU München, München, 1993.
66. SH Chen, P Gallo, F Sciortino, P Tartaglia. Phys Rev E 56:4231–4243, 1997.
67. RW Pastor, RM Venable, M Karplus, A Szabo. J Chem Phys 89:1128–1140, 1988.
68. SR Wassall. Biophys J 71:2724–2732, 1996.
69. EG Finer, A Darke. Chem Phys Lipids 12:1–16, 1974.
70. S Tristram-Nagle, R Zhang, RM Suter, CR Worthington, W-J Sun, JF Nagle. Biophys J 64: 1097–1109, 1993.
71. WJ Sun, RM Suter, MA Knewtson, CR Worthington, S Tristram-Nagle, R Zhang, JF Nagle. Phys Rev E 49:4665–4676, 1994.
72. HL Casal, RN McElhaney. Biochemistry 29:5423–5427, 1990.
73. R Mendelsohn, MA Davies, JW Brauner, HF Schuster, RA Dluhy. Biochemistry 28:8934–8939, 1989.
74. L Senak, MA Davies, R Mendelsohn. J Phys Chem 95:2565–2571, 1991.
75. E Lindahl, O Edholm. Biophys J 79:426–433, 2000.

Appendix: Useful Internet Resources

It is well known that the resources available on the Internet are in constant flux, with new sites appearing on a daily basis and established sites disappearing almost as frequently. This also holds true for the dedicated tools used in biochemical and biophysical studies. New tools are constantly becoming available, and established tools, obsolete. Such rapid change makes it difficult to stay current with the state-of-the-art technologies in the areas of bioinformatics and computational biochemistry and biophysics.

To help the reader keep abreast of these advances we present a list of useful WWW sites in this appendix. Realistically, this list should be updated on a daily basis as many of the tools offered on the Internet are made available not only by large organizations and research groups but also by individual researchers. The goal, therefore, has not been to provide a nearly complete guide to the WWW but rather to provide material representative of the tools useful to researchers in the fields of biochemistry and biophysics.

Most web sites listed contain links to other web sites. This "hyperconnectivity" is what makes the WWW a virtually unlimited information source, which we hope you will be able to exploit to expand the limited list of sites presented below. In addition, this appendix will be regularly updated at the following web site:
http://yuri.harvard.edu/~watanabe or *http://www.geocities.com/masakatsu_w/index.html*

A. Internet Resources for Topics in Selected Chapters

1. Force Fields (Chapter 2)
   - MacKerell group: *https://rxsecure.umaryland.edu/research/amackere/ research.html*
   - Popular empirical force field web sites:
     CHARMM: *http://www.pharmacy.umaryland.edu/~alex/research.html*
     AMBER: *http://www.amber.ucsf.edu/amber/*
     GROMOS: *http://igc.ethz.ch/gromos/gromos.html*
   - Force field evaluation suite: *http://www.ccl.net/cca/data/ff_evaluation _suite/*
2. Protein Dynamics (Chapter 3)
   - Quick guide to molecular simulations (see Online tutorials, below):
     http://www.tc.cornell.edu/Visualization/Staff/richard/Courses/biobm631
   - Molecular movements database:
     http://bioinfo.mbb.yale.edu/MolMovDB/db/ProtMotDB.main.html
3. Minimization and Conformational Analysis (Chapter 4)
   - Monte Carlo methods:
     http://zarbi.chem.yale.edu/programs/mcpro/mc_toc.htm
4. Structure Refinement Applications (Chapter 13)

- Software for structure determination and analysis:
  http://www.rcsb.org/pdb/software-list.html
- X-PLOR: A program for structure determination from crystallographic or NMR data: *http://atb.csb.yale.edu/xplor*
- NMR:
  Nilges group: *http://www.embl_heidelberg.de/nmr/nilges*
  Basics of NMR: *http://www.cis.rit.edu/htbooks/nmr/*
- X-ray crystallography:
  CCP4: *http://www.dl.ac.uk/CCP/CCP4/main.html*
  General site: *http://www.iucr.org/cww-top/crystal.index.html*

5. Comparative Protein Structure Modeling (Chapter 14)
   - Sali's group: *http://guitar.rockefeller.edu/sub-pages/programs.html*
6. Statistics in Molecular and Structural Biology (Chapter 15)
   Dunbrack group: *http://www.fccc.edu/research/labs/dunbrack/*
7. Computer-Aided Drug Designs (Chapter 16)
   - Tropsha group: *http://mmlin1.pha.unc.edu/~jin/QSAR/*
   - Molecular docking:
     http://www.scripps.edu/pub/olson-web/people/gmm/
   - Quantitative structure–activity relationship (QSAR):
     http://www.chem.swin.edu.au/modules/mod4
8. Protein Folding (Chapter 17)
   - Becker's group: *http://www.tau.ac.il/~becker/index.html*
   - Friesner's group: *http://www.chem.columbia.edu/cbs/protein/protein.html*
   - Okamoto's group: *http://konf2.ims.ac.jp/research.html*
9. Membrane Simulations (Chapter 21)
   - Tobias's group: *http://www.chem.uci.edu/research/faculty/dtobias.html*
   - van Gunsteren's group:
     http://www.nmr.chem.ruu.nl/%7Eabonvin/ToT/lukas/index_lukas.html
   - Feller's group: *http://persweb.wabash.edu/facstaff/fellers/*

B. Molecular Modeling and Simulation Packages

Listed is a collection of general-purpose molecular dynamics computer simulation packages for the study of molecular systems. The packages include a wide variety of functionalities for the analysis and simulation of biomolecules. In addition, they contain integrated force fields.

1. CHARMM (Chemistry at HARvard Molecular Mechanics): General-purpose molecular dynamics computer simulation package
   http://yuri.harvard.edu/
   http://master2.lobos.nih.gov/Charmm/
   http://www.scripps.edu/brooks/charmm_docs/charmm.html
2. AMBER (Assisted Model Building with Energy Refinement): General-purpose molecular dynamics computer simulation package
   http://www.amber.ucsf.edu/amber/
3. GROMOS: A general-purpose molecular dynamics computer simulation package for the study of biomolecules *http://igc.ethz.ch/gromos/welcome.html*
4. GROMACS (GROningen MAchine for Chemical Simulations)
   http://rugmd0.chem.rug.nl/~gmx/

5. TINKER: Software tools for molecular modeling
   http://dasher.wustl.edu/tinker/

6. NAMD: Object-oriented molecular dynamics code designed for high performance simulation of large biomolecular systems
   http://www.ks.uiuc.edu/Research/namd/

7. MMTK (Molecular Modeling ToolKit): Open Source Program library for molecular simulation applications *http://starship.python.net/crew/hinsen/MMTK/*

C. Molecular Visualization Software

Listed is a collection of visualization software packages that are widely used in academic and industrial research groups.

1. RasMol: A free program that displays molecular structure. It is easy to use and produces space-filling, colored, three-dimensional images.
   http://www.bernstein-plus-sons.com/software/rasmol/

2. MOLMOL (MOLecule analysis and MOLecule display): A molecular graphics program for displaying, analyzing, and manipulating the three-dimensional structure of biological macromolecules.
   http://www.mol.biol.ethz.ch/wuthrich/software/molmol/

3. WebMol: JAVA PDB viewer to display and analyze structural information.
   http://www.embl-heidelberg.de/cgi/viewer.pl

4. Swiss-Pdb Viewer: An application that provides a user-friendly interface allowing simultaneous analysis of several proteins.
   http://www.expasy.ch/spdbv/mainpage.html

5. WebLab ViewerLite (Freeware from Molecular Simulation, Inc.): A fully Microsoft Windows integrated program allowing for 3D molecular visualization and the generation of high quality rendered images.
   http://www.msi.com/download/index.html

6. Jmol: A free, open source molecule viewer and editor.
   http://www.openscience.org/jmol
   Jmol can be also used to animate the results of simulations that are in a multiframe XYZ format and to animate the computed normal modes from ab initio quantum chemistry packages.

7. VMD (Visual Molecular Dynamics): MD-generated trajectories can be read.
   http://www.ks.uiuc.edu/Research/vmd/
   VMD is designed for the visualization and analysis of biological systems such as proteins, nucleic acids, and lipid bilayer assemblies. It may be used to view more general molecules, as VMD can read several different structural file formats and display the contained structure. VMD provides a wide variety of methods for rendering and coloring a molecule. VMD can be used to animate and analyze the trajectory of a molecular dynamics (MD) simulation.

8. gOpenMol: Graphical interface of computational chemistry
   http://www.csc.fi/~laaksone/gopenmol/gopenmol.html
   gOpenMol can be used for the analysis and display of molecular dynamics trajectories and the display of molecular orbitals, electron densities, and electrostatic potentials from programs like the Gaussian.

9. Molscript: A program for displaying molecular 3D structures in both schematic and detailed representations. *http://www.avatar.se/molscript/.*

10. List of other free or public domain software that is relevant to visualizations:
    http://www.ahpcc.unm.edu/~aroberts/main/free.htm

D. Computational Biophysics Related at the National Institutes of Health (NIH)

1. General home page: *http://webasaurus.dcrt.nih.gov//molbio/*
2. Scientific resources: *http://helix.nih.gov/science/*
3. Sequence and molecular databases: *http://helix.nih.gov/science/databases.html*
4. Center for Molecular Modeling: *http://cmm.info.nih.gov/modeling/*
5. Molecular biology software list: *http://bimas.dcrt.nih.gov/sw.html*

E. Molecular Biology Software Links

1. Resource site for biotechnology—Molecular biology, bioinformatics, biophysics, and biochemistry—A well-organized web site:
   http://www.ahpcc.unm.edu/~aroberts/
2. Molecular surface package: *http://www.best.com/~connolly/*
3. Biotechnological software and internet journal:
   http://www.davincipress.com/bsj.html
4. Computational chemistry web site: *http://www.ccl.net/chemistry/*

F. Online Tutorials

1. Molecular dynamics: *http://cmm.info.nih.gov/intro_simulation/course_for_html.html*
   http://www.chem.swin.edu.au/modules/mod6.
2. Monte Carlo method:
   http://www.cooper.edu/engineering/chemechem/MMC/tutor.html.
   http://www.cooper.edu/engineering/chemechem/monte.html
3. Bioinformatics:
   http://www.iacr.bbsrc.ac.uk/notebook/wwwresource/bioinformaticcourses392.htm
   http://biotech.icmb.utexas.edu/pages/bioinfo.html.
4. Molecular modeling workbook:
   http://www.ch.ic.ac.uk/local/organic/mod/Chem99.pdf

G. Additional Resource List for Computational Chemistry and Molecular Modeling Software

1. The Center for Molecular Modeling at NIH:
   http://cmm.info.nih.gov/modeling/software.html
2. Laboratory of Structural Biology at NIEHS:
   http://dir.niehs.nih.gov/dirlmg/strFxn.html
3. BioMolecular research tools—A collection of WWW links to information and services useful to molecular biologists:
   http://www.public.iastate.edu/~pedro/research_tools.html
4. Center for Scientific Computing in Finland:
   http://www.csc.fi/~laaksone/docs/stuff.html
5. Rolf Claessen's chemistry index:
   http://www.claessen.net/chemistry/soft_mod_en.html

6. Software for structure determination and analysis at the Protein Data Bank: *http://www.rcsb.org/pdb/Modeling*
7. W. L. Jorgensen group at Yale: *http://zarbi.chem.yale.edu/*

H. Databases of Biological Molecules

1. Protein Data Bank: *http://www.rcsb.org/pdb/*
2. IMB Jena Image Library of Biological Macromolecules: *http://www.imb-jena.de/IMAGE.html*
3. Nucleic Acid Database: *http://ndbserver.rutgers.edu/NDB/*
4. Cambridge Structural Database: *http://www.ccdc.cam.ac.uk*
5. Molecules R Us: *http://molbio.info.nih.gov/cgi-bin/pdb*
6. ExPASy (Expert Protein Analysis System) molecular biology server: *http://www.expasy.ch*
7. Lists of useful databases (including the Genome database) *http://www.gdb.org/gdb/hgpresources.html*

Index